CELL BIOLOGY OF BRAIN

CELL BIOLOGY OF BRAIN

W.E. Watson

M.D., M.R.C.P., F.R.C.S.

Professor of Physiology
Medical School, University of Edinburgh

LONDON
CHAPMAN AND HALL

A Halsted Press Book
JOHN WILEY & SONS, INC., NEW YORK

First published 1976
by Chapman and Hall Ltd

Typeset by Santype Ltd (Coldtype Division)
Salisbury, Wiltshire
Printed in Great Britain at the
University Printing House, Cambridge

ISBN 0 412 11950 1

Distributed in the U.S.A. by Halsted Press,
a Division of John Wiley & Sons, Inc., New York

Library of Congress Cataloging in Publication Data

Watson, William Eric.
Cell biology of brain.

Bibliography: p.
Includes index.
1. Brain. 2. Neurons. I. Title. [DNLM:
1. Brain—Cytology. 2. Neurophysiology. WL102 W343c]
QP376. W27 611'0181 76–1921
ISBN 0-470-15042-4

CONTENTS

PREFACE

I have written this for Honours students and postgraduates. Some subjects have been omitted as such: developmental neurobiology, the synapse, nerve growth factor and prostaglandins. These subjects have been widely reviewed, and I have nothing fresh to add.

I wish to thank Catherine for lending me my typewriter, Julian for leaving me some paper, and Aline for not criticizing the manuscript. I am grateful to my publishers for their patronage.

Edinburgh
October, 1975

W.E.W.

1 BRAIN AS AN EPITHELIUM

Brain develops from an epithelium. A persisting pattern of epithelial organization is apparent in ependyma, and discernible in astrocytes (3072). Neurons and oligodendrocytes show this organization less readily. Four types of mesenchymal cell are found in normal brain: endothelial cells of blood vessels, perivascular cells, and cells of the pia mater and arachnoid mater.

Organization of an epithelium

A basal lamina can be formed between epithelial cells and mesenchyme, or between mesenchymal cells of different types: for example, between capillary endothelium and perivascular fibroblasts. It is not formed between different types of epithelial cell. A basal lamina contains collagen (139, 1570, 1571, 1572, 2286, 2287, 2288, 2798), which can be detected biochemically, immunologically and by X-ray diffraction (1741). Collagen cannot usually be seen in the basal lamina by electron microscopy because it is masked by glycosaminoglycans. Mesenchymal cells synthesize the precursors of collagen. The epithelial cells synthesize glycosaminoglycans and also precipitate the collagan precursors at their own surface (713, 1273, 1275, 1523, 1741, 2070, 2159, 2287, 2288). When visible, the collagen of some basal laminae is structurally organized; it is arranged in layers (1576, 2503, 3107), and each is orthogonally aligned. The newest side of a basal lamina is therefore the one next to the epithelial cells (1275) (Fig. 1.1). The epithelial cell cannot precipitate collagen, if it is prevented from synthesizing protein.

Mesenchyme and epithelium interact during differentiation. Mesenchyme alters the rate of proliferation of epithelial cells (60, 62) and allows their structural (60, 622, 1170) and biochemical (61)

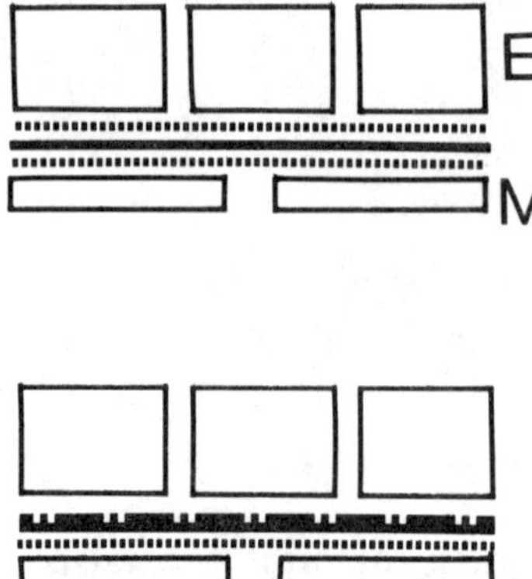

Figure 1.1 *Top.* Section of an epithelium. The epithelial cells, E, are shown at the top, and the mesenchymal cells below M, separated by a basal lamina which is indicated by a thick black line. Soluble precursors of collagen, synthesized by the mesenchymal cells and indicated by the two dotted lines, lie between the epithelial cells and mesenchyme. *Bottom.* Section of an epithelium. The soluble precursors of collagen are precipitated on the epithelial side of the basal lamina. The newly precipitated material is shown as —·—·, older basal lamina as ———, and soluble precursors as · · · ·.

differentiation. Some epithelia require mesenchyme from a closely related source for effective mutual interaction (1114, 1160, 1161, 1162, 1163, 1165), but others are less demanding (1164). In some tissues epithelium influences the differentiation of mesenchyme (1161, 1164, 3246). In these situations the part of the cell abutting upon the basal lamina can be considered as the cell's dominant pole, as its differentiation is directed to a considerable degree through this surface.

Several kinds of junction can join adjacent lateral surfaces of epithelial cells. Functionally they may be divided into three groups. In the first are junctions which hold the surfaces together, but do not allow the cytoplasmic compartments of adjacent cells to communicate; these also occlude the extracellular space between the cells. If such junctions pass around the whole circumference of the cell, an apical compartment of extracellular space is sealed from the basal, and molecules which are not soluble in fat cannot pass by diffusion from one to the other. In the second group are junctions which are more localized. These too do not allow the cytoplasmic compartments of adjacent cells to communicate, but, being more localized and not passing around the whole circumference the extracellular space round the apex of the cell is continuous with that around the base. The third group are also localized junctions. They permit communication not only between the different compartments of extracellular fluid but also between the cytoplasmic compartments of adjacent cells. These are called gap junctions.

Gap junctions are found in many epithelia. They allow direct electrotonic interaction between adjacent cells (200, 202, 203, 204, 205, 976, 1320, 1504, 2210, 2211). The junction has two systems of channels. One provides continuity between the cytoplasmic compartments of adjacent cells through channels about 1 nm in diameter and separated with a centre to centre spacing of about 10 nm. These channels are 15 to 20 nm long. The other system of channels, open to the extracellular fluid but closed to the cytoplasmic compartments, allows the extracellular and lumenal fluids to communicate through channels about 2 nm in diameter and 1 to 3 μm long. The intercytoplasmic channels form an hexagonal array (Fig. 1.2). Freeze-fracture studies show characteristic membrane inclusions, at the sites of gap junctions (459, 973, 975, 976, 1106, 1647, 1892, 2038, 2698). The particles are in hexagonal array, and their centre to centre spacing corresponds closely to that of the intercytoplasmic channels. A membrane fraction enriched with gap junctions has been obtained from liver cells of the mouse (1107). Preliminary biochemical analysis suggests that the composition is fairly simple. The structure of unfixed junctions determined by X-ray diffraction corresponds closely with that seen by electron microscopy. The fact that they can be isolated shows that they are stable. It is unlikely that the repeating pattern seen represents only a phase change in the membrane lipids. The structure of the membrane at a gap junction is not known, but is likely to be of considerable interest, as the 10 nm spacing of the intercytoplasmic channels must include the 1 nm channel and the 2 nm extracellular channel; insufficient space would therefore appear to remain for two unit membranes. Communicating gap junctions can be formed and lost rapidly both in tissue culture and *in vivo*. Although it has not yet been shown that these junctions can be formed between cells after their protein and lipid synthesis has been inhibited, the speed of their formation makes it very probable that the junctions are formed from

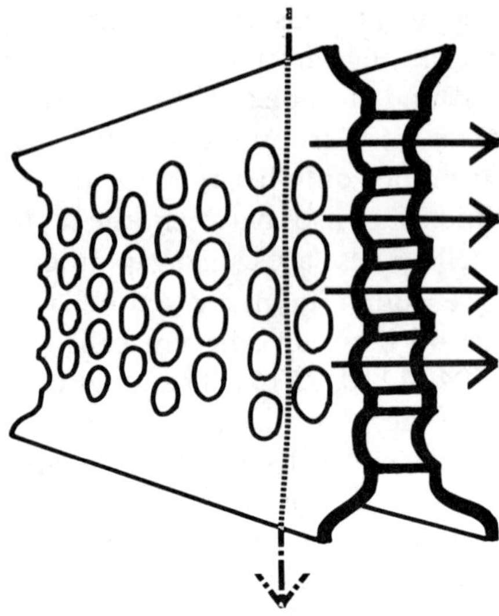

Figure 1.2 A diagram of a gap junction between two cells. It is viewed from the inside of one cell. The cell membranes are shown as thick black lines. The communicating channels appear as holes on the inner face of the membrane, and are in hexagonal array. Four black arrows indicate the lumen of the channels between the cytoplasmic compartments. These channels have been shown opened longitudinally. A dashed arrow indicates a channel which joins the regions of extracellular space separated by the junction: the arrow is dotted where the channel passes deep to the nearer plasma membrane.

specialized components already present in parts of the cell membrane, from 'potential half channels' (205). Nothing is yet known of the sequence of events that can lead to the formation of gap junctions, that determine the part of the cell at which they are made, and that regulate their individual and total areas.

Gap junctions are formed only between cells of certain types; for example, ependymal cells, astrocytes and nerve cells can all form junctions, but each does so only with cells of its own type (342). Such junctions between astrocytes and ependymal cells or between neurons and astrocytes either do not occur or are very rare. This could be due to important differences between the potential half channels themselves in different types of cell, or perhaps more probably a consequence of differences in the overlying glycocalyx which either facilitates or impedes the early stages of cell interaction.

Communication can be established between mesenchymal cells also, for example, between fibroblasts in tissue culture. Junctions are formed between fibroblasts from a normal mouse and fibroblasts from a mutant mouse, which are deficient in hypoxanthine-guanine transferase or in adenine phosphoribosyl transferase (561, 562, 2849); as a result of this communication the cells cooperate metabolically by the movement of the substrates and products of enzyme activity from one cell to the other probably by diffusion, and not by the transfer of enzyme. This cooperation occurs only between cells of the same species. Normal fibroblasts from man or from the hamster cannot cooperate with enzyme-deficient cells of the mutant mouse.

Tetraparental mice prepared from two fused morulae, one of a normal mouse and the other of a mutant deficient in a histologically demonstrable enzyme contain a mosaic of cells. Cells with a normal quantity of reaction product lie next to cells without any. These presumably communicate through gap junctions. There is no evidence to indicate that the enzyme has been able to pass from one cell to the other. The reaction product, being insoluble, cannot (2013).

Ions and small molecules can pass through these junctions. Nothing having a molecular weight greater than 80 000 daltons has been shown to pass. Very little is known of the restrictions of charge or of molecular shape which influence the intercellular movement of smaller molecules.

It is not known if gap junctions are normally labile *in vivo*. When cultured epithelial cells are injured the electrical coupling decreases considerably. This decrease is associated with an increased intracellular concentration of Ca^{++}. Other circumstances in which the coupling resistance is increased or restored to normal are those in which the intracellular Ca^{++} is likely to be high or low respectively. A small increase in membrane resistance is usually not accompanied by a structural change apparent by electron microscopy; this is however, a crude morphological measurement. The coupling between epithelial cells of the salivary gland of *Chironymus* decreases as the animal develops, and the area of the gap junctional membrane also falls (1828).

Such cell to cell direct contact is almost certainly concerned in a variety of biological processes including induction of the immune response (836), cell recognition during development (765), contact inhibition of cell movement (4), and of cell division (2829, 2932), electrical coupling (996, 1824), metabolic cooperation (561, 623, 974, 2848) and possibly the regulation of organ or tissue growth (1109).

Epithelial organization of brain

Ependyma

The ependyma lines the ventricular surface of the brain, except for certain bare areas (2171, 2579, 2588) and for small regions where it is regularly penetrated by the processes of neurons (345, 515, 702, 1468, 1765, 1766, 1767, 1768, 2195, 2515, 3011, 3144). These are usually dendrites and they often terminate within the ventricular cavity. It is not clear whether these processes act as chemoreceptors, whether they secrete into the cerebrospinal fluid, or whether they are there by accident. Not all such intraventricular cells are neurons. Occasionally the converse situation is found; ependymal cells lie beneath the surface of the ventricle and form subependymal rosettes. The lumenal aspect of these cells faces a small central cavity. Cultured ependymal cells behave in this way (1176). It is a pattern of organization that resembles that of implantation or sequestration dermoids. The nature of the lumenal surface varies from one part of the ventricular wall to another. This heterogeneity can be explored very well by scanning electron microscopy (2634, 2635, 2636). In some areas most of the cells are ciliated; in others the cilia are replaced by microvilli; in yet other regions, especially near and upon the floor of the third ventricle, there are bullous projections. The functional significance of these variations is far from clear. There are differences between the ependyma of the floor of the third ventricle of the male and female monkey and cat; the ependyma of the third ventricle of the skunk changes considerably during the breeding season (1215); after intraventricular injection of catecholamines the ependyma of the floor of the third ventricle changes its surface configuration.

The lateral surfaces of adjacent ependymal cells are joined by gap junctions; most ependymal cells are also joined by non-occluding zones of adherence. It has not yet been shown that these gap junctions form a low resistance pathway between adjacent ependymal cells, but it is very probable (208, 210, 344). In most parts of the brain, fluid and molecules up to at least 100 000 daltons can pass between the extracellular fluid and the cerebrospinal fluid, through the clefts between the cells (224, 312, 339, 340, 341, 343, 869, 921, 1286, 1779, 1780, 2315, 2391, 2392). Some parts of the brain have permeable capillaries (189, 2360, 3162). In these the ependymal cells are joined by occluding junctions as well as by gap junctions (1903, 3181). Here exogenous markers of 2000 to 70 000 daltons can enter the extracellular space from plasma, but not from the cerebrospinal fluid. Diffusion within the extracellular space from parts of brain having leaky capillaries into parts with sealed capillaries is not restricted (1372, 3164).

A process passing from the deep surface of an ependymal cell makes contact with a basal lamina (1350, 3068, 3072), often at a site remote from the cell body (Fig. 1.3). Usually the basal process is not seen in sections, but is evident in teased preparations. These processes are applied either to the basal lamina beneath the pia mater or to the outer layer of lamina surrounding vessels (920, 1015, 1666, 1764, 2031, 2871, 3068). In both situations the ependymal cell becomes related to mesenchyme through basal lamina; its relationship here is characteristically epithelial, and present from an early stage of development. No basal lamina separates the base of an ependymal cell from the neuropil; nor would one be expected, because both the ependymal cell and the bulk of the constituents of neuropil are epithelial. No ependymal cells have yet been seen that lack this relationship with mesenchyme. The basal lamina beneath the ependymal cells of the subcommisural organ is structurally specialized (2107, 2207). This might result from a leak from the base of the cells of the glycosaminoglycans, which are synthesized by these cells and secreted from the apex as Reissner's fibre; that such leakage occurs is indicated by the phagocytes found in the vicinity of the blood vessels. Commonly they contain much 'neurosecretory' material. Alternatively, the specialization might result from the basal secretion of these cells, which has been often suggested but never proved. Both suggestions, however, seem weak.

The development of ependyma has not yet been shown to depend upon contact with mesenchyme. The observation that presumptive ependyma differentiates only when in contact with brain (247) could indicate a dependence upon mesenchyme, for the cell bodies are almost certainly separated from mesenchymal influences as well as from others. The lamina at this interface may be viewed in several ways: as indicating a plane at which interaction between mesenchyme and epithelium once occurred (1939), as a barrier preventing or limiting interaction between these cells (626), as a consequence of continuing dynamic interaction between these cell types, and as a stable support to preserve the structure of brain efficiently. These possible roles are not necessarily mutually exclusive. For example, it seems almost certain that these cells interact dynamically both before and after the formation of the lamina, and that the nature of the interaction is different at these two stages. Or, even if the lamina were to act as a barrier to prevent mesenchyme from continuing to direct the further differentiation of ependyma, the lamina itself still results from the mutual interaction of these two types of cell. It is, incidentally, improbable that if it acts as a barrier to interaction at all, it does so by impeding the diffusion of molecules between the cells. Molecules as large as horse-radish peroxidase can pass across it in brain without significant hindrance. Furthermore, such a method of regulating mutual influence would be very inefficient; it is surely more probable that the synthesis of morphogenetic molecules, if any, would be stopped rather than that their synthesis should continue and their extracellular diffusion should be blocked.

Astrocytes

Astrocytes do not line an obvious surface; lacking a clear lumenal surface they are bathed on many aspects by the extracellular fluid of brain. They touch the subpial or the external layer of the perivascular basal lamina by at least one process (1650, 2066, 3068, 3072, 3172) (Fig. 1.3). As the vascular process may pass either to a distributing or collecting vessel, and not exclusively to an exchanging vessel, it seems to be unlikely that this process is concerned with facilitating or restricting exchange between blood and brain. Furthermore, astrocytes are found in parts of brain in which barrier properties have not been demonstrated (745). Capillaries which receive basal processes of ependymal cells also receive vascular processes of astrocytes (3068). The shark has leaky capillaries throughout its central nervous system; exogenous protein can pass

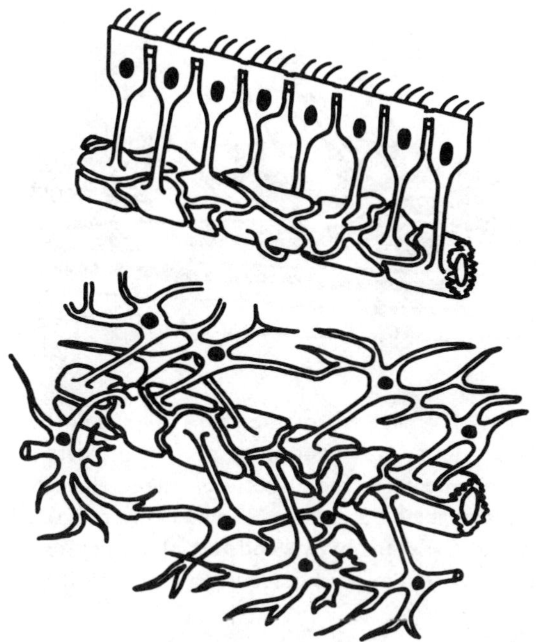

Figure 1.3 *Top.* Diagram of ependymal cells with basal processes lying upon a basal lamina around a blood vessel. Details of the basal lamina are not shown. Note the communicating junctions joining adjacent ependymal cells near their lumenal ciliated surfaces. *Bottom.* Diagram of astrocytes with basal processes lying upon a basal lamina around a blood vessel. Note the communicating junctions joining processes of adjacent astrocytes. The basal processes either of ependymal cells or of astrocytes could alternatively have made contact with the basal lamina beneath the pia.

from the blood to the extravascular space. The end-feet on the vascular processes of the astrocytes surrounding these capillaries are joined by occluding junctions; the protein markers cannot penetrate from the extravascular space into the extracellular space in these animals (Fig. 1.4).

Other processes of astrocytes ramify widely and form gap junctions with other astrocytes (344) but not with other types of cell. Not being confined to a surface, the organization of astrocytes is not planar. If positional information in brain is provided through intercellular communications, by a phase shift of conducted pulses

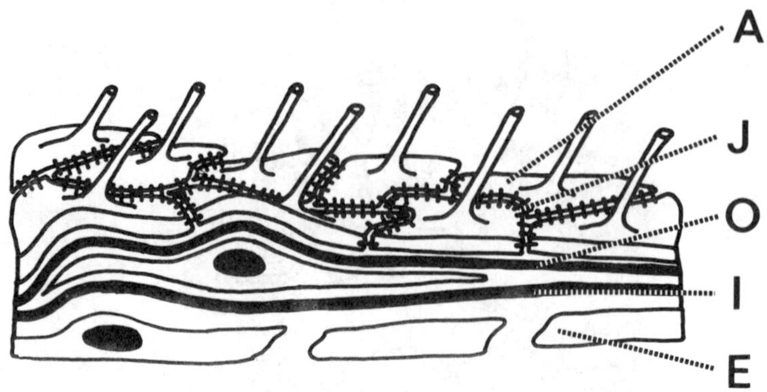

Figure 1.4 Diagram of vessel wall of the shark. The endothelial cells, E, are not joined by occluding junctions; exogenous protein markers such as horse-radish peroxidase can penetrate from blood between them, then through the outer, O, and inner, I, layers of the basal lamina. Their further progress is stopped in most parts of the brain of the shark by the end feet of astrocytes, A, which are joined by occluding junctions, J. The arrangements in mammals differ from this.

(1109), or in some other way, the code would be more difficult to interpret when the communicating cells are arranged in three dimensions rather than two. Perhaps it is for this reason that ependyma or ependymoglia rather than a system of astrocytes play an important role in guiding the regenerating nerve processes of long tracts within the central nervous system of those vertebrates in which such regeneration occurs (Chapter 7).

Mesenchymal cells of brain

Pia and pericytes share the capacity of interacting with ependyma or with astrocytes to form basal laminae, and may share a capacity to induce the differentiation of these cells at a critical period of development. Only one layer of basal lamina lies between pia and an astrocyte or an ependymal cell. The vascular pericyte is related to two layers of basal lamina, clearly seen when separated by a pericyte (1764), collagen (812, 1603) or a perivascular space (2676) (Fig. 1.5). Although interaction of pericytes and vascular endothelium almost certainly formed the layer that lies between them, these cells cannot have made the layer that is remote from the endothelium,

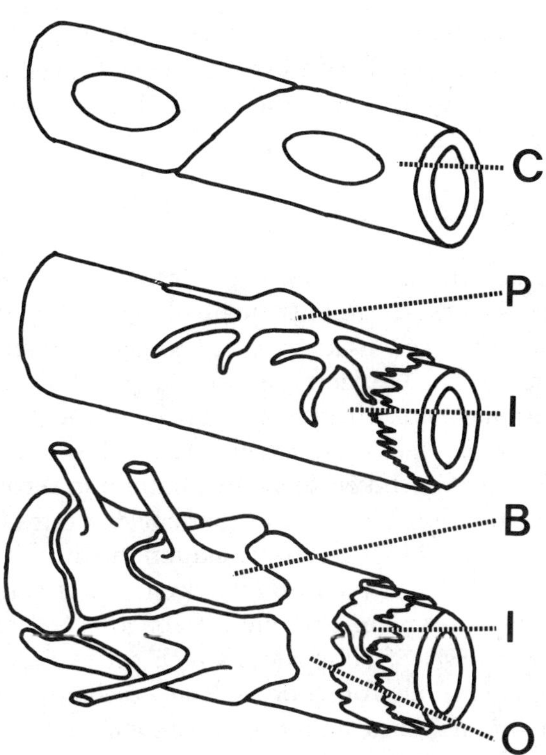

Figure 1.5 Diagram of the two layers of basal lamina surrounding a cerebral blood vessel. *Top.* A vessel showing two capillary endothelial cells, C. *Centre.* The capillary endothelium is shown overlain by the inner layer of basal lamina; this layer, I, lies between the capillary endothelium and the pericyte, P. *Bottom.* The pericyte and inner layer of basal lamina are now shown covered by the outer layer of basal lamina, O, which lies between the pericyte, now covered, and the basal vascular process, B, of astrocytes or of ependymal cells.

between the pericyte and the ependymal cell or astrocyte. No single type of mesenchymal cell has yet been shown to produce a basal lamina alone; it is therefore improbable that the pericyte produced this layer by itself. It seems very probable that it is formed through the interaction of the pericyte with the ependyma or astrocytes. In this context the astrocyte, ependymal cell and the Schwann cell have two roles; to synthesize and secrete glycosaminoglycans and to control the precipitation of collagen. Cytoplasmic specializations in the basal expansions of astrocytes close to the basal lamina (2102) are compatible with this role, but could equally well be a result of this relationship. Vascular endothelium and arachnoid mater share certain barrier roles (201).

Neurons

Although the neuron as a cell will be described in Chapter 2, it is considered briefly here as a constituent of an epithelium. A striking feature of the nerve cell is its capacity to generate and to conduct action potentials. This capacity is not unique. Pulses of electrical activity are propagated without decrement through much of the epithelium of *Hydra medusae*, except at the base of the tentacles (1882, 1883). Care was taken to avoid conduction through the nerve net. Similar conduction is found in other coelenterates (1517) and in the skin of tadpoles (2455, 2456). Conduction without decrement is also found in plants: in internodal cells of *Characea*, in the parenchyma of protoxylem and phloem of *Mimosa*, and in *Dionea* (2682). The synthesis of substances which act as neurotransmitters is not confined to neurons, or indeed to brain. For example, acetylcholine, indolamines and catecholamines are synthesized by the developing sea urchin at definite stages; it is not known whether these mediate intercellular communication in this preparation. Drugs which in mammals facilitate or antagonize the action of neurotransmitters cause disordered growth in *Hydra*.

Some neurons, such as primary sensory neurons, motor neurons, neurosecretory cells and many nerve cells of the autonomic nervous system make limited contact with a basal lamina. Virtually all nerve cells which have their processes confined to the central nervous system do not make such contact except for trivial, inconstant and probably coincidental and transient apposition. It is possible that neurons are not capable of synthesizing those substances that are necessary for the formation of a basal lamina, and that the apparent insulation of nerve cells from this structure reflects this inadequacy. The most likely and specific defect would then be of factors required to polymerize and precipitate the products of mesenchyme. The evidence summarized above indicates that mesenchyme normally influences the biochemical and structural differentiation of epithelial cells by a process which involves, or leads to, the formation of a basal lamina. It is therefore possible that the separation of a neuron from direct mesenchymal influence frees it from the restrictive aspects of such differentiation; this may permit its structure and biochemical properties to be determined more easily by other contacts, such as those with other nerve cells. The inductive influence of mesenchyme is such that it is very likely that when a part of the neuron is in contact with a basal lamina, this part constitutes its dominant pole and has an overriding influence upon the direction and intensity of the metabolic activity of the perikaryon. The response of such neurons to axotomy and to related procedures (Chapter 7) is clearly compatible with this hypothesis, but does not prove it. Many neurons are joined by gap junctions which may be between dendrites, between cell bodies, or between an axon and a dendrite or perikaryon (113, 126, 206, 207, 209, 777, 994, 995, 1095, 1158, 1218, 1348, 1627, 1628, 1648, 1787, 1933, 1968, 2128, 2210, 2212, 2233, 2732, 2766, 2768, 2797, 3057, 3059). Between particular nerve cells the position of the gap junctions is constant. They provide channels for electrotonic coupling, which can facilitate the synchronization of impulse initiation in cells which are already partially depolarized through the action of other synapses. The degree of coupling between cells depends upon the number of junctions, their size and their position on the cells. The degree of coupling can be altered physiologically if the cells are joined by a bridge of cytoplasm belonging to another cell (2797), or pharmacologically by many agents which change the intracellular concentration of calcium ions (102, 2210, 2482). The degree of coupling is reduced if the cells are depolarized, and then restored if the cells are hyperpolarized.

Oligodendrocytes

These cells, like most neurons, are not characteristically epithelial. They too fail to make contact with a basal lamina. In addition they do not form gap junctions with similar cells. They are heterogeneous. Some interfascicular oligodendrocytes form myelin (14, 388, 1664, 1665, 2251, 2252, 2253, 2254, 2257, 2625, 2626, 3112).

Other oligodendrocytes seen near cell bodies of neurons are

termed 'perineuronal oligodendrocytes'. This term has three shortcomings. Firstly, it is not possible by light microscopy alone to determine whether an oligodendrocyte really abutts on a nerve cell body. In some situations thin astrocytic processes lie between neurons and these oligodendrocytes. This is so in Deiter's nucleus, from which were taken cells for the microchemical analyses which led to hypotheses of extensive metabolic collaboration between nerve cells and glia (602, 1153, 1448, 1449, 1450, 1451, 1454, 1455). In some other regions oligodendrocytes are really next to neurons (1664, 1665, 1750, 2098, 2202). Secondly, some of these oligodendrocytes may have formed the myelin sheaths of presynaptic fibres, or of fibres passing between neurons to remote endings. If so their proximity to neurons is fortuitous. Serial electron micrographs (2771) could indicate whether their processes pass to myelin sheaths. Thirdly, the term 'perineuronal' is applied only to such cells as lie in the vicinity of the neuronal perikaryon and not to those in the neuropil which may lie equally close to dendrites.

Yet other oligodendrocytes may lie only transiently in white matter. Some of these are almost certainly migrating from the periventricular zone; some of these divide *in situ*. Some probably die in the fibre tracts, although obviously dying cells are seldom found. Others may migrate more remotely. Their further differentiation is uncertain. These cells have been called glia of the 'third type' to emphasize their uncertain nature. The term is conveniently non-commital, but unfortunately is numerically accurate only if ependyma is disregarded (3003). Glia of the third type might become perineuronal oligodendrocytes but despite autoradiographic study this has not been convincingly proved. Even if it were, it would show only that a cell of uncertain antecedents had become one of unknown function. They might differentiate into astrocytes, but this seems to be unnecessary as mature astrocytes can themselves divide (451). Demyelinated axons of the central nervous system can be remyelinated under some circumstances (164, 544, 834, 1065, 1085, 1102, 1349, 1611). The remyelinating cells are derived from others intermingled with myelinating oligodendrocytes in adjacent intact fibre tracts (1610, 1611, 1612). It is not known whether these are derived from the division of differentiated interfascicular oligodendrocytes, or from glia of the third type. The latter appears to be more likely.

Microglia

Cells defined as microglia by metal impregnation are certainly heterogeneous. Some of these are identical with glia of the third type, within fibre tracts. Macrophages in injured brain, which are derived from blood monocytes are also stained. Vascular pericytes are also commonly impregnated. The vagaries of staining by techniques of metal impregnation make them an insecure sole criterion for classifying cells (423, 2047, 3020). Because of this heterogeneity the term 'microglia' is avoided in this book as far as possible.

Glial cell turnover

Some glial cells of adult animals are labelled by tritiated thymidine. Many of these are subependymal, a region in which persists part of the periventricular layer from which neuroblasts and glioblasts migrated (65, 368, 620, 986, 987, 1343, 1603, 1637, 1642, 1710, 1711, 2043, 2620, 2740, 2818, 2819). Many of the labelled cells die without migrating or differentiating (2206). Some migrate (2736) especially along fibre tracts, in which they also divide. The division, migration, differentiation and death of these cells are presumably regulated, although nothing is known of the factors involved in such regulation, whether, for example, the repeated induction of spreading depression influences these processes. It is difficult to study quantitatively, as the stem cells, daughter cells, migrating cells and other glia which may or may not be related, all divide. Although the migrating cells appear to pass peripherally along fibre tracts towards the cortex, it is not known if they pass only towards the cortex; nor is it known if they are actually guided – whatever that may mean – by the tracts of parallel fibres and their sheaths. There is no indication that they are guided by any residual radially oriented system of fibres, such as guide neurons that develop late (2384, 2385, 2386, 2390).

Regional variation of glia

Protoplasmic astrocytes are found in grey matter and fibrous astrocytes predominate in white matter. The former have more microtubules and fewer aggregated filaments; they also contain less immunologically detectable glial acidic fibrillary protein (239, 240, 3001, 3002). Microtubules are concerned with creating cell assymmetry (411, 1058, 2321, 2900, 2905, 2924, 2926) or with the movement of cytoplasm or organelles (81, 138, 1379, 2321, 2368) (Chapter 2). In the developing astrocyte microtubules are present in growing processes. Their role in mature protoplasmic astrocytes is less clear. They might indicate continuing lability of the cell's processes or considerable transport along the processes.

The structure of astrocytes is as strikingly different in different parts of brain as is that of neurons. For example, Bergmann glia of the cerebellum, Müller cells of the retina and thalamic astrocytes each have a characteristic pattern of arborization. The distinctive patterns of the major branches are not secondary to an overriding previously determined pattern of neuronal processes. The finer terminal branches are influenced by the distribution of nerve processes, and in some parts of brain adopt characteristic arrangements around nerve cells and terminals (2255). In some parts of brain, processes of glial cells run as ordered bundles (922).

The matrix and glycocalyx of brain

The glycocalyx covers all aspects of most cells (1055, 1573, 1934, 2237, 2238, 2395, 2396, 3105, 3156). It is specialized at junctions and at basal laminae. It can be demonstrated histochemically in brain (277, 286, 287, 288, 1847, 2237, 2275, 2396, 2466, 2882). In brain it is largely composed of hyaluronic acid, chondroitin-4-sulphate and chondroitin-6-sulphate, with lesser amounts of heparan sulphate and dermatan sulphate (6, 503, 716, 1914, 1972, 2181, 2721, 2722, 2723, 2870). Cultured neuroglial cells synthesize hyaluronic acid, chondroitin-4-sulphate and chondroitin-6-sulphate (716). Synaptosomes have characteristic glycoproteins. The properties of the glycocalyx are likely to be the same in brain as in other tissues. It influences the social behavior of cells: their capacity to aggregate, to move upon each other or to form junctions (3, 1913, 2053, 2054). This influence is brought about in part through the action of glycoprotein-enzyme bridges (2510, 2511, 2512, 2513): one cell contains a glycoprotein, the end of which can act as a substrate for a glyco-glycoprotein glycosyl transferase. The enzyme is on the surface of the other cell, probably as an intrinsic membrane protein

(Chapter 3). The enzyme binds the substrate, so the cells adhere; the adherence can be broken if the reaction goes to completion. This occurs if the appropriate nucleotide-sugar is available. The sugar is added to the glycoprotein and the molecule is released from the enzyme. The glycocalyx presents a stereospecific array of sites on the cell surface (392, 393, 472) and masks deeper sites. It is not known if the glycosyl transferase system is the only way in which specific adherence is obtained or if there is an ordered array of glycoproteins on the cell surface which matches precisely a corresponding array of enzymes. Not all the glycoproteins need be on one cell and all the enzymes on the other. If the array is not an ordered mosaic, the individual molecules can probably float about like icebergs in the planar fluid of the membrane (Chapter 3), until they become more anchored by incomplete glycosylation. It is not known how this early adherence may pass on to the formation of specialized junctions or how adherence changes the patterns of movement or metabolic activity of the interacting cells.

The glycocalyx can also alter the concentration of ions in the immediate vicinity of the membrane. Although interaction between cations and polyvalent macromolecular anions conforms *in vitro* to the predictions based upon mass action in simple systems (298, 2559, 2560, 2561, 2562), the situation *in vivo* is complicated by other factors: by changes in hydrogen ion concentration, to which cation binding is very sensitive (169, 436), by chelation of divalent cations by nucleotides (2559) and by the ensuing change in molecular shape (130, 279, 504, 1981, 2527). The potassium permeability of some cells is reduced when they are treated with neuraminidase (1086), but their ability to transport sodium ions, glucose and lysine is barely affected. If hyaluronidase is applied to brain, seizure discharge patterns ensue (1214); it is not clear if this is due to the altered membrane properties of the neurons alone, to a possible reduction in the permeability to potassium of the glial cell membranes, which could reduce their capacity to sequester this ion passively when its extracellular concentration rises rapidly, or to altered efficacy of synapses of pathways of recurrent inhibition or excitation. The amount of extracellular material that can be demonstrated histochemically is greater near the axon hillock and node of Ranvier than elsewhere; these are sites of greater rates of ion movement across the membrane (548, 1709). The significance of such accumulations is demonstrably slight under experimental conditions in which the volume of the bathing fluid is great and in which measurements are made after imposed alterations in ion concentration have become stable. Predictions of changes in the resting membrane potential are confirmed experimentally when the ionic concentration of the bathing fluid is altered even though the predictions make no allowance for the binding of ions by molecules of the glycocalyx and matrix. On the other hand, the release and binding of ions by these molecules may be more important when the volume of extracellular fluid is restricted or when its composition is changing rapidly.

Extracellular fluid of brain

Extracellular fluid does not drip from the surface of a freshly isolated brain. Like the extracellular fluids of other organs (1904), that of brain is not easily seen, being held in the gel of the matrix. The composition of this fluid has been repeatedly studied, and its volume has been the subject of some controversy (660, 1251, 1255). Dilution methods (869, 1779, 2170, 2392, 3180, 3221), impedance measurements (957, 1257, 2397, 2398, 2399) and electron microscopy of rapidly frozen brain (1249, 1250, 1253, 1254) all indicate a volume of about 12 to 24% in common laboratory animals. Lower values derived by electron microscopists from conventionally fixed material indicate the shift of fluid from the extracellular compartment into the cell during asphyxia (1399, 2770, 3199).

Exogenous markers of molecular weight 20 000 to 70 000 daltons introduced either into the blood or into the cerebrospinal fluid cannot pass the occluding junctions of mammalian vascular endothelium (Fig. 1.6) (341, 343, 345, 346, 869, 1780, 2315, 2391, 2392, 2425). They can penetrate from the cerebrospinal fluid between the ependymal cells and through the neuropil as far as these junctions. If horse-radish peroxidase (43 000 daltons), or albumin (70 000 daltons) can diffuse freely past all surfaces of the mammalian astrocyte, it is likely that ions can also. It is therefore improbable that an astrocyte attempts the sisyphean task of maintaining actively an ionic gradient between different aspects of its surface (2213). It is more likely that differences between the

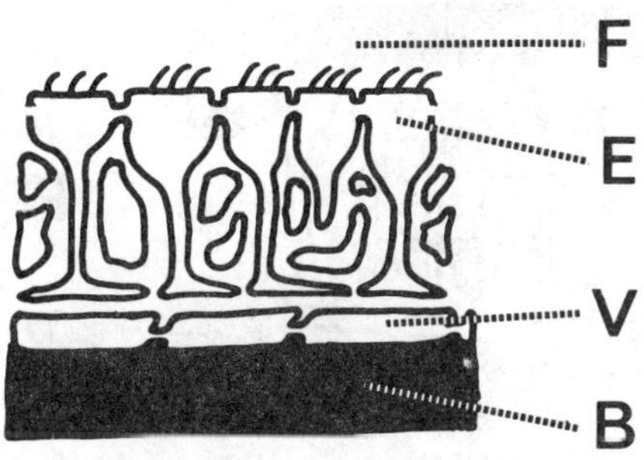

Figure 1.6 Diagram of a section of part of brain which has permeable capillaries, showing a line of ependymal cells, E, facing cerebrospinal fluid, F: and vascular endothelium, V, facing blood, B. *Top.* When horse-radish peroxidase is injected into CSF it diffuses past the junctions between the ependymal cells, through the extracellular fluid and basal lamina to the deep aspect of the vascular endothelium. It passes into the cleft between adjacent endothelial cells until its further progress is stopped by the occluding junctions between them; hence none enters blood. The distribution of horse-radish peroxidase is indicated in the diagram by a dense black 'reaction product'. *Bottom.* When horse-radish peroxidase is injected into blood it penetrates the clefts between endothelial cells only as far as the occluding junctions between endothelial cells: hence, none can enter the extracellular space by diffusion.

compositions of plasma and extracellular fluid are maintained by vascular endothelium, the cells of which are mostly joined by occluding junctions. The difference of potential between blood and cerebrospinal fluid (253, 422, 1288, 2961, 3198) is almost certainly sustained across the endothelium. The endothelium of brain transports some substances (107, 237, 589, 590, 591) and metabolizes others.

Those parts of brain which have permeable capillaries and impermeable ependyma (189, 1903, 2360, 3162, 3181) sustain the differences between plasma and cerebrospinal fluid across the ependyma (Fig. 1.7). Ionic differences between plasma and extracellular fluid that can be altered actively are brought about by electrogenic pumps at these sites. Such active transport need not of course occur at all sites of occluded passage. It is very difficult to measure the contribution of different regions or types of cell to

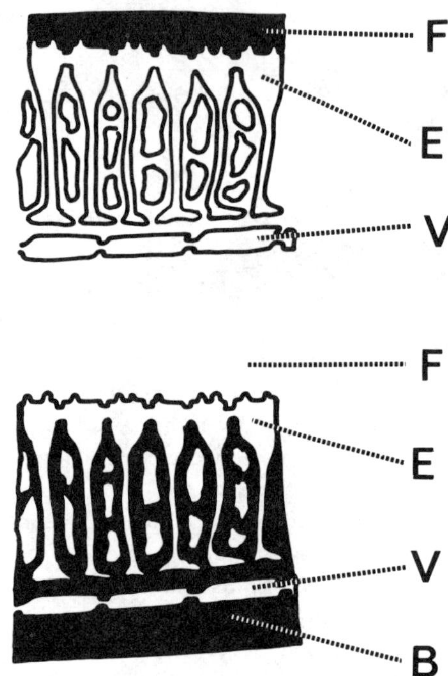

Figure 1.7 Diagram of a section of part of brain which has permeable capillaries, showing a line of ependymal cells, E, facing cerebrospinal fluid, F: and vascular endothelium, V, facing blood, B. *Top.* When horese-radish peroxidase is injected into CSF it cannot pass, in these parts of the brain, the occluding junctions between ependymal cells; hence the exogenous protein marker cannot enter the extracellular space. *Bottom.* When horse-radish peroxidase is injected into the blood it can diffuse between the endothelial cells to gain the extracellular fluid. It can diffuse through this as far as the occluding junctions between the ependymal cells, but no further. The 'reaction product' is shown here dense black. Compare with Figure 1.6.

active ion transport, to determine, for example the relative contributions of vascular endothelium, ependyma of the choroid plexus and of ependyma with tight junctions in other parts of the brain in transporting bicarbonate from brain fluid compartments into plasma during processes of acclimatization to high altitude. There is no significant barrier to the free diffusion of substances between cerebrospinal fluid and the extracellular fluid as a whole, except for the reservations expressed below.

The 'blood-brain barrier' is therefore a result of different factors, regionally heterogeneous and depending upon the substance examined and the circumstances of its investigation (1671). It is not a simple mechanical obstruction due to one type of cell or to the presence of particular occluding junctions alone.

The division of extracellular space into compartments by processes of astrocytes has been suggested to explain the prolonged activity of transmitter substances applied close to postsynaptic sites (605, 606, 607, 771). The prolonged activity has been ascribed to impeded diffusion. Some synaptic complexes are characteristically surrounded by astrocytic sheaths (100, 2255, 2256, 2454) (Figs. 1.8, and 1.9), but it is uncertain whether they retard diffusion. Exogenous markers can penetrate synaptic clefts (184, 339, 3012), but they were allowed to diffuse for a time that was very long by comparison with the duration of transmitter action; the marker molecules are larger and have a different charge density and distribution. The synapses to which exogenous markers have been shown to gain access are not those from which evidence has been obtained for restricted diffusion of transmitter. Labelled noradrenaline, 5-hydroxytryptamine, gamma-aminobutyric acid and glycine can gain access to some perisynaptic regions from the remote extracellular space (40, 278), and transmitter substances released from synapses and their degradation products can be detected in cerebrospinal fluid (775). It is uncertain however whether these substances have diffused through the extracellular space alone, as they can be taken up and concentrated by glial cells, and released from them. The diffusion of cationic transmitters could be impeded by polyanions of the matrix (1725, 1726, 1969), which are distributed irregularly, especially in relation to synapses (184, 277, 286, 288, 2396, 2466). The problem has certain similarities to the compartmentation, distribution and uptake of inspired gas in the lung.

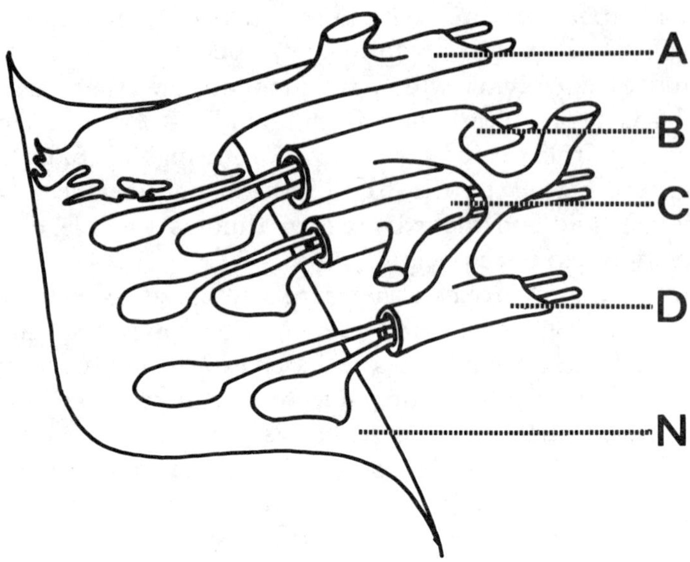

Figure 1.8 Astrocyte processes are frequently wrapped around axons and boutons. The astrocytic wrapping is shown at A, but the wrapping around the boutons has been removed in B, C and D, to show the boutons. N is a surface of the postsynaptic nerve cell. These astrocytic sheaths are shown intact in Figure 1.9. Note that more than one bouton or axon can be contained in one wrapping, exposing them to the same microenvironment, and possibly facilitating interaction between them. Note too that one astrocyte can envelop more than one group of fibres; for example, B and D are contained in different wrappings of one astrocyte.

Difficulties were met in attempting to reconcile the measured sodium content of brain (633, 800, 1255, 1804, 2929, 2965, 3203) with the apparently small extracellular space seen on electron micrographs (1399, 3199). These led to the hypothesis that some of the sodium must be intracellular (1559). As the electrophysiological properties of neurons indicated that it could not be contained in neurons in a sufficiently high concentration (766), it was suggested that some neuroglial cells should have a high concentration of sodium (1559, 2073). It is now evident that the electrophysiological properties of glial cells are incompatible with this hypothesis, and that the extracellular space of brain has been seriously underestimated by conventional electron microscopy. Furthermore,

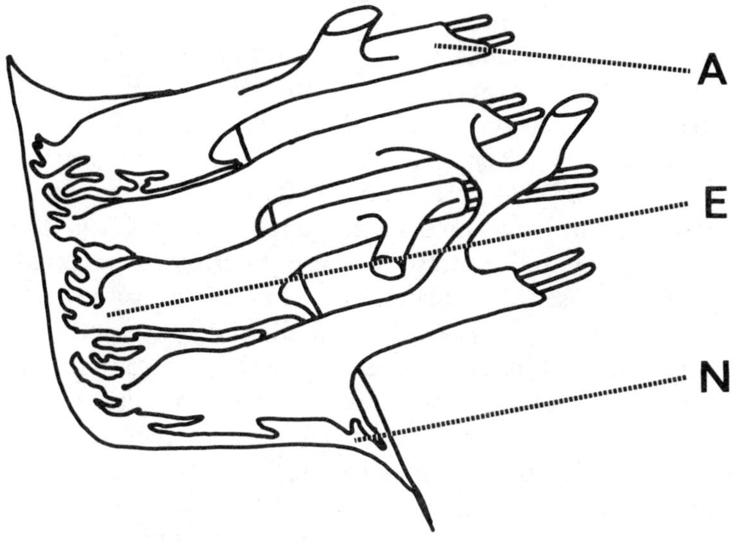

Figure 1.9 The processes of an astrocyte can cover boutons and much of the surface of the postsynaptic neuron. A is a process of an astrocyte ensheathing some preterminal axons; E is an expansion of an astrocyte which covers some synapses; and N is a surface of the postsynaptic nerve cell. Note that very little of the postsynaptic surface remains uncovered. An oligodendrocyte lying against these boutons and upon the expanded processes of the astrocytes would appear by light microscopy to be 'perineuronal'.

a greater proportion of sodium is bound in brain than in muscle (941), so the electrophysiological effect of this sodium is slight. The hypothesis that some glial cells are 'high sodium cells' is now not only unnecessary, but is also incompatible with their known electrophysiological properties (1672, 2127).

Electrophysiology of neuroglia

Large glial cells of invertebrates (520, 521) have been studied electrophysiologically (177, 178, 179, 1669, 1672, 1673, 2123, 2124, 2125, 2127). The measured resting potential of 60 to 75 mV (1674) is likely to be an underestimate, as many of the cell processes are fine and liable to be damaged easily. Technically it is easier to study glial cells of the optic system of *Necturus*; these have resting potentials of about 90 mV (1673, 2003, 2187). This is close to the equilibrium potential for potassium. When the external concentration of potassium is altered experimentally, the cells behave as potassium electrodes (177, 178, 1669, 1672, 1673, 1674, 2123). The glial cells are coupled to each other through low resistance junctions, but not to neurons (1673, 1674, 2185, 3019). It is difficult to calculate the specific membrane resistance of these cells both because of their complicated shape and because of these junctions. The considerable spread of imposed depolarization makes a value exceeding 1000 $\Omega\,cm^2$ probable (1674). The membrane resistance is substantially constant within the range +75 mV to −75 mV. The glial membrane potential can be used to measure the extracellular concentration of potassium and to determine the alterations that accompany induced nervous activity (177, 178). Such measurements made at different temperatures show that potassium leaves the extracellular space by diffusion in the invertebrate, and that the redistribution of this ion across the membrane of the glial cell is passive (311). This might not be true for the mammalian glia. The extracellular concentration of potassium was also measured in invertebrates from the undershoot of the neuronal action potential. This measurement made both before and after the removal of the investing glial sheath showed that the extracellular concentration of potassium fell more rapidly after cessation of neuronal activity when the glia was absent than when it was present. Far from acting as a potassium buffer, the glia appeared to impede rapid restoration of the extracellular concentration of potassium ions.

Most mammalian glial cells in tissue culture have a resting membrane potential of 40 to 70 mV (1328, 1329, 1330, 1331), or less (3047). The membrane potential is greater when the cell density is high (3047). The cells are smaller than those of invertebrates and are more likely to be damaged when impaled. It is likely that this damage is responsible for the low membrane potential. The very low value derived for the specific membrane resistance (1330, 1331) almost certainly results from current leak at the site of impalement, from underestimate of the area of the cell membrane and from electrotonic coupling between adjacent cells. The reported 'responses to stimulation' (1327) result from breakdown of the membrane dielectric (3047). Similar results are obtained if a cell membrane is injured in other ways (2893). Cultured ependymal cells behave similarly (1330). No differences in behaviour can be detected between cells of cultures that are almost exclusively astrocytic and those that are oligodendroglial.

'Idle cells' of mammalian brain have a high, stable membrane potential and lack an injury discharge when they are impaled. They have no spontaneous or evoked action potentials and no postsynaptic potentials (33, 440, 553, 950, 1166, 1167, 1169, 1538, 1662, 1796, 1932, 2208, 2209, 2283, 2759–2761, 2850, 2890, 2892, 2947, 3061). A few cells have a lower resting membrane potential (1168). There may be two classes of idle cells with different membrane potentials (1323, 3061), but the contribution of cell damage in producing this difference is uncertain. If the difference is due to the presence of two types of cell, they have not been correlated with morphological types. Idle cells in brain do not behave as ideal potassium electrodes (678, 1662, 2208, 2209), but this departure could be due to cell damage. When the external concentration of potassium is raised by direct or indirect neuronal stimulation, or by spreading depression, a slow depolarization of these cells occurs (440, 678, 1168, 1169, 1323, 1538, 2850). Iontophoresis of various transmitter substances affects the membrane potential of idle cells directly or indirectly (1662). Acetylcholine and high concentrations of gamma-aminobutyric acid depolarize idle cells without changing their membrane resistance; the way in which they do so is uncertain. The depolarization could be caused by release of potassium from excited neurons; if so, the glial potential change is indirect. The transmitters could be acting directly upon the glial cells and producing the potential change without alteration of the membrane resistance either by activating an electrogenic pump or by altering the surface fixed charge. Although both of these effects could be mediated in glia by cyclic nucleotides (Chapter 5), they are improbable causes, because transmitter substances have no such effect upon cultured glia. However caused, depolarization of idle cells, which are almost certainly glia, contributes significantly to recorded changes of surface potential (536, 1538, 2003, 2850).

The plasma membranes of many other types of epithelial cell respond similarly. In most the measured ratio of sodium to potassium permeability is higher than in glia (1061, 1827, 1941, 3123, 3135, 3179), but as in glia the severity of damage sustained by the plasma membrane is uncertain; it is certainly difficult to maintain a stable resting membrane potential in these cells. Their potential too is altered by the extracellular concentration of potassium; their failure to behave as ideal potassium electrodes can also be ascribed hypothetically to membrane damage.

Impedance of brain

The electrical impedance of a tissue depends upon a number of factors: the extracellular volume, the conductivity of the extracellular fluid and the resistance of the cell membranes limiting the extracellular compartment (Fig. 1.10). In an organ as heterogeneous as brain the results are difficult to interpret. Changes

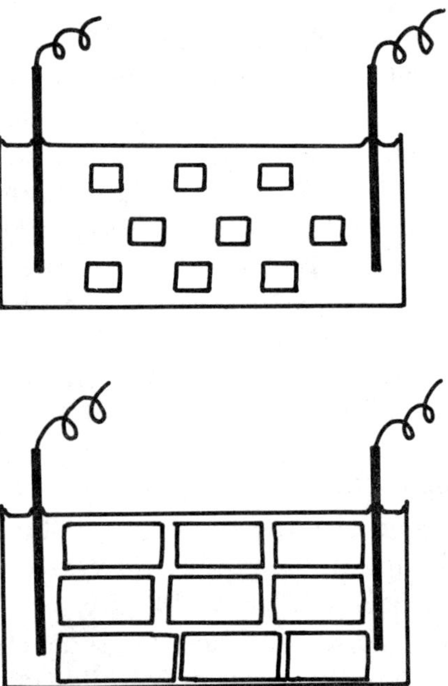

Figure 1.10 A tissue is represented here as a bath of extracellular fluid containing two electrodes and a fixed number of square 'cells'. *Top.* The cells are small and the volume of extracellular fluid is large; impedance is therefore low. *Bottom.* Fluid and electrolytes have entered the cells; the volume of the extracellular fluid has fallen and that of the intracellular compartment has increased. If the cell membrane resistance remains very high, the impedance of the tissue rises.

of impedence which have been measured in brain may be divided arbitrarily into three classes depending upon the duration of the change.

In the first (1604, 1605), impedance decreases by 0.1% or less for about 10 ms during an evoked cortical potential change. This decrease is greatest in the layer of cortex in which the phase of the evoked potential alters. This suggests that the permeability changes are associated with synaptic events, but nothing more is known of their significance.

The second type of impedance change lasts longer and is larger. An initial period of increased impedance lasting 200 to 1000 ms is followed by a period of increased impedance lasting between 4 and 5 s. In sympathetic ganglia these changes are produced by preganglionic stimulation (248). As the second phase can be prevented by hexamethonium it may result from changes in the ganglion cells or interneurons. Hypothetically this second phase could result from the synthesis of tri-phosphatidyl inositol under the direction of dopamine-regulated adenyl cyclase (Chapter 5). The first phase is resistant to hypoxia, fatigue and hexamethonium and has been ascribed to presynaptic activity. Although the total blood flow did not change at the time of the impedance shifts, changes in blood volume or in the distribution of blood within the ganglion were not sought. It is not known if the extracellular volume of the ganglion alters during the period of changed impedance. Similar alterations of impedance in the hippocampus can be correlated with gross changes in behaviour (31, 32), but not at present with precisely defined changes in neural activity. The changes have been interpreted in two ways, each speculative: firstly, as due to a change in the membrane resistance or volume of glial cells, and secondly, as a result of altered conformation of the extracellular polyanions of the matrix and glycocalyces. It has already been seen, however, that the membrane resistance of glial cells does not alter when they are depolarized by the activity of surrounding neurons. The changes cannot yet be explained adequately by altered conformation of constituents of the glycocalyx. These polyanions bind various cations with different affinities (504, 1555, 1724, 2559, 2560, 2561, 2562, 2637, 2997), but there is no evidence that one cation displaces another significantly during nerve activity. The variable changes of impedance seen after injecting calcium ions locally (26, 27, 28, 29, 3042, 3043,

3044) are less if the hyaluronic acid has first been hydrolyzed (3042, 3043), even though this is not the only polyanion present. The striking differences between the effects of magnesium and of calcium upon impedance (29) contrasts with their similar binding affinities and with the competition between these divalent ions for the same binding sites (436). The causes of these changes must still be considered as unknown. It is likely both that the changes which occur on arousal (30), and that the very large changes found in the subicular region during paradoxical sleep (2399) should be considered as belonging to this type of impedance change.

The third type of impedance change is far slower and very large; it is usually caused by asphyxia, and is also seen during seizure activity and during spreading depression (30, 248, 1258, 1259); by analogy with these states it is likely also to accompany migraine. Chloride moves into dendrites and into some astrocytes (1249, 1250, 1252, 1254, 1256, 1260, 1830), and their volume increases. As this change of volume closely resembles the response of cells of other epithelia to adversity (1252, 1830), it cannot be considered as a specialized response of brain. The part played by potassium-dependent enzymes in this active uptake of chloride, and the possible role of bicarbonate-dependent ATPase remains to be clarified.

Biochemical properties of glia

Good methods for obtaining a high yield of glial cells of a single type that retain their metabolic and structural integrity have yet to be developed. The glia should also be obtained from discrete parts of brain from animals of known neurophysiological and endocrine state.

Methods of bulk fractionation provide heavily contaminated mixed glial fractions (583, 859, 881, 918, 958, 1224, 1625, 2487, 2489, 2565, 2566, 2998, 3166). Glial cells isolated from white matter are usually assumed to be almost entirely oligodendrocytes (879, 880, 2369); it is difficult to reconcile this assumption with the observation that about 25% of cells present in fibre tracts are astrocytes (2047, 2048). Furthermore white matter contains several different kinds of oligodendrocyte, which probably differ biochemically.

Lumps of neuropil adjacent to neuronal perikarya have been analyzed as 'perineuronal oligodendroglia' (602, 1153, 1448, 1449,

1450, 1451, 1454, 1455). These fragments correspond to the contaminated glial fractions obtained by bulk separation. The method has the advantage that the brain area from which they are derived is known. Neither the relative contribution of each type of cell, cell process or fragment that they contain, nor the viability of each constituent has been reported. Without such measurements the results cannot be interpreted meaningfully. In other studies single identified glial perikarya have been obtained from discrete brain areas. Only very limited and non-specific measurements have been made (2859, 3068). The degree of cell damage and the extent of contamination by adhering fragments of other cells have not been assessed quantitatively.

The metabolic activity of cultured neuroglia (389, 390, 1236, 1400, 1742, 2319) is influenced by the composition of the medium (963, 3066, 3069). The state of the cells in culture is therefore unlikely to be normal. Measurements made on well-differentiated cultures where neurons and glia are tightly packed will probably indicate better the normal metabolic state. Unfortunately, however, this close packing will increase the difficulty of isolating glial cells for measurement. The growth of glia beneath the skin of newborn hamsters is an interesting attempt to circumvent the influence of the artificial medium, but the composition of the microenvironment of these cells is not known (2669). The use of fresh tumour tissue suffers from the abnormality of the starting material, cellular heterogeneity and the presence of reacting cells (598, 1748, 2162). Cultivating tumour tissue removes the influence of host reaction, but substitutes those of tissue culture. It is a very popular method, and several clones are now available. The results obtained are often clear and frequently interesting; nevertheless their application to the role of glia *in vivo* must be cautious. For example, some cultured glia have been found to synthesize nerve growth factor; while they might perhaps also do so *in vivo* this cannot be presumed (1834). It is by no means uncommon for epithelial tumours to synthesize active polypeptides, which cannot be detected in the normal epithelium from which the tumour arose. Or again, several clones of glia have an adenyl cyclase which can be activated by a neurotransmitter (Chapter 5). It would not be surprising if glia *in vivo* responded similarly, for this is a characteristic of some other epithelia also – such as liver; it must not, however be assumed from study of tumour tissue alone. A fundamental difficulty of using cultured or tumour tissue for biochemical studies is that the interaction between different types of cell, which is of overriding importance in brain, cannot be properly studied at present. The use of cultured tissue is however of great value for morphological studies. The interaction between different cells can be followed by light microscopy, and confirmed in the same cells after fixation, by electron microscopy. This technique has been useful in studying myelination, and is likely to be very valuable in the study of genetic disorders of the nervous system (Chapter 9), both in removing the abnormal tissue from the influence of remote organs of the host, and in allowing the interaction to be followed between cloned cells of different kinds, some of which are genetically abnormal; this could show if the genetic abnormality is confined in its expression to one type of cell, and whether the expression can be influenced by other cells or by the medium in which they have been grown.

Histochemistry and autoradiography have considerable limitations; only in the perikarya can the cell that is overlain by reaction product or silver grains be identified by light microscopy. Electron microscopy usually shows the reacting elements of the neuropil with low sensitivity and presents great difficulties of sampling.

The metabolism of ependyma and of astrocytes, so far as it is known, differs little from that of other epithelia; that of the myelinating oligodendrocyte is dominated by the myelin (Chapter 4); the metabolic activity of nerve cells differs because of the production and effects of neurotransmitters (Chapters 2 and 5), and is quantitatively unusual in some respects because of the peculiar geometry of the cells. While it seems possible that the brain might produce some polypeptides as a result of certain patterns of activity (2972, 2973, 2974, 2975), there is no substantial evidence that peculiar and unique molecules are responsible for 'memory'. It is becoming more and more likely that the modification of behavior by experience is achieved through cellular and biochemical interactions that are not in themselves unique, but are probably shared to a varying degree with other tissues at various stages of their development.

There is a large literature on the metabolism of neuroglia that reflects the severe methodological limitations.

DNA

Most glial calls are diploid. In some species astrocytes near Purkinje cells are tetraploid (1712, 1713). DNA is synthesized by some glia around stimulated or injured nerve cells (732, 3062); this is detected readily in autoradiographs after tritiated thymidine has been injected. Under these circumstances labelled DNA can also be found around the nucleoli of the adjacent neurons. It is unlikely that this indicates either metabolic collaboration between neurons and neuroglia or the synthesis of DNA by nerve cells; it is more probable that nerve cells ingest labelled DNA released from glial cells dying near them. Such uptake occurs in cultured cells (1333); the surviving cells take up labelled DNA from cells dying near them, and this ingested DNA also lies around the nucleolus. Some glial cells are labelled in normal brain, especially in certain subependymal regions and provide evidence for the continuing turnover of these cells.

RNA

RNA polymerase activity is as great in neuroglial cells as in hepatocytes (153, 1556). The types of RNA synthesized are unexceptional (1489, 1864, 1865, 2875); they appear in the same sequence and at about the same rate as in other types of cell (435, 703, 1142). Neuroglia synthesize at least one protein that is unique to brain (1808, 2985, 3244), and probably others (240, 821, 1564, 2381, 2382, 2680, 2863, 2984, 3049, 3050, 3051). Therefore there must be at least one RNA sequence that is also unique. Several tissues have been shown to synthesize certain proteins exclusively. Brain as a whole has several, which are present in some types of cell or cell process rather than in others; for example, some are concentrated in synaptosomes, others in myelin (Chapter 3), and yet another is limited to reacting or fibrous astrocytes. The bulk of the evidence indicates that RNA metabolism in glial cells and in neurons does not differ qualitatively from that of other cell types, except in the production of such messenger molecules.

Structural proteins

The protein metabolism of brain has been reviewed (734, 1704, 2679). Structural proteins of neurons, which are of interest in relation to the geometry of the cells and the problems associated with the movement of organelles, are considered later on in Chapter 2. Studies describing the incorporation of radioactively labelled amino acid by neuroglial cells are of three kinds. In the first, isotope is given to an intact animal and the cells are later fractionated (276, 1232). In the second, brain slices are incubated *in vitro* and then separated into 'neuronal' and 'glial' fractions (276, 1232). In the third, cells are first fractionated and then incubated in suspension (1230, 2488, 2931). All methods are subject to the extreme limitations of the methods of cell fractionation currently used. In addition, the first method suffers from the difficulty of introducing the labelled amino acid into brain, and of knowing the concentration in the vicinity of the cells synthesizing protein. In the third method, many of the nerve cells are dead and virtually all are injured before their period of measured protein synthesis begins. The second approach is probably the most useful available now. No firm conclusions can be drawn from the studies yet except that protein is synthesized by glial cells in the same way as in other cells. Subfractionation of organelles derived from a 'neuroglial' fraction is unlikely to yield preparations of greater purity than the starting material; in theory it could do so if, for example, a given type of organelle in glial cells of an impure 'glial' fraction had a density different to the same type of organelle in the contaminating cells. There is no evidence to suggest that this is a method that can be usefully employed. The incorporation of amino acids into mitochondria derived from a crude 'glial' fraction is unremarkable (1230). Except for a high rate of incorporation of amino acid into the nuclei of malignant astrocytes (1221), the synthesis of nuclear protein has no unusual features.

Enzymes

It is neither profitable nor possible to attempt to review the very large literature concerned with the activity of glial enzymes. Quantitative studies are confounded by the impurity of the preparations used, whether the cells are 'isolated' by bulk fractionation or by microdissection. Discrepancies abound between biochemical, cytochemical and histochemical studies. For example, in studying oxidative enzymes best agreement is obtained between measurement of mitochondria and of cristae (2352) on the one hand, and subjective histochemical assessment on the other. Both show high oxidative capacity in the choroid plexus (2162, 2357), less in ependyma (961, 962, 1292, 2094, 2162, 2232, 2772, 2910), and in neurons (962, 2056, 2057, 2162, 2910), and least in other glia (962, 2056, 2057, 3036). These observations correlate poorly with microchemical measurements of 'glial' activity measured on perineuronal lumps of neuropil. In these the oxidative activities have been described as exceeding those in nerve cells (1230, 1231, 1232, 1233), possibly because of the high density of dendrites and boutons in the neuropil (2767, 2768). The similarity between the oxidative capacity of the 'neuronal' fraction of brain and that of the 'glial' fraction (2487) probably results both from neuronal death (281) and damage and from contamination of the 'neuronal' fraction by glia. At present the most reliable method of estimating oxidative enzyme activity in different types of cell in brain is histochemical – despite all its disadvantages. Even with this it is difficult to interpret the results obtained; because cell processes often cannot be identified in the neuropil, because undue emphasis is easily placed upon a few prominent but uncharacteristic cells and because the kinetic state of the sequence of reactions leading to the assessed precipitate is usually not known. The greater oxidative activity of protoplasmic than of fibrous astrocytes (280, 2352) may be related either to their greater structural lability, if the greater number of microtubules in these cells can be assumed to indicate the active maintainance of cell assymmetry or the greater transport of organelles within their processes, or to the greater variability of their environment which results from neural activity. The wide range of activities demonstrated in interfascicular glial cells is compatible with their heterogeneity.

The measurement of ATPase activity also illustrates the unrewarding limitations of the present methods of cell fractionation. Bulk fractionation and microdissection (602, 1230) indicate that the ATPase activity of contaminated neuropil is as high as that in damaged neuronal perikarya, or higher (1960). Astrocytes grown subcutaneously have a much lower Na-K-ATPase activity than neurons (814, 815), but an activity 10 times greater than that of erythrocyte ghosts. The reported increase in 'glial' ATPase activity found when the brain of the rat matures (1961) is likely to be due at least partially to the progressive contamination of this fraction with growing dendrites and developing synapses.

Reactions and roles of glia, lacking reaction with neurons

Discussion of the 'function' of neuroglia has been dominated by the assumption that their role is secondary to that of the neurons. There is no doubt at all that neurons and neuroglia influence each other (Chapters, 2, 4, and 5). This is not the whole story. In view of their epithelial nature the ependymal cells and astrocytes most certainly have other roles. These may be considered briefly under two headings: firstly, the cells of any epithelium are involved at some time in regulating the growth, differentiation, repair and replacement of the epithelium as such, and secondly there are more specialized activities such as secretion or absorption. It is unnecessary to assume that the astrocyte has one function. It may appear histologically to be a rather dull sort of cell derived from an epithelium and to sit there without doing very much, but so too of course appears the hepatocyte: and the hepatocyte has many functions. Both the hepatocyte and the astrocyte are cells with a great capacity for reaction. There are, incidentally, many interesting parallels between the two types of cell. For example, both contain glycogen; both have beta-adrenergic receptors that form the regulatory subunit of an adenyl cyclase which activates phosphorylase through the action of a protein kinase, so that the glycogen content is diminished (Chapter 5). The astrocytes in some situations react markedly during prolonged hepatocellular failure, or following porta-caval anastomosis.

It seems most likely that the cell responsible for maintaining the shape and gross structure of brain is the neuroepithelial cell in the first instance, but is later either the ependymoglial cell in those vertebrates which keep them, or the ependymal cell. This is likely *a priori* because these cells maintain a planar epithelium, lining the ventricular lumen; the great importance of the ependyma of the spinal cord of the goldfish in guiding the regenerating fibres after injury (Chapter 7) is consistent with this role of organization. It is possible that some astrocytes also act in this way during development, and guide migrating neurons radially through the thickening neuropil. The way in which the shape of any organ is controlled by its constituent cells is not known, but it seems to require the interaction of mesenchyme and epithelial cells in epithelia, the formation of a basal lamina, and communication through junctions. Models have been proposed in which phase differences occur between propagated waves of metabolic change (1109); the hypothesis is too general at present to interpret for brain.

After direct injury astrocytes become larger and divide (451, 1621). This starts within 8 hours of a needle stab. Their glycogen content increases and the activity of phosphorylase and of branching enzyme is raised (960, 962, 1192, 1196, 2166, 2675). The part played by adenyl cyclase in producing this change is not yet known; certainly noradrenaline can accumulate in damaged brain. It is known to cause further damage from intense vasoconstriction. It might also activate the adenyl cyclase. Cells of other epithelia respond to injury similarly (588, 2183, 2191, 2192, 2193, 2534, 2585, 2586). The response of mesenchymal cells to brain damage also resembles the response of these cells in other epithelia. Macrophages pass from the blood into the brain (1222, 1443, 1620, 1621, 2474, 2475, 3035), and after phagocytosis return to the circulation. Vascular pericytes do not become migrating phagocytes. In other epithelia also, macrophages enter from the blood and are not formed from local fibroblasts (3023). In brain, pericytes divide only when capillaries are made to sprout, probably because more basal lamina is required. In other epithelia also, new fibroblasts arise by the division of those that are already there (2505, 2506, 2507, 2888). The interactions between reacting astrocytes and pericytes cannot be examined easily in brain because of its geometrical complexity. Study of interaction between pia and astrocytes should be easier. As the responses of astrocytes and of mesenchyme individually resemble those of the corresponding cells in other tissues, it is likely that their patterns of interaction are also similar.

There is no evidence yet for secretion by astrocytes *in vivo*, although active polypeptides are produced by clones of cultured astrocytes. Some ependymal cells unequivocally secrete. Ependymal secretion in mammals may be confined to the choroid plexi and to the subcommisural organ. These latter ependymal cells make Reissner's fibre (354, 828, 1513, 1599, 1734, 1769, 1950, 2177,

2178, 2179, 2203, 2388, 3010, 3161). This gelatinous column of neutral mucopolysaccharide is formed in the third ventricle and passes caudally through the aqueduct, fourth ventricle, and central canal of the spinal cord to the terminal ventricle (1513, 1615, 2108, 2178, 2179, 2820), where it is ingested by macrophages. Although changes have been seen in these ependymal cells when the hormonal or electrolyte state of the animal is altered (101, 940, 1734, 2523), these are not found in all species. Changes in the staining characteristics of the cells are not accompanied by significant differences in the rate of movement of mucopolysaccharide in Reissner's fibre, measured autoradiographically. There is no evidence that the rate of secretion is regulated, or varies. The cells often atrophy with advancing age (148, 667); in man the subcommisural organ usually atrophies during fetal life (2383).

The occurrence of ependymal secretion here is unequivocal, but the role of the secretion is less clear. It may contribute to the homeostasis of cerebrospinal fluid within the confines of the central canal of the spinal cord by acting both as a moving flypaper to remove adhering small debris, and as a migrating ion-exchange column to bind charged molecules (2821) and to buffer local ionic changes. As there is no evidence that the debris is more abundant in the caudal part of the fibre and no quantitative description of its ion-binding affinities, these possibilities are entirely speculative. The pineal synthesizes and releases melatonin. There is no certainty that other ependymal cells secrete. The striking changes that occur in the ependyma of the third ventricle of the skunk during the menstrual cycle (1215), the sex difference found in tanycyte ependyma of the floor of the third ventricle (1677, 1678, 1679, 1680), and the atrophy of the ependyma of the third ventricle found in some patients with Cushing's disease (1283, 1284) are of unknown significance.

The ependyma controls the composition of cerebrospinal fluid in several ways: by secreting cerebrospinal fluid, by stirring the layer of fluid that has equilibrated with extracellular fluid, by secreting into the fluid a mucopolysaccharide, which passes through some poorly perfused cerebrospinal fluid spaces, and by sealing the cerebrospinal fluid from the extracellular fluid compartment in those parts of brain which contain permeable capillaries.

Immunological status of brain

Traditionally, brain has been regarded as an 'immunologically protected organ'. Evidence indicates that the degree of protection is slight (2438, 2580, 2581, 2582), and is expressed when the implanted material is a weak antigen or when the implanted tissue is poorly vascularized. Some heterologous tumours (1143, 1144, 1145, 2074, 2075), probably of low antigenicity (1631), can survive in brain. Younger heterologous tumours, which have been serially transplanted fewer times or not at all previously, and which are therefore less likely to have been 'emasculated' antigenically, are rejected more briskly (2075).

Animals have been sensitized by skin grafts applied either to the dermis or implanted in brain. An accelerated rejection occurred of a second identical graft applied later to skin; the rejection occurred at the same accelerated rate both in animals which had received their first graft intracerebrally and in those which had been grafted onto the dermis (2580, 2581, 2582). Tissue implanted in brain can therefore cause sensitization. If a skin graft is implanted into the brain of an animal which has been previously sensitized, it is rejected as rapidly as a graft placed on dermis (1959). Brain therefore does not protect a recently transplanted tissue from the reaction leading to accelerated rejection. On the other hand, grafts of skin and other tissues implanted into the brains of animals which have not been previously sensitized can survive for a longer period than such grafts implanted subcutaneously (1959, 2438). It seems likely that under these conditions brain has recovered from the trauma of implantation by the time that the immunological attack is mounted, and that cells and plasma proteins can no longer gain such ready access to the transplanted tissue. It would be expected that if such a graft or the immediately surrounding brain were injured, rapid rejection would follow. It is probable that this slender degree of relative protection results both from the absence of lymphatics and from the relative impermeability of the cerebral capillaries to large molecules. If so, the survival of the graft is likely to be critically dependent upon the size of the graft, upon the degree to which it undergoes immediate necrosis, and upon the density of its vascularization. Grafts do not survive well in the cerebellum, nor if they encroach upon the ventricles, the choroid plexus or the meninges (2074).

Constituents of brain can act as antigens. It is likely that the human encephalomyelitis following vaccination against rabies is of this type (1332, 2215). Isolated basic protein of myelin is antigenic, and if injected with an adjuvant can cause experimental allergic encephalitis in a number of species. Antigenic fragments of the basic protein can be obtained by digestion with trypsin or pepsin (840, 843). Experimental allergic neuritis is produced similarly by sensitizing a recipient with whole nerve in the presence of an adjuvant, or with a purified basic protein extracted from nerve (3029). In some species both neuritis and encephalomyelitis can be produced by a basic protein.

During experimental allergic encephalitis at least two kinds of cell enter brain, accumulate around blood vessels and are associated with demyelination: mononuclear macrophages derived from bone marrow, such as are found in similar lesions of skin (1639) or at sites of chronic inflammation (3023), and small lymphocytes (2753). The progression of the disease can be suppressed by saturating sensitized lymphocytes with soluble antigen, or by L-aspariginase. There is no evidence to implicate serum or cerebrospinal immunoglobulins in the initiation of the disorder. In man, 88% of normal individuals have an immunoglobulin G which reacts *in vitro* with myelin. This appears between the ages of 1 and 4 years, a period of considerable myelination. Immunoglobulin A and immunoglobulin M are found only infrequently (782). Antibodies can be raised to other brain-specific proteins (197, 497, 821, 972, 1564, 2381, 2680, 2863, 2984, 3049, 3051), but a progressive cell-mediated immunological attack upon brain has not occurred. The myelin basic protein is synthesized in significant quantities only after the animal has become immunologically competent; some at least of the other brain-specific proteins probably appear earlier during the period of immunological tolerance (497). After brain injury or irradiation, the serum and cerebrospinal fluid concentrations of IgA, IgG and IgM alter (305, 1617); a cell-mediated reaction does not occur. Experimental allergic encephalitis can be transferred from a sensitized donor to a normal recipient by lymphocytes (2222). The serum of animals with experimental allergic encephalitis demyelinates axons of cultured cerebral or cerebellar cortex, a reaction involving complement-dependent antibodies (92, 2367, 2504). Many of the synapses in the cultured tissue cease to transmit and degenerate; it is not known if this is a direct effect of the serum or a consequence of the demyelination or other tissue reaction. Some synapses are unaffected.

Antibodies raised against brain and applied directly to the cortex cause an increased rate of discharge and convulsions (1984, 1986, 1987). An antibody raised against the hippocampus increases the rate of discharge in either the hippocampus or in the caudate nucleus of a recipient, according to the site of application. Similar antibodies raised against the caudate nucleus alone also affect either site. There is therefore no evidence from these studies that an antibody raised when a particular region of brain is used as an antigen reacts only, or predominantly against that region. A localized effect was not sought after attempted adsorption of shared antibodies. The observations that antibody raised against the caudate nucleus increases the concentration of histamine-like substances in this nucleus, but not in the hippocampus, and that antibody raised against the hippocampus increases the concentration of histamine-like substances in this nucleus but not in the caudate nucleus (1986, 1988) are unexpected, and the origin of this material is not known. If this finding is confirmed, it indicates that a remarkably localized effect can result from antibody raised against a heterologous and heterogeneous antigen. Antibodies injected into the caudate nucleus of monkeys impair their performance on tasks involving delayed alternation (1985). Lesions or stimulation of this nucleus cause similar impairment (2509). The cells primarily affected by the antibody and the areas of their membranes which are peculiarly susceptible, if any, have not been identified. Learning is impaired after the injection of antibodies to brain microsomes into the cerebral ventricles (1895, 1896). Synaptosomes can also be used as antigens; the resulting antibodies also cause convulsions when applied topically to the cortex (1550).

Conduction of action potentials by the whole ventral nerve cord of the lobster is disturbed when antibody prepared against this tissue is added to the bathing medium, The resting potential falls and the height of the action potential declines gradually (1989). A similar impairment follows the internal perfusion of the giant axon of *Dosidicus gigas* with an antibody prepared against its axoplasm.

These results can be compared with the effects upon other membranes of antibodies in the presence of the complement system. The resting potential of meningioma cells falls before immune cytolysis (2337). Myocardial cells transplanted to a previously sensitized recipient rapidly stop beating. Such reactions, occurring in the presence of the complement system, are associated with the formation of holes of 8 nm diameter in the plasma membrane (145, 1435). Such holes would reduce the membrane resistance in a manner similar to the effect of membrane trauma; the increased rate of neural discharge is likely to be a result of the sustained depolarization constituting the membrane injury potential. The immunological studies so far performed reflect the limitations of the methods available for preparing pure proteins or other antigens from discrete cells of known brain areas. The results obtained indicate clearly, however, that in immunological as well as in other ways, the cellular susceptibilities of brain are essentially the same as those of other tissues.

2 THE NEURON AS A CELL

Neurons are unusual in several ways, but not unique. Each of their unusual features is shared with other types of cell. Other cells conduct action potentials (1517, 1882, 1883, 2455, 2456, 2682), or synthesize, secrete and respond to some of the substances called neurotransmitters (299, 410, 1755, 1763). Neurons are formed from the same line of precursor cells which later form astrocytes, mature ependymal cells and oligodendrocytes (1343, 1344, 1345, 1710, 1711, 1918, 1974, 2735, 2736, 2737, 2738, 2739, 2741). It is therefore not surprising that clones of malignant glial cells cultured *in vitro* can synthesize, release or accumulate a number of neurotransmitters (Chapter 5). Neurons transport organelles and molecules for considerable distances along their processes, a property shared with some slime moulds and other extensive syncytia. The ends of these processes can move and extend, a feature shared with many, perhaps with most, cells. Both the electrophysiological and developmental aspects of the brain have been thoroughly and repeatedly reviewed (384, 766, 767, 1034, 1087, 1261, 1398, 1472, 2119, 2211). Such material will only be included here when of relevance to the topic being discussed, and will not be included as an entity in itself.

Differentiation of a nerve cell

In vivo

The differentiation of a neuron is evidently subject to a number of precise constraints, but little is known of their mechanisms. For example, most neurons of the anterior horn die during development (2333) and it is likely that those which survive do so because they have established and sustained a peripheral connection with muscle (1705, 2333). This contact directs the development of the cell away from the 'terminal differentiation' that leads to controlled cell death, and towards the sequence of morphological, biochemical and physiological changes, which characterizes a mature motor neuron (965). Such contact is between mesenchyme and an epithelial cell; this is an example of the small class of neurons which have the opportunity for such interaction (Chapter 1). A basal lamina is later formed at the junction between the nerve terminal and muscle, but it is not yet clear whether this structure has been formed by direct interaction between these cell types. It is also possible that overlying Schwann cells, or mesenchymal cells other than muscle may interact with the muscle fibre to produce the basal lamina in the cleft of the junction. Regardless of the type of cell that later forms the lamina at the site of contact, it is clear that this interaction of the neuron with mesenchyme causes the axon terminal to emerge as this cell's dominant pole. Throughout the life of the animal the axon terminal of the motor neuron directs the metabolic activity of the nerve cell, so that its pattern of synthesis is appropriate for growth, for the restoration or loss of the neuron's afferent connections, or for the secretion of neurotransmitter and perhaps other substances. This domination is clearly illustrated by the response of the mature or immature animal to axotomy (Chapter 7). The axons of most nerve cells are confined to the central nervous system, and do not have this opportunity to react with mesenchyme. The axon terminal of such a cell does not necessarily become its dominant pole; it may, as one of the sites at which the neuron reacts with another cell, or it may not if the differentiation is initiated through receiving a particular afferent connection. This potential variability may be a factor which contributes to the variation in the chromatolytic response of nerve cells of the central nervous system to axotomy; there are others (Chapter 7).

There are clear examples of the ability of afferent connections to regulate the differentiation of a nerve cell; this regulation is not confined to the receptive area beneath the regulating bouton itself, nor is its nature limited to the development of denervation hypersensitivity (p. 263). It can be illustrated by the afferent connections of the pyramidal cells of the cortex and by those of Purkinje cells of the cerebellum. In both, spines develop on dendrites only after afferent connections have been made. These connections do not themselves end upon the spines (1240) (Fig. 2.1). The spines of the tertiary dendrites receive connections overwhelmingly from parallel fibres; these are the axons of granule cells. Climbing fibres do not form synapses here. The granule cells and other local interneurons can be destroyed selectively by a virus (1587). The Purkinje cells then fail to receive contacts from parallel fibres; nevertheless, spines are still formed upon these dendrites. They

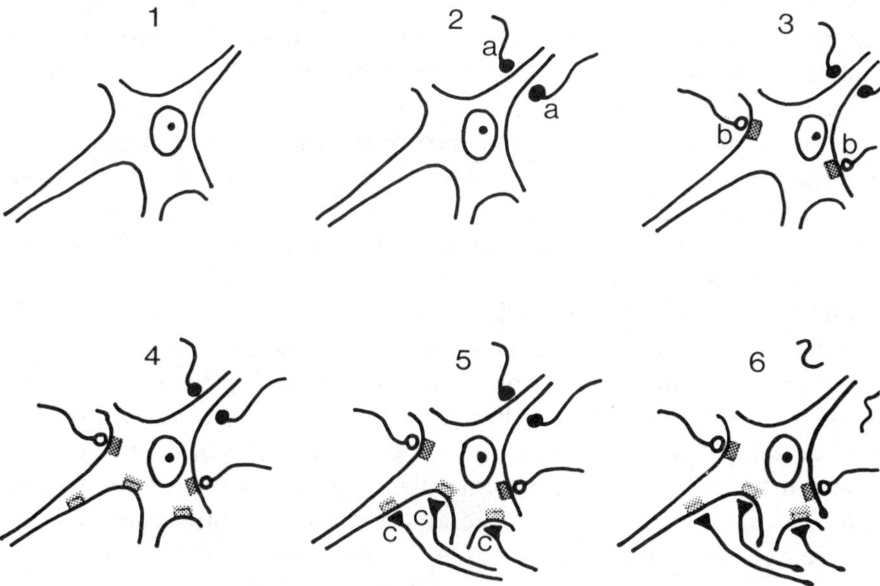

Figure 2.1 A possible sequence of heterosynaptic induction. In (1) the neuron is shown without contacts; in this state it is able to receive boutons of type 'a' and does so (2). The consequent changes in he postsynaptic membrane, which might or might not be a result of transmitter action cause other specializations of the nerve cell surface, shown shaded. These permit boutons of type 'b' to make contact (3). These acting alone or in concert with boutons type 'a' cause further and different specializations of the nerve cell surface (4), so that boutons type 'c' adhere (5). The combined influence of boutons type 'b' and 'c' may lead to changes in the membrane causing the detachment of boutons type 'a' (6). Heterosynaptic induction might therefore lead both to the sequential establishment of connections and to the loss of some contacts attained earlier. In the diagrams the nucleus and the nucleolus are shown as it is very probable that the sequential changes are a result of gene activation.

receive no contacts from parallel fibres, which are absent, or from climbing fibres, which terminate upon other parts of the Purkinje cell. This shows that the spines cannot have been induced locally by either type of fibre. If on the other hand, the climbing fibres are destroyed many of the spines are lost. Pyramidal cells of the sensory cortex also lose spines or fail to form them, if fibres ending on a remote part of the neuron fail to reach the cell (1091). It seems unlikely that the axon terminal is the dominant part of these cells; instead, one type of afferent connection determines the capacity of the cell to accept others. It is a process closely related to anterograde transneuronal degeneration and to the hierarchy of connections which can form collateral connections upon the surface of a partially denervated nerve cell (p. 259). The dominant pole is the receptive surface on which terminate the fibres responsible for heterosynaptic induction. If this is so, axotomy should not alter the differentiation of the neuron, judged by reorganization of its dendrites or by change of its afferent connections. From the rather slender evidence available at present, it is not possible to predict which part of a nerve cell will form its dominant pole, if it is a cell which does not interact directly with mesenchyme. It is not yet clear whether afferent connections can be divided into two distinct classes: those that induce other synaptic connections and those that are permitted only by heterosynaptic induction. Nor is it known if transmitter substances of a particular type are associated with this process. It is possible that the assumption that only one type of afferent connection is responsible for enabling a neuron to accept others is much too simple; different afferent connections might act sequentially in this way. A hierarchy of consecutive action would result. It is possible that temporary connections may act in this way; if so, it is unlikely that proper connections could ever be re-established should they be lost by a mature neuron as a result of injury. The nature of the changes which occur in the membrane of a neuron at the time that it becomes capable of accepting a new type of bouton is not known. It is likely that they include the synthesis of the appropriate receptor (Chapter 4), which may be distributed more widely initially than later; these changes may also include the synthesis of new constituents of the glycocalyx or their exposure by enzymatic hydrolysis. These constituents might be incomplete glycoproteins and glycolipids on the one hand, or glycosyl transferases on the other (p. 16). The axotomized nerve cell is an appropriate model in which to study this (Chapter 7).

As the differentiation of a neuron is influenced by the connections it receives and makes, the position it comes to occupy within the developing neuropil is evidently crucial. Difference in the final position of the nerve cell body, rather than an intrinsic difference in the phenotype, has been used to explain the morphological diversity of interneurons of the cerebellum (2387). The orientation of the dendrites of a neuron depend in part upon the position of the perikaryon, and in part upon the position of the afferent fibres which terminate upon the dendrites. This orientation is one of the most characteristic features of a neuron. The dependence upon the position of the afferent fibres can be seen if, for example, the dendrites of a pyramidal cell of the hippocampus are compared with those of nerve cells in auditory relay nuclei, or in the reticular formation. In the hippocampus the afferent fibres are adjacent to each other and laminated; the dendrites are contained in a single compact region. Neurons of the reticular formation often receive boutons from widely separated tracts in the brain stem. Major dendrites then pass from the perikaryon to each of these principal sources of afferent connection, and each ramifies independently of the others. The factors responsible for such differences of distribution are not understood. It seems likely, however, that two factors at least are operating. The first depends upon an initial direction of outgrowth of the major dendrites that is not random (see below). The second may be due to a process which confirms an appropriate contact, but does not support an inappropriate one. By the term 'appropriate' I mean a contact which can be established because the pattern of the glycocalyceal mosaic is appropriate for immediate or rapid adherence, and because receptor molecules are present in the membrane of the cell that becomes postsynaptic, which respond to the quantal release of neurotransmitter by the fibre that will become presynaptic. It is possible that the growing fibre releases transmitter molecules before any contact is established, or that release follows interactions of the glycocalyces. By 'appropriate' I do not mean a pattern of neural activity that is behaviourally correct. Not only is there evidence to show that synapses can be formed both *in vivo* and *in vitro* in the absence of detectable neural activity, but also many synapses are formed precisely long before they play their part in adult behavior: for example, synapses of the preoptic nucleus involved in the expression of sexually appropriate behaviour. The selection of such connections cannot be explained on the basis of a mechanism confined to one animal. It is only possible to say that pressures of selection have led to the more successful survival and breeding of such animals, so that their proportion in the population becomes very high indeed. In the example cited, of neurons of the preoptic nucleus, the relevance to survival of the species is immediate and obvious; animals which posses neurons with membranes or with glycocalyces in this region, which are inappropriate, will fail to form appropriate connections and hence may resemble female animals testosteronized at birth, or males which have been castrated at this time. Their endocrinological state and sexual behaviour will be abnormal, and their chance of producing offspring similar to themselves is negligible. Presumably similar pressures of selection act for all neuronal connections, although the degree of polymorphism is likely to be considerable. Such polymorphism has hardly been explored yet at the cellular level, although there is much evidence for it from behavioural studies.

In view of the regulation of extension of the processes of many types of cell through the action of transmitters, hormones or other agents upon membrane receptors which are themselves regulatory subunits of adenyl cyclase (Chapter 5), which indirectly alter the degree of polymerization of the microtubules or their capacity to extend cell processes, it is very probable that similar mechanisms act in the neuron. Dibutyryl cyclic AMP, a derivative of cyclic AMP which can enter cells more easily than the unesterified nucleotide, can make some cultured neurons extend processes.

In some situations at least, the final position of a neuronal perikaryon is determined by interaction between neurons and glia (p. 72). Little therefore is known of the factors which regulate the shape of a nerve cell.

The disposition of an axon in the vicinity of the perikaryon is determined in part by the cell, and in part by the surrounding neuropil (Fig. 2.2). Occasionally these determinants conflict. Such conflict is seen when neurons are orientated atypically (1835). About 18% of cortical pyramidal cells are aligned with their long axes more than 20° from a line drawn perpendicular to the pial surface of the brain; some of these cells are completely inverted. If it

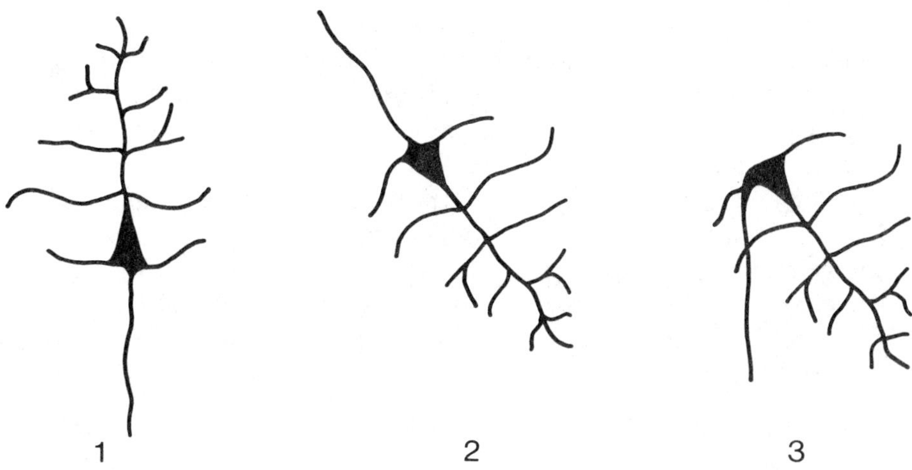

Figure 2.2 Some neurons are oriented inappropriately (1835). The normal orientation of a cell is shown in (1). If the direction of axon growth depends only upon the orientation of the perikaryon and not upon the surrounding neuropil, the axon would continue in a direction abnormal for the brain as a whole, but appropriate for the neuron (2). If the direction of axonal growth depends only upon the surrounding neuropil, the axon would pass in a direction appropriate for the brain, but inappropriate for the nerve cell as soon as it leaves the perikaryon (3).

is assumed that these cells are aberrant and do not constitute a morphogenetically distinct class, a conflict exists between the intrinsic, or neuronal factors, and the extrinsic, or neuropil factors, in regulating the course of the axon. Intrinsic factors would make the axon emerge from a point which is normal for the neuron, but abnormal for the neuropil; extrinsic factors would make it emerge from an abnormal part of the nerve cell, but proceed in a manner that is proper for the neuropil. The solution of this conflict is variable; in some cells the axon emerges from the normal basal part of the perikaryon which is itself pointing in the wrong direction. After the axon has passed a variable distance in the neuropil it changes direction and runs in a direction appropriate to the neuropil and parallel to similar axons derived from properly aligned nerve cells. The axons of other inverted pyramidal cells, however, do not emerge from the basal part of the perikaryon but from its apex or from a major dendrite, and pass at once in a direction that is appropriate for the neuropil. The solution probably depends upon the relative dominance of intrinsic influences and extrinsic influences in that part of the brain at the time of the initial outgrowth of the axon. It is not of course clear from this study to what extent the point of emergence of the axon from the neuron has altered after its first emergence as a result of remodelling. A most interesting point to emerge from this study was the constancy of the distribution of the dendrites with respect to the abnormally orientated cell body; the influence of the intrinsic factors was considerable, while that of the extrinsic factors was slight (Fig. 2.3). This indicates that the

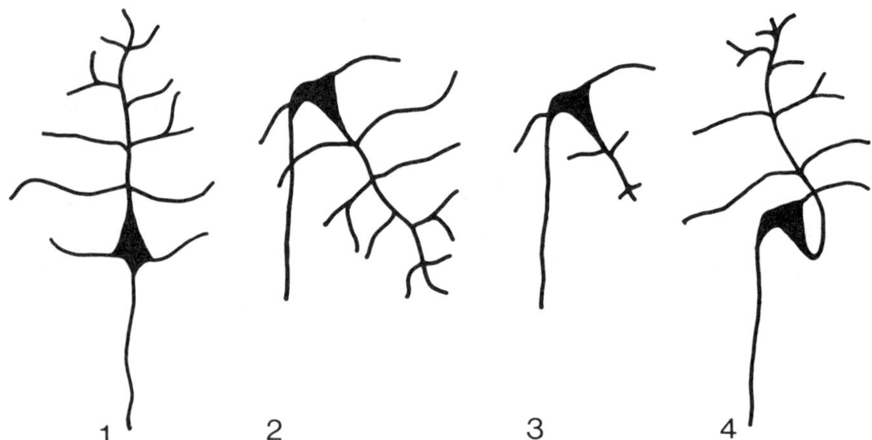

Figure 2.3 An appropriately oriented nerve cell is shown in (1). If the growth of dendrites depends upon the orientation of the perikaryon alone, an improperly oriented neuron would have an inverted array of dendrites (2). If the growth of dendrites is induced by the contact of afferent fibres, the dendrites of an inverted neuron should remain stunted, provided that the afferent fibres are confined to a layer of cortex into which the inverted dendrites fail to penetrate (3). If the growth of dendrites is determined not by the perikaryon but by the surrounding neuropil, their orientation should be appropriate for brain but inappropriate for the nerve cell body (4). It is likely that (4) cannot always be distinguished from (1), and so possible that this orientation by neuropil is important at certain stages of development. As (2) is frequently seen (1835), is is evident that the direction of the large dendrites is determined by the orientation of the perikaryon if the cell body becomes misaligned at certain stages of development.

disposition of the major dendrites of a neuron is not a result of random outgrowth followed by the selection of processes which make appropriate connections. It is a surprising result in view of the role of synaptic connections in permitting the establishment of others. It does however indicate that the two factors leading to the formation of synapses – dendrites in the correct position, and interaction between the presynaptic and postsynaptic membranes – progress independently, although presumably subjected to similar pressures of selection. Axons of monopolar neurons of the lamina of the butterfly *Pieris* behave in a rather similar way (2551, 2552). The direction in which the axon grows at first is inconstant, frequently being centrifugal instead of centripetal. After passing for a variable distance through the neuropil, the axons curve to enter a parallel centripetal bundle.

Most studies of the shape of nerve cells have used Golgi methods. The fine processes of neurons can be filled well, judged by light microscopy, after intracellular injection of Procion Yellow, a fluorescent dye (651, 1486, 2843, 3238). This method can be used to study the structure of a nerve cell from which intracellular electrophysiological recordings have been obtained. Large neurons of arthropods can be shown well by intracellular injection, diffusion or electrophoresis of cobalt.

In vitro

Explants of brain continue to differentiate in a suitable medium. Usually fetal nerve tissue is studied, but adult neurons and glia also survive (1584). Myelin is formed and synapses appear. The demonstration of the formation of synapses demands two things: that they are absent at the time that the tissue is explanted, and that they are later present. Their presence should be demonstrated both electrophysiologically and morphologically. Their formation has been demonstrated on many occasions in a number of different preparations. For example, synapses cannot be identified in the spinal cord of the 14-day fetal rat (390), and there are no reflex responses to stimulation (3145, 3147). After this tissue has been explanted for three days working synapses are found (387, 390). They appear rapidly at first, and later much more slowly; their number never exceeds a very small proportion of those that are formed *in vivo*, probably because of the absence of those that originate outside the explant, and possibly because these terminals, being absent, cannot exert an inductive effect to facilitate the formation of some synapses from nerve cells within the explant. It is also an example of the limitation of the formation of synapses to a very short period.

A neuron can accept boutons of certain types only. This has been shown clearly in cultures of sympathetic ganglia. These contain very few synapses if they are cultured alone; if they are grown with separate explants from the thoracic part of the spinal cord, many synapses appear in the ganglion. When the spinal cord is later removed from such an established paired culture, the synapses disappear from the ganglion (2176). Synapses do not appear within the ganglion if grown with an explant of cerebral cortex instead of spinal cord. Fibres from the ganglion do not form synapses in either explants of spinal cord or of brain. This indicates that synapses are only formed in the sympathetic ganglion by axons which originate from the part of the central nervous system which contains the nerve cells that form such synapses *in vivo*; it does not prove that the same neurons are responsible although this seems likely. It is possible, however, that nerve cells from the same part of the spinal cord may have a similar glycocalyx; cells of the retina and tectum adhere with some regional specificity. This preferential adhesion is not confined to neurons, but is shared by other cells in the same part of the nervous system (p. 173). Synapses are not only formed within explants; they also occur between dissociated neurons grown in culture, for example between dissociated cells of the spinal cord of the chicken (246). Neurons obtained from the superior cervical sympathetic ganglion by mechanical disaggregation (324, 2424) extend their processes rapidly and form an interconnecting network. This preparation has been used to study the extension of processes and the manner of movement of growth cones.

The fact that synapses are formed in cultures in a manner that obeys some rules shows that some determinants of the patterns of connection are independent of precise patterns of afferent stimulation, of neural activity or of experience. This conclusion is confirmed by the failure of low concentrations of local anaesthetic in the culture medium to inhibit the formation of synapses (2021). The concentration used was sufficient to prevent the conduction of action potentials.

Explants of brain commonly differentiate to form an organized tissue possessing many of the features of the part of the brain from which the explant is derived. The resemblance is often marked, but never complete. Some, but not all of the explants of cerebellum form a well laminated structure (1326, 1720, 2342, 2651, 3167, 3168, 3169), in which granule cells migrate to an approximately correct depth, and a layer of Purkinje cells is formed. In such a culture the cortical and subcortical groups of nerve cells can be distinguished. If the explant includes the vestibular nuclei, nerve fibres pass between these and the cerebellar neurons. The axons of neurons in culture form bundles (1176); it is not clear at present whether bundling only represents the fact that the direction of extension of axons which offers the greatest freedom from contact inhibition of growth is found when they pursue a parallel course, or whether it is a process that involves more precise types of interaction. Complicated patterns of electrical activity can be recorded from nerve cells in cultures in response to stimulation. Occasionally activity is recorded that appears to be spontaneous, but it is difficult to be certain that it does not result from the injury of a nerve cell during preparations for recording, or even from small changes of temperature (417, 570–580, 2138, 2259, 2260). Dispersed cells reaggregate; the tissue obtained often resembles the part of brain from which the cells were obtained (295, 670, 671, 1023, 1024, 1965, 2644, 2645, 2684), and functioning synapses are found. The structure has been studied both of explants and of reaggregates obtained from dispersed neurons of mutant mice, in which the migration of cells or the formation of synapses is abnormal *in vivo* (Chapter 9).

Dorsal root ganglion cells grow in culture, and their axons readily become myelinated (750, 1937, 1938, 2089, 2840). The growth and cytochemical differentiation of these cells is abnormal in several respects (2754, 2755). The growth and differentiation of dorsal root ganglion cells and of sympathetic neurons are greatly influenced by nerve growth factor. Under certain circumstances dorsal root ganglion cells bear junctions *in vitro* which resemble synapses in some respects. (2002); *in vivo* these cells do not accept synapses upon their cell bodies.

The formation of neuromuscular junctions *in vitro* has been widely studied. Neuromuscular junctions which have been made *in vivo* can be sustained in culture. (573, 901, 902, 2260, 2449, 2450, 2674). Working junctions are also established in culture between explants of spinal cord and fragments of muscle, which need not be of the same age (573, 1485, 2259, 2449, 2450). Synaptic potentials can be recorded and imitated by iontophoretically applied acetylcholine. Conduction is blocked by curare and by α-bungarotoxin. Direct electrical coupling was found between a small proportion of the nerve-muscle pairs examined (904). The terminals of the nerve fibres ramify upon the muscle (2259), and multiple innervation is not uncommon (2449, 2450). The formation of junctions alters the membrane properties of the cultured muscle fibres (849, 851, 852, 903, 904, 2150). Culture of spinal cord and muscle has been used to study the relative roles of nerve and muscle in producing muscular dystrophy in mutant rodents (Chapter 9). The scope of the study of factors which contribute to the formation of junctions between nerve and muscle has been extended by the use of dissociated neurons and muscle cells (902), and by using cloned nerve cells and cloned muscle (1264, 1265, 2814, 2815). Muscle activity is not necessary for junctions to be formed between neurons and muscle fibres, as they are established when the muscle is kept depolarized by a high concentration of potassium ions in the culture medium (2814); this finding is consistent with the appearance of synapses in cultures of brain grown in the presence of sufficient local anaesthetic to prevent the propagation of action potentials (2021). The successful formation of neuromuscular junctions in cultures from *Drosophila* (2642, 2643) increases the scope of genetic analysis of the factors leading to the establishment of working junctions.

Fragments of brain have been implanted in the anterior chamber of the eye as an alternative to culture *in vitro* (1368, 2175). There is less scope for the experimenter to alter the medium, and the situation is complicated by the presence of tissue in the eye, such as the iris, which can accept certain types of innervation from intraocular grafts, and by the presence of sympathetic nerve fibres, which can invade the graft. The period of survival of these grafts is usually limited by the invasion of macrophages, and by slow fibrous replacement. Orthotopic grafts of cerebellum containing cells with tritium labelled DNA have been studied (629). Most of the implanted cells die, but it is possible to follow the migration and differentiation of the survivors. It is not clear that this technique will provide more information than can be gained from tissue culture. While it might be of value in studying mutant animals, its value here may be limited both by the distortion of the recipient cerebellum by the trauma of grafting, and by the difficulty of seeing the particular synapses that the cells with labelled nuclei have made, if any.

Movement of neuronal processes

A neuron is a dynamic cell in several ways: during development and regeneration the cell extends and retracts processes, throughout life many organelles are transported along the processes, and it is possible that the cell forms new connections and loses old ones. Microtubules and microfilaments are closely associated with these movements; the involvement of neurofibrils is much less clear. Perhaps these are concerned in maintaining established form efficiently rather than in attaining new structures or in acting as a support for myosin. Aspects of these and of related features have been reviewed: cell movement (1447, 1619, 2653) cytoplasmic streaming (66, 2420), axoplasmic flow (154, 1719), microtubules and microfilaments (1917, 2172, 3196), and the role of actin in cells other than muscle (3118, 3119).

Microtubules

Most cells have microtubules (292, 366, 1917, 2172). They can be as long as 100 μm but are frequently shorter (3197), and have an outer diameter of 23 to 25 nm, a bore of 13 to 15 nm and a wall thickness of about 5 nm. It is not known if they are significantly longer in long processes, such as axons and dendrites. Microtubules have a longitudinal periodicity of 8 nm. Microtubules in different types of cell differ in the ease with which they can be isolated by homogenisation and in their stability in vivo (187). Those of the mitotic spindle, axons and dendrites are more labile than those of flagellae, and can be isolated only with difficulty, for example by treating them with hexylene glycol. Microtubules contain 13 protofilaments (2444, 2925), which can be accentuated in electron micrographs by Markham rotation (2444). The long axis of a protofilament lies parallel to, or nearly parallel to, the long axis of the tubule. Each protofilament consists of a chain of globular subunits 3.5 to 4.5 nm in diameter (3052).

The protein from the microtubules of brain is extracted in solution after homogenization and is called 'native tubulin'. Apart from the regularly repeating subunits of the tubule, the tubulins, there are at least two other proteins in native tubulin separable only with difficulty from the tubulins. Tubulin has a molecular weight of 110 000–120 000 daltons (866, 1594, 3102, 3142). It is a dimer, being composed of two similar but not identical subunits, α-tubulin and β-tubulin (366, 2173), bound by interaction that is not covalent but not yet fully described. These each have a molecular weight of 55 000 ± 2000 daltons (367). The molecular weight of α-tubulin appeared to be greater when it was initially isolated, because it reacted in an unexpected manner with the sodium dodecyl sulphate used to prepare it. The separation of α-tubulin from β-tubulin by electrophoresis depends upon their difference of charge; they differ in six amino acid residues (367, 866) and only β-tubulin contains phosphorylated serine. Equimolar amounts of α-tubulin and β-tubulin are always found, regardless of the source of the microtubules, of the manner of isolation, or of the degree of fractionation. It is very probable that microtubules are composed of heteropolymers (α—β). Homopolymers (α—α) or (β—β) have not been identified. It seems most likely that the globular subunits of the protofilaments are these heteropolymers. Only the dimer of native tubulin binds colchicine with high affinity; each mole of dimer binds one mole of colchicine and two moles of guanosine triphosphate (GTP). At one of the binding sites for GTP exchange occurs readily; binding at the other site is stable (3102). Freshly isolated native tubulin from brain contains bound GTP and guanosine diphosphate (GDP). The structure of native tubulin varies with temperature (3004). At 4° C it contains 22% α-helix and 30% β-structure. At 37° C, the temperature at which it binds colchicine with the greatest avidity, the protein is nearly all in the β and random coil configurations. Very little difference is found between animals of different orders in the composition of α-tubulin and β-tubulin. Probably the α- and β-protomers evolved from a single ancestral protein before the divergence of echinoderms and chordates (1845, 1846, 2173). It seems likely that the steric constraints for the production of effective heteropolymers are too small to have permitted animals with marked variants to have survived. It is therefore improbable that mutant strains will be found with abnormal tubulins. Membranes have some capacity to bind colchicine (1703). It is not yet clear whether this indicates that tubulin is present in some membranes *in vivo*; the evidence at present is inadequate. Antisera prepared against 'tubulin' precipitated by vinblastine cross-reacts with particulate fractions of brain containing membranes. It is known, however, that vinblastine precipitates a number of proteins other than tubulin; any of these could also be the cause of the cross-reaction. No microtubules have yet been identified in the crude particulate fraction of brain. The identification of tubulin in plasma membranes of organelles would be of interest in determining the way in which microtubules interact with organelles in facilitating their movement. In some cells microtubules are very close to the plasma membrane, for example in the erythrocyte, and in many plant cells.

Tubulin can be made to polymerize *in vitro* under well defined conditions (290, 291, 292). Polymerization occurs if a solution of native tubulin is warmed from 0° C to 37° C (290, 291, 292, 830, 3101). GTP is needed, and GMP or GDP do not suffice. Native tubulin does not polymerize if GTP is replaced by an analogue that cannot be hydrolyzed; nor can other nucleotides be substituted for GTP. Little is yet understood about the nucleation of microtubules, the initial stage of their formation. The initiation of microtubule formation differs thermodynamically from the elongation. The significance of discoid nuclei that facilitate the formation of microtubules and reduce the latency of their polymerization is controversial. They may themselves consist of short 'active' polymers. Even less is known of the factors which regulate the formation and loss of microtubules *in vivo*. The significance of the cyclic AMP-dependent protein kinase that can phosphorylate tubulin *in vitro* is uncertain. Such phosphorylation has not been shown to be a necessary stage of polymerization either *in vivo* or *in vitro*, and many other proteins can act as substrates. Microtubules can be formed rapidly by polymerization; a length of 100 μm can be reached in 30 min.

The microtubules of flagellae grow by adding new material at the tip (2490, 2491). It is uncertain whether the growth of microtubules of brain occurs only at one end. Native brain tubulin can extend *in vitro* microtubules obtained from flagellae; tubulin may then be added at one end only or at both, depending on the circumstances of incubation. Polymerization is prevented by a concentration of calcium ions that is significantly greater than that found inside cells

(2589, 2809, 3103). Colchicine prevents polymerization and causes disaggregation by altering the dynamic equilibrium between polymerization and depolymerization in favour of the latter. It does so by binding to the dimer. The disaggregation caused by colchicine is not complete; it ceases when the microtubules are very short (292, 293, 294). If this indicates that depolymerization proceeds in two stages, one from long tubules to short and one from short tubules to dimers, and if the two stages differ kinetically and thermodynamically, then it is consistent with the hypothesis that the discoid structures, which may act as sites of nucleation, are an 'active' form of tubulin oligomer. Filaments appear when microtubules disaggregate through the action of colchicine. These differ both from microfilaments and from neurofibrils (1242, 3197). The relationship of these filaments to the protofilaments of microtubules is uncertain. They are not themselves protofilaments; their diameter is 10 nm instead of 5 nm. The proportion of brain tubulin that is in a polymerized state in normal brain is not known; the size of the pool of microtubule protein cannot yet be measured. Such measurement is easier in organisms which have more stable microtubules that can be isolated without depolymerization, for example, the sea urchin embryo (112). The microtubules of a desheathed nerve can be disrupted if they are incubated under conditions which increase the intracellular concentration of calcium ions (2589). This has not been shown to be reversible.

The role of microtubules in brain is incompletely understood. As in other tissues they are formed in the mitotic spindle (1879, 1880, 3100), but the way in which they contribute to the movement of the chromosomes is uncertain; in brain glial cell division occurs by mitosis, despite early reports to the contrary. Microtubules are also concerned in the creation of cell assymmetry (2924, 2926). This morphogenetic role is seen clearly in epithelial cells of the differentiating lens and in the extending processes of neurons and glial cells (1859, 1860, 2235, 2748, 3165). Little is known of the factors which regulate their formation, orientation or movement. Regional differences must exist within a given cell, for microtubules can be formed in one part while they are disappearing elsewhere in the same cell. That such precise regional regulation must occur is evident in cells like neurons; the neurons of invertebrates can be individually identified. The variation in the branching pattern of the first two or three orders of dendrites is negligible. It seems most likely that the regulation is peripheral; that is, microtubules form under the influence of regional variations in the cell membrane or adjacent parts of the cytoplasm. Microtubules are also closely involved with the movement of organelles and cytoplasm (244, 1379, 3132) (see below) and, possibly indirectly, with the release of neurotransmitter substances. The movement of microtubules almost certainly depends upon the actions of at least one of the proteins other than tubulins that are present in native tubulin. Two high molecular weight proteins, dynein and nexin are associated with the microtubules of flagellae (1059, 1917). Two proteins of high molecular weight are present in native tubulin (292, 400). These are precipitated with tubulin when a cold solution is warmed, and remain closely associated through several cycles of polymerization and depolymerization. One high molecular weight protein has been studied in some detail. It may represent the dynein of brain, and is likely to be derived from the short sidearms found on microtubules. The ATPase activity of this protein is disappointingly low, however, when measured after isolation (400).

Many drugs affect the microtubules. Most of these have been studied because of their effect upon axoplasmic flow or upon cell division. Colchicine combines with tubulin dimers, a reaction which depends upon the conformation of the protein (3004, 3102, 3131). The reaction with colchicine is weak, and no covalent bonds are formed. Vinblastine and vincristine precipitate tubulin and a number of acidic proteins both *in vivo* and *in vitro*. The precipitate is crystalline. Although these crystals have been isolated from brain and used to stimulate the formation of antibodies to 'tubulin', the purity of the preparation is not adequate to allow the results obtained to be interpreted confidently. The *Vinca* alkaloids bind to the protein at a different site from colchicine (211, 212, 1649, 2590). There is some correspondence between precipitation by calcium and precipitation by these alkaloids (3143). Halothane, a general anaesthetic, alters the structure of microtubules (1347). This action is almost certainly not related directly to its anaesthetic effect. If it disrupts the microtubules of cilia or prevents their effective beating, it may help to limit the capacity of the respiratory tract to clear particles postoperatively. It seems likely that lidocaine, procaine and tetracaine may also impede the action or the formation of microtubules, as they can stop the rapid phase of axoplasmic flow. Indolamines and catecholamines also act in an interesting manner. Pargyline, an inhibitor of monoamine oxidase, can make many cells of the brain develop cilia; reserpine can impede the binding of colchicine to tubulin dimer, and interfere with the polymerization of microtubules *in vitro*. Both mescaline and melatonin can cause mitotic arrest by interfering with the mitotic spindle.

Microfilaments and actin

In many cells, filaments of diameter 4 to 6 nm form a network or bundle beneath the plasma membrane. Some of these have been identified as actin (25, 188, 2312, 2313, 2314, 2615, 2616, 2617, 2618, 2619, 2927, 3118, 3119). Three types of method have been used: the best is chemical extraction and identification by peptide mapping and by determining the amino acid sequence. The next is by histochemical or physiological identification: the substance believed to be actin is made to act *in situ* with heavy meromyosin. The complex formed can be seen by electron microscopy as having characteristic arrowheads, or by light microscopy if the heavy meromyosin is labelled with a fluorescent dye. This reaction can be blocked with ATP or magnesium pyrophosphate (96, 1467, 2312, 2313, 2734, 2799). Immunological demonstration is possible if a pure preparation of the actin is available. There is usually a degree of cross-reaction between brain actin and muscle actin if the species in which the antibody is raised is chosen with care. Physiological methods of demonstration are those in which the substance believed to be actin is mixed with known muscle myosin under conditions that produce superprecipitation. The third class of method for the detection of actin is not sufficiently reliable to be used alone. Cytochalasin B (2803) disrupts many microfilaments known to be actin, and prevents several activities of cells that are believed to be mediated through actomyosin. Such microfilaments are not disrupted in all cells. Some activities of the cell which are disrupted may be disturbed indirectly.

Actin has been identified both in neurons and in glia. It forms a filamentous network beneath the plasma membrane of cultured glial cells. When this is dissociated by cytochalasin B, movement stops (2801, 2802). In other parts of the glial cell similar microfibrils form bundles; these are not dissociated by cytochalasin B, and these presumably are playing no part in the movement that the drug stops, or can be paralyzed without structural change; alternatively these bundled fibres are not made of actin microfilaments and have no direct role in causing cell movement through this mechanism. Actin and tubulin have been extracted from nervous tissue of chick embryos (888): they form 20% of the total protein. Peptide mapping shows that the primary structures of actin and tubulin differ considerably. Very similar peptides maps are found for actin from muscle and actin from brain. It is probable that the differences between actomyosin of muscle and actomyosin of brain result mainly from dissimilar structures of the myosins, just as the dissimilar properties of actomyosins obtained from different types of muscle are due to different myosins (Chapter 8).

Actomyosin has been isolated from brain. It has Mg^{++}- and Ca^{++}-activated ATPase activity, which is inhibited by cytochalasin B, and it can be superprecipitated (225, 2348, 2349, 2350, 2351). Both superprecipitation and ATPase activity are inhibited by p-chloromercuribenzoate and mersalyl, but not by ouabain (2350). The actomyosin isolated from synaptosomes does not cross-react immunologically with that from vascular smooth muscle (2349). Some cross-reaction is found between actomyosins from the brains of different species. The actin resembles that of skeletal muscle, but is not identical with it, and has a molecular weight of about 47 000 daltons. The half life of brain actin in a rat aged 3 weeks is 4.4 days, about the same as that of tubulin. Myosin obtained from brain has a molecular weight of about 240 000 daltons and differs from muscle myosin. The brain myosin is probably mixed and has not been extensively studied. Between 8% and 10% of the protein of synaptosomal fractions is actomyosin; none has been found in mitochondria, microsomes or myelin (2351). When the synaptosome is subfractionated, actin is found to be concentrated in the membrane fraction and myosin in the fraction containing synaptic vesicles. If this does not result from the redistribution of these proteins during isolation, it may be of relevence to the methods involved in moving the vesicle to or from the synaptic membrane. The myosin has not been identified ultrastructurally yet. In the contractile ring of HeLa cells actin may be present as an oligomer which forms fine fibrils. In smooth muscle, ribbons of myosin are attached to a fibrous cytoskeleton. Muscle actin can stimulate the ATPase activity of brain myosin (225), and brain actin can stimulate the ATPase activity of muscle myosin (2349). Tropomyosin has been extracted from the brain of an embryo chick; its molecular weight is smaller than that of muscle tropomyosin (887).

In cells other than muscle, actin and myosin are probably involved in a number of activities: movement (5, 1447, 2799), cytoplasmic streaming under circumstances in which tubules cannot be acting, morphogenesis and exo- and endo-cytosis (67, 188, 2246, 2615, 2616, 2617, 2618, 2619, 2799, 2800, 2801, 2802, 3118, 3119). Movement has been analyzed in the ruffling membrane of the fibroblast (1098), which has many microfilaments beneath its plasma membrane. The morphogenetic role is well shown in the differentiating epithelium of salivary glands (2801), in the oviduct of the immature chick, in which differentiation has been induced with oestrogen (3190), and in the contractile epithelial cells of *Distaphia occidentalis* during tail resorption (507). Both in amoeba and in nitella it is probable that cytoplasmic streaming is brought about by bundles of microfilaments between 5 and 6 nm in diameter rather than by microtubules (2085, 2312, 2313, 2314, 2420). Phagocytosis and endocytosis are inhibited in leucocytes and macrophages by cytochalasin B; micropinocytosis is not (67). The discharge of exocrine secretions from some glands is similarly inhibited.

Microfilaments are concerned with the movement of growth cones and filipodia in neurons. Glial cells migrate in brain and probably do so in a way similar to the fibroblast. The morphogenesis of brain has been extensively studied, but the basic mechanisms are understood no better than in other tissues. It is very probable that the neural groove and the neural tube are formed by the contraction of microfilaments arranged beneath the luminal surface of the neuroepithelial cells, a mechanism very similar to that seen in developing salivary glands. The relative contributions of actomyosin and of tubulin in moving organelles and cytoplasm in the various types of cell found in brain are not known. Cytoplasmic flow is not confined to neurons, but is most easily studied in them. Cytochalasin B does not affect the accumulation of noradrenaline storage vesicles within an axon proximal to a site of constriction (138), but colchicine prevents it completely. Cytochalasin B also prevents the release of catecholamine and of dopamine-β-hydroxylase from postganglionic nerve terminals of sympathetic nerves (2909). It also blocks conduction through the synapses of sympathetic ganglia; this second effect is prevented by pyruvate. It is difficult to interpret this interaction of pyruvate and cytochalasin B; the action of the drug may not be directly upon the actin of the terminals, but possibly on the energy-producing systems of the presynaptic bouton. Cytochalasin B disorganizes epithelial differentiation brought about by the actions of microfilaments: the development of salivary epithelium and the differentiation of the oviduct. Cytochalasin B also causes microspikes or filipodia to retract and growth cones to round up. *Vinca* alkaloids precipitate actin and other acidic proteins. Unlike the precipitate formed by interaction with tubulin, the precipitate of actin and the alkaloid is not crystalline, but amorphous.

Neurofilaments

Unlike microfilaments and microtubules there is no evidence yet to suggest that neurofilaments are involved in causing or directing the movement of cell processes or organelles. They have received much less attention. They have a diameter of 10 nm and are of uncertain length (3197). In cultured neurons they are very straight and are frequently more than 200 μm long. They are usually most prominent in axons, but are also common in other processes of the neuron and in the perikaryon (3194). The extraction and concentration of the constituent proteins is difficult; the protein binds avidly to other substances and particles; it adsorbs strongly onto cellulose and its derivatives. Neurofilament protein constitutes 25% of the protein extracted from the giant axon of *Dosidicus gigas* (657, 658, 1436, 1437, 2603). The acidic protein, filarin, isolated from it has a molecular weight of 80 000 and an unusual amino acid composition in that 12% of its amino acid residues are serine. Filarin from *D. gigas* cross-reacts with extracts of nerve or brain prepared from the octopus *Loligo pealii*, and with extracts from the nerve cord of the lobster. Filarin differs in molecular weight, charge, amino acid composition and immunological reaction from both tubulin and actin. Neurofilaments cannot therefore represent a different state of aggregation of microtubule protein or actin (2258). When microtubules are dissociated at low temperature, fibrous material is formed; this is not related to neurofilaments. Furthermore, neurofilaments are not affected in this way by cold (830). The purest filarin yet obtained forms two bands when studied immunologically by double diffusion. These could result either from the presence of two substances, or from one in two states of aggregation.

Little is known of the structure of filarin, or of its mode of organizaton into filaments. X-ray diffraction studies are hampered by the unsuitable state of the extracted material for analysis. Protofilaments cannot be clearly seen by electron microscopy; globular subunits 3.5 nm in diameter are seen and may correspond to a repeating subunit presumably largely or entirely composed of filarin. Between 3 and 6 of these protofilaments are arranged helically or longitudinally around a lumen 3 nm in diameter. It is not known whether there is a pool of dissociated filarin *in vivo*, or whether neurofilaments are in a state of dynamic equilibrium, like microtubules. Some filarin is found in solution after axoplasm has been homogenized, but this may be an artefact of preparation. The function of neurofilaments is unknown. They may serve either as a cytoskeleton, sustaining an established assymmetry without expending energy, or supporting molecules which must be retained in precise alignment, such as myosin. The protein of neurofilaments differs from the acidic protein derived from fibrous astrocytes; the filaments in these two types of cell also differ when examined by electron microscopy.

Axoplasmic transport

That axoplasm flows is not itself remarkable. Movement of organelles can be seen in many kinds of cultured cell. In neurons and in other cells with long processes it is the scale of the flow that is impressive. Axoplasm moves in two ways: by bulk flow (871, 3106, 3108), and by the transport of organelles. Axoplasm moves by bulk flow at a rate of about 1 mm/day, or 700 nm/min in mature animals. In young animals it moves at about 5 mm/day, or 3.5 μm/min (1300). Two factors may contribute to this slowing with increasing age: firstly, the reduction of rapid growth of the axon when its terminals have established contact with the periphery, and secondly, a later and lesser reduction when the animal stops growing as a whole; for, if the animal is growing its nerves are also likely to be getting longer. A similar high rate of bulk flow is present in regenerating nerves of adult animals (428). Many proteins pass along the axon at the rate of bulk flow; tubulin has been identified (872, 1121, 1122, 1546).

Axoplasmic transport is both centrifugal and centripetal. Fast transport in both directions can be seen in cultured neurons (1593), and in suitable preparations *in vivo* (1889). It is frequently saltatory and different organelles can be seen passing in each direction simultaneously (2420). The rate of centrifugal transport varies considerably, between species, between different nerves of the same animal, and according to the organelle or protein studied.

Transport is faster than bulk flow. Rates are commonly between 50 and 500 mm/day, or 1 μm/sec to 6 μm/sec, although faster and slower rates have been reported (313, 1544, 1545, 1546, 1547, 1716, 1717, 1718, 1719, 1762, 2155, 2156, 2157, 2158, 2725, 2727, 2728). Transport requires energy (2155). The rate of transport is no greater in stimulated axons, but the amount of protein carried increases (1854). Many rapidly transported organelles and substances have been identified. Neurotransmitters and associated substances such as neurophysin, vesiculin, chromogranin, synthetic enzymes and ATP are carried distally at the same rate as synaptic vesicles (613–618, 1045–1048, 1535, 1536, 1537, 1616, 1821, 2899). These and mitochondria are frequently seen within axons, closely applied to microtubules. Other vesicles of the endoplasmic reticulum are also seen; it is likely that some of these are responsible for the transport of adenyl cyclase and phosphodiesterase (330). Sulphated mucopolysaccharides and glycosaminoglycans (736, 803, 804, 946, 947, 1613, 2773), phospholipids (1123, 1975, 1976) and cholesterol are transported mainly at the fast rate. Some of the phospholipids exchange with those of the axon as they pass distally. The S-100 protein, formerly believed to be confined to glia, is transported in axons (1101, 1977, 1978). Many other transported proteins have not been characterized. (735, 737, 1289, 1299, 1300, 1716, 1717, 1854, 1875, 2725, 2727, 2728, 2901).

Rapid transport of labelled proteins has been used to study the anatomical projections of nerve fibres (558, 1547). Adenosine also seems very suitable for this purpose (2621, 2622, 2623); it is carried distally, released and then incorporated by the postsynaptic cell into RNA. There is no evidence that this release and transsynaptic uptake has a transmitter or trophic role.

The mechanism of centrifugal transport is not clear. Swellings are frequently seen on the processes of cultured neurons. When the axon is growing these sometimes appear to pass along the axon or to pulsate (3106, 3108). The rate of pulsation is about once every 16 min. It is not certain whether this is an activity confined only to growing fibres, or whether the pulsation or movement actually aids the longitudinal passage of organelles; nor is it known whether cytochalasin B or calcium-free media prevent this activity. Similar swellings are seen on long processes of other types of cell in culture, such as neuroglia. Colchicine and vinblastine impede transport (615, 1545, 1655, 2135, 2726). Flow ceases if the microtubules are physically disrupted. Vinblastine can also prevent flow without disrupting the microtubules (3143). It is not clear if the motion is imparted through one microtubule moving in relation to another, perhaps deriving the necessary energy through the ATPase activity associated with dynein, or if an organelle moves with reference to a related microtubule; if so, the nature of the dynamic coupling between the organelle and the microtubule is not understood. Cytochalasin B can block axonal transport (594, 785, 873), without preventing the bulk flow. This by itself is not adequate evidence that an actomyosin-like system is concerned in the transport, but suggests that such a mechanism might be involved. Perhaps colchicine destroys the railway and cytochalasin B damages the engines. Chlorpromazine and lidocaine block fast axoplasmic transport (784, 897). It is reversibly inhibited by D_2O (78).

There is the problem of recoil. If an organelle is in contact with a microtubule, and as a result of interaction the organelle is thrust distally, the microtubule will be thrust proximally. There is no direct evidence that this balanced acceleration of mass occurs; especially there is no evidence that organelles are accelerated both proximally and distally in the same region of the axon simultaneously. A balanced system could result if organelles of the same mass were transported centripetally and centrifugally. The evidence indicates, however, that the amount of material that passes proximally is much less than that passing distally. It would therefore be expected that the propelling organelles, almost certainly the microtubules, would pass proximally. In fact they pass centrifugally at the rate of the bulk flow. It is therefore likely that they are anchored when they thrust. Possibly one of the functions of the neurofilaments is to provide an anchoring framework.

The functional significance of centrifugal axoplasmic flow is qualitatively self-evident but quantitatively uncertain. The perikaryon is not the only site of protein synthesis. Protein can be synthesized in the axon of invertebrate neurons (783, 786); these contain ribonucleic acid (RNA) (783). Protein is also synthesized by the axons of mammals, but the amount is small (1608, 1609). A few ribosome-like particles are in mammalian axons (3234). The bulk of the labelled RNA extracted from a mammalian nerve after a radioactive nucleoside has been injected in the region of the cell body is not axonal, but glial (116, 289, 329, 1490). Probably the nucleoside, or a derivative of low molecular weight, passes centrifugally and leaves the axon. It is then taken up by a sheath cell and incorporated into RNA. Synaptic boutons do not depend completely upon the perikaryon for their protein and glycoprotein. The synthesis of both can be demonstrated in preparations of synaptosomes. The demonstration has to be interpreted cautiously, both because autoradiographic study of preparations of synaptosomes shows that only half the radioactive activity of some such preparations is associated with identifiable synaptic components, and because the postsynaptic membrane nearly always adheres to the presynaptic bouton. The effect of re-uptake of transmitter is quantitatively unclear. It is known that dopamine-β-hydroxylase, chromogranin, neurophysin, vesiculin and ATP are released with neurotransmitters. The molar ratios of noradrenaline, chromogranin A, and ATP that are released vary very little. Some of the transmitter is taken up again. Unless an amount of the other substances is also taken up that is equivalent in molar ratio, some rearrangement must occur within the vesicles of the bouton. If this occurs the vesicles of the bouton are not simply equivalent to the vesicles passing down the axon.

Some labelled amino acid appears in the postsynaptic cell after it has passed along the axon. This would be expected if some labelled proteins are hydrolyzed at the presynaptic terminal, and their amino acids are released into the extracellular space. It is not by itself evidence for the transsynaptic passage of protein (725, 3130). Better evidence for this is available. If labelled amino acid is injected into the eye of a goldfish, it is carried to the opposite optic tectum. Some labelled protein appears there in the postsynaptic cell, and is carried in its axon to the ipsilateral tectum. This could represent either the transsynaptic transport of protein or the re-utilization of amino acid. If, however, intracerebral protein synthesis is almost completely stopped by cycloheximide, this uptake by the second neuron still occurs. It seems likely that some protein passes from the presynaptic terminals into the postsynaptic cell. It is easy to overestimate the significance of such an observation. There is no necessary relationship between such transfer and neurotrophism (Chapter 8). A neurotrophic action could occur without the effective agent entering the postsynaptic cell; it could, for example, act upon a receptor in the plasma membrane of the postsynaptic cell and activate an adenyl cyclase or a guanyl cyclase. Furthermore, a substance that enters the postsynaptic cell does not necessarily act as a trophic agent.

Within a given nerve cell material is directed along particular processes. For example, vesicles containing noradrenaline, dopamine-β-hydroxylase, chromogranin A and ATP pass along the axon, but not along the dendrites of an adrenergic neuron, even though dendrites possess an intrinsic transport system (1657, 2623). Neurophysin is transported along the axon of a neurosecretory cell, but does not enter the dendrites unless the axon is divided and considerable retention follows (1492). A dorsal root ganglion cell has a single process which divides into central and distal branches. The protein passing along one process differs from that passing along the other both in amount and nature (79). The way in which organelles, proteins or other materials are directed into one process of a neuron rather than another is not understood.

Centripetal transport (1578) is now well confirmed (2, 391, 1727, 1816). It can be demonstrated most easily with horse-radish peroxidase injected into the vicinity of the nerve terminals (1321). The role of centripetal transport is not known, but is likely to be concerned with regulating the metabolic state of the nerve cell body according to the state of its remote processes. In a cell like a motor neuron, in which the axon terminal is the dominant pole of the cell, such retrograde transport would be expected to play an important part in such regulation.

The axon terminal

Structural lability is evident in a chemically transmitting synapse. Despite considerable variation these share common features. The pre-

and postsynaptic membranes do not fuse, but remain separated by a uniform cleft about 25 nm wide, which contains extracellular material including glycosaminoglycans. The junctional area of the presynaptic membrane is well defined, and has features that can be demonstrated clearly if the preparation is stained with ethanolic phosphotungstic acid, especially if combined with three dimensional reconstruction from serial sections. The cytoplasmic aspect of the presynaptic junctional area is then seen to have a number of dense projections extending into the cytoplasm. These are 50 to 60 nm thick at their bases and about 60 nm high; they have a centre-to-centre spacing of 80 nm. Between them lie channels about 25 nm in diameter, extending from the interior of the presynaptic terminal towards the membrane. These dense projections are in hexagonal array; fine fibrils pass between them close to the cytoplasmic aspect of the membrane. The chemical nature of these projections and fibres is not known. Within the synaptic cleft, stained extracellular material appears to be more concentrated next to the dense projections. Synaptic vesicles lying near the junctional area often lie opposite the channels passing between the dense projections, suggesting that they may pass along these to gain the plasma membrane. A dense sheet, the subsynaptic web, lies close to the cytoplasmic aspect of the postsynaptic membrane and parallel to it. In material stained with ethanolic phosphotungstic acid it has a substructure of units 2 nm in diameter. Fine filaments extend from the subsynaptic web to microtubules. Other vesicles, tubules and mitochondria are also found in the presynaptic terminal. The origin, fate, composition and function of coated vesicles is controversial. The frequency of these can only be measured quantitatively in serial sections. It is likely that they have a role in the cycling of synaptic vesicles.

Discharge of transmitter occurs by exocytosis (717, 718, 1964, 2969). In many types of synapse, perhaps in all except the cholinergic, much of the transmitter re-enters the presynaptic terminal and new vesicles are formed from the plasma membrane by endocytosis (1376, 2608). The uptake of neurotransmitter is a major mode of terminating its activity. It is very improbable that the transmitter re-enters the bouton only within the newly formed vesicles, but likely that it enters through a wide area of membrane and is then concentrated within the vesicles. It is not yet certain whether the transmitter enters the bouton by diffusion and is then actively transported into vesicles, or whether the transmitter is also transported through the plasma membrane. Different types of bouton may differ in this respect. Freeze-fracture studies show pits in the plasma membrane of the bouton, where vesicles are discharging or forming; these can also be seen, but less well, in sectioned material. More are found when a nerve is stimulated. Endocytosis has been demonstrated by adding an easily identified protein to the extracellular fluid; horse-radish peroxidase is incorporated into vesicles of the neuromuscular junction of the frog, and into synaptic vesicles of brain. This uptake is increased by stimulation (1321). Similarly, bovine serum albumin has been identified within synaptic vesicles concentrated by centrifugation after it has been injected into cisterna magna. The rate of accumulation of the exogenous protein, and the rate at which the marker is lost again if the tissue is placed in unlabelled medium, is surprisingly low. While the former is compatible with the hypothesis that a small proportion of the vesicles turns over more rapidly than the remainder, the latter is not. There is more evidence from autoradiography that vesicles recently added to the presynaptic terminal from the axon lie closer to the junctional membrane than the remainder.

The extent to which cycling of presynaptic vesicles leads to the exchange of membrane protein and phospholipid between the vesicles and the plasma membrane is not clear, nor is it easily studied. Such exchange could be slight if a vesicle opened into the cleft, released its contents and then closed before departing again towards the interior of the bouton (2245). On the other hand such exchange could be considerable if exocytosis occurred at one site on the presynaptic membrane, say at the junctional zone, and uptake occurred at another, say at the edge of this zone. In this latter circumstance the membrane of the vesicle becomes part of the plasma membrane. The composition of the plasma membrane of synaptosomes differs from that of the vesicles. Little can be concluded from this difference, however, because membranes of the synaptosome include the postsynaptic membrane and membranes of the presynaptic terminal remote from the junctional zone, because some of the lipid of the plasma membrane may be bound to its fixed structures, just as certain phospholipids are found close to particular membrane proteins (Chapter 3) and are sometimes uniquely necessary for the normal functioning of these proteins, and because different vesicles appear to differ in their rates of turnover and so may differ chemically from each other either as a cause or as a result of this difference of behviour. If exocytosis occurs within the boundaries of the junction and endocytosis more peripherally, the system has an interesting parallel in the growth cone (Chapter 7), where there is very probably a similar 'source and sink' for vesicles, which are also cycled through the cytoplasm. If this model for a synapse is correct, lipid must move past the fixed anchored points of the membrane; it seems likely that the dense projections are some of these fixed points. Such a movement would not be difficult because of the fluidity of the lipid constituents of the membrane, which contains a high proportion of unsaturated fatty acids (Chapter 3). The fate of membrane retrieved by endocytosis is not clear. If retrograde transport is important, the membrane should accumulate in parts of the neuron away from the axon terminal. As the half life of the lipids of vesicles far exceeds that of neurotransmitters it is very probable that the lipid is reused many times at the axon terminal.

Each bouton is believed to contain only one kind of transmitter, although evidence is growing that some neurons can release different transmitters from different nerve terminals. Many transmitters are cationic, for example dopamine, acetylcholine and noradrenaline. A polyanionic protein is found within these vesicles: vesiculin in cholinergic vesicles, chromogranin A in noradrenergic vesicles, and neurophysins in vesicles of the posterior pituitary, which contain vasopressin or oxytocin. These acidic proteins differ from each other and from other acidic proteins such as tubulin, S-100, and actin, in their molecular weight and amino acid composition. When first isolated both chromogranin A and vesiculin are bound to a nucleotide. In the presence of ATP both chromogranin A and vesiculin are predominantly monomers; in its absence they aggregate. The roles of these proteins are not known for certain. Vesiculin has sufficient anionic groups to bind the acetylcholine of a vesicle; if 36% of the protein of a vesicle is vesiculin, one molecule of the vesiculin–nucleotide complex would have to bind only 19 molecules of acetylcholine. There is no evidence to suggest that the binding of acetylcholine to vesiculin, of nordrenaline to chromogranin A, of neurophysin to oxytocin or to vasopressin, or of histamine to heparin is specific. It is improbable that a molecule could possess so many specific binding sites, unless the protein–nucleotide–transmitter complex within the vesicle is crystalline. If the binding is not specific then the binding of the 'wrong' transmitter must be avoided; this is likely to be achieved by precisely controlled transport systems in either the plasma membrane or in the vesicular membrane. Noradrenergic vesicles contain dopamine-β-hydroxylase; it therefore seems likely that noradrenaline can be synthesized within or upon the vesicle; in contrast choline acetylase is found predominantly in the soluble cytoplasmic fraction of the synaptosome. It is unfortunate that it is so difficult to raise antibodies to vesiculin; this prevents this technique from being used at present to localize cholinergic synapses.

All these contents of the vesicle are discharged. The molar ratios of the different substances found in the superfusate are close to the molar ratios found in preparations of vesicles. The vesicle therefore appears to discharge its contents as a whole. If an adrenergic nerve is stimulated, noradrenaline, chromogranin A, dopamine-β-hydroxylase and ATP are all found in the medium. *In vivo* it is likely that their fates differ. Much of the noradrenaline re-enters the terminal. Some at least of the ATP crosses the postsynaptic membrane perhaps as adenosine, and is then incorporated into RNA within the postsynaptic cell. It is not known how much of the chromogranin A and of the dopamine-β-hydroxylase re-enters the presynaptic bouton. The cholinergic synapse differs in that the transmitter is hydrolyzed rather than removed from the cleft by the terminal; choline is re-utilized. The situation in the posterior pituitary differs in that there is no postsynaptic neuron. Neurophysin exists in two pools with different rates of turnover. These may correspond to the contents of vesicles in different parts of the bouton. It is by no means certain that substances other than the recognized neurotransmitter released from the vesicles are without significance as messengers; there are after all, many polypeptide hormones which are released in a manner that is almost identical. The ATP released may influence the postsynaptic cell; extracellular adenosine facilitates the activation of adenylate cyclase by neurotransmitters (Chapter 5).

Transmitter is released after calcium ions enter the presynaptic terminal (711). Release can be initiated by injecting calcium ions into an axon terminal when the surrounding fluid is calcium-free. The action of calcium is mimicked by strontium, but not by magnesium or manganese. After entry, calcium ions are sequestered rapidly by mitochondria, and perhaps by tubules of endoplasmic reticulum. It is not clear how the entry of calcium ions initiates exocytosis. There are many resemblances between the factors affecting release of neurotransmitter and those causing muscle to contract. The involvement of actomyosin in this process has not been proved. Although cytochalasin B blocks the release of acetylcholine at the superior cervical sympathetic ganglion, this is not an adequate criterion; it is also difficult to explain the ability of pyruvate to prevent this action of cytochalasin B. Three aspects of the release of transmitter must be explained, which may be related: movement of the vesicle towards the junctional membrane, fusion of the membrane of the vesicle with the membrane of the junction, and release of the neurotransmitter from the protein to which it is bound. If the synapse is considered as a variant of the pattern of terminal organization which also forms growth cones, it is probable that the ordered movement of the vesicles is directed by aligned molecules of myosin both on the cytoplasmic surface of the plasma membrane and on the cytoplasmic surface of the vesicles. The sites at which vesicles are formed by endocytosis will be a site of radiating or of converging myosin molecules. Aligned molecules of myosin should also be present on the surfaces of the dense projections. Unfortunately, however, actin microfibrils have not yet been identified either over the surface of the projections or beneath the plasma membrane of the synapse, although these have been clearly demonstrated in the growth cone. The ability of neuraminidase to prevent the release of neurotransmitter after the enzyme has been injected into the presynaptic terminal may have more to do with the process of membrane fusion than with vesicle mobilization. The findings indicated that the enzyme was probably not acting through disturbing the synthesis of transmitter or its transport to the site of release. Again speculatively, it is possible that neurotransmitter is released from the binding protein as soon as the vesicle opens; the contents are then exposed to a high concentration of calcium ions. The ability of ATP to chelate these may cause its release from the protein–neurotransmitter–nucleotide complex, a loss that could alter the capacity of the protein to bind molecules of transmitter, perhaps by causing it to aggregate.

The adherence of a presynaptic bouton to the postsynaptic membrane appears to be influenced rather by the state of that membrane rather than by the condition of the terminal in the adult animal. Boutons can be shed, for example, by a motor neuron or sympathetic ganglion cell after axotomy (Chapter 7). The factors which normally retain stable adhering boutons are not known. Stability might be maintained through the dense projections, material in the synaptic cleft and into the intrinsic proteins of the postsynaptic membrane or into the subsynaptic web. The glycocalyces of synaptosomes have been analyzed. They are rich in fucose, galactose and derivatives of sialic acid. The glycosaminoglycans have been fractionated by affinity chromatography, using bound lectins. It is not yet possible to say how the glycocalyx of synapses affects their capacity to make or retain connections with other cells, their electrophysiological behaviour, or whether it facilitates the processes of endocytosis or exocytosis.

Interactions between neurons and neuroglia

Morphogenesis

Neurons can survive and differentiate in tissue culture without neuroglia (490, 2167). Neuromuscular junctions develop without the aid of astrocytes, oligodendrocytes or ependymal cells. There is no ground for assuming that glia are essential for the formation or function of simple synapses.

The granule cells migrate radially through the developing cerebellar cortex from the external to the internal granular layer. In doing so, they pass through the molecular layer, which is about 200 μm thick, and which contains many neuronal and glial processes orientated in different directions. Electron microscopy has shown that the granule cells are directly apposed to radially oriented processes of glial cells during this migration, and that these processes pass across the whole thickness of the molecular layer (2384). They are perpendicular to the pial surface and have an expanded foot, which ends upon the basal lamina. The processes have a number of spiny or flattened protrusions and contain electrolucent cytoplasm with many microtubules oriented longitudinally. It seems likely that these fibres provide a framework which guides the migrating nerve cell precursors. Such a role is compatible with the hypothesis that the differentiation of brain is determined to a considerable degree by interaction between epithelial cells and mesenchyme, and that astrocytes play a significant part in this organization (Chapter 1). A similar relationship is found in the neocortex of the fetal monkey (2385, 2386); neurons here pass from the ventricular surface towards the pia, pursuing a direct radial course through a complicated tissue. Most migrating cells lie throughout their course very close to a glial fibre, which is radially oriented and ends beneath the pia in contact with the basal lamina. Each neuron usually has several processes which encircle the fibre. Each fibre is in contact with several neurons.

The simplest interpretation of this relationship is that the glial fibres guide the nerve cells. It has not been shown that only these fibres act in this way, nor that the fibres determine the direction of migration; it is possible that the fibres provide only a convenient route for a neuron to follow. There is also no evidence that indicates that the glial fibre provides the motive force for the migration of the cell. The late migration of neurons of the mutant mouse *weaver* is disorderly (Chapter 9). The radial orientation of the glial fibres is disorganized in this mutant (2389, 2390). This correlation does not prove that the neurons migrate along the fibres, nor that they determine the direction in which the nerve cells move. Both the disordered migration and the disarray of glial processes could equally well result from a common immediate cause.

Cultured neuroblastoma cells differentiate better if they are grown in a medium in which neuroglial cells have previously been cultured (2029, 2030); other cells influence differentiation less. This enhanced differentiation is not associated with an increase in the cell's content of cyclic AMP, nor with an altered growth rate. Glial cells also facilitate the chemical differentiation of these cells (229). Neuroblastoma cells are not normal neurons, however, and the demonstrated interaction may be a consequence of this abnormality: equivalent for example, to the capability of normal fibroblasts to support normal metabolic activity in mutant fibroblasts deficient in an enzyme, by transferring substrates and products of reaction (561). Nor can a normal neuron of the adrenal medulla be considered a typical cell; it resembles neurons of the posterior pituitary and many nerve cells of the median eminence in lacking immediate contact with a responding postsynaptic membrane. If the axon terminal of these cells is the dominant pole, the state of the dominant pole may be expected to be determined by adjacent cells; that is, by the pituicytes, the tanycytes and by the satellite cells of the adrenal medulla. Some cultured astrocytoma cells synthesize and release nerve growth factor. This might have physiological significance, or it might be another example of a tumour secreting a polypeptide that is biologically active, but which is not made by normal cells of that epithelium. If nerve growth factor is secreted it might support the growth of catecholaminergic nerve cells, but it is improbable that it supports the growth of all.

It is debatable whether the glia should be considered as forming an epithelium that is innervated, or one virtually lacking innervation. In only a very few situations are nerve fibres found which end upon the surface of glia and bear no relationship to any other kind of nerve cell. At least one example is present in the brain of mammals; many axons in the median eminence end upon tanycytes, and can be demonstrated by serial sections to lack contact with other cells. It is probable that these terminals discharge releasing factors into the extracellular fluid, from which it diffuses into the hypothalamo-hypophyseal portal system. There is no evidence at present to indicate that the material released acts upon the postjunctional cell. The membrane of the tanycyte is not obviously specialized at the site of contact, and the 'synaptic' cleft contains little demonstrable material (1179). On the other hand the epithelium can be considered as responding to neural influences, in that the glial cells are depolarized by increase of the extracellular potassium resulting from neural activity. The glia also possess – at any rate in culture of malignant derivatives – receptors which respond to neurotransmitter by metabolic changes (Chapter 5). It seems to be very probable that neurotransmitter could gain access to such receptor sites if they exist on the surface of the normal glial cell *in vivo*.

Morphogenetically it appears that glial cells are not essential for the survival and interaction of nerve cells electrophysiologically, but they are essential for the organization of nerve cells into a tissue.

It is interesting to note that regenerating nerve fibres seem to be guided similarly. Regenerating optic nerve fibres in amphibia pass across the surface of the optic tectum before plunging radially. It seems probable that these fibres are guided by radial processes of ependymoglial cells. It would be interesting to know whether these cells react metabolically or in any other way at the time of re-innervation of the tectum. The apparent ability of the ingrowing axons to react with the basal processes of these cells, and possibly with the cells most appropriate to guide the axon terminal to the correct region of the tectum, is consistent with the ability of non-neural cells as well as nerve cells to adhere with some specificity to particular cells of the retina (143).

The reaction of astrocytes around an axotomized neuron may be related to the remodelling of connections; the astrocytes respond by increasing their mass both when boutons are withdrawn from the surface of the axotomized cell and when they are restored (597, 597(a), 2857, 3068, 3072). If the neuron reacts without such remodelling the astrocytes do not respond. It is not known if the astrocytes are guiding the moving cell processes (Chapter 7).

Glial responses to neuronal activity

The apparent migration of neuroglial cells into the cytoplasm of a stimulated neuron (1675, 1683, 1684) is largely due to the movement of some of the nerve cell's cytoplasm into its processes from the perikaryon. The cause of this movement is not known. The early and transient reduction of the RNA content of the nerve cell body during prolonged intense stimulation (1153) may be similarly caused. Cells divide within the sympathetic ganglia following prolonged transsynaptic stimulation (732). These are probably satellite cells, although invasion by blood macrophages has not been excluded. The division was increased by the administration of neostigmine and reduced by atropine (2631, 2632, 2633). These pharmacological studies suggest that the glia may be responding to postsynaptic consequences of stimulation, rather than to changes in the presynaptic endings (Figure 7.11). The effects of antidromic stimulation were not studied.

Protoplasmic astrocytes hypertrophy near stimulated neurons (3067, 3072). This initial hypertrophy declines despite progressive metabolic change in the nerve cells. Since astrocytes respond briskly to changes in their environment and sustain their response until the original environment is restored (2184, 3066, 3067), the transient hypertrophy is almost certainly a response to a transient environmental change. After axotomy astrocytes respond to presynaptic rather than to postsynaptic events (3068, 3072). If this is also true of the stimulated neuron, it is possible that the transient change in the environment results from levels of presynaptic activity which are greater initially than later. This could result from progressive reduction in the use of those presynaptic fibres that are active initially, but which do not contribute significantly by summation to the activity of the postsynaptic neurons. Neither an active role for the astrocyte in 'selecting out' active but ineffective terminals, nor a structural remodelling of synapses under these circumstances has yet been demonstrated. Such selection is presumably essential to prevent a progressive increase in neural activity with increasing experience.

It is unlikely that astrocytes support presynaptic terminals metabolically, and very improbable that they aid the postsynaptic cell. There is no evidence for the transfer of organelles, nucleic acid, protein, enzymes or substrates. The membrane of the astrocyte and the adjacent cytoplasm are not specialized in the vicinity of a neuron. It is improbable that phosphorylated nucleotides could be transferred without hydrolysis at the cell surfaces. Ultrastructurally the astrocyte does not appear to be a cell of great metabolic activity, although possessing a great capacity for reaction. By analogy with muscle it is more probable that nucleotides are lost from neurons during periods of intense activity. Glial cells can concentrate a number of neurotransmitters (1306, 2006). They seem to do so with some specificity: glycine is an inhibitory transmitter in the spinal cord, but not in the brain; conversely, γ-aminobutyric acid is an inhibitory transmitter in the brain, but hardly in the spinal cord. Glial cells of the spinal cord can accumulate glycine much better than γ-aminobutyric acid, and glial cells of brain can accumulate γ-aminobutyric acid better than glycine. The glia may play a significant role in removing many neurotransmitters from the synaptic cleft, but the presynaptic terminals are very efficient themselves at this task. The capacity to accumulate these substances may be unrelated to synaptic transmission; glial cells around dorsal root ganglion cells accumulate γ-aminobutyric acid, but the perikaryon receives no synapses.

Glia react also to more complicated patterns of stimulation. Rats kept in an enriched environment (198, 1653) have a greater brain weight (870), cortical thickness (693, 694), neuron size (694), altered synapses (2083), and more neuroglia (692, 693). They are better at solving problems (944, 1653), and the rate of glial cell division is higher (68). The type of glial cell affected is not known: but from the position of the cells which have divided, they appear to be migrating and hence might be of the third type. The final distribution of these cells and the duration of their survival have yet to be determined. These reactions result more from continuous exposure to new experience rather than from repetitive activity. It is very likely that these rats have varying patterns of neuronal activity, with patterns and intensities of presynaptic activity. By analogy with the changes in astrocytes around stimulated neurons (3067), and in glia of the third type around stimulated axons (2078), the glial cells may be responding repeatedly or continuously to these changing states.

During exercise the nuclear diameter of ependymal cells of the central canal of the spinal cord close to active nerve cells increases (1676). This may be a response to a change in the composition of cerebrospinal fluid, for its 'sink action' here may be slight, as its volume and rate of flow are small.

Biochemical studies of the interaction between neurons and glia are very difficult technically. Good methods for obtaining a high yield of glial cells of a single type that retain their metabolic and structural integrity have yet to be developed. The glia must also be obtained from a limited and functionally homogeneous part of the brain, from animals of a known neurophysiological and hormonal state. It seems likely that the early history of the animals will also be important.

The extracellular concentration of potassium ions increases when neurons are active (954, 2185, 2341). Glial cells influence this change in two ways: by acting as potassium buffers, redistributing this ion across their membranes, and by restricting diffusion. They may both keep the concentration of potassium within fairly narrow limits (2316, 2947), but facilitate extrasynaptic interactions between those nerve processes which share a common microenvironment.

Such interaction may be greater during transient states of increased sensitivity of the membranes, for example during after-hyper-polarization. (177, 178). The membrane potential of neuroglial cells falls when the activity of adjacent neurons increases, and this contributes to the recorded change in surface potential (2405, 2406, 2407). The membrane resistance usually does not alter during this depolarization; this does not necessarily indicate that the depolarization is secondary to the release of potassium from the neurons although this is the most likely explanation. It could also result either from activation of an electrogenic pump, or from change in the membrane phospholipids, which alter the fixed charge of the membrane. As neither of these possible causes has been demonstrated, and as the degree of depolarization is compatible with the glial cell behaving as a potassium electrode damaged by penetration, the membrane potential of the glial cell should for the present be taken as indicating the extracellular concentration of potassium.

The role of the glial cells in generating slow potential changes in brain, and the possible effects of these upon neurons are uncertain (2347). There is no evidence at present to suggest that glial cells, divorced from neuronal influences, can act as primary generators of varying electrical fields. The membrane potential of glial cells falls when the activity of adjacent presynaptic fibres or of nerve cell bodies increases, and this depolarization contributes to the surface potential change. It is much less certain that a glial cell by depolarizing can generate a field potential sufficient to alter the excitability of an axon terminal (584). It is naturally even less certain that glial cells can provide a degree of specific coupling between boutons of different fibres, so that discrete patterns of hetero-synaptic facilitation can occur.

Glia in special situations are likely to be useful in clarifying aspects of the interaction between neurons and glial cells. The large glia of some invertebrates have been valuable in studying the electro-physiological responses to neuronal stimulation, and the membrane properties (1669, 1672). Pituicytes are related to the terminals of the neurosecretory cells of the posterior pituitary. There is no local postsynaptic membrane. When the release of vasopressin or oxytocin is increased, the pituicytes become larger and divide (245, 1053, 1054, 2187, 2429, 2652, 2847). Synthesis of RNA is increased. The granules in the cytoplasm become larger and more numerous. Pituicytes facilitate the release of hormone in an unknown way (245, 2429). It is not known whether the granules of the pituicyte contain material ingested from the extracellular space, or derived from the nerve terminal, or whether they result from alteration of the metabolism of the pituicyte itself (1774). Tanycyte ependymal cells of the median eminence behave very similarly under circumstances in which the liberation of releasing factors is enhanced. The changes are less striking than those seen in pituicytes, and the cells have not been seen to divide.

Pigment cells of the retina are modified ependymal cells. They lie close to the nerve cells that form the photoreceptors. Photoreceptor discs of the retinal rod outer segment (Chapter 3) are synthesized throughout life and pass distally along the rod. At the end of the rod they are shed, ingested by pigment cells and degraded (1225). There is no direct evidence for the phagocytosis of material shed from other nerve cells by other glia, although it is possible that some of the granules or vesicles in pituicytes or in tanycytes result from phagocytosis. In the mutant mouse 'retinal degeneration' this interaction is disordered both in the outer segments of the retinal rods and in the pigment cells (Chapter 9).

Satellite cells around the dorsal root ganglion cells are likely to be useful in studying interactions between neurons and glial cells remote from synapses.

3 MEMBRANES

Biological membranes are very varied (2525); they differ in composition and function. Different parts of the plasma membrane of a single cell are not identical. A plasma membrane differs from the endoplasmic reticulum or sarcoplasmic reticulum. During the cell cycle the plasma membrane changes. Membranes from different types of cell are dissimilar. Despite this variation, recent study (24, 473–481, 892, 893, 1298, 1900, 2189, 2463, 2464, 2516, 2517, 2717, 2719, 2720) supports the hypothesis that different membranes are composed to a large extent of a bimolecular phospholipid leaflet (661).

Membrane structure

Phospholipids

This confirmation has been obtained in a number of ways (131, 132, 1298). The surface area of a compressed monolayer of phospholipid indicates that the phospholipid is orientated with the long axis of the fatty acid chains being largely vertical to the plane of the membrane. This orientation is energetically favourable, as the polar head is related to the aqueous medium, and the fatty acid chains form a sequestered hydrophobic phase. The alignment of spin probes attached to the fatty acids indicates that this orientation is adopted by phospholipid films dried upon a glass slide (1797), and by phospholipids in hydrated membranes of sarcoplasmic reticulum, packed into pellets by centrifugation (810, 811). The periodicity of multilayers of bimolecular phospholipid leaflets measured by X-ray diffraction indicates that the width of the central zone of low electron density, occupied by fatty acids, is consistent with this orientation (896, 1784, 1785, 3133). Multilayers of barium stearate have a similar structure (1772). In neither preparation do the terminal $-CH_3$ groups appear to overlap. Although electron micrographs obtained from conventionally fixed and sectioned tissue seem to provide further evidence for such a structure, the membrane seen in fixed tissue (2462, 2463, 2464, 2465) is not identical with the bimolecular leaflet of the living cell. X-ray diffraction studies show clear changes in the electron density and the periodicity of the repeating unit on fixation and dehydration (46, 47, 889, 895, 2993). Furthermore, most of the lipid can be removed from membranes without altering their histological structure (919, 2095, 2906). Many of the neutral lipids and phospholipids can be lost during fixation (1626).

Electron microscopy of freeze-fractured tissue shows aspects of membrane structure (1387, 1888, 1892, 1893, 2276–2281, 2694–2698, 2923, 3092); this method can be augmented by freeze-etching (1607, 2036, 2037, 2812). The fractured membrane preferably splits along the plane occupied by fatty acid chains (320, 321) (Fig. 3.1), because frozen water offers the greatest resistance to

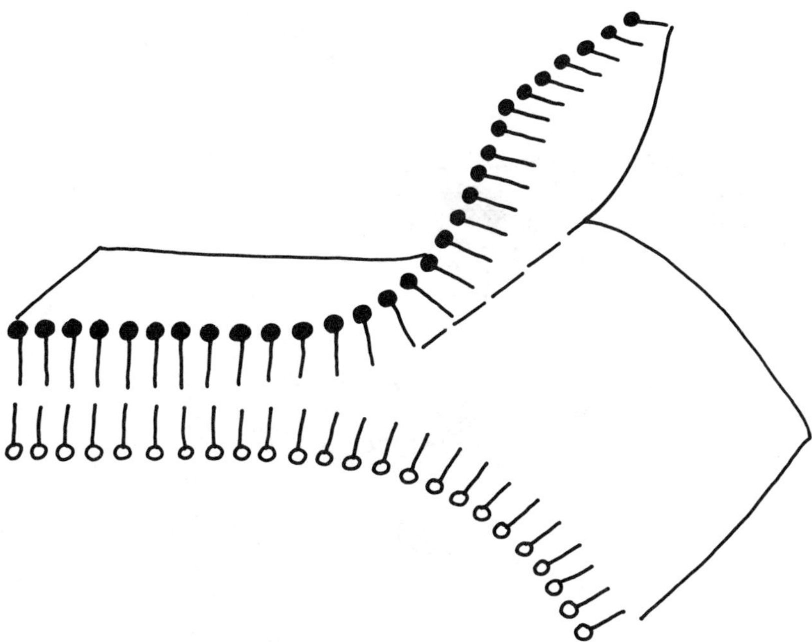

Figure 3.1 Diagram of a bimolecular lipid membrane to show the plane of fracture within the hydrocarbon phase.

fracturing in such a tissue, and is largely excluded from this plane by the hydrophobic molecules. The fractured surfaces of artificial membranes of phospholipid in water are featureless (664, 665). If hydrophilic protein is added to the suspension of artificial membranes, or if it is added before they are prepared, the protein is not seen on the fractured surface (Fig. 3.2). On the other hand, the addition of hydrophobic proteins causes pits or particles to appear on the fractured surface (2036, 2695) (Fig. 3.3). Such particles are seen in many biological membranes; in myelin they are unusually sparse. The number of particles seen in the plasma membrane varies during the cell cycle (2641). The particles are not uniformly distributed. Frequently, a greater density of particles is found near gap junctions or synapses (2277, 2278, 2280, 2281, 2554). If the two split faces of a given membrane are examined, complementary patterns of pits and particles are seen (459). The external surface of

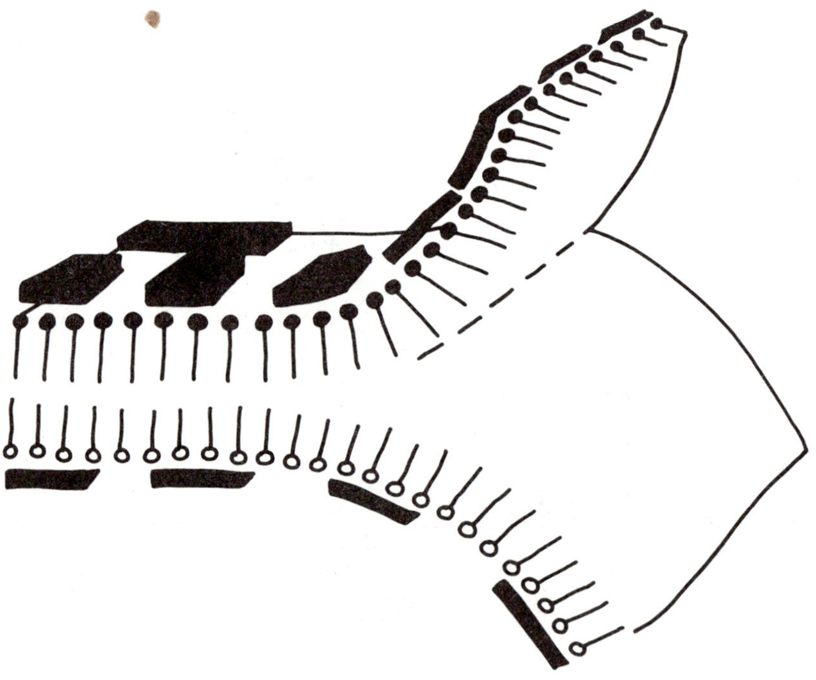

Figure 3.2 Diagram of a bimolecular lipid membrane with hydrophilic extrinsic protein. When the membrane is fractured the protein is not seen on the fractured surface.

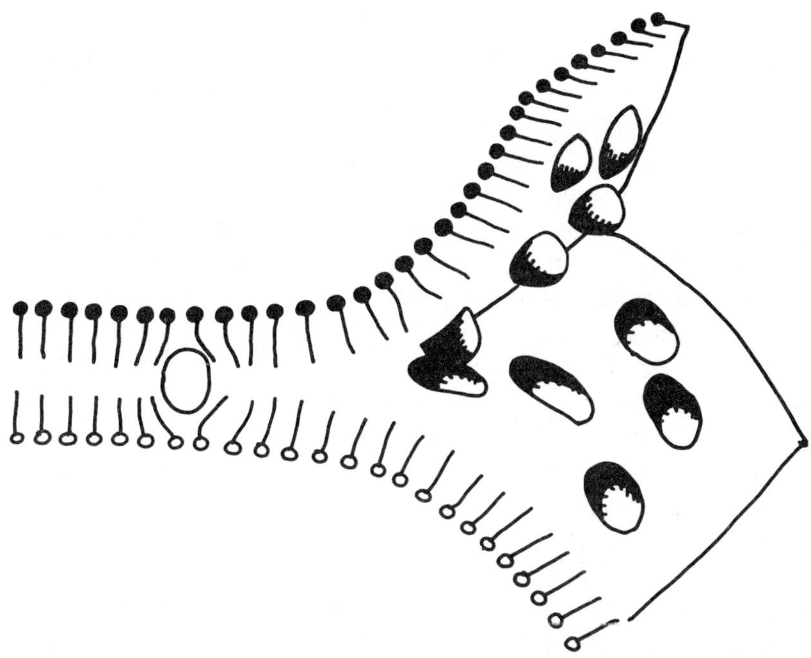

Figure 3.3 Diagram of a bimolecular lipid membrane containing hydrophobic intrinsic protein. When the membrane is fractured the protein is revealed; the two fractured surfaces are complementary. Projections on one correspond to craters on the other.

the membrane can be revealed from the fractured surface by deep etching. These methods are not free from artefact. The membranes of tissue prepared by rapid supercooling are not necessarily better preserved than those of fixed tissue. The measured relaxation times of proteins and lipids in intact membranes indicate that significant molecular movement is likely during freezing. Cooling also causes phase change in lipids (1855, 1856). Further artefact may be caused by perfusing the tissues with glycerol before cooling. Preparation of tissue by drying for negative staining or shadowing (1081, 1340) is also likely to produce artefact. Artificial lipid-water mixtures undergo phase transition during drying, which can resemble an altered substructure (134).

The lipids of a biological membrane are fluid; the fatty acid chains can move significantly within the membrane. This can be detected by measuring the motion of spin labels (1155, 1518, 2522, 2647, 2750) attached rigidly to assymetrical molecules of the cell membrane. The electron spin resonance spectrum changes when the rotational mobility of the label varies. The label provides information only about the membrane in its immediate vicinity. This environment is distorted by the presence of the label (413, 2930). It is probably for this reason that chemically different labels bound to the same part of the fatty acid chains indicate dissimilar degrees of movement (712). Spin labels attached near the methyl end of the fatty acids move more than those attached near the polar end. Nuclear magnetic resonance, a technique which does not require the insertion of artificial chemical probes also indicates that membranes are fluid (1397). The fluidity of phospholipid dispersions is increased when unsaturated fatty acids are included (811, 2646, 2945); this effect can be more marked at low temperatures than at high (2646). The rigidity of the double bond presumably impedes the close packing of the hydrocarbon chains at low temperature; at high temperature the increased interaction between adjacent molecules caused by the presence of a double bond is just sufficient to compensate for the gain in freedom of molecular motion. This could in principle stabilize the fluidity of membranes of poikilotherms when their temperature changes considerably. Both in goldfish and in bacteria, the proportion of unsaturated fatty acid increases when the external temperature is lowered. This adjustment could help to stabilize the fluidity of the membrane. Another interesting correlate of the unsaturated fatty acids in membranes is found in yeast; if yeast is grown anaerobically, the fatty acids of its membranes contain fewer double bonds. But, although the fatty acids are more saturated they are also shorter. The reduction in the proportion of unsaturated fatty acid would make the membrane less fluid; the presence of more short chain fatty acids would decrease the fluidity. These opposing trends would facilitate stabilization (2082). There is no direct evidence that membrane fluidity is measured and regulated by the cell *in vivo*, nor is it clear how such regulation could be achieved. It would be interesting to know if the lipids of mammalian nerve behaved similarly; the temperature of the cutaneous nerves of a rat tail can be increased considerably by burying the tail in the abdominal cavity. It might be expected that the proportion of unsaturated fatty acids would decrease. In contrast to the fatty acid chains of phospholipids, measurement of the nuclear magnetic resonance spectrum of the phosphorus of the polar head indicates that its movement is severely restricted (644, 645). This restriction is consistent with the lesser degree of movement shown by the adjacent part of the fatty acid chain.

X-ray diffraction also indicates that lipids of the cell membrane are fluid. Hexagonally packed crystalline paraffin chains provide a sharp reflection at 4.2 Å. Liquid paraffin chains give a broad band at about 4.5 Å (682, 823, 831, 1784, 1785). The most plausible interpretation of the wide angle 4.5 Å band found when biological membranes are examined (543, 823, 896, 3182, 3183) is that a significant proportion of the paraffin chains is fluid.

The plane of polarization of emitted light derived from a fluorescent probe inserted into the lipid part of an artificial or natural membrane is not precisely related to the plane of polarization of the incident light. This indicates that the fluorescent chromophore is able to move within the lipid environment. While this too shows that the paraffin chains are probably liquid in the immediate vicinity of the chromophore, the effect of this bulky molecule may not be negligible (480).

Raman infrared spectroscopy of phospholipid-water mixtures within the range $1200\ cm^{-1}$ to $1000\ cm^{-1}$ has been used to indicate that the hydrocarbon chains of a membrane are fluid (374, 375, 376, 1814, 1815, 2263). Crystalline chains produce intense bands at $1130\ cm^{-1}$ and $1060\ cm^{-1}$ and a weak band at $1090\ cm^{-1}$. Melted chains have a broad single band at $1080\ cm^{-1}$. Such a broad band is found in biological membranes.

It therefore appears that phospholipids are organized into artificial bilayers and into natural membranes in such a way that movement of the polar head and of the adjacent part of the fatty acid chain is severely restricted, while the remainder of the chain can move more freely. A model that is consistent with this (1870, 1871, 1872) is shown in Fig. 3.4. The chain is angled at about the tenth carbon atom. The oblique part of the chains near the polar head lies closer to its neighbours than the vertical part of the chain remote from the polar head. If the angle is 30°, the difference in density would be about 12%, which is the order of the difference between liquid and solid hydrocarbons at the temperature of transition.

Despite the fluidity of the artificial and natural membranes,

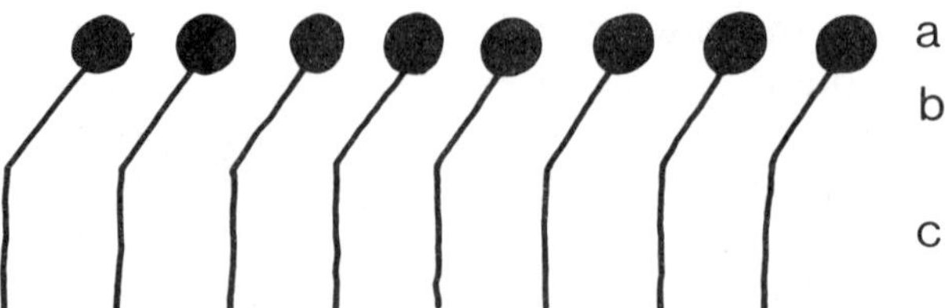

Figure 3.4 Diagram of phospholipid molecules to show angled hydrocarbon chains. The polar heads in layer 'a' have little mobility. The adjacent part of the chain in layer 'b' is more tightly packed than the part in layer 'c'. The part of the chain in 'b' is less mobile than that in 'c'.

phospholipids seldom pass from one lamella of a bilayer into the other (1629) (Fig. 3.5). A move in which the polar head must be buried transiently within the hydrophobic lipid interior would require the expenditure of energy. This limitation would not be predicted by a micellar model for membrane structure, nor for a membrane that undergoes frequent transition between lamellar and micellar forms (1842, 1843, 1844, 1845). The absence of significant exchange makes possible membranes having an assymmetrical composition of lipid (335). By contrast, diffusion laterally of lipid-soluble substances within the plane of the membrane should be less restricted. This is so; phosphatidyl choline can move laterally at a rate of about 0.5 μm/min in phospholipid vesicles. As a spin label was used in this study, it is possible that the fluidity of the membrane in the immediate vicinity of the labelled molecule was increased (1630). Different phospholipids can diffuse laterally within the two dimensional fluid domain of the sarcoplasmic reticulum (687). This lateral flow of material within the membrane can be demonstrated more directly (2694, 2696). Particles in a membrane can be made to aggregate by reducing the pH of the medium to 3.5; they can be dispersed again by raising the pH to 9.5. This aggregation and dispersion occurs within 2 to 4 min and can be studied readily by freeze fracture (Fig. 3.6). A similar but slighter change occurs physiologically. Rhodopsin moves within the fluid domain of the disc of the outer segment of the retinal rod when the charge of the protein is altered by light (p. 112). Immunoglobulins of the surface of cells from the mouse spleen also move (2264, 2265, 2266). These can be shown by univalent antibodies to be distributed diffusely;

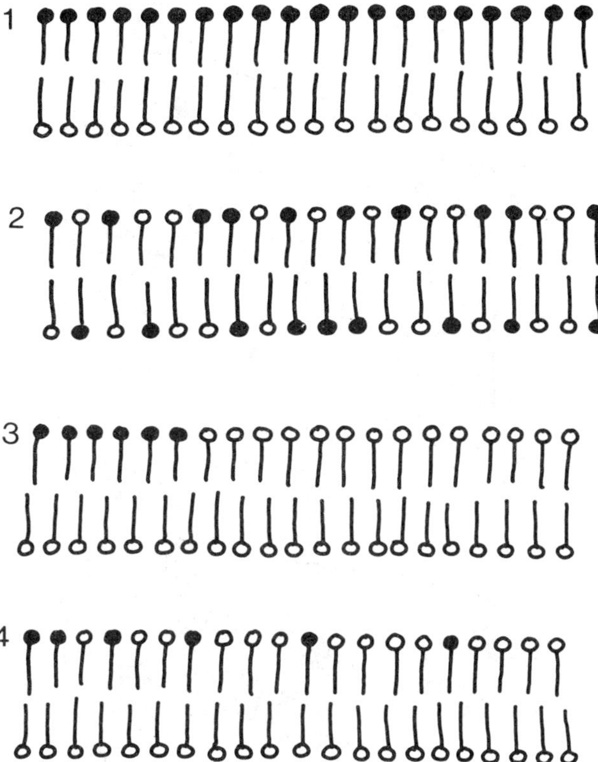

Figure 3.5 When labelled phospholipids are inserted into one layer of a lipid bilayer (1), they do not become rapidly redistributed between the layers; the situation shown in (2) does *not* occur quickly. When labelled phospholipids are inserted into one part of one layer of a lipid bilayer (3) they can diffuse laterally within that layer (4).

they are aggregated by divalent antibody. These aggregates can then be moved actively to the autosomal region of the cell, where they form a cap and are eventually phagocytosed. Cell surface antigens move similarly. The surface antigens of hybrid heterokaryons of mouse and man soon become mixed (982, 983). It is possible in these immunological studies that the molecules were not actually moving like icebergs within the lipid sea, but were rather skating over the surface. In all examples it is not clear whether the protein moves in relation to the immediately surrounding phospholipid, or whether the protein moves with some phospholipid so that the plane of

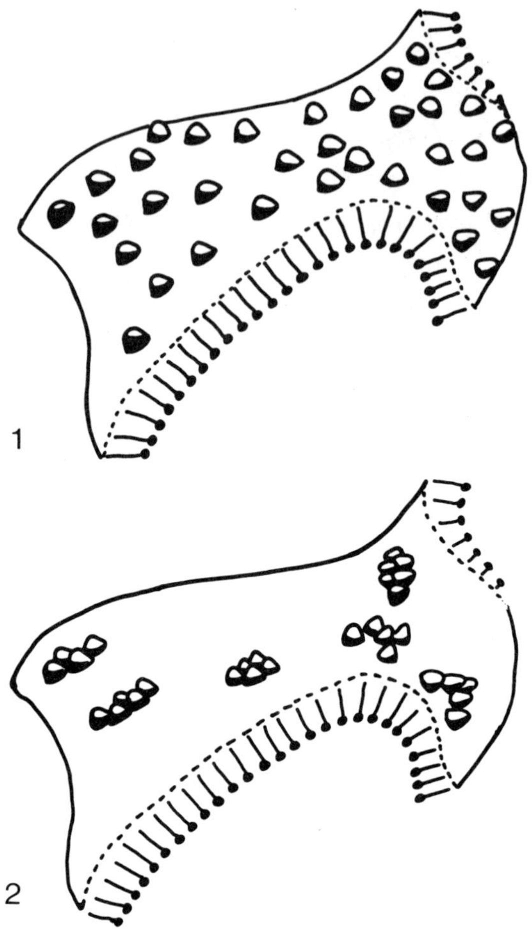

Figure 3.6 The distribution of bimodal proteins can be seen at the fractured plane of a membrane. If the pH of the aqueous phase is changed before the membrane is frozen, the proteins can move and may aggregate (2).

movement is not between protein and phospholipid, but between phospholipid molecules.

Steroids

Cholesterol influences the fluidity of phospholipid membranes in two ways. At temperatures above the melting point of the phospholipid hydrocarbon chains it reduces their mobility (1413, 1414, 1415, 1663, 1701, 1832). If the temperature is lowered, cholesterol delays freezing by impeding the ordered array of the side chains (480, 1814). These influences help to stabilize the fluidity of the membrane over a range of temperatures close to the melting point. If a biological membrane is loaded with cholesterol the structure as assessed by infrared spectroscopy does not change (1663). Cholesterol can be removed from isolated sarcoplasmic reticulum without altering the capacity of the system to accumulate calcium ions (722). Red cells from which the cholesterol has been removed are more fragile than normal (363); red cells that have been loaded with cholesterol are less fragile than usual (554); both the diffusion of sodium through the membrane and the transport of sodium are reduced.

Other steroids do not necessarily enter the membrane in the same way as cholesterol (2071). Because of their greater polarity, testosterone, progesterone and deoxycorticosterone form monolayers with the molecules lying almost flat on the surface; hydrocortisone appears to lie on the edge. Their orientaton within or upon biological membranes is uncertain.

Proteins

The proteins of membranes may be classified as intrinsic or extrinsic (1082, 2719, 2994). The former are associated with the membrane only by hydrophilic interaction, usually with the polar head of the phospholipids. These proteins can be extracted from the membranes by mild treatments, usually by using aqueous solvents at high ionic strength and at an appropriate pH. Such proteins are soluble in water and have an amino acid composition that is not exceptionally rich in hydrophobic amino acids. The basic proteins of myelin are examples. Chelating agents can remove more than just the extrinsic proteins; they can remove most of the proteins from the red cell membrane (2431). Intrinsic proteins lie partly within the lipid phase of the membrane. They are associated through hydrophobic interaction with the fatty acids of the phospholipids. (1138, 2546). Lipoprotein aggregates can be extracted by detergents or in organic solvents, such as chloroform-methanol. After the lipid has been removed, usually with difficulty, a very insoluble protein remains; in water the protein molecules reduce their surface energy by aggregation (2813, 2988). They contain a high proportion of hydrophobic amino acids (1751, 2995) often arranged in sequence, or close together (1751). The proteins are globular (2994, 3037, 3038, 3039) and contain a significant amount of peptide in α- and in β-conformation. The hydrophobic groups form the exterior part of the molecule (213).

Many intrinsic proteins also have externally oriented polar groups; this usually makes them bimodal. This structure causes the molecule to lie across the surface of the membrane (316, 1138). The hydrophobic part is then exposed to the central lipid lamella, and the hydrophilic part is related to the aqueous medium. Some proteins extend through the lipid phase to emerge on both aspects of the red cell membrane (334). Such molecules are stabilized both by hydrophobic and by hydrophilic interactions (1180, 1563, 2720, 2866). If the structure of the protein changes, then its orientation within the membrane may alter. Rhodopsin is an example of a molecule that moves in this way. Conversely, if a protein is removed from the membrane, its structure is altered, as an external placement of hydrophobic groups is energetically unfavourable in water. This change of conformation has been studied in a defined artificial system of methyl stearate and haemoglobin, using infrared spectroscopy (980). This denaturation is sometimes reversible. The renaturation of bimodal protein is illustrated by some enzymes, which are reactivated when they are recombined with phospholipid (482, 483, 1172, 1503, 1530, 2118, 2322, 2762, 2865, 2883, 2884, 2885). Ionic interaction, hydrogen bonding, and hydrophobic interaction are all involved in the renaturation of some enzymes (2762, 2876), and different phospholipids differ in their ability to facilitate the process. The structure of intrinsic and of bimodal proteins is often maintained within a membrane only in the presence of divalent ions; these proteins can be extracted from red cell membranes by chelating agents. It is of course possible that these ions are not acting directly upon the protein. Rhodopsin cannot be renatured by phosphatidyl choline if the fatty acids contain ten carbon atoms or fewer (1386), but renaturation occurs in didodecanoyl-, in ditetradecanoyl- or in dihexadecanoyl-phosphatidyl choline. This defect is not due to the closer packing density of the short chain molecules, as the pattern of restitution was hardly changed when the temperature was varied. The simplest interpretation is that insufficient room is available in the lipid phase when the chains are short.

The proteins of membranes also influence the phospholipid. If the rhodopsin content of a bilayer is increased the membrane becomes less fluid (1386). Spin labels indicate that this reduction affects the fourth, eighth and tenth carbon atoms from the polar surface. As the condensing effect of rhodopsin is nearly independent of pH and of ionic strength, it seems to depend upon non-polar forces, and to resemble the effect of cholesterol. Some intrinsic proteins bind certain phospholipids preferentially (2518). They can therefore influence their immediate environment, and may alter the physical state of the membrane in their immediate vicinity.

Some bimodal proteins contain sugars, and probably form part of the glycocalyx. At least one glycoprotein of the membrane of the red cell passes right through the membrane (334). Because the distribution of each type of protein within the membrane depends upon its tertiary and quaternary structures, it is very likely that it will move when denatured by fixation or by dehydration either by substitution or by freezing.

No ubiquitous 'structural protein' (2435) has been found which is an integral part of all membranes (730, 1278, 1588, 2298, 2880). While it is possible that each type of membrane has its own unique structural protein, there is no evidence that this is so. A lipid bilayer may be the only feature common to all membranes. Many diverse intrinsic proteins are enzymes or have other well-defined physiological roles (2537). In many membranes such proteins form most of the extractable intrinsic protein. It can be argued that a residue of protein remains to which a definite function cannot yet be ascribed; such an approach defines a structural protein only by what it fails to do. There is no evidence which suggests that a membrane cannot be formed from the lipid components of membranes because of the absence of one protein component to which no other function can be ascribed.

Subunit structure within the plane of the membrane

Some membranes have an ordered substructure within the plane of the membrane. Technically it is possible to detect the pattern of distribution of proteins but not of the lipid constituents. It is probable, however that a mosaic ordering is not confined to protein, because certain phospholipids lie adjacent to particular membrane proteins (1651) and because protein influences the physical state of adjacent phospholipids (1386). A repeating substructure can result from the dispersion of the molecules of one type of protein; this is seen in the membranes of the discs of the retinal rod outer segment. Rhodopsin is the preponderant bimodal protein. The charge it bears at a normal pH repels adjacent molecules of rhodopsin. As the concentration of this protein is high and the domain of the membrane is limited, the mutual repulsion results in a distribution of rhodopsin molecules within the plane of the membrane that is uniform, and a pattern of packing that is optimal. This has been shown in hydrated membranes of these discs by X-ray diffraction (p. 110). Cytochrome oxidase also becomes evenly dispersed throughout artificial phospholipid bilayers (2996); the type of lattice structure seen by negative staining in this preparation must be accepted with caution as the membrane is dried before examination.

Subunit structure also results from a balanced packing and dispersion of a single type of protein when the unit is made of ordered oligomeric aggregates of a constant number of molecules. It is probable that this form of substructure occurs in the membrane of the electroplax, which contains the acetylcholine receptor (433) (Chapter 4). In this preparation the oligomeric units are formed of six protomers. The hexamers themselves form a hexagonal lattice with a centre-to-centre spacing of 9 to 10 nm. This distribution was found both by deep etching of the freeze-fractured surface and by negative staining of the natural surface of the membrane. Both techniques are liable to artefacts from drying. Pharmacological studies of the intact membrane are consistent with this structure as they show a high degree of co-operativity.

Subunit structure can also be determined by the close packing of aggregates of protein molecules of different types. An example of such a situation is seen in the organization of enzymes which are closely associated both functionally and structurally and which act serially to separate charge across a membrane, in chloroplasts (322, 959, 2067, 2068, 2069, 2216, 2217, 2218), and in mitochondria (874, 875, 876, 1139, 1140, 2445). This type of substructure, based upon the ordered and close packing of protein subunits has some affinity with the capsules of viruses and some marine bacteria.

A regular subunit structure and the formation of oligomeric aggregates are almost certainly necessary if the responses of the membrane are to show co-operativity, that is, a facilitation of the response of one type of molecule through the direct interaction between adjacent molecules. Although co-operativity is a feature of many membrane reactions, its molecular basis is not understood in any, except in principle. It is not known, for example, if the co-operative binding of acetylcholine to the receptors of the electroplax of *Torpedo* is due to interaction between single protein molecules within an oligomer, or to interaction between adjacent oligomers (470, 471, 1971).

Membrane synthesis

Membranes are labile structures; phospholipids and steroids of some membranes can exchange with sonicated dispersions of these substances in the surrounding medium (363, 554, 1127), and the composition of the membrane can be altered in this way. The incorporation of lipids into membranes *in vivo* is increased greatly by certain soluble proteins of the cytoplasm, which facilitate the transfer of phospholipids (1, 48, 1891, 3157–3160). At least one of these proteins facilitates this process specifically, increasing the exchange of phosphatidyl choline between mitochondria and microsomes. It is likely that this facilitation is fundamental to the regulation of the phospholipid composition of a given membrane and to maintaining the differences in composition of different membranes in a given cell.

The synthesis of membrane lipids and of membrane proteins proceeds independently (1522, 2007, 2139, 2140). It is therefore unlikely that lipoprotein subunits are synthesized as such and are then incorporated into the membrane. More probably, the lipids and the proteins proceed independently to the membrane and are assembled *in situ*. Membranes can be reassembled *in vitro* from previously separated phospholipids and proteins (2418). Reconstitution must be in a fixed sequence. The UDP-galactose-lipopolysaccharide-α-3-galactosyl transferase system of *Salmonella typhimurium* can be reconstituted in this way from protein, phospholipid and liposaccharide (818, 2519).

Artificial membrane bilayers

Artificial bilayers of lipid (1307, 2064, 2065) have some similarities to biological membranes. They are 7 to 12 nm thick, have a low specific capacitance (82, 83), a similar interfacial tension (2920), and a similar permeability to water, urea, glycerol or mannitol (86, 438, 439, 1136, 1241, 1417). Unlike most biological membranes, bilayers have a very high specific resistance and obey Ohm's law (145, 441, 1390, 1391, 1773, 1804, 2019, 2032, 2033, 2060, 2062, 2064, 2065, 2656, 3245). In bilayers of charged phospolipids this resistance has two components: one interfacial and the other within the lipid (85, 1241), which can be distinguished experimentally. The interfacial resistance may be due to the ability of the phospholipid molecule to rotate as a counterion, to possible dehydration of the ions before they enter the lipid phase of the membrane, or to the different properties of the layer of ordered water next to the membrane (642).

Although the composition of the lipid mixture from which the membrane is made can be determined, the composition of the bilayer itself is not precisely known. Bilayers of fairly well defined composition can be made by apposing the hydrocarbon chains of two lipid monolayers (2033).

Bilayers made of different phospholipids differ in their permeabilities for water. The permeability is less in membranes made of longer fatty acid chains; the permeability is increased if the number of double bonds in the chains is increased (898, 1062, 1136). The permeability is reduced if cholesterol is added (898, 1124). If the charge borne by the polar groups of the phospholipid is increased by changing the pH of the medium, the permeability is increased (1733). This results from increased separation of the molecules due to enhanced coulombic repulsion. Factors which increase the fluidity of the membrane therefore increase its permeability. This may result from the accommodation of water in transient cavities formed within the lipid part of the membrane by the increased movement of the chains. The permeability of the membrane and its fluidity are also influenced by the concentration of cations in the medium (409). Univalent cations reduce the coulombic repulsion of adjacent charged phospholipids by acting as counterions; they do not act in this way on uncharged phospholipids; divalent cations may also act as stabilizing bridges (1885, 1886).

The surface of a membrane is covered by a layer of ordered water molecules and by a layer of unstirred water about 110 μm thick (86). This unstirred layer is not altered if the membrane is made porous by adding amphotericin. This layer of unstirred water offers a significant resistance to the diffusion of substances through the membrane; thus, the unstirred layer next to a porous membrane contributes 84% of the resistance to diffusion for water, 75% of the resistance for urea, and 34% for glycerol. These layers of unstirred water must be distinguished from the bound water. About 20% of the mass of a phospholipid membrane is due to bound water. If it is removed the structure of the membrane changes: the lamellar phase may become hexagonal (133, 813, 896, 1699, 1700).

The surface of a membrane composed of charged phospholipid attracts ions of opposite charge. Thus, more sodium ions than chloride ions will be present in the vicinity of a membrane which is negatively charged. Conversely, in the fluid remote from the membrane, the concentration of the chloride ions will be higher. Because of this, if sodium chloride is added to one side of a negatively charged membrane, but not to the other side, a potential will develop across the membrane. The fluid containing the sodium chloride will be the more negative. The magnitude of the potential is described by the Gouy-Chapman equation (1873, 1885, 1886). The potential difference in such a situation is not caused by the passage of ions across the membrane.

Ionophores

The conductivity of a bilayer can be increased considerably by adding certain substances (802, 2059–2065, 2920–2922, 2944). Well defined additives include cyclic peptides, cyclic polyethers, antibiotics and some proteins and peptides. Less well defined additives include proteinaceous extracts derived from bacterial membranes, or from various tissues including brain. They share the property of facilitating and controlling the passage of ions which could otherwise cross the membrane only in minute quantities (801, 802, 1276, 2063, 2921). There are several ways in which these substances can act. These are still controversial, despite some studies of the structural changes in an ionophore when it binds an ion (710, 865, 1586, 2292), of the interactions between ionophores (1277), and of the interactions between ionophores and phospholipids (1575, 1652, 2869).

Valinomycin

When valinomycin is added to a membrane its conductance increases. At concentrations of 100 pg/ml to 1 ng/ml its action is just detectable. At 100 ng/ml it increases the conductance of a bilayer between two solutions of 100 mM potassium chloride from 10^{-8} mho/cm^2 to 10^{-3} mho/cm^2. Membrane capacitance is not altered (2061). Valinomycin facilitates the passage of cations only. The permeability of potassium is increased about 400 times more than that of sodium (84, 1773). Hence if a bilayer containing valinomycin is placed between two equimolar solutions, one of sodium chloride and the other of potassium chloride, the compartment containing the potassium chloride will become the more negative. Bilayers of different phospholipids alter the ability of valinomycin to increase the conductance; through their charge they influence the concentrations of cations in the immediate vicinity of the membrane. They do not alter the capacity of the drug to facilitate preferentially the transfer of potassium. Valinomycin can act only in fluid membranes (1652), and the increase of conductance is less in membranes that are made less fluid by the incorporation of cholesterol (2869). Valinomycin increases the conductance not just of thin bilayers, but also of films up to 5 μm thick. The action of valinomycin is increased by a rise of temperature. Within wide limits the conductance is directly proportional to the concentration of the drug, which need be applied to one side of the bilayer only.

Valinomycin is a lipid-soluble cyclic depsipeptide with many carbonyl groups. Unlike some other ionophores, the ring structure is closely related to its action. If valinomycin-like molecules are synthesized, in which the ring is either greater or smaller, they are less effective; hydrated alkali metal ions can no longer fit into the smaller ring, and a larger ring holds them so loosely that the stability of the drug-ion complex is reduced. If valinomycin is made more lipophilic by substituting leucine for alanine, it becomes more effective. The lipid solubility results from the externally directed hydrophobic groups. Its ability to accept hydrated cations is due to proton-accepting carbonyl groups directed centrally within the ring (1469, 2165, 2292, 2332, 2670).

The resistance of a membrane has two components: that due to its surface and that due to its central lipid phase. Valinomycin reduces the surface component of membrane resistance (85). The affinity of valinomycin for cations depends upon its lipid environment; in water it has almost none (1575). At a lipid-water interface the molecule changes its shape considerably when it binds potassium. Nuclear magnetic resonance spectra do not show a tendency for valinomycin molecules to aggregate in a lipid film to form complexes (1277). This confirms the probability that molecules act singly, previously shown when conductance was found to be directly proportional to the concentration of the drug. The carriage of potassium or rubidium ions rather than sodium or lithium ions may be related to the energy needed to dissociate the ion from the carrier at the surface of the membrane.

It seems very probable that valinomycin acts as a cationic carrier with a high affinity for potassium. The properties described are not consistent with the formation of a stable channel or a transient pore. Qualitatively, the ionophores non-actin and enniatin behave similarly.

Alamethicin

Alamethicin increases the conduction of cations through a phospholipid bilayer. Its ability to do so is greatly increased if a potential is applied across the membrane (479, 492); the curve relating current to voltage varies with the charge on the cation. Conductance is proportional to the sixth power of the concentration of the drug. The conductance increases relatively slowly when the voltage is applied, but falls rapidly when it is removed. Alamethicin acts in membranes made of phosphatidyl serine or of lecithin; its action therefore does not depend upon the charge of the phospholipid. When the drug is added to a membrane the movement of the fatty acid chains of the phospholipid is restricted. This restriction could result from the formation of a phospholipid-drug complex, and perhaps from a co-operative change in other phospholipids.

According to one hypothesis, monomers of the alamethicin-cation combination remain on the surface of the bilayer, so that conduction is not increased until an applied potential drives them into the hydrocarbon lamella, in which they aggregate (2062). Alternatively (479, 492), alamethicin readily penetrates the lipid layer and causes considerable reorganization of the hydrocarbon chains; when a potential is applied the alamethicin is redistributed between the two surfaces and then aggregates to form a conducting system. The dependence of the conduction upon the sixth power of the concentration indicates that the active form of the drug is probably a hexamer. The structure of this aggregate is controversial.

Gramicidin A

Gramicidin A increases the cationic conductance of a bilayer. This increase also is dependent upon the presence of a voltage across the membrane. The conductance varies with the square of the drug concentration, suggesting that the active form of the drug is a dimer (1105). When the concentration of gramicidin is low, discrete 'quantal' changes of conductance are seen (1359, 1360), corresponding to a current flow of 2×10^7 monovalent ions/sec. The frequency of these steps increases with the applied voltage.

Gramicidin A is probably in the form of a left handed helix around an aqueous lumen. A variety of models have been considered, differing in length and pitch (2976, 2977, 2978). A channel through the membrane could be formed by the head-to-head union of two of these molecules, one within each layer of the membrane. It is unlikely that gramicidin A acts as a mobile carrier, as it can continue to support conduction when the membrane is no longer fluid. It also acts only in bilayers and not in thicker films. The difference in the energy of activation between the diffusion of cations through water and the passage of ions through the gramicidin channels is very small; the transfer of the ions cannot involve the formation of hydrogen bonds. The most probable model is conceived as being a channel containing a continuous file of water molecules through which ions can pass by diffusion. The selection of cations and the rejection of anions could result from a potential of about −60 mV within the pore.

Amphotericin B

The channels formed by amphotericin B have very different properties. Up to 12 molecules may form one channel, for under some circumstances the conductance varies with the twelfth power of the concentration of the drug. Cholesterol must be present, but the conductance does not vary with its concentration in a straightforward manner. Cholesterol cannot be replaced with α-tocopherol. Amphotericin must be applied to both surfaces of the membrane; this suggests that the drug, like phospholipids cannot move easily from one lamina of the membrane to the other. The large complexes necessary to facilitate conduction are easily disaggregated by thermal agitation. This is indicated by a striking decrease in conduction when the temperature is raised. Amphotericin has a low lipid solubility. Formed channels continue to conduct when the membrane is no longer fluid.

The channels pass anions and not cations; they discriminate between anions according to size (87, 438, 439). The selectivity is insensitive to pH within the range 3 to 10 but is reversibly lost in more alkaline media. Small electroneutral molecules can pass through the channels; small molecules pass more easily. Urea, glycerol and glycol can pass through, although at different rates. Glucose, which has a Stokes-Einstein radius of 400 pm, cannot. Nystatin increases the conductivity in a manner similar to amphotericin B.

Cyclic polyethers

These have been studied less intensively (1884). They are lipid soluble and produce a large increase in the conduction of cations. Conduction increases linearly with the concentration of the metal cation, and with the square or cube of the concentration of polyether. Probably they act as dimers or trimers. The channel might be formed in a manner similar to that of gramicidin A, by stacking pores concentrically.

Excitability-inducing material

Excitability-inducing material is extracted from bacteria (2059). It contains basic protein and ribonucleic acid, but has not been analyzed in detail. Separately the constituents have no action, but on recombination they are effective again. When added to a lipid bilayer, its conductance increases by several orders of magnitude, but only when a polarizing potential is applied. At low concentrations of the material the conductance changes in discrete steps. In membranes made of certain lipids each step represents the complete opening or closing of a single channel (797, 798); in othe membranes a channel can conduct at less than its maximum (181, 182). In some membranes the channels behave as a homogeneous class; in other membranes they are clearly heterogeneous. The voltage applied determines the period for which a given channel stays open. Cations are selected and anions are excluded (2062). As the relative mobilities of different ions through a channel are very close to their relative mobilities in free solution, it appears that a channel does not discriminate between them. It is likely that the channel is filled with water, and that its wall bears a fixed negative charge (1723). The conductance of a channel is greatly influenced by other added substances, such as multivalent cations and proteins (181, 2062): its capacity to reject anions may be modified considerably by changing the fixed negative charge lining the pore.

Miscellaneous materials

Synthetic amino acid homopolymers such as polylysine and polyglutamic acid also influence the conductance of artificial bilayers. Polylysine increases the conductance for cations and polyglutamic acid for anions (2032, 2033). It is not at present clear how they work; polylysine can bind ions, but it also affects the surface packing of lipid monolayers (2657, 2658), and alters the mobility of fatty acid chains of phosphatidyl serine. Haemocyanin from the keyhole limpet also increases conductance. The demonstration that a protein extracted from a tissue can alter the conductance of a lipid bilayer therefore is not evidence that it acts in this way *in vivo*. The S-100 proteins (416) can increase the conductance. Other extracts of brain have been prepared; these are not pure, but contain molecules or complexes which facilitate the transfer of potassium ions or of sodium ions across the bilayer. It is not known whether these substances are related to 'sodium channels' or to 'potassium channels'. Cerebroside can reduce the conductance of a bilayer (513), and ganglioside can increase it (514).

Many enzymes *in vivo* are within or closely associated with membranes. A crude preparation of Na^+-K^+-ATPase extracted from brain has been studied in an artificial bilayer (1480); the hydrolysis of ATP by the enzyme is accompanied by a fall of the resistance of the membrane, which can be prevented by ouabain. Changes in the conformation of proteins caused by reactions between antigen and antibody in the absence of complement increase the conductance of a bilayer (145, 441). Agents which uncouple oxidative phosphorylation also reduce the resistance of artificial membranes (1390, 1391).

Photosensitive responses can be elicited in artificial bilayers. The simplest preparation consists of a membrane in the presence of iodine and iodide (1951). Current is carried by polyiodides, which are sensitive to light. The conductance changes by about 1%, with a very short latency. Photosynthetic pigments facilitate electron transfer across a membrane between aqueous compartments of differing Redox potential. The photosensitive effect (1462, 2121, 2920, 2922, 2957, 2958, 2959) is of very low quantal efficiency compared to electron transfer in the chloroplast; this poor performance may be due to the virtual absence of subunit organization of the proteins of the artificial membrane. Thermal effects from light absorption without electron transport give a smaller increase in conductance (2959). Successful photosensitive responses using visual pigments have not yet been reported.

Black lipid membranes may be useful for studying the action of other biological transducers. If, for example, the chemoreceptors for taste and smell are proteins, the responses of a membrane upon which they are absorbed could be studied when odourants and tastants are applied. Receptors for well-characterized insect pheromones might be very suitable for such a study.

Biological membranes

Myelin

Lipids

Lipids of mature myelin differ from those of most other membranes in possessing a high concentration of cholesterol and cerebroside. They are also exceptional in that many of their conjugated fatty acids are very long, and relatively saturated (136, 170, 619, 656, 1614, 1958, 2151, 2152). Cholesterol is the only sterol present in large quantities in the myelin of mature animals. The concentration of galactolipids is lower than in other membranes of brain (2860, 2861).

The fluidity of membranes is low when the alkyl chains are long and saturated, and when the membrane contains cholesterol or cerebroside. The hydrocarbons of myelin are fluid, but the extent of their freedom is small (478, 1427, 1495, 2604, 2605). The isolated phospholipids of myelin have a clear transition temperature; in the presence of cholesterol this is not seen (1699, 1700, 1701, 1736). Without cholesterol some of the hydrocarbons are crystalline at 37° C. Cholesterol therefore acts both to reduce the fluidity and permeability of myelin, and to impede crystallization. The stability of the laminar phase is lost if the bound water is lost (478, 481). Detailed analyses of the lipids of myelin are available (699, 700, 1505, 2152). Apart from observing that the composition is well suited to a membrane which acts as an insulator, it is not yet possible to make further observations on the physiological, pharmacological, biophysical or pathological significance of the precise composition of the myelin lipids.

Protein

Myelin contains an unusually small amount of protein, only 20% by weight (115). This low proportion is consistent with the smooth fractured surface seen by electron microscopy (320, 321). Water soluble basic proteins make up about 30% of this protein (1927–1931), and proteolipid about 50% (115, 3174). In the peripheral nervous system a different proteolipid and different basic proteins are found (1220, 3173, 3174, 3175).

Basic protein

Myelin of the central nervous system of the different species, studied contains a very similar basic protein (1927–1931). It has an unusually great proportion of basic amino acid, and an isoelectric point greater than 12 (431, 432, 820, 839, 840, 841, 842, 843, 844,

1103, 2090, 2091, 2092, 2190). This composition is compatible with the absence of α- or β-structure demonstrated by optical rotary dispersion and by circular dichroism. Histologically also, basic protein is localized to myelin (1632, 1633). The basic protein acts as an antigen to produce experimental allergic encephalitis in susceptible animals (431, 1848, 1927–1931, 2468, 2753). The part of the molecule which acts in this way is not the same in all species (841, 842). In some species, including the rat, the basic protein has two components. The second is smaller and is only weakly antigenic (1930); it differs through deletion from the larger protein. Other fractions of basic proteins may also be present. The half-life of the basic proteins is about 35 days (3178).

Proteolipid

Proteolipid (937, 938, 939, 2023, 2673) forms about half the total protein of myelin of the central nervous system (3174) and rather less of that of the peripheral nervous system (3175). Unlike lipoprotein it is insoluble in water and can be extracted with chloroform-methanol (937). Lipid can be removed totally only with difficulty; the separated apoprotein, which has a molecular weight of 34 000 daltons, and which may be trimeric, is water-soluble and susceptible to enzymatic hydrolysis (2022, 2904). In a two-phase system of water/chloroform-methanol the apoprotein remains in the water. When an acidic lipid is added, a protein-lipid complex is formed which enters the chloroform-rich layer. The protein in such a complex has appreciable α-structure; the apoprotein in water has none (1459, 1460, 1465, 2928). It is probable that the change of conformation includes the placing of the non-polar residues on the outside of the molecule. The apoprotein has a high proportion of non-polar amino acids (820, 1103, 2904, 3174, 3176). The rate of turnover of the apoprotein is very low (2024). In several species the molar ratio of proteolipid: basic protein is about 1 : 1 (1962). The significance of this is far from clear, especially if the molecular weight of the proteolipid is that of an oligomer. Proteolipids are not confined to myelin. Some evidence suggests that different proteolipids in synaptic membranes might be involved as receptors in chemical transmission (Chapter 4).

Wolfgram protein

This is less well characterized (820, 3175, 3176). It is extracted by acidic chloroform-methanol (3173), but is precipitated at pH 5. It is rich in dicarboxylic amino acids (820, 1103, 3173) and can be hydrolyzed by trypsin.

A number of other proteins, for example peptidases, and phosphodiesterases are also present in myelin extracts from brain.

Structure

Early optical studies of structural birefringence in myelin indicated molecular order (2604, 2605). These have been fully confirmed and extended by X-ray diffraction and by electron microscopy. From an X-ray diffraction pattern the electron density profile of a repeating unit, or 'unit cell' can be determined. The resolving power of the method increases as higher orders of diffraction are analyzed. One of the simpler calculations is that of the length of the repeating unit. Study of artificial lipid-water phase systems (896, 1784, 1785, 1855, 1856, 3133) has validated the application of the method to biological membranes. The length of the repeating unit in myelin of unfixed peripheral nerve is 17.8 to 18.5 nm (46, 47, 269, 270, 275, 889, 890, 891, 895, 2605). Crystallographic centres of symmetry occur at the internal and external membrane boundaries. Five major orders of diffraction and a number of minor orders are obtained. The presence of odd orders of low amplitude indicates that the two layers forming the basic unit are not symmetrical. The electron density profile within a unit cell, that is, the variation of electron density along a line normal to the plane of the membrane, is calculated by Fourier synthesis. This requires that the correct phase sign is given to each order of diffraction (890, 894, 3188). There is substantial (270, 894, 2034, 3188), but not universal (46, 47) agreement about the signs to be allocated to the first five orders. This concensus emerged from a study of the changes in diffraction pattern which accompany osmotically-induced swelling of the myelin sheath, and represents the best solution provided no gross molecular rearrangement takes place within the membrane of the swollen nerve. The resolution possessed by an electron density profile based upon the first five orders alone is poor. The phase signs ascribed to further orders are at present based largely upon finding the combination which fits best a 'Patterson plot', that is, a topographically scrambled representation of the distances separating points of equal electron density. Diffraction patterns studied up to the 11th order for peripheral nerve myelin, and up to the 10th order for myelin of the central nervous system, have provided electron density profiles with a resolution up to 800 pm. Further analysis to the 15th order (437) provides a potential resolution to 570 pm. Extension of the analysis to the 18th order (270) increased the resolution 450 pm. It is doubtful if further useful resolution will be obtained by this method, for it provides only a mean profile for many different unit cells, not all of which may be identical.

Electron density profiles proposed for the myelin of peripheral nerve (3182, 3184, 3186, 3188, 3189) agree in showing that the unit cell is asymmetrical; they show a central region of low density and a high density peak on either side. The 'shoulders' seen on the slopes ascending from the central valley may be asymmetrical. The peaks may be divided. The profile was calibrated by data obtained from swollen myelin (269, 270).

Molecular models based upon such estimated electron density profiles (1595, 3185, 3186, 3189), upon the known molar ratios of lipids, and upon the probable patterns of interaction between the lipids, are provisional and hypothetical. The asymmetrical 'shoulders' might result from an asymmetrical arrangement of cholesterol (437). Similar 'shoulders' are seen in artificial mixtures of phospholipid and water when cholesterol is added (1785).

X-ray diffraction studies of myelin do not show where the protein is within the unit cell. A diffuse meridional ring only indicates its presence. It is usually assumed that the highly polar basic proteins are distributed on the surface of the lipid bilayer (1595) in the manner originally proposed for all membrane proteins (661). This is an unlikely site for the hydrophobic proteolipids. These cannot lie within the central hydrocarbon zone of the membrane, for the electron density of the protein would then exceed that found in this zone. It is probable that the proteolipids contribute to the density of one or more of the peaks of high electron density. The absence of proteolipid from the central region which contains the fatty acids of the phospholipids is consistent with the striking absence of globular subunits seen by electron microscopy within the fractured plane of the myelin (249, 250, 320, 321), and with the absence of significant neutron scattering by this region, which nitrogen would cause. Both

X-ray diffraction and neutron diffraction indicate the presence of water both in the extracellular and intracellular aspects of the lamellae (1595). X-ray diffraction patterns of myelin change during fixation, dehydration and freezing and during post-mortem degeneration (2045, 2046).

Myelin of the central nervous system differs from that of the peripheral nervous system in composition (2152, 3174, 3175). X-ray diffraction indicates that the radial unit cell is shorter in the myelin of the central nervous system, being 16 nm, and that the degree of asymmetry is different.

Biochemical stability

Myelin is not synthesized as a whole, nor is it degraded in this way (135, 654, 1958). It is therefore not possible to speak of its stability and turnover, but only of the turnover of identified constituents. In myelin, as in other membranes, the synthesis of lipid can proceed independently from that of protein, and *vice versa*. Different phospholipids turn over at different rates (2152, 2751, 2752) and the rates for cholesterol and for galactolipids are again different. Fatty acids and cholesterol are synthesized at an appreciable rate in the mature brain, and incorporated into myelin (654, 1958). Lipids of myelin may be in two compartments: one turning over rapidly and the other slowly (1519). Proteolipid has a half-life of at least 100 days; that of basic protein is 35 days.

Cholesterol and fatty acids are reutilized locally in the nervous system. This has been demonstrated clearly in regenerating peripheral nerves (1280, 1281, 2152, 2412–2416, 2751, 2752). If radioactive cholesterol is injected into a growing rat, it is incorporated into the myelin sheath of peripheral nerves, where it persists. If the nerve is crushed, the myelin degenerates; after the axon has regenerated, remyelination occurs. The 'new' myelin is still heavily labelled. Much of the original cholesterol has been re-utilized; a little exogenous cholesterol is also incorporated. The cholesterol is not totally lost from the Schwann cell; rather the cell membrane of the Schwann cell has been taken up in part by this cell, the cholesterol has presumably been stored within the cell and used again when the cell reforms the expanded membrane.

Structural stability

The diameter of an axon can change without distressing the myelin sheath. The way in which the myelin accomodates to an axon of changing size is not known (964); 'slippage' between adjacent lamellae has been suggested. The myelin sheath of a living nerve can be seen to move (370); such movement is most evident at the Schmitt-Lantermann incisura. The significance of the movements and the degree to which they affect the structure of the compact myelin are not known.

Myelination

Myelination results from an interaction between an axon and an oligodendrocyte or a Schwann cell, or exceptionally, between a dendrite and one of these cells (2290). Artificial fibres cannot yet be made to induce myelination in tissue culture (883); it is not known if this failure indicates only that the surface of the particular fibres used are inappropriate, or that other patterns of interaction occur between the neuron and the myelinating cell (1634). Myelination in mammals is critically dependent upon the size of the axon (508, 964, 967, 968, 969, 970, 1030, 1942, 2163, 2530, 2612, 2613, 2630, 2851, 2902), especially in the peripheral nervous system. The increased myelination of an axon that occurs when the functional state of the nerve increases may be due to an increase in the diameter of the axon, or to a more specific action of the neuron.

Myelination in the central nervous system is preceded by cell division in the nerve tracts (2609). The stimulus for this division is not known. Myelination begins with the formation of loose spiral folds of membrane around the axon; this membrane is continuous with the plasma membrane of the myelinating cell. Compaction follows (171, 1052). During myelination the concentration of myelin constituents rises; sphingomyelin, triphosphoinositide, phosphatidic acid, ethanolamine plasmalogen, galactosyl diglyceride, sulphatide and proteolipid. Initially sterol precursors and esters are incorporated; later, only cholesterol is found (15, 16, 17, 609, 652, 653, 654, 1983, 2136, 2137). Before compaction, myelin possesses much less cerebroside and basic protein (136, 609, 619). It is not known whether their appearance is a cause or a consequence of compaction, or only coincidental. During myelination two fractions of myelin can be isolated from brain. One resembles mature myelin. The other has less cerebroside and more phospholipid (136), and the fatty acids are shorter and less saturated. This is immature myelin. The rate at which myelination proceeds appears initially to be related to the area of myelin already present; it has not been shown that the rate limiting step for myelin synthesis is, in the first instance, the area available for inserting new material.

The activity of phosphodiesterase increases in the central nervous system during myelination, but not in the peripheral nervous system. It is found in those parts of the brain where myelination is proceeding. The enzyme activity is much reduced in mutant animals with little myelination. The enzyme might be concerned with the regulation of interaction between adjacent membranes by cyclic nucleotides (1695, 1696, 1697, 2168, 2169).

Retinal rod outer segment

The outer segment of a retinal rod is a photoreceptor. It contains a stack of discs, which lie normal to the long axis of the outer segment. Each disc is a separate organelle within the cytoplasm. Each is formed by invagination of the plasma membrane at the perikaryal end of the outer segment, at a rate of about one every 40 minutes. They pass distally and are finally shed, being then engulfed by retinal pigment cells (689, 719, 1807, 3219). Little is known of the regulation of their production, movement and loss. The stacked discs float within the cytoplasm, each containing a closed fluid space, bounded by membranes (338); *in vitro* they are osmotically sensitive (1290). The outer segment has strongly positive intrinsic birefringence due to orientated lipid, and weakly negative form birefringence caused by the orientation of the stacked discs (2035, 2604, 2605). Isolation of the discs is usually accompanied by irreversible changes in birefringence. These can be delayed by careful handling and by ATP (2463, 2464). The membranes have an unusually high proportion of polyunsaturated phospholipids and very little cholesterol (731, 799, 2306, 2309), suggesting that the hydrocarbon phase of the membrane is very fluid. Of the membrane dry mass, 60% is protein, and 38% is lipid (799, 2724); of the protein,

80% is rhodopsin (309). The lipid composition of the membrane is not altered by prolonged illumination (731).

Rhodopsin is found both in the membranes of the discs and in the plasma membrane of the outer segment (689, 727), but not in the plasma membrane of the rest of the cell. It is a conjugated protein, molecular weight 40 000 daltons (610, 1287, 2469), containing the prosthetic group retinal, which is attached stereospecifically probably at a hydrophobic site (307, 1420–1425, 2044, 3030). Rhodopsin is insoluble in water, and has a high proportion of hydrophobic amino acids (2467). It cannot be separated reversibly from associated lipids (3243).

Illumination isomerizes 11-*cis*-retinal to an all-*trans* form (1425, 3031, 3032) and a very large change in conformation results (1422, 2194, 2365, 2366, 3033, 3034). The rhodopsin chromophore is highly dichroic and light is absorbed most readily when the electric vector is parallel to the long conjugated chain of retinal (680, 681, 1805, 2601, 2602, 2783, 3034). Each chromophore lies for most of its time parallel to the surface of the disc membrane (680, 681). This limitation of movement has been used to demonstrate that rhodopsin is free to undergo Brownian rotation (359, 360, 361, 1216) within the plane of the fluid membrane. Its considerable mobility is indicated by a relaxation time for this movement of about 20 μsec (2949).

Structure

X-ray diffraction patterns have well-defined reflections at 30 nm, 14.7 nm, 9.4 nm and 4.3 nm, indicating respecting the 1st, 2nd, 3rd and 7th orders of a 30 nm repeating unit. Values of this order have been found in a number of species (271, 1128, 2464, 3079, 3187). Two different models have been used to calculate the electron density profile across the membrane, a symmetrical model (271), and a centrosymmetrical model (1128).

Using the first model, the results of the first ten orders of the lamellar diffraction pattern have been analyzed (271). Fourier synthesis provided only one solution consistent with the Patterson plot. The low electron density in the centre was ascribed to the hydrocarbon tails of the lipids; as this valley is only 1.8 nm wide, it is argued that the tails must be curved, or be obliquely oriented, or overlap; these solutions are compatible with the low concentration of cholesterol. The 4.0 nm separation of the peaks of high electron density is too great for these to result from the polar heads of the phospholipids alone, and it is suggested that protein must also contribute to these peaks. Meridianal diffraction at 470 pm indicates the likely fluid nature of the hydrocarbons, and a further meridional diffraction at 1.0 nm is due to protein, largely rhodopsin. This solution has the advantage of being consistent with the symmetrical appearance of the membranes in electron micrographs; but the peaks are not themselves sufficiently broad to accommodate a rhodopsin molecule if it is spherical. If rhodopsin is flattened, and about 1.0 nm thick, then it can be accommodated.

Using the second model, the first 11 orders of the lamellar diffraction pattern were analyzed (1128). The Patterson plot obtained from the first eight orders fitted the results obtained by Fourier synthesis better when the centrosymmetrical model was used than when the symmetrical model was employed. The electron density profile calculated shows a narrow central zone of low density occupied by hydrocarbon chains. One peak of high density is 4.0 nm wide, and the other only 1.85 nm. This model has the advantage that a molecule of rhodopsin can be placed in the wider zone of high electron density. It is difficult to reconcile with results obtained using transmission electron microscopy to study sections of fixed material. The centrosymmetrical model provides a picture that is consistent with results obtained by freeze-fracturing membranes of discs. In the dark-adapted preparation particles 5.0 nm in diameter project at least 2.0 nm beyond the internal hydropholic surface into the central hydrophobic zone; they cannot be seen on the external membrane surface. Many of these particles might be rhodopsin.

X-ray diffraction can also provide information about the ordered distribution of zones of high electron density within the plane of the membrane. A regular spacing of protein molecules can be sought in this way. The diffraction pattern obtained from fixed outer segments indicates a square array of spherical particles 4.0 nm in diameter with a centre-to-centre spacing of 7.0 nm. This is compatible with the size and spacing of particles seen by electron microscopy on the surface of dried isolated lamellae negatively stained with phosphotungstate or with silicotungstate (266). Membranes can undergo marked phase changes on drying and on fixation (134) and an ordered array could be an artefact. However, low angle X-ray diffraction patterns obtained from wet, unfixed photoreceptor discs confirm the presence of these particles (264, 268), but indicate that the spacing is less uniform. The mean distance between adjacent centres of high electron density remains about 7.0 nm. This spacing, in which one molecule occupies 49 nm^2, is compatible with the measured concentration of rhodopsin. This is about 2.5 mM (1805), and corresponds to a planar concentration of one molecule to every 45 nm^2. The density of the particles seen in the fractured surface of the membrane is of the same order: one to every 54 nm^2. X-ray diffraction indicates that antibodies to rhodopsin which are applied to the discs are separated by a similar mean centre-to-centre spacing, after they react with the membrane, and unattached antibody is washed away. Other serum proteins do not behave in this way (267, 689). It is therefore to all intents and purposes certain that these regions of high density are due to spaced molecules of rhodopsin. Further analysis has allowed the degree of submersion of the rhodopsin molecule within the lipid membrane to be calculated. The diameter of the rhodopsin molecule is 4.2 nm; in the unbleached membrane 1.4 nm is immersed in the membrane and 2.8 nm projects into the aqueous medium. This disposition is consistent with the data from freeze-fracture studies that the molecule emerges at least 2.0 nm.

A sequence of changes occur on bleaching. Light isomerizes the chromophore 11-*cis*-retinaldehyde to an all-*trans* form in several stages (1217, 1423, 3031, 3032). The 11-*cis* isomer is less stable than the 9-*cis*, 13-*cis*, and 9,13-di*cis* isomers, because the double bond in 11-*cis*-retinaldehyde is twisted by steric hindrance; the hydrogen on carbon 10 overlaps with the methyl group of carbon 13 (2230, 3233). Not all unstable *cis* stereoisomers of retinaldehyde can act photochemically (119). Retinal is the chromophore of visual pigments of vertebrates, cephalopods, crustaceans and insects (2231, 3031, 3032); it is the chromophore of both rods and cones of the primate retina, taking part in the transduction of colour vision (361). Retinal is not itself optically active, but activity is induced when it is bound to opsin (119, 1598, 2878). The aldehyde group of retinal is bound to an ϵ-NH_2 group of lysine by a Schiff base linkage (307, 611, 1159). Synthesized optically active pigments bind to the same part of opsin. It is probable that the optical activity of the chromophore is not just due to the twisting of the 11-*cis* linkage seen in crystals of 11-*cis*-retinal (1063), and indicated also by the spectra of solutions (1388).

Large changes of heat and entropy indicate that opsin changes its configuration when the conjugated protein is illuminated. This occurs in two stages: during the conversion of lumirhodopsin to metarhodopsin I and during the conversion of metarhodopsin I to metarhodopsin II. The hydrolysis of metarhodopsin II to all-*trans*-retinaldehyde and opsin might not produce a further change of shape (1949). The ultraviolet spectrum of opsin alters during bleaching. The absorbance at 235 μm and at 280 μm decreases both when metarhodopsin I is formed from lumirhodopsin and when metarhodopsin II is formed from metarhodopsin I. Optical rotary dispersion shows that part of the unbleached protein is an α-helix, and that this becomes less on bleaching (1423). During bleaching two sulphydryl groups are revealed (3033). When lumirhodopsin is converted to metarhodopsin I one group with a pK of 6.6 is exposed, probably belonging to histidine (2365, 2366).

When the retina is bleached the X-ray diffraction pattern also alters. This indicates that the molecule sinks about 700 pm further into the lipid matrix, so that 2.1 nm lie within the membrane and 2.1 nm protrudes into the aqueous medium. The molecules also come closer together. As rhodopsin bears a negative charge at physiological pH, and as the mean separation of rhodopsin molecules is related to coulombic repulsion caused by this charge, it is likely that bleaching is associated with a reduction of this charge. This is consistent with the exposure by bleaching of two polar groups which bear positive charge (7, 265, 361, 1949, 2307, 2308, 2365, 2366). This motion of rhodopsin is permitted by the highly fluid nature of the lipid membrane. Freeze-fracture studies confirm the greater immersion of bleached rhodopsin. The hydrophobic fractured plane is smooth in the dark adapted retina. After bleaching many particles 12.5 to 17.5 nm diameter project 2.0 to 3.0 nm above this cloven plane. These particles, however show no ordered arrangement. They are also considerably larger than the single molecules of rhodopsin studied by X-ray diffraction. If they represent aggregates of rhodopsin molecules (308, 1806), then this arrangement is not compatible with the data obtained from X-ray diffraction. It is, however possible that the aggregation occurs on freezing; if this is the explanation, it is fortunate that the movements of artefact occur only in the plane of the membrane and do not cause the rhodopsin to sink more deeply into the lipid lake, or to float out of it altogether.

The conversion of molecular changes in rhodopsin into hyperpolarization of the retinal rod involves a number of changes that are not well understood. Firstly, it is not clear how rhodopsin alters the permeability of the membrane within which it is embedded, whether, for example it forms part itself of a regulated pore. Illumination decreases the permeability of the retinal rod outer segment to sodium, but does not alter the flux of other solutes. The light-induced fall in permeability is directly related to the number of molecules of rhodopsin bleached (1624). This indicates a low degree of co-operativity, and suggests that the formation of oligomers or aggregates may not be involved in the formation of pores. The X-ray diffraction data also give no indication of oligomerization on bleaching. Rhodopsin may differ in this respect from the acetylcholine receptor (Chapter 4). Rhodopsin reduces the mobility of spin-labelled phosphatidyl choline in an artificial membrane made of this phospholipid. It would be interesting to know if this inhibition is increased in the natural membrane on illumination. Even if it were, however, it is unlikely that this could explain the selective reduction in the permeability for sodium. There is evidence that the discs can accumulate calcium ions and release them on illumination (356). If this is so, it is possible that the calcium ions released act upon the interior surface of the plasma membrane to diminish its sodium permeability or to increase its potassium permeability. There is not yet a clear role for cyclic nucleotides in visual transduction (Chapter 5).

Rather less is known of the process of reconstituting active chromophore. Bacteria can convert photolyzed rhodopsin to isorhodopsin with 9-*cis*-retinaldehyde (2520, 2521). A saline extract of bovine retina can promote the formation of rhodopsin from opsin and all-*trans*-retinaldehyde (2520, 2521); the constituents responsible may not be enzymes. All-*trans*-retinaldehyde cannot be isomerized in the absence of opsin. In the presence of opsin, only endogenous all-*trans*-rhodopsin can be isomerized, suggesting that an additional factor is also necessary. Dihydroflavine (1002) and an ammonium sulphate fraction of water-soluble protein from bovine retina facilitate this isomerization (1421).

Excitable membranes

It is less easy to study excitable membranes in this way. An axon plasma membrane is difficult to separate from contaminating membranes. Impure preparations of axolemma have been obtained from the retinal axons of the squid (907, 908, 3170, 3171, 3228), and from the giant axon of this animal (421, 3013). These preparations have a low concentration of cholesterol, and the phospholipids contain a large proportion of unsaturated fatty acids. The mobility of spin labels indicates that the hydrocarbon chains are very fluid (1427, 1428, 1429). Bilayers formed from lipids derived from these membranes and stabilized with α-tocopherol have a low ionic conductance, and a linear resistance in the range ± 120 mV.

Electrophysiological evidence indicates that channels exist in membranes through which cations can pass selectively (952–956, 1100, 1335–1339, 1361–1367, 1811, 2199, 2889, 2890, 3014, 3015, 3017), but does not prove that these channels exist as molecular entities. The behavior of antibiotics upon artificial bilayers indicates that such discreet molecules may well exist. Cation-conducting channels behave very similarly both in the giant axon of the squid and in the node of Ranvier of the frog. In myelinated nerve conduction across the axon membrane only occurs at the node, and the movements of different ions have been followed precisely. As drugs can affect the conduction of sodium alone or of potassium alone (1335, 2096, 2097) the ions probably move through the membrane at different sites. Each channel can probably accept other ions; for example, an action potential can be generated by a nerve when certain small organic cations replace sodium in the external medium. The ability of these cations, and of some metal cations other than sodium to pass through these channels has been studied (2113). Sodium channels at nodes of Ranvier, unlike channels of gramicidin A, or of excitability-inducing factor in artificial membranes, conduct ions at rates which are not related simply to their relative mobilities in free solution, suggesting that the sodium channel may be more restrictive. The relative mobilities of cations is not related simply to the size of the hydrated ion, or to the energies of the hydrogen bonds formed between the hydration shells of the passing ions and the inner wall of the channel (1337, 1338). The channel has a negative charge, with a pK of 5.2 (733, 1336), probably a carboxyl or phosphate group. The presence of this group permits suitable organic ions which can form hydrogen bonds to pass through a channel slightly smaller than the van der Waals radii of the ions; this explains the observed rejection of methylated ions by this channel (1338). The necessary transverse dimensions of the sodium channel are 300 x 500 pm.

It is unlikely that channels having such properties and such precise conformation represent only a change of state of surrounding membrane constituents; this precision, together with the specificity of drugs to block the passage of certain ions while leaving others unaffected, makes it likely that channels in nerve are discrete molecules or molecular complexes, having a hydrophobic exterior and a central polar cavity of closely defined shape, size and field strength (461, 1336, 1337, 1338, 1339). If such molecules are present it should be possible to extract, demonstrate and characterize them upon artificial lipid systems. Extracts have been obtained both from rat brain and from the electroplax which increase the ionic conductance of a bilayer, probably with some specificity. Two types of 'channel' have been demonstrated, differing in their stability and voltage dependence. Nevertheless, these results must be interpreted with great caution in view of the ease with which changes of conductance can be produced with a wide range of natural and synthetic molecules. Tetrodotoxin has been used to identify 'sodium channels' during attempted isolation from the unmyelinated olfactory nerve of the long-nosed gar fish (220, 221, 457), and from the electric organ and brain (1581, 1786). The material extracted has a molecular weight of about 229 000.

The mechanisms which make the permeability of the sodium channel voltage-dependent are not known. Nor is it known whether the channel alters the surface component or lipid phase component of the membrane resistance, or both. The former would appear more likely in view of the highly fluid nature of the hydrocarbon interior. The structural basis of the asymmetrical sensitivity to drugs is not understood. For example, tetrodotoxin prevents the conduction of sodium ions only when applied to the outside of an axon, and tetraethylammonium ions only block potassium channels when applied to the inside. This pharmacological assymmetry could result from assymmetry of the molecule or molecules forming the channel, from possible asymmetry of the phospholipid membrane, or from assymmetry of the aqueous phase next to the membrane.

Optical changes

Changes in the optical properties of membranes cannot be related directly to the state of channels which may conduct individual ions (111, 430, 525–534, 1579, 1580, 2229, 2891, 2894, 2895, 2896). These properties do not change in a way that can be correlated closely with the currents carried by either sodium ions or potassium ions during an action potential. Optical retardation, measured as change of birefringence, decreases during the passage of an action potential along a squid nerve; as this change persists after most of the axoplasm has been removed, it probably originates in the membrane of the nerve, in the immediately subjacent cytoplasm, in the plasma membranes of the sheath cells, or in the cleft between the axolemma and the plasma membrane of the sheath cell. Similar changes are seen in the electric organ of the eel (526); as this lacks a Schwann cell sheath, it is likely that the changes are taking place in the membrane of the axon or very close to it. The decrease in retardation follows the voltage change by about 40 μsec, and could be caused either by alteration of the orientation of the molecular dipoles within the membrane, or by reduced membrane thickness. The latter is compatible with the change of membrane capacitance (121, 538). The former would alter birefringence if the dipoles were optically anisotropic. The molecular nature of the changing dipole is not known; some experiments suggest that they are not lipid (532), while studies of the infra-red absorption spectrum indicate that phospholipids of conducting nerves undergo conformational change (2672). Although the membrane potential reverses, the change in membrane retardation is monophasic, and not triphasic; a triphasic change would be expected if the greatest membrane thickness occurred when no potential gradient was present across the membrane. It is therefore necessary to suggest additionally that the membrane is asymmetrical, and possibly biased by a standing charge on one surface. A structural assymmetry is compatible with the results obtained by X-ray diffraction and from the action of drugs. (433, 3185).

The amount of light scattered by a cell membrane also alters during depolarization; this variation has several components which may be distinguished by their different time courses and by the direction in which the light is scattered. One component is closely related to the charge that has crossed the membrane and is ascribed to the accumulation of ions and water in a fluid compartment close to the excitable membrane, either in a cleft between the axon membrane and the plasma membrane of the satellite cell, or in the tubular system beneath the plasma membrane of the electroplax. A second change of shorter duration and of unknown cause could be correlated most accurately with a combination of both voltage and current. A third component, also of obscure origin, is unrelated to anything so far studied.

Dyes which fluoresce only in a hydrophobic environment (120, 318, 549, 643, 1897, 2426, 2528, 2891, 2895, 3081) have been introduced into excitable membranes. X-ray diffraction studies show that these molecules lie only partially within the region of low electron density, corresponding to hydrocarbon chains. The extrinsic fluorescence of the membrane decreases during an action potential (430, 2896), or when the membrane potential is altered with a voltage clamp (533). The fluorphore 8-anilino-1-sulphonic acid is a rigid planar molecule which absorbs the exciting energy maximally when the absorption oscillator is parallel to the electric vector of polarized incident light; the emitted light is also highly polarized. The excited state lasts about 10 μsec. If the molecule alters its orientation within this period the plane of polarization of the emitted light will be less sharp. During an action potential the degree of polarization of emitted light decreases. This is not a direct effect of the voltage change across the membrane upon the electron configuration of the fluorphore. Fluorescent dyes indicate three properties of an excitable membrane: that the resting membrane is highly organized in the immediate vicinity of the fluorphore, that the membrane becomes less hydrophobic during an action potential, and that the membrane becomes more fluid during an action potential. The mobility of spin labels within nerve trunks, taken to indicate that the hydrophobic part of the membrane is fluid in the resting state, is not compatible with the relative immobility of the large planar fluorphores, or with the high degree of orientation of the fluorescent dyes. A number of points have to be considered in resolving this incompatibility. Different nerves were used; it would, however be surprising if the excitable membranes of different nerves differed greatly in this respect. It is also not clear how much of the spin label was within the axolemma itself, and how much was in other membranes. The degree to which the incorporated spin label, or the incorporated fluorphore, altered the physical state of the immediately adjoining hydrocarbon chains is not known, and is most unlikely to be negligible (413). With either method, the state of only those chains in the immediate vicinity of the label is studied. Although the fluorphore increases its rate of tumble during an action potential, the freedom of movement of spin labels does not change measurably when a nerve is stimulated (418). This difference may reflect only a difference in sensitivity of the two methods, and the diminution of any change in electron spin resonance by the signal coming from other membranes. Contrasting estimates of the fluidity of the red cell membrane have also been obtained from spin labels on the one hand, and Raman infra-red spectroscopy on the other.

These studies show clearly that an excitable membrane undergoes structural change when the membrane potential alters. It appears likely that these changes include the altered orientation, packing and mobility of its molecules, including phospholipids, and a change in the degree of hydration of the hydrophobic part of the membrane. This correlation agrees well with information obtained from artificial membrane systems. The results do not yet offer information about changes associated with the movement of particular ions, and especially concerning the existence and nature of specialized molecular channels.

4 CHEMICAL TRANSDUCERS

Receptors are recognized by their capacity to bind ligands; this capacity is shared by enzymes and to a lesser extent by many other proteins, lipids and gangliosides. The identification of a receptor requires that the binding should meet certain criteria (796, 1226, 2225, 2305).

Criteria

1. The binding must be of high affinity. The reaction of most presumptive receptors has a dissociation constant (K_d) between 100 pM and 10 nM. Some enzymes also bind ligands with high affinity; they can usually be distinguished by the action of specific inhibitors which attenuate the activity of the enzyme and its capacity to bind the ligand, but which do not reduce the ability of the receptor to bind the ligand. Enzymes differ from receptors, in that although they too may undergo a change of shape when they react with the ligand, this always results in alteration of the natural ligand, for example, its hydrolysis. Receptors almost certainly undergo a change of configuration, which is the basis of their role as transducers, but the ligand is not altered by this combination.

2. The affinity of the receptor for the ligand *in vitro* should be equal to the affinity *in vivo* measured from dose-response curves. This condition is limited by several constraints. Physiological responses, for example the depolarization of the electroplax, or the contraction of muscle, do not directly measure the affinity with which a receptor binds a ligand. The parameter is influenced by several stages of the response, which are not necessarily related linearly to the rate of turnover of the agent at the receptor's binding site (2223, 2224). In view of this, it is surprising how close on

occasion the agreement is (2225). An isolated receptor reacts with a ligand in a manner that is not affected by surrounding molecules of a different type, for example by the surrounding lipids of the membrane. If these influence *in vivo* the access of the ligand to the receptor site, then the apparent affinity *in vivo* will differ from the measured affinity *in vitro*. Isolation of the receptor probably changes the molecule itself; it may increase its degree of aggregation. If so, the affinity might change. Binding may be measured in solvents which are different from the natural phospholipids of the membrane. Binding to an isolated proteolipid is usually measured in organic solvents, whereas physiological binding occurs at the interface between a membrane and an adjacent aqueous phase, both of carefully controlled composition.

3. Antagonists should compete for the binding site in a manner that is quantitatively equal to the antagonism *in vivo*.

4. Inhibitors of specific binding of high affinity to sites which are not receptors should not alter the binding of the ligand to the isolated receptor. Conversely, inhibitors of receptor binding should not alter the activities of enzymes concerned with the destruction of the ligand. This symmetry of specific inhibition is seen in studies which distinguish acetylcholine acetyl hydrolase (cholinesterase) from the cholinergic receptor, or monoamine oxidase from the receptor for 5-hydroxytryptamine.

5. Substances that are chemically similar to ligands, but which lack pharmacological action either as agonists or as antagonists should not themselves bind to the putative receptor, nor influence the binding of the ligand. This lack of response is seen in the absence of reaction between possible receptors for opiates and isomers of narcotics.

6. The binding to a receptor site should indicate that binding occurs with a single affinity. On the other hand, binding to a receptor molecule may reveal several sites of differing affinities. The isolated proteolipid which bears high affinity sites for cholinergic ligands behaves in this way. This might indicate the presence of active sites and regulatory sites upon the molecule; such sites have been proposed as the cause of receptor desensitization observed with high concentrations of ligand, and of 'metaphilic' facilitation of the binding of antagonists (2402, 2403, 2404). It is usually then assumed that the site of greatest affinity is the specific receptor, and that later

occupation of sites of lower affinity reduces the affinity of the high affinity site. It is of course also possible that occupation of the regulatory sites could influence also, or instead, the interaction of the receptor molecule with other constituents of the membrane. A mundane cause of the presence of a number of sites of differing affinity in a preparation of 'receptor' is the presence of contaminants.

7. If the interaction of a labelled antagonist and a receptor is measured in vitro, and the rate of appearance and disappearance of the antagonism is determined pharmacologically in the tissue, then these rates should correspond. In a study of the action of atropine on the smooth muscle of guinea-pig ileum this correspondence could not be established, although other criteria suggested that binding to receptor sites was being measured (2225).

8. The greatest specific binding capacity (moles bound/mg protein) must be found in the fraction of brain containing the parts of the sensitive cell upon which the ligand acts. Thus, high affinity binding sites for noradrenaline, 5-hydroxytryptamine, glycine and acetylcholine have their greatest specific binding activities in membranes of synaptosomes. These organelles include both the pre- and the postsynaptic membrane. It should be noted in passing that a synaptosome is likely to have a number of molecules bearing a high affinity for the transmitter: the receptor site itself, regulatory sites, reactive sites of hydrolytic enzymes, regulatory sites on these enzymes, carrier molecules possibly concerned with the re-entry of transmitter molecules into the presynaptic terminal, and binding molecules of the synaptic vesicles.

9. The distribution of a receptor between different parts of the brain should correspond to the distribution of the transmitter, or preferably to its rate of turnover, if this can be measured reliably. For example, receptors reacting with strychnine, an antagonist of glycine, are found predominantly in the spinal cord; fewer are found more rostrally. The distribution of glycine is similar.

Nicotinic cholinergic receptors

Cholinergic receptors are of two kinds: nicotinic and muscarinic. This division, which is based upon the capacity of drugs to act as agonists or antagonists, might also correspond to a fundamental difference in the nature of the response of the membrane. The nicotinic receptor appears to act as a regulatory subunit of an ionophore, and the muscarinic receptor in some situations is a regulatory subunit of a guanylate cyclase. It is not yet clear if this distinction is universal, or even usual.

Nicotinic receptors of the electroplax and of skeletal muscle have been studied biochemically (795, 2153, 2305). Their extraction and identification have been greatly facilitated by the availability of constituents of snake venoms which bind to these receptors with very high affinity: the α-toxin from *Naja nigricollis*, or α-bungarotoxin (466, 468, 1738, 1994, 2703, 2962).

Electroplax

The electroplax of *Torpedo*, and of *Electrophorus*, has been used as a source of acetylcholine receptor (467, 468). Identical results are not obtained from these two species (524, 3082, 3083, 3084). Far more receptor protein can be extracted from *Torpedo* than from *Electrophorus*; this difference corresponds to the difference in area of the subsynaptic membranes, which is very large in *Torpedo* but small in *Electrophorus*. The α-toxin of *Naja nigricollis* and α-bungarotoxin bind only to the innervated face of the electroplax (302, 1994, 1552, 1553, 1554).

It is difficult to measure the molecular weight of the native receptor protein for two reasons: first, because it depends on how many of the associated molecules and subunits are included, and this varies according to the method of separation and measurement and second, because the cholinergic receptor protein has an anomalous hydrodynamic behavior (1434). The most reliable estimates so far suggest a molecular weight in the range 320 000 to 275 000 daltons (1434). In an aqueous medium the receptor protein has a highly ordered structure; ultraviolet circular dichroic spectra indicate that 34% is in the form of an α-helix and 29% is β-structure (2041). The protein does not have an exceptionally high proportion of hydrophobic amino acids (1971). When denatured, the molecular weight of the protein's subunits can be determined in SDS gel; there is some variation in the values cited. All agree that there are two principal subunits of molecular weight about 40 000 and 50 000 daltons (1282, 1434, 1971, 1982), and a smaller quantity of material with a molecular weight in the range 65 000 to 90 000 daltons. These larger fractions may be aggregated forms of the smaller two. The subunit of 40 000 daltons binds α-bungarotoxin, that of about 50 000 daltons does not; the subunit that binds α-bungarotoxin can also be labelled by affinity-alkylation. Relatively more of the smaller subunit than of the larger is isolated. Hypothetically, it has been suggested that the receptor might be composed of 3 of the smaller subunits and 2 of the larger (1434). It is still not clear which part of the molecule acts as the ionophore. It cannot be assumed that, because the smaller subunit binds cholinergic ligands, the larger subunit must therefore be the ionophore.

A number of ligands have been studied (1552, 1553, 1554): acetylcholine, *d*-tubocurarine, hexamethonium, decamethonium, carbamylcholine, gallamine triethiodide and phenyl trimethyl ammonium ions. These are known from pharmacological studies to activate or inhibit the receptor reversibly. The venoms α-bungarotoxin and the α-toxin of *Naja nigricollis* bind the receptor specifically and almost irreversibly. In the presence of a large excess of unlabelled toxin, the radioactive toxin is displaced with a half-life of about 60 hours (1554, 1994). A fluorphore attached to a tetra-ammonium ion has also been synthesized and successfully used (3080). An interesting ligand has been prepared by coupling a chromophore to a nicotinic agonist (167). This reacts with receptors on an intact electroplax without depolarizing it; the dissociation constant is rather high (K_d = 150 nM). When the chromophore is illuminated the electroplax is depolarized. Alkyl affinity labels have also been attached to reduced disulphide bonds in or near the ligand-binding site (1539–1542, 3094).

If the membrane fraction of the electroplax is used as the source of the receptor, it has only one site that binds decamethonium (K_d = 20 nM), and acetylcholine (K_d = 60 nM), at concentrations of ligand that are less than 1 μM (1971), and no evidence is found for co-operative binding. If the crude homogenate of the electroplax is studied more than one binding site is usually found. The affinity of the site which binds the ligands most avidly resembles the affinity of the receptor site found in membrane preparations (805, 2948). The capability of nicotinic ligands to interfere with the binding of α-bungarotoxin to membrane fragments of the electroplax agrees with the degree of interference predicted from measurement of their individually measured dissociation constants (468, 3083); they also interfere with the binding of labelled decamethonium. Displacement of one drug by another does not necessarily mean that both are bound to the same site. Allosteric transformation could also produce this effect, but only if there are at least two binding sites on each molecule, or intermolecular co-operativity. These measured dissociation constants for decamethonium and for acetylcholine can be compared with the apparent dissociation constants measured on the intact electroplax. The apparent dissociation constant for carbachol is 44 μM, and for *d*-tubocurarine is 240 nM (1324). Apparent dissociation constants for acetylcholine are 240 nM to 720 nM (1498, 1866).

There is considerable agreement concerning the amount of receptor that is bound by high-affinity sites, that is, by sites which are saturated by low concentrations of the ligand. Acetylcholine binds to the electroplax to the extent of 1.1 nM/g (2948) or 930 pM/g (805); the capacity to bind α-bungarotoxin is saturated at 1.1 nM/g (1994). High-affinity binding sites for muscarone, nicotine and decamethonium become saturated at ligand concentrations of 1 to 2 nM/g (807). If higher concentrations of ligand are used, they bind to other sites. It is possible that at least one of these is associated with the receptor. When the concentration of acetylcholine exceeds 1 μM, the binding is reduced. The receptor site of high affinity may be altered by a regulating site of lower affinity (809). This effect is likely to be allosteric, and may be related to desensitization. Most ligands will bind to a number of proteins if the concentration is sufficiently high; this non-specific binding has been studied with 28 arbitrarily selected proteins (807).

The affinity of the nicotinic receptor is reduced by pronase, trypsin, chymotrypsin and phospholipase C (805–808), and by attempted isolation with deoxycholate (2948). Nucleases, neuraminidase and hyaluronidase do not change the affinity (498). The receptor is thermolabile (468).

The nature of the proteolipid extracted by organic solvents (2153, 2937) is unclear. It can bind cholinergic ligands. The preparation has at least two kinds of binding site, with dissociation constants of 100 nM and 10 μM. The extracted material has a constituent with a molecular weight of 37 000 daltons which binds α-bungarotoxin (2300, 2301). This may resemble the toxin-binding polypeptide obtained from the membrane fraction of the electroplax, but further characterization of both is required before any conclusions can be drawn. The dissocation constant of 100 nM is rather higher than that found with receptor isolated with detergents, but closer to the apparent dissociation constant measured by the depolarizing response of the cell (2300).

Vesicles have been prepared from the lipids of the electroplax. They lose a significant amount of radioactive sodium (1982). This efflux is increased a little when acetylcholine receptor is also present in the membrane, and is further augmented in the presence of nicotinic agonists. The action of these ligands is prevented by the addition of α-bungarotoxin. The ligands do not cause an increase in the rate of loss of sodium if the acetylcholine receptor is not included in the membrane. As the 40 000 dalton subunit can act in this way, it is possible that the ionophore is also formed by this subunit or by an aggregate of them. Some caution is needed in accepting this conclusion because a variety of pure proteins studied on black lipid membranes can alter the permeability of the membrane when the structure of the protein is altered in the absence of a known ionophore. The proteolipid that might act as a receptor also increases the permeability of black lipid membranes in the presence of ligands.

Electron microscopy of negatively stained preparations of the 40 000 dalton subunit from *Electrophorus* or from *Torpedo* shows rosette-like particles about 8 to 9 nm in diameter; each is composed of 5 or 6 subunits of 4 nm diameter (1971). Electron microscopy of deeply etched freeze-fractured membranes of the electroplax shows a highly ordered hexagonal array of particles, with a centre-to-centre spacing of about 9 nm (3056).

X-ray diffraction of membranes of the electroplax of *Torpedo* has several points of interest (751). The membranes have a well ordered lamellar structure. The intense peak on the Patterson plot found at a spacing of 3.5 to 4 nm in most lipid bilayers is absent. Its place is taken by a peak at 6 nm. This suggests that the zones of high electron density on either side of the central low trough due to the hydrocarbon chains are further apart than usual. These peaks represent in part the polar heads of the phospholipids and the associated protein. It therefore seems probable that, as there is no evidence that the phospholipids are peculiar in any way that would cause their polar heads to be separated by a greater distance than usual, the protein is distributed predominantly towards the outer aspects of the membrane. This is compatible with the unexceptional amino acid composition of the protein, and its hydrophilic nature (2041). It does make its role as an ionophore rather more difficult to interpret, unless its principal action is to alter the interfacial component of the resistance rather than that of the bulk lipid phase. A reflection at 450 pm indicates that the hydrocarbons of this membrane, like others, are fluid. X-ray diffraction also indicates the presence in the plane of the membrane of a lattice of particles of high electron density, with a centre-to-centre spacing of 9.1 nm. This is identical with the spacing obtained by electron microscopy of freeze-fractured membranes. This spacing also corresponds closely to that calculated from the number of molecules of α-bungarotoxin bound, and from the area of the subsynaptic membrane. If it is also assumed that the protein has a gravimetric density of 1.37, the low molecular weight protomer might be a globular protein of 4 nm diameter. If so it will occupy about 5 to 10% of the available subjunctional area. The native protein is larger. This concentration is of the same order as that of rhodopsin in the discs of the outer segment of the retinal rod.

Skeletal muscle

When a motor end-plate binds a nicotinic agonist, the conductance of the postjunctional membrane increases simultaneously for both sodium and potassium ions (868, 1008). This suggests that the receptor may activate two types of channel.

The neuromuscular junction has been studied with the same range of drugs as the electroplax (3055). Radioactive α-bungarotoxin is bound almost irreversibly to the motor end-plate and not to extrajunctional membrane of innervated muscle (51, 52, 222, 462, 1739, 1995). The frog sartorius has 10^9 receptor sites at each end-plate (1995); an end-plate of rat diaphragm has 4.7×10^7 (1995), 3.98×10^7 (851), or 3.0×10^7 (149) and a mouse diaphragm has 1.6×10^7 (1995). The concentration of receptor sites beneath an end-plate of frog sartorius is about $10^5/\mu m^2$ (1995), and of rat diaphragm $1.2 \times 10^4/\mu m^2$ (851), values which are critically dependent upon the extent to which it is believed that the receptors occupy all the subjunctional surface, or are just confined to the tops of the folds in this membrane. Recent histochemical evidence indicates that the receptors are not uniformly distributed but are confined to the tops of these folds. The values calculated are close to those obtained from the electroplax. After denervation the extrajunctional membrane of muscle binds α-bungarotoxin (1739, 1995). *D*-tubocurarine and decamethonium are less specific ligands of lower affinity (1494); they bind to more than one site and not exclusively to the neuromuscular junction. The distribution of these two ligands on the muscle membrane is not so clearly altered by denervation.

Cultured muscle binds the α-toxin of *Naja nigricollis* (2868, 3021). Myoblasts from a chick embryo develop this capacity between the 7th and the 12th days of culture; this corresponds to the period in which applied acetylcholine begins to depolarize the fibres (2227). Fibroblasts in the same culture do not bind the venom. Even though the cultured myoblasts have never been innervated, the distribution of bound venom is not uniform; clusters of receptors are found autoradiographically. The average concentration of receptors on such a membrane is $9 \times 10^2/\mu m^2$. The average concentration in a cluster is $9 \times 10^3/\mu m^2$ (537, 2865). Some myotubes have as many as 20 clusters. It is unlikely that this clustering results from migration of the receptors within the fluid membrane after the ligand has become attached, as the ligand is monovalent. A similar clustering of receptor sites has been detected by iontophoresis of acetylcholine (537). It is possible that the clusters result from movement of the membrane, for example from the withdrawal of an extended process. Such a hypothesis implies that the receptors move less freely than the membrane in which they are contained. The action of drugs which bind to tubulin has not yet been studied in this context.

Two types of receptor constraint require explanation: first, the limitation of regions of high density of receptor sites to the subjunctional zone of the innervated muscle fibre, and second, the limitation of maximum receptor density, suggested by the similar values obtained for receptor concentration found at different neuromuscular junctions and at the innervated zone of the electroplax. The greatest density found in the membrane of cultured muscle is only a little less. The limitation of the zone of high receptor density to the subjunctional zone results from loss of extrajunctional receptors rather than from concentration at the subjunctional membrane. The cause should therefore be sought in the extra-junctional membrane rather than at the junction. The most attractive hypothesis is that propagation of action potentials through the membrane limits the concentration. The way in which it does so is not understood. The spread of sensitivity on the surface of a denervated muscle can be delayed by electrical stimulation (1863). It is attractive to speculate that membrane ionophores become converted into acetylcholine-regulated channels as a result of disuse. Such conversion cannot yet be understood because not enough is known about the structure of sodium and potassium channels, about the ionic channels activated by nicotinic ligands, about the interaction between the chemosensitive regulatory subunit and the ionophore, and about the components of the membrane which can bind nicotinic ligands with low affinity, but which are not normally necessary either for the conduction of action potentials or as chemoreceptors (see below). Extrajunctional chemosensitivity fails to occur if protein synthesis is prevented in denervated muscle, and recently synthesized proteolipid appears in the membrane of denervated muscle; this proteolipid binds nicotinic ligands (1851, 1852). This cannot yet be converted into a coherent story. The second constraint, that of maximum density of the receptor molecules can really only be discussed when it is known precisely what the maximum density really is. The information obtained from the electroplax is far better than that from skeletal muscle, as data from freeze-fracture and from X-ray diffraction are also available. The evidence indicates that the density of packing is similar to that of rhodopsin. At present it is reasonable to assume that the concentration is limited similarly by the mutual repulsion of adjacent molecules by the charge they bear. This assumption is made, however, in ignorance of the molecular charge, and despite some striking differences between the two molecules: for example, rhodopsin is hydrophobic but the acetylcholine receptor is quite hydrophilic.

The molecule that binds α-bungarotoxin can be extracted with detergents; it is polymeric and disaggregated into denatured subunits with SDS (1995). Proteolipids have been extracted from the membrane of muscle with organic solvents (1851, 1852); one fraction obtained by column chromatography binds acetylcholine and hexamethonium (2093). This proteolipid has also been extracted from muscle that has been denervated for 20 days. If radioactive leucine is given *in vivo*, the specific activity of this proteolipid is greater in denervated than in normal muscle. These findings suggest that denervation promotes an increased synthesis of a particular proteolipid, or group of closely related proteolipids, which might form part of the receptor complex. The relation between the proteolipid extracted with organic solvents and the receptor extracted with detergent is still not entirely clear.

Rabbits can produce antibodies if injected with acetylcholine receptor from *Electrophorus*. An autoimmune response to their own receptors can follow, and the condition closely resembles myasthenia gravis (1812). Patients with myasthenia gravis often have a complement-fixing serum globulin that binds to muscle (2841), and reduces the capacity of normal end-plates to bind α-bungarotoxin. The capacity of motor end-plates of patients with myasthenia gravis to bind the venom is low (850).

Brain

Brain contains both nicotinic and muscarinic receptors (296, 835, 862, 2481, 2547). The former bind α-bungarotoxin (835, 2481). The binding is less in the presence of acetylcholine, *d*-tubocurarine, carbachol, gallamine, decamethonium and hexamethonium, but it is not influenced by atropine or choline. The receptor appears to be nicotinic. The lack of influence of atropine makes it unlikely that this receptor is the one that is assayed when labelled atropine is used as the ligand. The venom is bound at two sites (K_{d1} = 66 nM: K_{d2} = 240 nM). The concentration of the site of higher affinity is 12 nM/g wet weight, and of the site of lower affinity is 26 nM/g. The complex of cholinergic receptor and venom can be extracted with detergents and purified by centrifugation and chromatography. The complex has a molecular weight of 94 000 daltons indicating a probable molecular weight for the receptor of 86 000 daltons.

The brain of the housefly binds muscarone, nicotine and decamethonium (K_d = 200 nM to 3.2 mM). The concentration of receptor sites is 2.2 to 3.2 nM/g wet weight (807).

Muscarinic cholinergic receptors

In brain, receptors identified by their capacity to bind ligands with high affinity would be expected in the subsynaptic membranes. Rough calculations (2154) that assume that the brain of a rat has 10^7 neurons, that there are 1500 synapses on each neuron, that 15% of these are cholinergic, and that each cholinergic synapse has 10^7 receptors, provide a plausible estimate of receptor concentration of 18 nM/g wet weight. The greatest specific activity of binding in rat brain is found in the fraction containing isolated synaptic membranes. Atropine reacts with two sites in these membranes ($K_{d\,1}$ = 600 pM : $K_{d\,2}$ = 900 nM). There were 89 pm/g of the high affinity muscarinic receptor (862). The binding was reversible, and could be reduced by scopolamine, but not by pilocarpine, carbamylcholine, succinylcholine, dimethyl tubocurarine, decamethonium, hexamethonium, or tetraethyl ammonium iodide. The dissociation constants of two atropine-binding sites in guinea-pig ileum are comparable ($K_{d\,1}$ = 1.1 nM : $K_{d\,2}$ = 500 nM) (2225). The dissociation constant for the higher affinity site in ileum agrees very closely with that determined pharmacologically.

Several proteolipids have been extracted from brain (2453). Preliminary concentration of the synaptic membranes increases the yield of those able to bind ligands (2452), and removes much of the greater quantity of myelin proteolipid, which also binds but with a low affinity. They are fractionated further by chromatography. The interaction between ligand and proteolipid has been studied mainly in organic solvents. The light scattering, measured at 90° and at 430 μm, of a solution of proteolipid increases when atropine is added. The dose-response curve is sigmoid in the range of atropine concentration 10 nM to 10 μM. Saturation occurred at 7 μM. Acetylcholine (10 μM to 1 mM) reduced the maximum scattering without changing greatly the co-operativity of the response; dimethyl-*d*-tubocurarine (100 μM to 500 μM), hexamethonium (500 μM to 1 mM) and succinylcholine (100 μM to 1 mM) reduce the maximum scattering and considerably increase the co-operativity of the response. The polarization of fluorescence of proteolipid was increased by atropine sulphate (2471). This response also was sigmoid in the dose range 10 nM to 10 μM. Saturation occurred at 10 μM. Acetylcholine (1 mM) reduced the maximum increase of polarization and reduced the degree of co-operativity of the reaction. Homatropine (500 μM to 1 mM) reduces the maximum increase obtained, but alters the degree of co-operativity less than acetylcholine. The effect of atropine sulphate is to alter the size and perhaps the shape of the aggregates of proteolipid (2999), so that the amount of light scattered alters, and the rotational relaxation time of the aggregate is increased. The maximum effect produced is reduced by other agents, possibly by competing with atropine for the binding sites. The change in co-operativity is probably not due to such competition, as it is also produced by other amine sulphates, such as eserine, amphetamine, dibenzylamine and strychnine. This group of studies is difficult to assess. The relevance to the action of acetylcholine on a muscarinic receptor is not necessarily close. The purity of the starting material is uncertain and the significance of the observed changes in the degree of aggregation is not obvious.

The circular dichroism of synaptic membranes does not change in the presence of high concentrations of acetylcholine, eserine, decamethonium or tetramethyl ammonium ions (100 μM to 1 mM) (2042). It is unlikely that a specific change in conformation of a receptor would be detectable in this way. The particular receptor studied is only a very small proportion of the total protein present, and the concentrations of ligand employed far exceeds the range in which specific binding only would occur.

Acetylcholine receptors in nerve

Acetylcholine is released from resting isolated nerves of vertebrates (463) and invertebrates (684, 685, 686); it is possible that this may be a feature of nerves damaged during preparation. This release is augmented by stimulation and also by depolarization by increasing the external concentration of potassium ions (686). High concentration of acetylcholine (100 μM to 10 mM) can depolarize an axon and reduce the height of an action potential (684, 685, 2446, 2447). Acetylcholine cannot initiate an impulse. This action of acetylcholine is nicotinic and is mimicked by metacholine (5 mM), carbachol (1.6 mM) and nicotine (600 μM). Tetraethyl ammonium ions, choline, and pilocarpine are inactive. It is most improbable because of the concentrations necessary that this action of acetylcholine has any physiological significance. The failure of acetylcholine to initiate a response indicates that it is not acting as a regulatory subunit of an ionophore.

A fraction can be obtained from the nerve bundles of the walking legs of the lobster, which is rich in membranes (672, 673, 674). The fraction cannot be assumed to be composed of the axolemma alone. It binds nicotine (K_d = 420 nM) (1838), and the concentration of binding sites is 700 pM/g of wet nerve. The affinity is altered by cation and anion concentrations (673). Acetylcholine is not bound, but can inhibit the binding of nicotine (K_i = 43 μM). Binding was also inhibited by curare, atropine, decamethonium, benzoylcholine, tetraethyl ammonium ions, eserine (all at 10 μM) and by α-bungarotoxin (3 to 40 μM), but not by 5-hydroxytryptamine, noradrenaline and γ-aminobutyric acid (10 μM). The binding of nicotine was reduced by pronase, trypsin, chymotrypsin, lipase and phospholipase, but not by papain, collagenase or nucleases.

Material can be extracted from peripheral nerve (2111) or from an electroplax that binds acetylcholine with low affinity; it also binds carbamylcholine, decamethonium, prostigmine, physostigmine and atropine. It can be localized by immunofluorescence to the membrane of the electroplax, but is seen on both the innervated and the uninnervated surfaces (791–798). It is improbable that this material has any receptor role, and there is no evidence that neurotransmitters are involved in its physiological role, whatever that may be.

Acetylcholinesterase and the acetylcholine receptor

Acetylcholinesterase and the acetylcholine receptor each have at least one stereospecific receptor of high affinity for acetylcholine. These are different molecules; they can be separated by affinity chromatography (468, 743). Inactivation of the acetylcholine receptor does not alter the activity of acetylcholinesterase (2701, 3220). Irreversible acylation of the catalytic centre of acetylcholinesterase does not reduce the capacity of the receptor to bind specific ligands with high affinity (469). Inhibition of acetylcholinesterase by organophosphates or other agents (805, 2304, 2305, 3177) does not prevent the receptor binding cholinergic ligands. Clones of neuroblastoma cells can be selected for their low activity of acetylcholinesterase; this value may be only 1–2% of normal. There is no concurrent reduction in the concentration of receptor (2701). Membrane fragments rich in acetylcholine receptor have a greater gravimetric density than fragments rich in cholinesterase (302). Although the acetylcholine receptor and acetylcholinesterase are different molecules, it is likely that in some studies the binding of ligands to both was measured (468, 1557). Such binding may be either to the catalytic site of the enzyme or to its regulatory sites (465, 3117). Decamethonium and *d*-tubocurarine lack specificity and bind to both enzyme and to receptor (465, 806, 807, 1554). The snake α-venoms appear to show the greatest capacity to distinguish the enzyme from the receptor.

When acetylcholinesterase is applied to a black lipid membrane, the impedance remains high. If acetylcholine is then added to the fluid of one compartment the impedance falls (441). Carbamylcholine and acetyl-β-methylcholine do not reduce the impedance, although these drugs can depolarize *in vivo* by interacting with the receptor. Nicotine reduces the impedance, although it is not hydrolyzed by the enzyme. Decamethonium and hexamethonium do not reduce the impedance, but prevent acetylcholine from doing so. It is likely that the reduction of impedance is due to steric change of the enzyme. It cannot be taken as evidence that the enzyme also acts as a receptor.

Receptors other than cholinergic

5-hydroxytryptamine

5-hydroxytryptamine is bound in a homogenate of brain with at least three affinities (K_{d1} = 500 nM : K_{d2} = 20 μM : K_{d3} = 2 mM). The greatest specific activity of high affinity binding is found in the preparation of synaptosomes. This binding is inhibited by *D*-lysergic acid diethylamide but not by harmine, an inhibitor of monoamine oxidase. Synaptosomes from the hypothalamus, basal ganglia and mid-brain bind more than those from elsewhere. The material is concentrated further in synaptosomal membranes and can be extracted from these with butanol. A proteolipid in this extract binds 5-hydroxytryptamine. Most binding of medium affinity (K_{d2}) is to a constituent of the mitochondrial fraction, and very little to synaptosomes. It is inhibited by iproniazid (100 μM) and by harmine, but not by *D*-lysergic acid diethylamide. The inhibition of the binding of 5-hydroxytryptamine by harmine (K_i = 180 nM) is very similar to the inhibition of monoamine oxidase (K_i = 120 nM). The amount of 5-hydroxytryptamine bound with medium affinity to different fractions of brain is related closely to their monoamine oxidase activities. The low affinity binding ($K_{d\,3}$) lacks specificity and probably indicates only organic cation exchange (914). It is plausible that the high affinity binding site is a receptor, and possible that part of this may be a proteolipid; the medium affinity site is likely to be the catalytic site of monoamine oxidase. A soluble protein can also be obtained from brain and has a high affinity for 5-hydroxytryptamine; this is associated with the presynaptic neuron (1911, 2879).

Noradrenaline

The receptor has not yet been identified even tentatively (1744). Noradrenaline is bound by preparations of synaptic membranes (913, 1737). There are two binding sites ($K_{d\,1}$ = 23.4 nM : $K_{d\,2}$ = 100 nM). Binding to the site of greater affinity ($K_{d\,1}$) is prevented by treatment with neuraminidase, or with trypsin, but not by phenoxybenzamine or propranolol. Binding to the site of lower affinity ($K_{d\,2}$) is prevented by propranolol but not by treatment with neuraminidase. The α-blocking drug, Sy28, also becomes bound to isolated nerve endings, and to preparations of their membranes (K_d approx. 3 μM). Noradrenaline interferes with this binding in a range of concentrations 100 μM to 1 mM.

The capsule of the spleen has been studied as a tissue more likely to contain a lot of receptor. A proteolipid has been extracted (915, 916), which has a high affinity for noradrenaline ($K_{d\,1}$ = 330 nM : $K_{d\,2}$ = 18 μM). The site of higher affinity is present at a concentration of 15 μmol/g of proteolipid, and the site of lower affinity at a concentration of 135 μmol/g proteolipid.

In none of these examples has characterization proceeded sufficiently far to allow the receptor to be identified even tentatively (596).

Glycine

Strychnine, an antagonist of the inhibitory transmitter glycine, is bound to membranes of synaptosomes prepared from the spinal cord. The binding has high affinity (K_d = 4 nm) (3214, 3215, 3216). The amount bound was increased by 25% when the membranes were washed with Triton X. Strychnine is displaced by glycine; maximal displacement was found with a concentration of glycine of 1 mM, and half maximal displacement at 25 μM. Strychnine bound to receptor was defined as that part of the total that could be displaced by glycine. Most binding to receptor was found in the spinal cord, and decreased progressively in more rostral parts of the brain stem. None was found in the cerebellum, cerebral cortex, hippocampus or corpus striatum, although strychnine is not without action in these parts of the brain. The distribution resembles that of glycine in the nervous system, and the distribution of neurons which respond to glycine applied iontophoretically. It is unlikely that strychnine is binding to material concerned with the presynaptic uptake of glycine, as this process is not inhibited by strychnine (1831). Glycine and strychnine appear to bind to distinct sites which interact (3215). Anions that are able to pass through the ionophore regulated by glycine impede the binding of strychnine. Anions which cannot pass through this ionophore do not interfere with the attachment (3216). The receptor has not yet been characterized further.

Opiates

Opiates are bound especially in the anterior amygdala, periaqueductal grey matter, thalamus, hypothalamus and the head of the caudate nucleus of the rat, of the monkey and of man (1099, 2248, 2249, 2250). Stereospecific binding of levorphanol or of naloxone indicates that there are 20 to 30 pmol of binding site/g brain. Binding is of high affinity (K_d for naloxone = 20 nm). No recognized neurotransmitter inhibits the binding of these substances. Prostaglandins E_1 and E_2 did not inhibit the binding despite their ability to compete with opiates in some pharmacological preparations (2702). The binding of opiates is not affected by electrolytic lesions which destroy cholinergic, indolaminergic, or cholinergic tracts. The ability of other opiates to inhibit the binding of naloxone is related closely to their potencies as analgesics. The receptor has not been isolated or characterized.

Others

Proteolipids have been isolated from crustacean muscle, which bind glutamate or γ-aminobutyric acid (2301).

Taste receptors

The gross structure of lingual papillae, and the histology of taste buds have been described (144, 853, 856, 1303, 2079, 2862). A circumvallate papilla of a cow may have 1500 taste buds (2182). Taste buds receive afferent and sympathetic nerve fibres. The gustatory afferent fibres run in the chorda tympani and glossopharyngeal nerves (2149). These fibres branch repeatedly within the tongue especially near the base of the papilla (1520, 2411). Not all the afferent branches necessarily enter the taste bud. Branches of single axons of the chorda tympani often pass to several papillae (2000, 2001, 2411). One papilla probably receives branches from several different axons (942). Afferent nerve fibres pass through the basal lamina beneath a taste bud and pass between its cells. Two types of innervation are seen (2079). One resembles a synapse; vesicles are found in the epithelial cell adjacent to the nerve terminal. The other is more unusual. No well defined nerve terminals are found, but the fibres end close to the plasma membrane of an epithelial cell, which may have a subsurface cistern; no vesicles are seen. It is likely that the processes receiving a synaptic contact are gustatory; the role of the others is not known. It is possible that they are temperature-sensitive fibres, or they might be collateral sprouts formed because of the continual replacement of the epithelial cells of the taste bud. They could be sympathetic fibres, or even efferent fibres regulating receptor sensitivity (832). There is no evidence to implicate them in a trophic role (Chapter 8).

Sympathetic axons enter the gustatory papilla (1007) and not all of these are obviously related to blood vessels. It is not clear to what extent these fibres penetrate taste buds or end upon their cells (1007, 1303). These axons in which primary catecholamines can be made to fluoresce, can be distinguished from other nerves, presumed to be sensory, which contain cholinesterase. Epithelial cells of the taste bud are replaced in the taste buds of adult rats, rabbits (193, 194) and mice (547). The cells have an average life of 250 hours, but several different populations of cell are involved, which might behave differently. This turnover means that contact between the gustatory nerve fibres and the epithelial cells is being repeatedly lost and re-established. There is no information to indicate the extent to which the nerve terminal is structurally dynamic, or to show the extent to which an axon retains contact with other epithelial cells when one connection is lost.

The functional relationship between the taste bud and the gustatory fibre is not fully understood. Single taste buds produce a receptor potential with more than one taste quality (1589, 2197, 2897). A single gustatory axon usually responds to more than one taste quality (535, 827, 911, 2086, 2271, 2272), but not always (2161). Receptors sensitive to sucrose and receptors sensitive to sodium chloride share an afferent fibre less commonly than receptors sensitive to sucrose and receptors sensitive to warming; receptors sensitive to acid, to quinine and to cooling commonly share a fibre. It seems likely that certain fibres tend to convey information concerning certain modalities, but this is far from absolute. It is one degree less definite that certain fibres tend to contact particular types of receptor cell. If the receptor cells are being replaced, if each taste bud responds to several modalities and if each axon passes to several taste buds, it is unlikely that a stable pattern of connection in all branches of an axon is maintained.

The nerve responds with a very short latency after a tastant is applied to the tongue, indicating that the responding site is superficial (190). The permeability of the lingual epithelium is low, and it is likely that tastants enter only through the pore of the taste bud. Taste buds are not essential, however, for the correct identification of tastants. Gustatory papillae and taste buds are absent in familial dysautonomia (2746). The threshold for taste is high (1305); if these patients receive methacholine the threshold is reduced within ten minutes to normal (1301–1305). Histologically, only unmyelinated axons are found in the epithelium of the tongue in these patients. The capacity of axon terminals to respond to chemical stimulation has been discussed (191, 192). If the epithelium is removed from a normal tongue, the responses obtained when tastants are applied resemble those found when the tastants are placed on terminals of the sciatic nerve.

It is likely that molecules tasted normally diffuse into the pore of a taste bud and act upon the plasma membranes lining the pore and perhaps upon the microvilli within the pore (2080). As inward diffusion is prevented by tight junctions, the tastants presumably react with constituents of the membrane at the apical part of the cell (2015). However, saccharine injected intravenously can be tasted, and this presumably gains direct access only to the basal parts of the cell; stimulation through secretion of saliva was prevented in this experiment as far as possible (192). Perhaps reacting components of the cell membrane are also present towards the basal parts of the cells.

Before a receptor can be sought with prospect of success, it is necessary to identify the molecule with which it reacts. The evidence for four classes of tastant is fairly strong; it has some electrophysiological support, in that a few single fibres of the chorda tympani or glossopharyngeal nerve respond only to one class: sweet, salt, sour or bitter. Other fibres respond to more than one class; they can adapt to tastants of one class without showing any diminution of their response to tastants of other classes applied immediately afterwards (951, 3045). It is technically difficult to study the apparent association between a receptor and a tastant, as the cells of the taste bud are small, and any changes of potential cannot be measured easily. Furthermore, nothing is known of the effects of a possible efferent innervation, which might be derived from collateral branches of other afferent nerves, or from elsewhere. Little direct information can be obtained from single unit recording from the gustatory nerves, which concerns the direct reaction between the receptor and the ligand. The coding in these fibres is complicated. Individual nerve fibres come from several taste buds and usually respond to a number of classes of tastant and also to temperature change (1585, 2001, 2086, 2160, 2274, 3045, 3208). Not all tastants stimulate in the manner that may be expected; saccharine, for example, stimulates in a manner both like sucrose and like salt if the sodium salt of saccharine is given (2567, 2568).

The binding of a ligand is not a sufficient criterion to identify a receptor; nevertheless, it is a necessary one, and is the usual starting point in such studies. A protein has been extracted from papillae of bovine and porcine tongue, which changes its ultraviolet absorption and refractive index when sugars or saccharine are added (630, 631, 632). The changes are greater with substances subjectively assessed as sweeter. The response has been attributed to the formation of specific complexes between a 'sweetness receptor protein' and sugar. Most of the proteins of the papilla do not form such complexes. Similarly, a protein binding with bitter tastants has been extracted from porcine taste buds. The 'sweetness receptor' fails to bind bitter tastants, and the 'bitterness receptor' fails to bind sweet tastants. These proteins, isolated by aqueous homogenization, are unlikely to be intrinsic constituents of the membranes. These proteins are also present in similar concentration in the lingual epithelium outside the taste bud (1640). It appears either that the reacting protein is not unique to the taste buds, or that the concentration of the specific binding protein is too little to be detected by the methods used, that is, less than 10 pg in each papilla (2971). The mode of action of zinc, copper and thiols upon the gustatory thresholds is obscure (1301, 1304). Glucocorticoids raise the threshold for all classes of tastant (1302); the low threshold found in patients with chronic adrenal insufficiency is not for salt alone, nor is it raised by DOCA even though this restores the serum electrolytes to normal; the low threshold is raised by prednisolone.

Attempts to define the receptor structure through complementarity by structural analysis of tastants (1668, 2659, 2660) is a classical pharmacological approach, but like all such analyses is unlikely to lead to a precise solution. If interaction between an enzyme and different substrates is taken as a guide, a receptor may adopt a different configuration with different ligands, and these may not all bind in precisely the same way (612). An alternative approach has sought the chemical identity of the receptor by studying the action of oral hydrolytic enzymes on the acuity of taste (1075). If buffered proteolytic enzymes are held in the mouth for 5 minutes, the thresholds for the recognition and identification of all four taste qualities are raised for between 1 and 12 hours. Thresholds for light touch and for two point discrimination are unaffected. The results might indicate that a protein essential for maintaining taste acuity is exposed in the mouth, and is inactivated by proteases. It is necessary to be sure, however that the hydrolysis affected neither the access of the tastant to the receptor, nor the capacity of the cell to respond to a normal change in the receptor.

Some molecules are themselves tasteless but can modify taste (1688). Gymnemic acid, obtained from the leaves of *Gymnema sylvestre* (1687), suppresses the sweet taste of cyclamate, *d*-amino acids, beryllium chloride and lead acetate, but not that of chloroform. It does not alter the appreciation of salt, bitter and sour substances. It is a basic glycoprotein with a molecular weight of about 44 000 daltons (1688, 1690, 1691). The active group is likely to be a *d*-glucuronide of hexahydroxytriterpine esterified with acetic, isovaleric and tiglic acids. It would be interesting to compare its molecular shape with the sweet tasting isomer of a diterpinoid from pine tree rosin, $4\beta,10\alpha$ dimethyl 1,2,3,4,5,10 hexahydrofluorene-4α,6 dicarboxylic acid.

Miraculin is a basic glycoprotein of molecular weight 42 000 to 48 000 daltons. It has no taste of its own, but 100 μg makes substances that usually taste sour, taste sweet (1688, 1690). It may bind near the receptor for sweet tastants, and increased acidity might bring the part which contains sugar closer to the receptor.

Many substances containing methyl xanthines taste bitter; they might act by inhibiting phosphodiesterase (1689), which is present in significant amounts in taste buds (1693). If cyclic nucleotides are intermediaries in transduction, it is possible that the receptor is a regulatory subunit of adenyl cyclase or guanyl cyclase.

It seems likely that the technique of equilibrium dialysis could be profitably applied to a study of the receptor. The affinity of binding could be studied and the capacity of different tastants to compete for a site of given affinity studied. The binding affinities of different tastants could be compared with the subjective intensity of the sensation. Affinity chromatography could also aid the isolation of a particular receptor. The evidence that these specific receptors exist is not yet compelling, however.

Smell receptors

The psychophysical classification of odorants is not agreed, and cannot be used as a basis for the identification of receptors (70, 186). It seems probable that olfaction has two clearly defined roles:

one as a general exteroceptor, providing information about a large number of different odorants, and the other as a highly specific exteroceptor regulating precisely aspects of the social behavior of the animal.

The first class includes many molecules, the olfactory threshold of which can be predicted from their absorption coefficients at an oil-water interface, and from the cross-section areas of the molecules (638–642). These molecules penetrate and distort the membrane, reducing its impedance. Red cell membranes leak in the presence of such odorants. Usually, the smellier the odorant, the greater the leak. Odiferous anaesthetics act similarly. If this action is a general one upon cell membranes, it would not be expected that the effect would be confined to olfactory neurons alone. Odorants depolarize some fibres of the trigeminal nerve; the consequences of peeling onions are well known. The cause of the low threshold of the bipolar receptor neurons of the olfactory mucosa is not known; perhaps the superficial part of the cell, including its cilium, has a lipid composition that makes it easily perturbed. Possibly it is very fluid. If this mode of action is correct for this class of odorant, no saturable binding sites of high affinity should be detected. If this is the whole explanation, it is difficult to understand why different odorants do not all smell the same. Electrophysiological studies (195, 1056) show that not all receptor cells are stimulated by all odorants, but that most respond to several. If odorants are arranged in an order of increasing potency, different receptor cells have different orders. It is not known if the lipid membrane of all olfactory nerve cells of the mucosa is the same. There is no evidence that one odorant of this class interferes directly and specifically with another.

The situation is entirely different with respect to odorants of the second class. These are substances of great and precise biological action. Pheromones (1732) are secreted by one member of a species and act upon another. They have been identified in many species and play an important part in the social life of invertebrates (72, 315, 1527, 1791, 2485, 2486); commonly they act as sex attractants or as alarm signals. Most insect pheromones are secreted by glands of the mandible, sting or anus. Many have been identified and synthesized. They are perceived by receptors on the antennae (2606). The degree of development of the receptive apparatus differs considerably from species to species, and between the sexes of one species. The sensory area of the dendrites is bathed in fluid, the composition of which might influence the access of airborne molecules to the receptor. Receptors are present only at the stage of the life cycle at which a response to the pheromone would be appropriate. The giant branched antenna of the male *Polyphemus* moth bears 60 000 sensilla, and has 150 000 receptor cells (2606); most of these respond to the sex attractant released by the female. Very few receptors need to be stimulated. A concentration of 10 molecules/mm^3 in air moving at 60 cm/s elicits a behavioural response within a few seconds. The antennae of the female *Polyphemus* moth has no long sensory hairs bearing receptors for the sex attractant.

Isomers of an identified pheromone differ in their capacity to make the receptors respond; this can be seen by measuring the electro-antennogram, or by studying the behavioural response (404, 1481, 2472, 2473). Very small amounts of isomers can also inhibit a response that a pheromone would otherwise produce (1476, 2473). This is difficult to explain if it is believed that a response to an odorant depends not upon combination with a receptor, but upon distortion of a lipid membrane. The interference caused by one isomer upon the action of another is elegantly shown in two gelechiid species of *Lepidoptera* (2473), in which it seems to provide an example of sympatric evolutionary saltation. In one species *cis*-9-tetradecenyl acetate acts as a sex attractant; in the other, *trans*-9-tetradecenyl acetate acts in this way. Each inhibits the action of the other.

The presence of a specific receptor protein for pheromones has not yet been firmly proved. A protein has been extracted from the antennae of the male silk moth *Antheraea pernyi* (2436), which is absent from the antennae of the female. The protein binds labelled, but unidentified, material synthesized and released by the female; this material might be a fatty acid. Such a receptor protein, if present, is likely to have considerable specificity. It is not yet clear which features of a receptor-pheromone interaction alter the permeability of the receptive cell. Study of substances related to the alarm pheromone of the ant suggest that certain frequencies of molecular vibration, indicated by infra-red spectroscopy, are important. The great sensitivity of many pheromone receptors suggests that the receptor cells might contain an amplifying system;

by analogy with other chemoreceptors, adenylate cyclase or guanylate cyclase are likely to be concerned. It might be of some interest to study the ability of insect pheromones to regulate the activity of these enzymes in extracts of antennae.

Vertebrates differ considerably in the degree to which pheromones regulate their social behaviour (1509, 1732, 2845, 2846). The sex attractant found in the vagina of some primates during oestrus is produced through the action of bacteria upon a secretion. This pheromone in the Rhesus monkey, and perhaps in other primates also, is a mixture of short chain fatty acids (1979, 1980).

Classes of odorants can be studied in subjects with genetic partial anosmia (70, 71, 1177, 1178). While it is attractive to equate an inability to smell certain substances with the absence of a particular receptor protein, this equation is not necessarily true. It is possible to argue that an enzyme defect has lead to an altered composition of the membrane lipids of the olfactory bipolar receptor neurons, or to an altered composition of the bathing mucus, so that the access of particular odorants to the receptive cell is hindered. A biochemical analysis of insect mutants in which the response to pheromones was absent or abnormal might perhaps be of greater value; there at least the primary odorant might be known. It is however of considerable interest that some subjects cannot smell isobutyric acid and closely related aliphatic carboxylic acids of a homologous series (71), as these have been identified as sex attractants in some primates.

It therefore appears that olfaction serves two roles. It is likely that pheromones of great chemical specificity act upon defined and specialized receptors to produce a precise behavioural response of social importance. These receptors must lie in an exposed position. This exposure makes them susceptible to depolarization by volatile lipid-soluble agents which possess only the capacity to enter the lipid membrane and distort it. That these agents smell is an artefact of reception that has been exploited biologically. These agents injure the cells rather than stimulate them. This injury is usually reversible, but it is interesting to note that prolonged exposure to an odorant can lead to the degeneration of neurons in the olfactory bulb of the rat (2289); this might be anterograde transneuronal degeneration (2290) resulting from the death of the primary receptor neurons. Within this context, it is not surprising that most nerve fibres passing through the cribriform plate discharge when many odiferous substances are applied (195); it is an injury discharge. Pheromones would be expected to evoke a discharge of much greater precision.

If the primary and original role of the olfactory system is to receive specific stimulants of social importance, which influence such behaviour as mating, defence and the ordering of rank, then it is not surprising both that the rhinencephalon is large, and that it influences considerably the neural and endocrine behaviour of the animal. Such connections and consequences seem bizarre if the primary role of the olfactory system is to detect substances such as amyl nitrate.

5 EXCITATION-METABOLISM COUPLING

Stimulated cells react biochemically. Some of these reactions have been studied in brain as 'the formation of molecular memories', perhaps wrongly. Stimulation can cause a large number of interacting changes in the metabolism of a neuron. Some of these are described here. Not all the metabolic responses necessarily affect directly the capacity of the nerve cell to respond to stimulation, or are concerned with learning. This interaction between excitation on the one hand, and metabolism on the other, is termed here excitation-metabolism coupling. Only in very few examples is it yet possible to trace far the sequence of reaction. It seems to be most logical to begin the sequence with consequences of interaction between receptor and ligand.

Receptors as regulatory subunits of enzymes

Many hormones influence tissues by activating adenylate cyclase (621, 659, 2393, 2469, 2470, 3104). Neurotransmitters can act similarly. Brain contains much cyclic adenosine monophosphate (cAMP) and rather less cyclic guanosine monophosphate (cGMP). The activities of adenylate cyclase, phosphodiesterase, protein kinase and phosphoprotein phosphatase are high; proteins are also present which can be phosphorylated by protein kinase. Much more is known about the factors which influence the formation of cyclic nucleotides in brain than about the ways in which these substances act.

Adenylate cyclase of brain is activated by adrenaline (486, 487, 488, 788, 2087, 2624), noradrenaline (1438, 1439, 2393, 2597, 2598), 5-hydroxytryptamine (2106), histamine (487), and dopamine (357, 1393, 1438, 1439, 1568, 1569, 2004, 2005, 2671), but not by acetylcholine, glycine or γ-aminobutyric acid. The enzyme in the cerebral cortex cannot be activated with thyroid stimulating hormone, luteinizing hormone, adrenocorticotrophic hormone or glucagon. Experiments in which different stimulating neurotransmitters are given concurrently indicate that they stimulate the activity of different enzymes. The activation sites of the enzymes have all the features of pharmacological receptors (Chapter 4). It is now certain that some pharmacological receptors are regulatory subunits of adenylate cyclase. Activation by dopamine can be blocked by haloperidol; the activation by adrenaline is antagonized in many parts of brain by drugs which block β-receptors, and in fewer situations by α-blocking drugs. Prostaglandin E_1 antagonizes the action of noradrenaline in the cerebellum (2690, 2691). An increased amount of cAMP in brain does not necessarily result from increased synthesis. The actions of apomorphine and bulbocapnine provide examples of this. These drugs probably react with cerebral dopamine receptors, but in addition they reduce the activity of phosphodiesterase in homogenates of brain (2087). It is not known if they also inhibit this enzyme *in vivo*. Among other factors, this depends upon whether the drugs can gain access to the enzyme.

The sensitivity of adenylate cyclase to noradrenaline responds to denervation in a manner appropriate for denervation hypersensitivity (2014). After noradrenergic terminals have been destroyed with 6-hydroxydopamine, the augmentation of adenylate cyclase activity that noradrenaline can produce increases (1416, 1526). The augmentation is also increased in animals treated with reserpine (705). Interneurons of sympathetic ganglia release dopamine which activates adenylate cyclase of the ganglion cell (1148, 1568). A slow hyperpolarization results, which is potentiated by theophylline and antagonized by prostaglandin E_1. The hyperpolarization is mimicked by exogenous dopamine, or by dibuteryl cAMP. The activation does not occur if the postsynaptic cell is stimulated directly and antidromically. The activation can be blocked by drugs which block the cholinergic depolarization of the interneurons or by drugs which prevent the dopamine released by the interneuron from combining with the receptor on the ganglion cell (1867).

Dopamine and 5-hydroxytryptamine activate adenylate cyclase of the abdominal ganglion of *Aplysia* (454, 455); carbachol, glutamate, noradrenaline and histamine do not. The activation does not depend upon the movement of ions across the membrane, which normally follows the binding of these neurotransmitters to receptors in this tissue, as the activation also occurs in the presence of isotonic sucrose without extracellular ions. The enzyme activity also increases when the nerve is stimulated at 1 Hz, a normal frequency. This activity fails to increase when synaptic conduction is blocked by a high concentration of magnesium ions in the extracellular compartment. Depolarization by ouabain or glutamate does not increase the amount of cAMP in the ganglion.

The adenylate cyclase activity of brain is also increased by extracellular adenosine (2569); AMP, ADP and ATP are less effective, possibly because they enter cells less easily (454, 486, 487, 488). It is not known if adenosine acts upon an extracellular receptor, or if it has to enter the cell. Adenosine acts synergistically with neurotransmitters (1524, 1525, 1877, 2344, 2345). Adenosine, inosine and hypoxanthine are released from superfused slices of brain (2344, 2345), and this release is increased by electrical stimulation. The physiological significance of this is unknown. The source of the adenosine is not clear. First, it is possible that adenosine results from dephosphorylation of ATP released as a specific transmitter substance (401, 402, 1380, 1381). It would not be the only substance known which is utilized both as a neurotransmitter and metabolically; for example, glycine and γ-aminobutyric acid are accepted as probable transmitters. If the activation of adenylate cyclase were brought about in this way, however, it is surprising that ATP, the substance released, is not as effective or more effective than adenosine. Second, it is possible that the ATP is released in conjunction with catecholamines or with acetylcholine (2327); it is present in the vesicles of both these types of synapse and very probably in others also. Third, the adenosine might be released from the postsynaptic cell as a result of increased metabolic activity; the nucleoside can pass through the membrane more easily than its phosphorylated derivatives. This situation has affinities with the release of adenosine, hypoxanthine and nucleotides from active muscle; here too, it acts as a minor transmitter, contributing to the vasodilatation. The formation of cAMP by asphyxiated brain is greatly increased, possibly by the release of adenosine. This third hypothesis is unlikely, as the adenosine released is derived from a small fraction of the total intracellular pools of adenosine and related nucleotides. When cells in brain slices are depolarized with ouabain, veratridine or by increasing the extracellular concentration of potassium ions, more adenosine is released and the synthesis of cAMP is increased (3232). The cells of the abdominal ganglion of *Aplysia* do not respond in this way.

Homogenization of brain in detergents or in non-physiological buffers reduces the capacity of adenylate cyclase to be regulated by neurotransmitters (251, 252, 3247). Similarly, when the enzyme is removed from cell membranes, it ceases to respond to hormones (1775, 1776, 1777). The response to glucagon of adenylate cyclase from heart is restored by adding phosphatidyl serine, but not by phosphatidyl inositol. Conversely, the response of the enzyme to noradrenaline is restored by phosphatidyl inositol, but not by phosphatidyl serine. The response of the enzyme from liver to glucagon was similarly restored by phosphatidyl serine. These studies indicate that certain phospholipids must be present to allow the enzyme to be activated by hormones (2299). The phospholipids could act either upon the protein, altering its configuration, or upon the hormone, adjusting its alignment with respect to the enzyme and facilitating its approach to the regulatory subunit. These enzymes provide an additional example of the association of particular phospholipids with intrinsic proteins. The sensitivity of adenylate cyclase of brain homogenates to neurotransmitters is increased by phosphatidyl serine and by phosphatidyl inositol (1438, 2132). The complex of membrane proteins which is necessary for the formation, action and destruction of cAMP can be preserved if vesicles are prepared from brain by homogenization in a physiological salt solution (486); probably this neither removes the enzyme from the membrane, nor dissociates it from particular adjacent phospholipids.

The presence of an adenylate cyclase that responds to a putative transmitter is not adequate evidence that the substance is in fact a specific transmitter acting at that site; examples in which this may not be so include the turkey erythrocyte, which has an adenylate cyclase sensitive to adrenaline, and the lymphocyte, which has a guanylate cyclase sensitive to muscarinic drugs (2844, 3247). On the other hand it is possible that even in these cells the action really is a specific response to a messenger. It is possible for example that the turkey 'receptor' regulates the synthesis of some constituent of the red cell.

Cyclic AMP produces its physiological effects by combining with a protein kinase, and so altering its activity. Protein kinase of brain is found in the same regions as adenyl cyclase, in the same subcellular fractions, and it appears at about the same stage of development (1004, 1005, 1006, 1488, 1499, 1898, 2242, 2243). Protein kinase phosphorylates a protein substrate; the transferred phosphate is derived from ATP and not from cAMP itself. Very low concentrations of cAMP activate protein kinase; between 10 and 100 pM produces half maximum activation. The enzyme is not activated by other nucleosides and nucleotides, except in far higher concentrations (1147). Different kinases are activated by cAMP and cGMP.

Many proteins can be phosphorylated *in vitro* by protein kinase: histones, protamine, tubulin, and protein mixtures from brain microsomes and synaptosomes, from secretory granules of the anterior pituitary, and from reticulocytes, adrenal cortex or liver. The phosphorylation of a protein *in vitro* is not evidence that it acts as a natural substrate in brain. Once protein kinase has appeared during development, its activity does not change much. It is likely that the endogenous protein substrates change, and so produce the appropriate variation of response.

Some membrane proteins of brain can be phosphorylated more easily than histones (2017, 2018), especially one minor protein of the synaptic membrane with a molecular weight of about 100 000 daltons. These proteins are distributed in brain and among its subcellular fractions in a way very similar to adenylate cyclase, phosphodiesterase and protein kinase. It seems probable that some at least of these act as natural substrates. Only in a very few cases is the physiological significance of the phosphorylation understood. In chick brain a cAMP-dependent protein kinase converts phosphorylase *a* into phosphorylase *b* (788), and the conversion is accompanied by rapid glycolysis. This response is very similar to that of the liver. It does not occur in the cerebral cortex of the rabbit (1525).

A ribosomal protein can be phosphorylated (2600). If it serves as a substrate *in vivo*, cAMP might play a part in controlling translation. Such a control in brain has not yet been shown. If it occurs, it would allow the rate of synthesis of protein to be regulated locally; the synthesis could be confined to the region of the membrane possessing the complex of enzymes, adenylate cyclase, protein kinase, phosphodiesterase. It is an attractive but entirely speculative possibility that a subsurface cistern acts in this way (Chapter 7). It is properly situated both to be influenced by cAMP produced in response to a neurotransmitter or other extracellular messenger, and to donate protein for insertion into the membrane near the site of enzyme activation.

Cultured cells commonly produce processes when dibuteryl cAMP is added to the medium (582, 993, 1412, 1502, 1592, 2186, 2476, 2477). These processes contain microtubules, organelles which create cell asymmetry (Chapter 2). Although it has been claimed that tubulin can act as an effective substrate for protein kinase (1702, 2076) and that it possesses protein kinase activity itself (1702, 2756), it has been possible to separate the substrate of the enzyme from tubulin (2410). The impurity is very closely associated with the tubulin, and is not removed by several cycles of aggregation and disaggregation of isolated microtubules. cAMP is not required for the formation or destruction of microtubules *in vitro*; it is possible that it may be involved in determining the sites of nucleation *in vivo*, either by regulating the efficacy of hypothetical sites of nucleation, or by controlling the concentration of calcium ions in the regional intracellular compartment. In muscle, protein kinase is very closely associated with the phosphorylase that it regulates.

Brain has a cAMP-dependent phospholipid kinase, which catalyzes the phosphorylation of diphosphatidyl inositol. The cAMP-sensitive subunit of the kinase can be separated from the catalytic subunit. Dopamine, noradrenaline, or acetylcholine and eserine increase the incorporation of ^{32}P into phosphatidic acid, triphosphatidyl inositol and phosphatidyl choline (2933–2936). *In vitro* many different kinds of membrane are affected; the diphosphatidyl inositol is phosphorylated in microsomes and in mitochondria as well as in synaptosomes. *In vivo* the response might be more limited by the distribution of the enzyme synthesizing cAMP. The synthesis of triphosphatidyl inositol by this method is necessary for the hyperpolarization of the slow inhibitory postsynaptic potential (S-IPSP). The S-IPSP is produced from the activation of the dopamine-sensitive adenylate cyclase in the postsynaptic membrane of the ganglion cell by dopamine released from the interneurons. It is potentiated by theophylline. Exogenous dopamine causes a similar hyperpolarization. Both the hyperpolarization caused by exogenous dopamine and the S-IPSP which follows stimulation are antagonized by prostaglandin E_1 and by phentolamine. It is not known how the triphosphatidyl inositol produces hyperpolarization. The membrane resistance does not fall, but commonly rises; it might alter the fixed charge on the membrane. A similar increased incorporation of ^{32}P into triphosphatidyl inositol is a feature of the response of many glands to secretion; it is also produced in heart muscle through the action of noradrenaline (1377).

The activity of protein kinase in brain can remain at an increased level for 30 min after the synthesis of the activating cAMP, an observation which is probably of great relevance to the persistence of a 'trace' after synaptic activity. Protein phosphorylated by a cAMP-dependent protein kinase is dephosphorylated by phosphoprotein phosphatase. This is measured in brain by its capacity to hydrolyze phosphorylated histones (3111). This phosphatase has a distribution similar to that of protein kinase.

Adenylate cyclase is found in many tissues and in many types of cell; it is improbable that it is localized in brain to a particular type of cell. Synaptosomes, which include the subsynaptic membrane, contain all the enzymes necessary to form, degrade and use cAMP. Similarly, cultured astrocytes contain adenylate cyclase, phosphodiesterase and cAMP-dependent protein kinase (499, 500, 2242, 2243, 2587). The adenylate cyclase is activated by noradrenaline and by histamine; the activity is further augmented by adenosine. The action of noradrenaline is prevented by the action of drugs which block β-receptors (1064, 1524, 1525, 1877, 2344, 2345). The hormone sensitivity of the adenylate cyclase is greater in cultures in the log phase of growth than in cultures near their terminal density (1064). This is of interest in two respects. First, it is uncertain whether the observation is of great relevance to astrocytes *in vivo*; these are not growing rapidly, but form a fairly stable epithelium (Chapter 1). Second, a similar dependence of the response of hormones upon cell density and growth is found in other cultured epithelial cells: HeLa cells, hepatoma cells and cells of the lens (1899). This indicates again the unexceptional nature of the behaviour of astrocytes. It is improbable that the sensitivity of these cells to noradrenaline results from their normal innervation. Noradrenaline can activate phosphorylase in cultured glia, and catabolism of glycogen results (90).

Less is known about guanosine cyclic 3′ : 5′ monophosphate

(cGMP). It is formed by guanylate cyclase, which can be regulated by neurotransmitters. Acetylcholine activates this enzyme in the sympathetic ganglion cell; the regulatory subunit is pharmacologically indistinguishable from a muscarinic receptor (347). The cGMP formed is responsible for the slow excitatory postsynaptic potential (S-EPSP) (1798, 1799, 2129, 2130, 2131). The S-EPSP is due to a reduced permeability of the membrane to potassium, and an increased permeability to sodium and to calcium (1667, 3093). Exogenous acetylcholine also leads to the formation of cGMP by acting upon muscarinic receptors in the cerebral cortex, heart, ileum, lung, and fat cells (1051, 1463, 1685, 1743, 2837). In most situations the formation of cGMP is inhibited by atropine. γ-Aminobutyric acid increases the amount of cGMP found in the cerebellum (1909, 1910); depolarization by raising the extracellular concentration of potassium, or through the action of ouabain or veratridine also increases the cGMP content of the cerebellum (877). Extracellular adenosine increases the synthesis of cAMP but not that of cGMP in the cerebellum. Several phosphodiesterases destroy cGMP; they have different substrate specificities. Some destroy cGMP more readily than cAMP, and others cAMP more readily than cGMP (1243). Just as it appears that β-adrenergic drugs may act upon cells which do not receive an adrenergic innervation (3247), so it is likely that cholinergic drugs can act upon muscarinic receptors of cells which do not receive a cholinergic innervation; for example, cholinergic drugs enhance lymphocyte-mediated cytotoxicity almost certainly by activating guanylate cyclase (2844).

Cyclic nucleotides and the associated enzymes are found in primary receptor cells: within the retinal rod outer segment, and probably within the chemoreceptor cells of the olfactory mucosa and the taste buds. The mammalian photoreceptor (Chapter 3) is hyperpolarized by light. Light inactivates adenylate cyclase of the outer segment (251, 252), and the degree of inactivation is proportional to the bleaching of the photoreceptor. It is likely that the adenylate cyclase is tightly coupled to opsin. The eye of *Limulus* is depolarized by light. Light activates adenylate cyclase in the eye of this invertebrate. Depolarization of a photoreceptor therefore seems to be associated with activation of the enzyme, and hyperpolarization with inactivation; it is not yet clear how this process interacts with the release of calcium ions from the discs (2117). Outer segments also possess considerable activity of guanylate cyclase (1110–1113, 2204, 2205). It is uncertain whether this is also sensitive to light. Protein kinases are also present, some of which are sensitive to cAMP and some to cGMP. Adenylate cyclase activity is also considerable both in the olfactory mucosa and in taste buds (1693). No information is yet available about the capacity of odorants, such as pheromones, or tastants to regulate the activity of this enzyme. The enzyme has not yet been localized to a particular type of cell in these epithelia.

cAMP or its dibuteryl or monobuteryl derivatives can sometimes mimic the action of noradrenaline in brain, but not always. In the cerebellum it can. The rate of discharge of Purkinje cells of the cerebellum is reduced by iontophoresis of noradrenaline or cAMP (2690–2693). Prostaglandin E_1 antagonizes this action of noradrenaline but does not affect the inhibition caused by γ-aminobutyric acid. The cerebellum normally receives noradrenergic innervation from the locus coeruleus. Noradrenaline also reduces the rate of discharge of many cells in the cerebral cortex; this cannot be mimicked reproducibly by cAMP or its derivatives. The response to prostaglandin E_1 is also very variable.

cAMP has presynaptic as well as postsynaptic actions. Both cAMP and phosphodiesterase are found in the subcellular fraction containing synaptic vesicles (1500), but there is no strong evidence that they are released. It is likely that this phosphodiesterase is the same as can be detected accumulating in vesicles proximal to the site of nerve constriction (330). Dibuteryl cAMP increases the store of transmitter that can be released from a neuromuscular junction, but does not affect events in the postsynaptic cell (3140). It is probable that adrenaline augments the release of acetylcholine from a motor nerve in this way. Theophylline, cAMP, adrenaline and propranolol increase the rate of quantal discharge at a motor end-plate, but not the amount of acetylcholine forming each quantum (332, 333, 1096). The action of cAMP in imitating the action of noradrenaline in the cerebellum is not produced presynaptically, through increasing the release of noradrenaline from the presynaptic terminals ending upon the surface of the Purkinje cell; the cAMP continues to act after the presynaptic noradrenergic terminals have been destroyed with 6-hydroxydopamine. cAMP also increases the rate of synthesis of noradrenaline in brain by increasing the activity of tyrosine hydroxylase, a rate-limiting step in the reaction sequence (1267); more enzyme is not synthesized under these circumstances, as the activation occurs *in vitro* under circumstances in which such synthesis cannot take place. Its mode of action is at present obscure. The rate of synthesis and turnover of 5-hydroxytryptamine is also increased by cAMP (2872, 2873), which increases the rate of entry of tryptophan into brain. The uptake of tyrosine is also increased. These augmented uptakes are a direct result of the injection of cAMP and not due to the hyperexcitability that follows such injection; increased penetration is also found in tissue slices (948).

Some of the actions of cAMP may be mediated by altering the intracellular concentration of calcium ions. cAMP can restore electrophysiological activity to cultured nerve tissue after calcium ions have been removed (579, 580). Dibutyryl cAMP inhibits uptake of calcium ions by the sarcoplasmic reticulum (984). The activation of adenylate cyclase in the outer segments is associated with the release of calcium from the discs.

Previous electrophysiological activity may therefore influence the response of the nervous system to later stimulation in at least five ways through the actions of cyclic nucleotides. These influences have differing latencies and durations. First, the synthesis and release of transmitter substance from the presynaptic terminals can be altered. Second, the efficacy of the depolarization produced in the post-synaptic cell by the presynaptic release of a given amount of transmitter is altered during the S-EPSP and the S-IPSP. Third, the protein synthesized in the immediate vicinity of the subsynaptic membrane through altered translation might alter the response of the postsynaptic cell. Fourth, altered transcription through the action of phosphorylated histones or through activation of RNA polymerase may change the electrophysiological response or the types of synapse which can terminate upon the cell. Fifth, augmented glycolysis might alter conduction or response through the supply of energy.

There is therefore no lack of possible markers to indicate recently active synapses; there is no need to postulate the initial stage of memory or learning as a reverberating neuronal circuit, which somehow holds the information until something else happens. Beyond this point, however, the trail peters out, lost amidst the many uncertainties of defining learning precisely, of measuring it when it can only be inferred from behaviour and not witnessed directly, of designing experiments that are not made ambiguous by making the animal learn, and of determining which of the many changes seen, if any, constitute the process of learning rather than the execution of the learned response.

Metabolic changes in active nerve cells

Without neural activity it is probably impossible for learning to occur, or for a learned act to be executed. This view can be held with fair confidence for vertebrates, and probably for nearly all metazoa, but obviously fails abruptly when faced with protozoa that show some patterns of behaviour that are acquired. In vertebrates, the sensation that provides the information, and the executed response are both mediated as action potentials.

If a relationship between the electrophysiological and metabolic activities of a nerve cell is to be established clearly, each must be defined and measured. Few studies have attempted this. When the state of the animal is altered it is often assumed that the rate of discharge of a group of neurons has increased or decreased. While such a guess might often be correct, it is an insecure assumption even for motor neurons, neurosecretory cells, or for primary sensory neurons. For example, a muscle can work harder if it hypertrophies (2583); it may then do more work in response to a given rate of motor nerve stimulation; or again, the sensitivity of many sense organs can be altered peripherally by efferent nerves passing to the receptor cells. Even if the rate of discharge of the neuron is measured, this is not an adequate measure of the relevant electrophysiological activity. The generation of inhibitory or excitatory post-synaptic potentials are capable of inducing metabolic change even when they do not lead to a regenerative discharge. It is evidently difficult to measure the total activity in a cell with the complexities of structure and connection of the average vertebrate neuron. In assessing metabolic change, it is commonly assumed that the synthesis of RNA or of protein has increased because more of a radioactive precursor has been incorporated. Only occasionally has the altered rate of synthesis been measured; only rarely has an altered rate of formation of a particular class of RNA or protein been measured. In studies made upon intact animals the effect of 'stress' in producing all or part of the response is frequently discussed, but seldom either defined with precision or measured.

It is difficult to make meaningful measurements of RNA and protein histochemically in sections of nerve tissue; by light microscopy only the perikaryon can be identified clearly and measured (2555). If the amount of RNA or protein changes, this can be interpreted in several ways: as an altered rate of synthesis, as a changed rate of destruction, or as a changed distribution between the perikaryon on the one hand and the processes of the nerve cell on the other. It is also possible that the perikaryon is one of the least rewarding places in which to seek rapid metabolic responses to transsynaptic stimulation; it is remote both from its own axon terminal and from much of its plasma membrane which receives the boutons of other neurons. Protein can be synthesized both in dendrites and at the axon terminal; if this alters as a result of stimulation there would not necessarily be an associated change observed in the cell body. Microchemical studies of single nerve cells, and measurements made upon neuronal perikaryal fractions of brain suffer from the same defect; in all only the perikaryon is studied and the large basal dendrites. Studies of single isolated nerve cells have two advantages: the part of the brain from which the nerve cells have been obtained is known, and it is evident whether all neurons in this region are behaving similarly. When neurons are isolated in bulk a greater number of biochemical studies can be made. The study of neuroglia is similarly limited. Some measurements have been made upon nuclei or nucleoli. Provided that the neuron studied is not multinucleate, like the ganglion cells of the superior cervical sympathetic ganglia of rabbits, it is certain that the entire organelle of one nerve cell is being examined (1310).

Difficulties of interpretation increase even more when interneurons rather than motor neurons are examined. It is very difficult to guess how their rate or pattern of discharge alters, and it is probable that when conditions change, at least as many cells are inhibited as are stimulated. Furthermore, the proportion of cells that are stimulated is likely to change with repeated stimulation.

The metabolic changes that accompany electrophysiological events have been studied in a giant neuron of *Aplysia californica* and in mammalian sympathetic ganglia. An identifiable neuron, R2, of the abdominal ganglion of *Aplysia* slowly increases the rate of synthesis of RNA when excited transsynaptically by stimulating ganglionic nerves (2262); it is sufficient that postsynaptic potentials are produced and unnecessary that action potentials are generated in the postsynaptic cell. If action potentials are produced in the postsynaptic cell by direct stimulation and not transsynaptically, the synthesis of RNA does not increase. One of the features of this experiment is the demonstration that the passage of action potentials as such does not require an appreciable increase in the synthesis of RNA within the period studied. This is in accord with the ability of an axonal membrane to transmit action potentials after much of the axoplasm has been extruded (123), and with the ability of the stretch receptor of the lobster to transmit action potentials after the synthesis of RNA has been much diminished by actinomycin D; furthermore, no additional synthesis of RNA can be detected when the normal receptor is stimulated by stretching for 12 hours. Neurons of a sympathetic ganglion synthesize more RNA after sustained stimulation of the preganglionic nerve (1076). This is prevented if synaptic conduction is blocked pharmacologically. The effect of sustained antidromic stimulation of the postganglionic neuron in the presence of drugs which block synaptic conduction was not studied. Prolonged application of acetylcholine increases synthesis; it is not known if desensitization occurs. This action of acetylcholine does not require the generation of action potentials, but is blocked by tetrodotoxin. Depolarization by potassium chloride does not increase the synthesis. These results suggest that more RNA is made as a result of stimulation and that this results not from the generation of action potentials but from the increased number of postsynaptic potentials. It would be of interest to know if dopamine could also increase the rate of synthesis through causing an inhibitory postsynaptic potential. It is not known if the increased synthesis of RNA is caused only by receptors which act by synthesizing cyclic nucleotides, or whether receptors which act as regulatory subunits of ionophores, and which do not form cyclic nucleotides, can also act in this way. It is possible that the generation of action potentials as such, might increase the rate of synthesis of RNA if stimulation continued for a longer period. Over a relatively short period, there might be insufficient time for a signal to return from the axon terminal. The axon terminal is not completely dependent upon the perikaryon for all new protein synthesis; some can be made within the terminal (114), and this might be sufficient to sustain the axon terminal over short periods of increased activity.

Many other studies have sought to relate increased synthesis of nucleic acid and protein with altered neuronal function (1482, 1483, 1484). In most it is not certain that the metabolic change observed is associated with an increased rate of discharge of the cells observed; it is also possible that the synthesis is a response to an increased rate of generation of postsynaptic potentials generated through the activity of local or of remote cells. It is very probable that a population of cells in a part of brain taken for analysis contains neurons responding in different ways.

Biochemical correlates of behaviour can be measured, but are almost impossible to interpret. There are several reasons for this difficulty. Experience not only elicits appropriate responses as a result of learning, but also determines the way in which information is later handled, determining, for example, which aspects of sensation are inhibited and which are accepted. This determination of cues is much more striking in young than in old animals (Chapter 6), possibly because the predominant act is the rejection rather than the facilitation of information: a procedure that is evidently self-limiting. It is uncertain whether this is a process that continues throughout life to a diminishing extent, or whether it is a process that can be completely or partially reversed under certain unusual conditions, for example, when information of a type that has previously been rejected must now be accepted to learn an appropriate response. To learn such a response might be expected to require much greater adaptation than a response not requiring this reversal of cue. There are interesting relationships is this context between three apparently unrelated conditions: relearning to a high criterion of performance after a lesion of the central nervous system, learning in animals which were deprived when young of experience of the type of sensory stimulation needed to solve the problem now set, and hysterical reactions seen in some animals when they are required to discriminate to a degree that may exceed their capacity. A learning task is not an isolated entity: isolated either from concurrent interference from factors that an experimenter cannot control and of which he may be unaware, or isolated from the previous experience of that animal. The neurophysiological basis of these variables has been only partially explored (Chapter 6).

Learning is usually inferred when the behavioural response to a stimulus becomes more appropriate in a given animal as a result of experience. If learning is different from the changing response by which it is recognized, it is not possible to study it without the response. It is therefore not possible to ascribe biochemical changes to a process of learning rather than to its expression. The biochemical correlates of a response once it has become established do not make a satisfactory control to compare with the biochemical changes occurring while learning is taking place. The patterns of neural activity employed during an act of learning may be different from those used in the execution of a learned act; the situations may differ for example in the intensity of the cue needed to provoke a response, in the latency of the response and in the precision of its execution. Each of these features indicates clearly the changing pattern of synaptic action; little is yet known about the changing turnover of neurotransmitter in different brain areas during learning. Any biochemical changes may only reflect this altered turnover as such.

Some learning involves monoaminergic tracts as part of the 'reward' system; that is, as part of the system that facilitates responses that are appropriate, or rewarding. Virtually nothing is known about the mode of action of this system, and surprisingly little about the connections which excite or inhibit it. Presumably many of its actions are mediated through activation of protein kinases in the postsynaptic cells.

Much attention is focussed upon models in which learning is viewed as a process that increases the efficiency of certain synapses. There is remarkably little evidence to show that this process occurs, or where it takes place, to indicate the extent to which the synapses used to execute a response after the period of learning is complete are the same as those involved in the process of learning, to show how many different learned responses a given synapse or synaptic system can partake in, or to indicate the necessary minimum extent of the path shared when the same response is produced on two different occasions either by repetition of precisely the same stimulus or by a given stimulus presented differently – for example by a visual stimulus presented in two different parts of the visual field; nor does such a view of learning explain the related ability of an animal to execute a learned response in many different ways. There is no evidence to indicate that the mean level of neural activity in the brain increases as the number of responses that have become learned increases. It is not known whether there is a transient increase in activity during the act of learning, which is due to the learning itself. The activity of nerve cells does not appear to increase with age. This is very difficult to measure as such, but two lines of evidence indicate that it is so. First, there is no evidence from the many single units studied in different parts of the nervous systems of many species that the mean rate of discharge increases with age. Insufficient recordings of the resting membrane potentials of neuroglial cells have been made to show that they do not decrease during life; if the mean activity of nerve cells steadily increases, the mean extracellular concentration of potassium ions should also increase, and depolarize these cells (Chapter 1). Second, the oxygen consumption of brain does not increase progressively with increasing age.

Biochemical correlates of changing behaviour are sometimes ascribed to 'stress'. In so far as a natural or experimental situation requires a new response to be learned, a stress is in one sense experienced. A new state must replace the old state of equilibrium in brain. This is a stress in the same sense that distortion of bone constitutes a stress that leads to structural remodelling; as a result of remodelling the distortion is minimized. Both in brain and in bone the stress is a localized distortion. Biochemical correlates of such a stress are a proper part of the learning process. It is less certain that perturbations which cause endocrine changes should also be considered as part of the learning process. In some situations the hormonal response is not due to the learning task as such, but to the pressures applied to the animal to make it learn; this is a feature of many experimental situations in which failure to act appropriately results in punishment (1783). In other situations, the hormonal response is produced as part of an attempt to solve a problem that the animal finds difficult or insoluble, either because of the nature of the problem itself, or because of the animal's inappropriate earlier experience. In this situation the hormonal response may be considered as part of the learning process. This situation is complicated, however, by the necessity of making the animal persevere. Trophic hormones released by the pituitary, and hormones released by the endocrine glands have both biochemical and electrophysiological effects on brain; it is certain that some of these are independent, and others interrelated. Some aspects of learning under conditions which are hormonally unusual may have affinities with drug-dependent learning. 'Yoked' control animals are commonly used in an attempt to exclude effects of stress. It is improbable that this is sound. Hypophysectomy or adrenalectomy can prevent the action of certain hormones on brain. Sometimes the removal of a target organ has an effect upon learning brought about by increasing the secretion of trophic hormone, for example ACTH, which then acts directly on brain at several sites.

For all of these reasons the interpretation of biochemical correlates of behavioural change is difficult. Several models have been studied in some detail, but none is ideal.

If the food of rats is placed so that it can be obtained by using only one paw and the container is placed in a position that allows the rat to use either paw, then most rats use one consistently more than the other; this one is the 'preferred paw'. If the food container is now placed so that the food can only be obtained by using the other paw, the rat soon learns to do so. Four days after changing the position of the food container biochemical differences are found in the hippocampus (1778, 1779, 1780). If attention is confined to the neronal perikarya of the pyramidal cells of the region CA3, preferential increased synthesis is found of the S-100 protein and two other acidic proteins (1452, 1453, 1790). Interpretation of this finding combines all the difficulties of understanding the significance of changes when only the perikaryon is studied with the difficulties following our ignorance of the electrophysiological correlates of the behavioural change. If the synaptosomes of the hippocampus are examined, the labelling of many different proteins is found to have increased. The changes in the hippocampus are present on the fourth day but absent on the fourtieth even though the animal continues to use the 'non-preferred paw'. Rats which have been trained to use the 'non-preferred paw' can be retrained to use the 'preferred paw' (1453). Again, there is an increased synthesis of these protein fractions. These findings indicate that the synthesis of protein increases when the animal uses the unaccustomed paw. It is in many ways similar to the older observations that a group of neurons which are subjected to a sustained increase in their stimulation respond with a transient increase in their synthesis of RNA and protein (3062(a)); this characteristic response was seen in supraoptic neurons, anterior horn cells and primary sensory neurons. When the rat reverses the hand it uses to obtain food, cortical neurons described as being in the 'hand area' increase their rate of synthesis of RNA; it is not known if neurons in other parts of the hemisphere also do so. The additional synthesis occurs only in the hemisphere opposite to the hand which is now being used. This finding is unexpected in two respects: firstly in that changes might be expected in the ipsilateral hemisphere because the rat is learning not to use its preferred hand, and secondly because a decreased rate of synthesis of RNA has been reported for nerve cells of the cortex during learning (see below). The base content of the total extractable RNA is altered during training.

Metabolic responses of the brain to avoidance conditioning have been sought in mice. A mouse is trained to jump from a grid floor to a platform to avoid an electric shock; a light and a buzzer are used as a conditioning stimulus. Under these circumstances more ribosomes and polysomes are synthesized in brain, but not in liver or kidney (11, 12, 825, 1083, 1084, 3236, 3237). The increase is found as early as 15 minutes after training. If the mice are trained daily, the increase of RNA follows the training period on the first two days, but not later (1084). The increased synthesis is found in the 'diencephalon' – a part of brain removed which includes the entorhinal cortex, hippocampus, amygdala, thalamus, hypothalamus and mammillary bodies (1521). Synthesis is reduced in the cerebral cortex. All fractions of RNA are affected (1084), and there is no evidence of a specific and new RNA being formed during learning. Extinction of a learned response is also accompanied by a similar response (542). In this respect the response of motor nerve cells, supraoptic neurons and primary sensory neurons is different; although the initial response of these to a sustained increase of activity is only transient (3062(a)), no additional transient increase is seen when the activity returns to its former level. In learning it appears that there is a metabolic response that is associated both with learning and with unlearning, but if increased activity alone is studied, an increased metabolic activity accompanies only an increased electrophysiological activity. Either the two processes are fundamentally different, or much more probably, both learning and unlearning are associated with an increased electrophysiological activity, which may be transient. The changes seen in mice during avoidance conditioning also occur if the gonads and pituitaries have been previously removed (541).

Each trained mouse has a 'yoked' control mouse; this has a shock at the same time as the trained mouse but there is no shelf to which it may escape. The increased synthesis in the brain of the trained mouse always greatly exceeds the changes found in the 'yoked' control. Usually the values for the 'yoked' mouse are the same as those of normal animals. This study shows clearly that 'learning', in the sense of expecting something to happen, is not accompanied by the synthesis of additional RNA. The 'yoked' control mouse does not remain naive; it learns that it will receive a shock shortly after hearing the buzzer and seeing the light; it behaves differently, freezing, or showing some other evidence of alarm. More RNA is synthesized only under circumstances in which the mouse executes a response that is more appropriate than previously; as soon as the response ceases to be more appropriate, that is, when the response is part of the established repertoire of the mouse, synthesis again becomes more normal. But a more appropriate response is one that is more rewarding. It is virtually certain that the noradrenergic system is involved in establishing responses as rewarding and perhaps in confirming them. It is therefore very probable that the increased synthesis of RNA results from activation of this system, and that this effect is brought about by altering the activity of cyclic nucleotide-dependent protein kinases.

The buoyancy of goldfish has been altered by small floats. The fish float upside-down until they learn a new pattern of swimming (2666, 2667, 2668). Fish which learn successfully synthesize more RNA, and the base composition of the total extractable RNA changes. The fish which learn the task most slowly synthesize the most RNA. Some fish with large floats cannot learn to swim in a way that allows them to be the normal way up; these usually struggle for a short time and then lie still. The RNA in the brains of these animals does not change. As in the mice learning an avoidance conditioning task, the synthesis of additional RNA is transient in those fish that successfully learn the new task; 24 hours later the synthesis is normal. It is not known whether restoring the normal pattern of swimming by removing the float on the third day causes a similar increase in synthesis. If the situation is analogous to that of the mice described above, such an increase would be expected. Other studies in which goldfish have been trained using techniques of conditioned avoidance suggest that the results described here should be interpreted cautiously (168). In such experiments the pCO_2 of the water rises; such a rise can itself cause changes in the synthesis of RNA similar to those described. Nevertheless, the results obtained show a close and interesting correspondence to those obtained in the mouse, and perhaps may be interpreted similarly. If dibuteryl cAMP is injected into the cerebral ventricles of the goldfish, the synthesis of RNA is increased crudely in the same way (2668); although this is clearly consistent with the process of metabolic stimulation suggested above, the increase could also be due to the hyperactivity that frequently follows such an injection.

The evidence on the whole suggests that the synthesis of RNA and protein found when an animal learns to execute a response that is rewarding but not when it learns only to expect something unpleasant, is a result of activation of the system in brain that is concerned with the recognition that a reward has been received, and with the unknown consequences of that activation. Many subsidiary questions follow. For example, if the synthesis of protein or of RNA is prevented in the period of learning is the confirmation process or the consolidation of memory affected, or is the rate of extinction of the transient trace altered? If the activation of the reward noradrenergic system is impaired pharmacologically or by lesions, can a new response still be learned, and if so can it be consolidated without the activation of protein kinases. Can new tasks be learned better under conditions in which the activity of the noradrenergic reward system is increased, or in which the action of cyclic nucleotides is facilitated, albeit diffusely. These questions can only be answered in a rather unsatisfactory way by methods which add a dimension of pharmacological and technical imprecision to the uncertainty of interpreting the basic observations.

Inhibitors of nucleic acid and protein synthesis

Many studies indicate that memory is stored in two ways; these are usually called short-term memory and long-term memory. Short-term memory arises directly from the primary sensory experience; it lasts for a period variously estimated at between 15 s and 4 hours. It might last even longer if the animal is deprived of the phase of sleep characterized by rapid eye movements (910). Although long-term memory is commonly said to 'follow' short-term memory the evidence for this is sparse; it certainly outlasts short-term memory but may not be built upon it. Initially, short-term memory is the only form to which access can be obtained; this does not exclude the parallel but independent development of long-term memory. It is also uncertain whether short-term memory is extinguished at precisely the time that long-term memory develops or becomes accessible, and if so whether this extinction is a necessary part of the consolidation. Short-term memory can be disrupted by electroconvulsive shock, by fits caused by drugs, by head injury or by spreading depression. The degree of amnesia is related to the intensity and duration of the process designed to produce it (491). It is usually assumed that these methods act by disorganizing the electrophysiological consequences of recent sensory information; unfortunately, there is no direct evidence for the persistence of such electrophysiological consequences as such in the normal nervous system, whether they are considered to be action potentials chasing their own tails in hypothetical reverberating circuits, or a change of firing pattern of such subtlety that they cannot be recognized. Both explanations appear to be rather unlikely. An alternative is available. It is now clear that the activation of many receptors leaves several types of biochemical trace of different durations (Chapter 4): cyclic nucleotide concentrations, adenylate cyclase activities, phosphodiesterase activities, protein kinase activities and the phosphorylated protein; a similar sequence of events involves the system beginning with guanylate cyclase. The rate of decay of the induced alteration of activity of protein kinases which follows nervous stimulation matches well the decay of short-term memory. Each of the methods of disrupting short-term memory could disturb the transient biochemical trace by masking it with the 'noise' produced through the release of transmitter substances in a disorganized manner during, and after, the shock, fit or depolarization.

Electroconvulsive shock, fits produced by drugs and spreading depression can each alter the rate of synthesis of protein by the brain (199, 749, 1659, 1660). The polysomes may also be disrupted. The reduction of total protein synthesis is less than is necessary to disrupt consolidation of memory if acetoxycycloheximide is used (see below). Charateristically, drugs which inhibit the synthesis of protein leave the short-term memory unaffected.

The period in which electroconvulsive shock can disrupt memory is short. It does not represent the total duration of short-term memory, but only the period in which it is the only record to which access can be obtained. After this period short-term records may still persist, but if obliterated artificially learned tasks can still be performed through access to long-term traces. During this period long-term memory might be developing, but it may not be accessible. The period delimited by the use of shock and related procedures is therefore not the period of short-term memory, but the latent period of accessible long-term memory.

Some drugs which inhibit the synthesis of protein impair the development of accessible long-term memory. These are often given by intracerebral injection, a procedure which can itself cause amnesia in mice (285, 3110).

Acetoxycycloheximide prevents the synthesis of protein by inhibiting the movement of amino acids from transfer RNA into peptides (64, 348, 824, 925, 2689). After it has been injected intracerebrally protein synthesis in brain is reduced by about 95% for 6 to 10 hours. If the drug is given before training, so that the synthesis is reduced during learning, the recall is normal for up to 3 hours after training, and sometimes for rather longer, and is then lost. This effect is found in mice, rats, chicks and goldfish (41, 42, 156, 928, 929, 930, 931, 1924, 2355, 2356, 3074–3076), and can be dissociated from an impaired capacity to express the learned action (2648). Under some circumstances the loss of memory is then permanent (156, 157). These findings are usually interpreted as indicating that the consolidation of memory requires the synthesis of protein. The amnesia is more marked if mice are trained in a single short session than if they are trained in a long one (156, 157). In other circumstances the loss of memory produced by acetoxycycloheximide is temporary (929, 930, 1794, 2654, 2655, 2804); such animals appear naive if tested between 3 and 6 hours after learning, but can recall the training in a normal manner between 24 hours and 7 days after training. Four factors may be crucial in determining whether the loss of memory is permanent or temporary: the intensity and duration of training, the duration of short-term memory, the period for which the synthesis of the relevant proteins is prevented, and the state of arousal of the animal when synthesis of protein recommences. The interaction of these factors has been studied (156, 157, 158, 159). Cycloheximide given subcutaneously to mice inhibits protein synthesis for about 3 hours; a residue of short-term memory is usually just detectable at this time but not at 6 hours, or later. This indicates that the residual and just detectable short-term memory cannot lead to consolidation after so long a period. If the state of arousal of the animal is increased either by stimulation or by drugs 3 hours after training, long-term memory can then be consolidated; consolidation is not facilitated in this way 6 hours or more after training. The effect of enhanced arousal at 3 hours is absent if protein synthesis is again blocked. These results show that an aspect of short-term memory, defined here as a memory independent of the synthesis of new protein, persists for at least 3 hours and that it can induce consolidation if the state of arousal is enhanced in the presence of active protein synthesis. Augmented arousal might increase the dwindling effect of short-term memory in at least two ways, depending upon the model of short-term memory that one favours. It might raise the dwindling activity in reverberating circuits to a critical threshold for consolidation. It might augment the activity of the reward system so that the almost vanished trace represented by transient chemical changes can be consolidated. Perhaps both systems operate. As would be expected, cycloheximide given before learning and electroconvulsive shock given just after learning together abolish the learned response (88).

Although it is likely that the effects of cycloheximide and acetoxycycloheximide upon brain are produced by reducing the synthesis of protein, these drugs have other actions; the current threshold for producing electroconvulsion is raised (1878). It is not known if this is a result of reduced protein synthesis. The penetration of some drugs into brain from blood is also facilitated (89). Adrenergic drugs can reduce the efficacy of acetoxycycloheximide if given before training or within 2 hours (2654, 2655). It seems likely that this effect is due to the increased arousal described above, and strongly supports the suggestion that the biochemical effects are brought about through the activation of the adrenergic reward system.

Puromycin prevents the synthesis of protein by releasing peptides containing puromycin prematurely from ribosomes (2105, 3139). Injected intracerebrally it disrupts memory (155, 927, 930, 931, 932, 933). Part of this action is unrelated to the inhibition of protein synthesis during the period of consolidation. If puromycin is injected before training commences it acts in a way similar to acetoxy-cycloheximide (41, 42, 650). It can also impair recall if injected between 1 day and 43 days after training (930, 931), long after consolidation has ceased. This late action of puromycin can be prevented by the simultaneous administration of acetoxy-cycloheximide; it is therefore likely that this effect of puromycin is due to the release of abnormal peptides which contain puromycin. These peptides can be isolated from synaptosomes and from mitochondria (933). By autoradiography they can be seen both in the perikarya of neurons and in the neuropil (1014).

Puromycin has other actions also. It reduces the respiration of brain slices (1506). Mitochondria become swollen in dendrites and in neuronal perikarya but not in axons or in neuroglial cells (1012, 1013). It is uncertain whether this variability reflects the uneven penetration of the drug into these different cell processes, or whether it indicates a different response of the mitochondria themselves. Puromycin, but not cycloheximide, also causes epilepiform activity in the hippocampus (522, 523, 2607) and facilitates the action of some convulsive drugs and of electroconvulsive shock (1878). The significance of this abnormal discharge is unclear, as the drug appears to be without effect upon short-term memory. It seems probable that the effects upon mitochondria and upon excitability are produced through increasing the permeability of the membranes in a manner that lacks specificity; if so it would be expected to have a similar action upon a black lipid membrane *in vitro*. Puromycin might also affect neurotransmission through its O-methyl-tyrosine group (2459, 2460). The action of puromycin is clearly less straightforward than that of cycloheximide. It is evidently necessary to know how these drugs influence the turnover of neurotransmitters in the brain, and synaptic transmission both in brain and at peripheral junctions. An effect related to the turnover of amines is also suggested by the capacity of these drugs to lessen some behavioural changes caused by inhibitors of monoamine oxidase (1125). This effect too is consistent with the hypothesis that the activity of the noradrenergic 'reward' system regulates the consolidation of transient traces through facilitating the action of adenyl cyclase and so influencing the synthesis of protein.

A number of procedures reduce the effect of puromycin upon memory. Bilateral intracerebral injection of saline lessens its effect (926), and releases some peptidyl-puromycin from synaptosomes but not from mitochondria. This action might be produced either by the saline or by spreading depression caused by the act of injection itself. Adrenalectomized mice are less susceptible than normal mice to the action of puromycin (927). As corticotrophin also provides a similar protection in adrenalectomized mice (928), it is possible that the effect of adrenalectomy is due at least in part to the augmented release of pituitary hormones. This action of corticotrophin gel is due partially to contamination with vasopressin (1707). Both vasopressin and corticotrophin are released in greater quantities from the pituitary of an adrenalectomized mouse. Puromycin apart, it is known that pituitary peptides facilitate the acquisition of a conditioned avoidance response (1077, 1078, 1079, 1152, 3124, 3125), and delay its extinction (3124). Both vasopressin and corticotrophin act in this way (1708). Pituitary peptides facilitate the synthesis of RNA and protein in the brain stem (1077, 2610).

Actinomycin D prevents the transcription of DNA-primed RNA. If given to a goldfish before conditioned avoidance training, the acquisition of immediate appropriate responses is not affected, but the consolidation of long-term memory is blocked (43). Actinomycin D does not prevent a mouse from learning a maze, or from recalling the correct route two hours later; because of the systemic toxicity of the drug, only a few animals can be tested 24 hours after learning. These show no significant impairment of memory. Apart from its action upon the synthesis of RNA, actinomycin D has a profound effect upon the metabolism of phospholipids (1973). Injections of 8-azaguanine leads to the initial synthesis of abnormal RNA by brain, followed by decreased synthesis (704). Animals which have received this drug appear to learn a water maze more slowly than normal in the first training session started 15 min after injection. This drug is more than likely to disrupt short-term memory systems also, however, if it interferes with the activity of guanylate cyclase.

It would be idle to pretend that these studies in which the synthesis of protein is inhibited fail to raise more questions than they answer.

Although RNA is synthesized while an animal is learning to execute an appropriate learned response, the inhibition of synthesis of RNA does not always prevent the response from being learned. This indicates that a new RNA is not needed as an entity that is itself required for the registration of the experience; it follows from this that a new protein is also not required. Furthermore, as learning can continue despite cessation of RNA synthesis, the half-life of either the RNA or of the protein must be long. If the synthesis of protein or polypeptides is blocked, but the synthesis of RNA continues, consolidation does not occur. A protein of some sort is therefore necessary for consolidation, but it is not a novel one. An imposed experiment is not the only event that happens in the life of an animal. Unless it is assumed that an experimental animal learns nothing and seeks to do nothing new other than the wishes of the experimenter, then this protein, or class of proteins and polypeptides must be present before the start of the period of specific learning imposed by the experimenter. The role of the synthesized RNA is unknown; it might be responsible for the replacement of the proteins or polypeptides required for consolidation to occur. If so, its absence would produce an effect only after a longer latency, which might vary between species.

Inhibitors of protein or polypeptide synthesis can inhibit learning; as it is probable that the protein or polypeptide is present before the formal learning session begins, then the molecule must either have a very short half-life once formed, or it must be required in far greater quantities than normal for the particular circumstances of the experiment, or only the nascent protein or polypeptide is effective. There is at present little to choose between these alternatives. It seems unlikely that all the different models studied should share the characteristic of requiring a far greater synthesis of the protein or polypeptide than normal; it is, however, possible, because many of the experiments are designed with the interests of the experimenter in mind, rather than the natural inclinations of the rat or mouse. If a particular protein is required to execute certain learned tasks, pressures of selection might well have acted to produce the quantities needed to allow the animal to meet problems of the type or intensity that its ancestors have experienced. If inbred strains of rodent have been studied that have been kept in the laboratory under standard drab and underprivileged conditions (Chapter 6), then the pressures of selection might produce animals in which very little of this particular protein is available before the start of the experiment. For such animals, the experimenter's task might well constitute the intellectual high-spot of its restricted life. It would be interesting to compare the sensitivity of different strains of rat or mouse to the effects of inhibition of protein synthesis upon the consolidation of memory; for example strains selected for their problem-solving capabilities, and kept as individuals within an enriched environment that allows them to excercise this ability, could be compared with strains selected for their dullness.

It is also necessary that the newly formed protein or polypeptide can interact with the trace left by short-term memory, which includes activation of protein kinases. It is tempting to suggest that the nascent protein acts as a substrate for protein kinase, and that the phosphorylated protein is the first step in the process of consolidation. The later steps are still unknown.

It is difficult at present to interpret the possible role of certain polypeptides which might be formed in brain under certain conditions (2824, 2972, 2973, 2974, 2975). A polypeptide, scotophobin, has been extracted from the brain of rats trained to avoid the dark. When injected intraperitoneally, an untrained recipient has a greater chance than a normal rat of avoiding the dark. The amino acid sequence of the pentadecapeptide is known and the substance has been synthesized. The synthesized material acts similarly. Another polypeptide has been extracted from the brains of rats which have become habituated to a loud noise regularly repeated. When administered intraperitoneally to naive recipients, they become habituated to a repeated noise more rapidly than normal animals. The materials act upon species other than that from which they are extracted. It has not yet been shown that the polypeptides enter brain. These observations may be related to those concerned with the possible transfer of learned responses from trained animals to naive recipients by brain extracts (917). These experiments with scotophobin and with the other polypeptide require independent verification.

6 PLASTICITY

The structure of the nervous system may alter in various ways. This chapter discusses some of the factors which may influence the formation or loss of synaptic connections: hormonal and genetic influences, environmental factors, and perhaps some processes determined by experience. In so far as use may influence the structure of the nervous system the contents of this chapter cannot be dissociated firmly from those of the last; in so far as injury has been used to study aspects of plasticity, the contents of this chapter are related to those of the next.

Some experiments clearly indicate that genetic, humoral and environmental factors interact. For example, a mature greenfly can develop wings when short of food; this development is associated with considerable change in the thoracic ganglia. The response is really that of one stage of development proceeding to the next, because of environmental influence.

Development

During development axons grow to particular parts of the brain. For example, optic nerve fibres of the mammal pass from the retinal ganglion cells to the lateral geniculate nucleus, the pretectal nucleus and the superior colliculus, and a few fibres enter the hypothalamus. Fibres originating from the retina have an ordered projection in these nuclei. Such an ordered projection has been found in other vertebrates also (1033–1042, 2784–2794). Fibres from certain parts of the retina project reproducibly to certain regions of the recipient nuclei. The resolution of the methods used have not been able to demonstrate that the reproducibility is perfect. Other studies indicate that afferent fibres from a particular source end upon a defined region of the postsynaptic neuron.

This process of orderly arrangement of cells is not of course only found in brain, and it is unlikely that the mechanism producing it is unique to brain (603, 2510–2513). It is only the length of the axon that makes the nerve cell unusual and the large number of connections that one cell makes and receives. Cells from discrete parts of the retina preferentially adhere to cells from those parts of the recipient nucleus to which they normally project. For example, axons from the nasal part of the retina of the pigeon innervate the posterior part of the tectum; axons from the temporal retina pass to the anterior part of the tectum; axons from the dorsal retina go to the ventral tectum, and axons from the ventral retina go to the dorsal tectum (1239). If a cell suspension made from the dorsal part of the developing retina is placed in a dish with the separated dorsal half and ventral half of the optic tetum, they adhere preferentially to the ventral half (143). Similarly, cells derived from the ventral part of the developing retina adhere preferentially to the dorsal half of the tectum. This mimics the pattern of innervation of the tectum by the fibres from the retina. When the adhesion is studied soon after the cells of the retina have been separated by the action of trypsin, cells from both halves of the retina adhere preferentially to the ventral half of the tectum; if the cells are incubated in nutrient medium for 9 hours after the action of trypsin before their capacity to adhere is studied, those from the ventral half of the retina lose their preference for the ventral tectum and stick to the dorsal half. This may indicate that cells from both ventral and dorsal halves of the retina have molecules upon their surfaces which make them adhere to the ventral tectum; these molecules may be relatively insensitive to trypsin, or they may be regenerated rapidly after hydrolysis. Cells from the ventral half of the retina may have in addition surface molecules which make them adhere to the dorsal tectum; the results suggest that they are sensitive to trypsin and replaced only slowly.

This study has other aspects of interest. It is very improbable that each adhering retinal cell is stuck to the precise site with which its axonal tip would later normally make synaptic contact; it is also unlikely that only that part of the retinal cell which ultimately forms the axonal tip adheres to the tectum, asuming that such a part of the cell's surface is even represented at this early stage of development. It is much more likely that all parts of the surface of retinal cells from a wide area of retina have a capacity to adhere to a large part of the tectal surface. The property of adhering to particular parts of the tectum is not confined to retinal ganglion cells; other cells, including glia, adhere preferentially in the same way. It therefore appears that all cells of a given part of the retina at this stage of development share similar surface properties. The types of cell to which retinal cells adhere have not been identified precisely. It is improbable that they are mesenchymal, as the pia is removed. They could be precursors of nerve cells, glia or neuroepithelial 'ependymal' cells. Probably, by analogy with the retina, all cell types from one part of the tectum might share a capacity to bind preferentially cells from one region of the retina. The specific adherence does not depend upon previous direct contact between the cells of the retina and of the tectum. These results indicate that preferential adherence is present, but do not indicate the precision with which it might determine the formation of stable contacts. It is probable that if the tectum could be divided further, and challenged with cells obtained from smaller areas of the retina, that a finer scale of attachment with differing affinities would be found.

A number of other studies have shown that cells of a particular type in brain adhere preferentially to other types of cell, and that these patterns of adherence are 'organotypic', that is, the patterns of organization resemble the pattern of the organ from which the cells were obtained (1023, 1024, 1967).

The adhesion of one cell to another has been ascribed to a complementarity of the surfaces. Much evidence indicates that the glycocalyx is involved in such interaction. If some constituents of the glycocalyx are hydrolyzed, the adherence of cells is disturbed (2336, 2511, 2512, 3041); it is also impaired by the presence of some free hexose and pentose sugars, by amino sugars (1022), and by lectins (2524). It is likely that one mechanism of adhesion is arrested intercellular transglycosylation. The evidence for this has been obtained from non-neural cells (2512, 2513). During synthesis of glycoproteins, a sugar is added to the terminal sugar of the glycoprotein. If we assume that the added sugar is galactose, it is transferred to the terminal sugar from UDP-galactose through the action of galactosyl transferase. Both galactosyl transferase and galactosyl acceptors are present on the outer surfaces of cultured Balb/c3T3 cells. The transferase transfers galactose from UDP-galactose to a galactose acceptor; but the transferase is on the surface of one cell and the acceptor is present on the surface of another. The transferase binds to the acceptor site, and the binding can only be released if the process can pass to completion, and this only occurs if UDP-galactose is available. If the nucleotide sugar is present, the sugar is transferred to the acceptor and the glycoprotein is released from the enzyme; the acceptor is then in a state different from that before the contact: it possesses a terminal galactose. Other factors are certainly involved; for example, the capacity of the plasma membrane of liver cells to bind circulating glycoproteins is reduced if they contain sialic acid, and increased if they are treated with neuraminidase. The ability of neuraminidase to alter the adherence of malignant cells to the endothelium of vessels is a related observation.

Ordered connections may develop in two ways: either because of total and unique complementarity, so that each cell can adhere to one target cell only and to no other, or because each cell can adhere to a large number of target cells, but to each with differing strength. There is no evidence to support the first, and much evidence to support the second type of adhesion.

When the concept of varying strength of adherence or of varying affinity is employed, a model can be developed which is very similar to that which describes enzyme-substrate and drug-receptor combinations. The formation of connections is considered to be a reversible interaction, which occurs with certain forward and backward velocities, and which therefore reaches a certain calculable equilibrium. When a connection of high affinity is made, the rate at which the connection is spontaneously broken is far lower than the rate at which it is formed. The formation of 'specific' connections is competitive, not because one terminal of high affinity somehow pushes a terminal of lower affinity off the postjunctional cell, but because the rate of spontaneous loss of connection by terminals of low affinity exceeds the rate of spontaneous loss of connection by terminals of high affinity. The affinity of a connection can be defined as the ratio of the rates of the reactions forming the connection to those of the reactions breaking the connection. In applying such a mode of analysis to the affinities of connections between cells, a number of factors must also be considered. It seems likely that a connection does not follow the creation of one molecular link between two cells; it seems probable that many are formed and their formation may well show co-operativity. It is also probable that the interaction consists of several types of molecular interaction occurring in parallel; for example, several different glycosyl transferases may be acting in parallel, and each type may be binding a number of different incomplete glycoproteins with different affinities. If the model of intercellular transglycosylation is accepted as one method of adherence, connections are not broken only by reversal of the primary reaction, but also by a second reaction, which is practically irreversible.

A practical technique for measuring the affinities of cell interaction rapidly *in vitro* is required, so that it may be measured and the actions of agents such as nucleotide sugars studied. It is possible that a sonicated dispersion of plasma membranes could be used in this way, and the rate of interaction studied by following the increase in light scattering that would be found as the particles mixed from two different sources aggregated. It would be interesting, for example, to repeat in this way the experiments described above which investigated the interaction of cells from the retina and tectum.

Effects of early experience

Effects upon behaviour

Experience gained at certain stages of development of brain has a persisting influence upon behaviour. Such influence may in general be assigned to one of four categories. The first is imprinting (434, 1491, 1837, 2025, 2026, 2120, 2548, 2677, 2733); early experience leads to a response which persists and which is not conditioned, as the animal receives no reward. Imprinting is an impressive example of the ability of the brain to register experience. Very little is known about the structural or biochemical correlates of this registration as such. It is known that the roof of the forebrain of the chick synthesizes more protein and more RNA and that RNA polymerase is activated. These changes occur in the appropriate sequence (172, 173, 1395). But it is not known how these observed changes are related to the registration of experience in this situation any better than we know how similar changes in a mature animal are related to the learning of a response. The study of the immature animal during imprinting has the advantage that there is less possible interference from previous exteroceptive experience, and that the learning is independent of an immediate obvious reward. It has the disadvantage that the results must be interpreted within the context of a rapidly growing and differentiating brain. At present we have no clearer idea about the mechanisms of imprinting than about those of memory in the mature animal.

In the second category is the early social experience that is required for the later expression of mature patterns of social and sexual behaviour (560, 675–677, 989, 990, 1154, 1171, 1245–1248, 1341, 1342, 1926, 1935, 1936, 2011, 2012, 2016, 2039, 2120, 2538, 2539, 2638, 2639, 2640). The young of many species need maternal contact, or contact with another mature and tolerant female for a certain period of life. This period undoubtedly corresponds to a certain stage of brain development in any one species, but it has not been clearly shown that the experience is necessary for certain aspects of the biochemical or structural differentiation to proceed. If the experience is not gained within this period, but later, it either fails to facilitate the later development of normal social or sexual behaviour, or else a far greater intensity and duration of stimulation is needed to produce a much smaller effect. There is evident similarity between this and the dependence of proper development of the visual cortex upon early visual experience. Options for determining certain patterns of behaviour do not remain open indefinitely, even though these patterns may not emerge until a later stage in the life cycle.

The third category is the influence of complete deprivation of sensory experience of one modality, upon the ability of the adult animal to discriminate. If the sensory deprivation occurs throughout a well-defined period the impairment of discrimination is considerable; the resulting defect is either irreversible or reversible partially only if a carefully graded training programme is used (1017–1021, 2261). Electrophysiological consequences of partial or complete visual deprivation have been studied in the cat.

Neurons of the striate cortex of an adult cat respond optimally to well-defined stimuli; they respond well to bars or edges moving across the receptive field in particular directions (1430–1433, 2629, 2795). The cortical nerve cell studied only produces its greatest response if the edge is precisely oriented and moves in a direction at right angles to its long axis. The response is often greater if the edge or bar moves in one direction rather than in the opposite direction. The response of such a nerve cell is described as having orientation specificity and direction specificity. The nerve cell responds only to a stimulus presented in a small part of the visual field.

The newborn kitten lacks visual experience. Its eyes are closed at birth and do not open for several days. Until its eyes open it can gain no effective experience of patterned vision. The neurons of the striate cortex of a newborn kitten have large receptive fields and show little or no orientation specificity or direction specificity (147). If a kitten is reared in total darkness for 3 months, its nerve cells fail to develop the characteristically specific responses of neurons of the adult cat (147, 261, 262, 263). If the total visual experience of the kitten consists of bars oriented in one direction, then the responses of the nerve cells of the primary visual cortex show orientation specificity, but the response is exclusively to bars with an orientation close to that experienced (261, 496, 1352, 1353, 2268, 2269, 2270). Neurons are not found which have an orientation specificity remote from that experienced. This limited specificity is not associated with a reduction in the total number of cells which respond to oriented stimuli (2270). Hence it appears that nerve cells which respond to bars oriented in directions which are various, respond in such a cat only to bars having the experienced orientation. The specificity of many neurons has developed along lines that it would not have done in the absence of restricted experience. The organization of the cortical cells of such a cat cannot be considered as impoverished; it is, on the contrary, highly appropriate for the experience of the animal. The 'tuning' of the orientation specificity may even be sharper than that of a normal cat, a possible consequence of lateral inhibition by closely similar adjacent stimuli (263). It is only when tested outside the context of the animal's experience that the pattern of organization is defective, for example in the failure of cortical units to respond to bars that are not orientated precisely in the manner of bars previously experienced, or in the failure of the cat to see or to respond to a stick held at right angles to the direction of bars previously experienced (261). It is behaviourally blind to such inappropriately orientated stimuli. On the other hand, and entirely hypothetically, it is possible that if there was a world that was really composed just of bars oriented in one direction, the cat reared just with exposure to bars of this orientation would fare better than a cat reared in the normal manner.

It is not clear whether the development of direction and orientation specificity is due entirely to the acquisition of certain patterns of inhibition, which are virtually irreversible, or whether there is an additional structural remodelling, with the loss and formation of different synapses. The exposure of a visually naive cat of the appropriate age for one hour to bars of one orientation is sufficient to determine the orientation specificity of most of the cortical nerve cells in the primary visual cortex. It is not yet clear how soon after such an exposure the specificity emerges (262). The orientation and direction specificity of a neuron of the visual cortex of a normal adult cat can be partially reversed by applying bicuculline, an antagonist of γ-aminobutyric acid (2267). GABA is an inhibitory neurotransmitter in the cerebral cortex (608, 1661). This observation confirms the importance of active surround inhibition in maintaining the characteristic specificity of the recorded neuron (585), and shows clearly that the development of orientation specificity is not experimentally irreversible, as it would be if due to the loss of synapses during development. The striking feature of the changing response pattern of a neuron of the primary visual area during development is that of progressive restriction: restriction both with respect to the size of the receptor field, and with respect to those features of a stimulus that elicits an optimal response.

In the adult cat the properties of neurons of the striate cortex and of the lateral geniculate nucleus are not absolutely constant. The size of the receptor field and the degree of orientation specificity vary. This variation can be produced by drugs (183, 2564), by alteration of the animal's state of arousal (183, 517, 1970, 2568), and possibly in other ways. It seems probable that as these properties of a neuron are variable, they may also be regulated, so that in any given state of the animal they are optimal. Hypothetically, it is possible that the degree of analytical precision achieved by the cortex during the performance of difficult discrimination tasks is enhanced by reducing the size of the receptor fields and by sharpening the degree of orientation and direction specificities. Conversely under conditions in which a response is urgently required, the degree of such specificities could be reduced, so that the chance of the information getting through, even though less well filtered, is enhanced. This variation of the acquired restriction of response is probably brought about by varying the degree of surround inhibition. This may be under the control of ascending pathways from the reticular formation.

Many neurons of the striate cortex of the normal cat discharge when an appropriate stimulus is presented through either eye (1430, 2442, 2795). The stimulus comes from the same part of the visual field of each eye. This is binocular convergence. Not all cells respond with equal intensity to stimuli presented through each eye. One eye is frequently dominant. A kitten can be reared with one eyelid sutured shut, so that one eye is deprived of patterned vision, and receives only weak diffuse light. The pathways from the other eye gain normal visual experience. Single cortical neurons of such cats respond normally to stimuli presented to the normal eye; simple, complex and hypercomplex cells are found (1430). Stimuli presented through the eye that has been occluded evoke much less response; this response is easily fatigued, and the responding cells show very little orientation or direction specificity and have large receptor fields (1019). A given cortical cell can be stimulated normally through the normally experienced eye, but abnormally through the eye that was previously occluded. The normal eye is clearly dominant in the cells tested (3126). The situation is different in animals reared with both eyes covered, or in animals that have been reared in darkness. In these the balance of dominance remains essentially normal, and the responses are not so easily fatigued; only the development of orientation specificity and direction specificity is lacking. An occluded eye is therefore represented less well in the cortex if the opposite eye is normal than if it is occluded.

Cats have been reared under conditions in which each eye can see form as well as light, but in which the information presented to each eye is not congruent with that of the other. This has been done in three ways: by occluding each eye alternately, so that binocular vision cannot occur, by exposing one eye to vertical stripes and the other to horizontal stripes (2270), and by moving the relative positions of the retinal images, either by making the animal squint or by placing a prism in front of one eye (2678). The results of all the studies are concordant. The visual cortex of such a cat has many simple, complex and hypercomplex cells. Nearly all of them can be driven through one eye only. Adjacent cells tend to be driven through the same eye to a greater extent than normal. It appears that a given cortical cell only responds to stimuli presented through both eyes if they are congruent.

Ocular dominance is not irreversible. It is possible to alter the relative dominance of the eyes measured in a cortical neuron of a normal cat by conditioning while recording from the cell (2269). A cell is selected which is clearly dominated by one eye, say the left. Then, a conditioning stimulus is presented to the right eye shortly before one is presented to the left. The response to stimulation of the right eye increases steadily. Only a few cortical cells behave obviously in this way, but many show a lesser tendency to act similarly. The study clearly has some affinities with that demonstrating heterosynaptic facilitation in *Aplysia.*

If a cortical neuron is preferentially excited by congruent stimuli, it might be expected that slight binocular disparity caused by a prism would make a cortical neuron accept information from points on the two retinae that are matched with respect to the position of their images, but not with respect to their absolute position. A cat was reared for four months with a prism causing vertical disparity; the cortical neurons showed a tendency to behave in this way, but the degree of compensation obtained was incomplete (2678).

Studies of *Xenopus laevis* (867), and of the frog (2729), give results that are essentially consistent with those obtained in the cat.

Morphological aspects of some of these electrophysiological changes have been studied (546, 1026–1029, 1543). Simple cells are almost certainly stellate and complex cells are pyramidal. Too few hypercomplex cells have so far been identified by injection of dye to allow them to be identified confidently; they might be either stellate or pyramidal (833). It is probable that the type of physiological response given by a certain morphological type of cell varies according to the layer of cortex examined; this has so far not been studied sufficiently to give firm conclusions.

Kittens that have been deprived of vision in one eye since birth have atrophied cells in the layers of the lateral geniculate nucleus which receive fibres from the deprived eye; the sizes of the neuronal perikarya, cell nuclei and nucleoli are reduced. The total number of nerve cells is not reduced (3127); atrophy does not follow the closure of one eye in the adult cat. Atrophy is less in kittens that have had some visual experience before the eye is closed. If a kitten is reared in the dark, some atrophy of all layers of the lateral geniculate nucleus is present, but is less than that found after monocular deprivation. When kittens are reared in the dark and then exposed to patterned vision for several hours, structural changes occur in the synapses of the visual area. Before visual experience many of the boutons have few synaptic vesicles and these are of low density. One hour of patterned vision increases both the number and density of the synaptic vesicles. This is more marked 14 hours after the visual experience and is still present 8 days later. The affected boutons are small, and found on dendrites and on dendritic spines. They have round vesicles and assymmetrical junctions. Kittens reared with a divergent squint have no atrophy of the lateral geniculate neurons (1028, 1432).

Fibres carrying information from the two eyes first converge upon a single nerve cell in the visual cortex. Several lines of evidence indicate that the electrophysiological and anatomical changes result from competition between fibres carrying information from one eye and the fibres carrying information from the other. The responses to stimuli given through the previously deprived eye are more easily fatigued only when the other eye is normal and not when that too has been deprived of patterned vision. There is considerable imbalance of ocular dominance when only one eye is deprived, but not when both have been deprived. The cells of the lateral geniculate nucleus are atrophied when one eye has been deprived, but not if both have been treated in this way. Not all parts of the visual field are seen binocularly; hence those parts of the temporal field which are seen only by one eye project to the contralateral lateral geniculate nucleus without an associated projection from the contralateral eye. This means that these cells of the lateral geniculate nucleus cannot be represented binocularly in the cortex; hence, their terminals do not compete for representation; these cells of the lateral geniculate nucleus are spared from the atrophy which affects other cells of the same lateral geniculate nucleus receiving information from the previously deprived eye. These findings support the hypothesis that the fibres containing information from each eye compete at the surface of the cortical neuron for representation (1432, 3129). The nature of the competition is unclear, but it seems to be confined to the period in which the cortical connections are being established.

Structural changes are found in the brains of other species if the animal is reared in the dark (885). Pyramidal cells of the visual cortex of mice have fewer dendritic spines in layer IV (2991, 2992); cells of the auditory cortex also atrophy, but later become larger than normal (1212, 1213). The spines of dendrites in layer IV of the visual cortex are not reduced in number in rabbits that are reared in the dark, but they are deformed (1090). These changes are not precisely equivalent to those found in the cat and do not correlate with the interpretation of the electrophysiological results obtained from the cat and described above. The mice and rats show atrophy of the cortical cells under conditions of binocular deprivation of vision (1088); this is the state in which the electrophysiological behaviour is well preserved. No information is available concerning the electrophysiological activity of the afferent nerve fibres to the cortex in these mice and rabbits during the period of visual deprivation; it cannot be assumed that the mean level of activity is less than normal, although it might be.

The simplest explanation of the results is perhaps obtained if the formation of all connections is viewed as a reversible process, in the same sense that drug-receptor and enzyme-substrate interactions are reversible (p. 175). The affinity of a terminal for a receptive site depends upon the ratio of the rates of the forward and backward reactions. If the response of the postsynaptic membrane to transsynaptic stimulation increases the rate of the forward reaction, or diminishes the rate of the backward reaction, then the connection is made more stable. Such connections would 'compete' effectively against other terminals which are either inactive or which cause a lesser response in the postsynaptic cell. The relevant postsynaptic response is likely to be a postsynaptic potential rather than a propagated action potential: the size of such a potential is not fixed, but can be altered by heterosynaptic facilitation. Such a model is compatible with the observed 'conditioning' of cortical neurons to alter ocular dominance.

Even less is known of the biochemical reactions that underlie this suggested mechanism, but it is possible that the molecular 'trace' left by activity is the sequence of reactions started through the activation of adenylate cyclase or guanylate cyclase (p. 145).

The consequences of early deprivation of patterned vision have certain points of resemblance to early social deprivation. In both, the experience has to be gained within a certain crucial period, and in both a very short exposure to appropriate stimulation is adequate for the development of normal behaviour (566, 2911, 2912, 3128). Animals of some species, if deprived of social experience in infancy remain 'socially blind' when adult. It seems as though they too cannot analyze the information in a way that leads to the normal pattern of social behaviour. Physiologically, it is possible that the defect is of the same type that is present in animals deprived of patterned vision; it is likely to be more difficult to study because many modalities are involved, and the nature of the electrophysiological changes which develop are not known.

Deprived animals of some species are hyperactive, either when stimulated or when behaving naturally (2012, 2441). Different strains of a given species vary considerably in this respect. It is probable that this results from a failure to form systems appropriate for the analysis of sensory information, and especially appropriate for the rejection of information that is not relevant to the current situation; the capacity to select appropriate cues is impaired. The picture is therefore one of susceptibility to easy distraction.

The fourth category of early experience which produces effects upon the adult brain includes three types of influence which might be inter-related: handling of the infant animal by the experimenter (1591, 2533, 3095, 3096), the action of certain hormones administered or withheld during certain phases of development, and malnutrition during the period of brain growth (636).

If infant mice are exposed to cold, they develop earlier the capacity to regulate their body temperature (2572). Cold is probably also a significant factor promoting the differences of development which follow early handling; the eyes open earlier and the adult pattern of response of the adrenal to adversity develops earlier. Adult rodents, which are handled in infancy, can withstand deprivation of food and water, exposure to cold, and prolonged immobilization better than animals which are not handled by the experimenter (1782, 3095, 3096). The observations are not easy to interpret, both because not all the variables of experience are in the hands of the experimenter, and because the behavioural observations lack thorough electrophysiological and biochemical study; the maternal care received by a handled infant mouse exceeds that received by a littermate that has not been handled. Although increased 'cortical activity' and 'reticular activity' have been invoked as causing the changes, evidence for either is slender and very difficult to obtain. The precise hormonal changes are not known, and some of the changes in brain may be produced by the action of hormones, as well as inducing altered hormonal secretion. The increased incorporation of cholesterol into the brains of developing mice which are being handled indicates earlier myelination.

The sexual behaviour of adult rodents is influenced by the hormonal state of the animal shortly after birth (161–163, 336, 362, 789, 924, 1117, 2864, 2919). In the absence of androgens a hypothalamus with female potentiality develops; in their presence a potentially male hypothalamus is formed.

The structure of the preoptic nucleus of the hypothalamus of a mature male differs from that of a mature female. Some boutons ending upon neurons of this nucleus come from neurons in the amygdala (884, 2373, 2374, 2378, 2380). Others come from elsewhere. The differences so far reported concern only those boutons which are not derived from the amygdala. In the male, these boutons almost all make contact with the shafts of the dendrites. In the female, a significant proportion ends on spines (2378, 2380). If the male is castrated at birth, the proportion of these boutons ending on spines increases to become the same as in the normal female. If testosterone is given to the female rat when it is 4 days old, the number of boutons found upon the spines of dendrites is negligible; the proportion becomes the same as that of a normal male. If the male is not castrated until he is 7 days old, the disposition of boutons is the same as in a normal male, if testosterone is given to a female aged 16 days, the final disposition of boutons is the same as in a normal female. No differences are found in the ventromedial nucleus of the hypothalamus. The types of cell upon which the changed pattern of connection are seen have not been identified. It is likely that the axons of neurons of the preoptic nucleus influence neurosecretory activity through the arcuate and ventromedial nuclei. Physiological studies implicate the preoptic area in the regulation of ovulation. It is very probable that the changes seen are related to the persisting alteration of the hormonal state of the animal. It is not yet clear whether the altered pattern of connection is caused directly and immediately by the imposed endocrine change; nor is it known if the altered pattern of connection is the immediate cause of the changed pattern of endocrine control. They could alternatively, for example, be related to a hormonally determined difference between the sexes in the response of the hypothalamus to olfactory stimulation by pheromones. It is probable that other differences between the sexes found in regions of the nervous system remote from the hypothalamus are not concerned with causing persisting endocrine change, but are a consequence of it; for example, differences in the number of preganglionic sympathetic nerve cells in the thoracolumbar intermediolateral nucleus (415) may be caused by the endocrine difference and might be a cause of the difference between the sexes in vascular reactivity.

If bovine growth hormone is injected into pregnant rats, the brains of the litter contain more nerve cells (1442, 3229, 3231); the perikarya are large (506). Cells of the cortex are affected more than those of subcortical nuclei. The dendrites of cortical neurons extend more widely than usual, even though they develop postnatally, after the series of injections is completed. Estimated histologically, the number of glial cells does not increase (3231). The glial cells of the rat also develop to a large extent postnatally. This finding suggests that the proliferation of glial cells might not be linked precisely to the number of neurons formed earlier. Changes in the behaviour of these rats are slight and are detected only in the performance of simple tasks (275, 3048, 3231). Certain normal reflex responses are found marginally earlier than normal. The rats have a greater rate of extinction of a conditioned avoidance response. They do not run through a complicated maze more rapidly than normal rats. The rats appear to derive but little benefit from their neuronal excess. Nor should any be expected. A rat and its behaviour is the product of a considerable period of development in the face of pressures of selection. No information is yet available about the connections established by the additional neurons, nor about the electrophysiology of the large dendrites.

The brains of tadpoles treated with growth hormone synthesize more DNA while the hormone remains available.

If the pituitary is removed from rats aged 4 or 5 days, the brain increases in weight normally. The basal dendrites of pyramidal cells of the somatosensory cortex are of normal size and spread (1149, 1150). There is no change in the thickness of the cortex (691). The acetylcholinesterase activity is reduced in the visual cortex, the somatosensory cortex and in the ventral cortex. While this enzymatic change might reflect alteration in the number or activity of synapses, it need not, for much of the acetylcholinesterase activity of nerve cells is found not at synapses but in the endoplasmic reticulum. Early after hypophysectomy the acetylcholinesterase activity is also reduced in the hypothalamus.

The effects of thyroid deficiency on the development of brain has been most studied in the rat; its brain is immature at birth. Thyroid function has been depressed by drugs, by surgery, or by radioactive iodine. Thyroidectomy on the day of birth leads to a reduction in the level of circulating thyroid hormone, which reaches its lowest level about the sixth day. The consequences have been studied anatomically, biochemically and behaviourally.

The differentiation of brain is not only retarded but is also altered (755–759, 1394). Judged by light microscopy, radial thalamocortical fibres appear to be affected to a greater extent than tangential association fibres; most thalamocortical fibres enter the cortex after the sixth day; many association fibres are already present. The nerve cells remain tightly packed and the perikarya are small. The dendrites become stunted and their pattern is abnormal (757). The duration of cell division is extended. The changed shape of the head almost certainly results from altered bone growth and is not a consequence of the altered development of brain.

The high concentration of DNA and the low RNA/DNA ratio (128, 129, 1043, 1044, 2220) are consistent with this histology. The rate of synthesis of protein is low (1043, 1044). The activity of a large number of enzymes is reduced. There is no evidence to indicate that any of these enzymes is particularly important within the context of altered development; they are just easily measured: hexokinase, phosphofructokinase, pyruvate kinase, succinic dehydrogenase, mitochondrial aspartate aminotransferase, GABA transaminase, Mg-ATPase, and Na-K-ATPase (94, 2220, 2627, 2628). The rate of conversion of (^{14}C)-glucose into (^{14}C)-amino acids is reduced (516). The concentrations of sodium and of water in the brains of these hypothyroid animals are high (1043, 1044); this may result from an increase in the volume of the extracellular space associated with deficient myelination (127). Thyroidectomy does not affect the activities of adenyl cyclase or of phosphodiesterase, and does not alter the increase in the concentration of cAMP induced by noradrenaline. The activity of cAMP-dependent protein kinase is not altered (2599).

The effects of thyroidectomy are prevented if physiological concentrations are maintained artificially. Bovine growth hormone can also prevent some of these changes (1643). Thyroid hormone administered later in life cannot reverse the effects of early deprivation.

Corticosterone suppresses mitosis when administered to a neonatal rat. Cell division in brain is considerably reduced. The effect begins rapidly and persists for as long as the concentration of exogenous hormone remains high. If cell division is suppressed in this way, the number of cells remains low in the adult rat. The cerebellum is affected by a greater extent than the cerebrum, because a greater proportion of the cells of the cerebellum divide postnatally. Cell division ceases at the same age as in normal animals. The differentiation of the nerve cells formed is not affected (128, 557, 1404, 1405, 1406).

Some reflex behaviour appears later than normal in animals treated with corticosterone, but learned behaviour does not seem to be affected (2575, 2576, 2577). The way in which corticosterone acts to produce these permanent changes is not known. It is unlikely that the permanent defect is a consequence of the persisting impairment of glucose homeostasis (2903).

There is unequivocal evidence that inadequate nutrition of a litter of rats up to the age of weaning is associated with retardation of development of the brain and with altered behaviour. Many of these changes cannot be reversed by later adequate feeding (34–38, 238, 412, 655, 706–709, 819, 909, 1009, 3151, 3153). Undernourished pigs are similar in many respects (699, 700, 1722, 2825). The period of vulnerability in the rat lasts from birth to about the 21st day. Animals underfed throughout this period have small brains; the reduction in total wet weight is proportionately less than the reduction in total body weight. Not all parts of the brain are equally vulnerable; regions growing by cell division are more vulnerable than regions growing by increase of cell size. The striking susceptibility of the cerebellum is related to the late division of the granule cells. The brains of rats undernourished in this way have fewer cells than those of normal litter mates (485, 912, 1404, 3230), and less myelin in the central nervous system (484, 485, 707, 1049, 2529). Less myelin is present because less is formed (216, 373). Undernutrition delays the appearance of certain patterns of behaviour (216, 373, 559, 2700, 2743, 2744, 2745). Rats which have been underfed until weaning remain clumsy throughout adult life (2744, 2745), a factor probably related to the small cerebellum (485, 707, 708).

If rats are undernourished for only the first 9 days of life and are then allowed to feed *ad libitim* the weight of the brain and the cell number are restored. If rats are undernourished after weaning, the biochemical changes are reversed when free access to food is again permitted (3150, 3152).

Information obtained from each of these four types of manipulation of the developing brain indicates that the pattern of development can be affected. By gross hormonal insults the development can be disrupted or arrested; more precise hormonal studies show that changes of great specificity can be produced. More subtle changes follow the alteration of the sensory experience of the animal. All these influences share two important features: their characteristic action can be brought about only by acting within or throughout a defined period of development, which is not the same for each agent studied, and the effect produced cannot be reversed once the period of susceptibility is completed.

Plasticity and insect metamorphosis

The nervous system of the moth *Galleria mellonella* changes at metamorphosis. The meso- and metathoracic ganglia fuse, and the terminal abdominal ganglion is formed from ganglia 6, 7, and 8. The nerve cord of the moth is 15 to 20% shorter than that of the caterpillar (2293, 2294, 2296, 2297). Early in metamorphosis adipohemocytes invade the basal lamina that lies between the glial cells and the perineural connective tissue, and probably ingest it. It is interesting that the first detectable change should be seen at the junction of epithelium and mesenchyme; this would be expected if interaction at this boundary is of primary importance in regulating the differentiation of the nervous tissue (Chapter 1). These two cell layers become separated. The shortening of the connectives takes 30 to 40 hours, but the axons shorten more slowly over 140 to 150 hours. Because of this difference the axons become coiled (2296). After shortening is complete a new basal lamina is formed; this attains adult thickness by 120 to 130 hours.

If the brain of a pupa is removed, the connectives shorten more slowly. Further delay results if the corpora cardiaca and the corpora allata are removed (2295). The delay caused by removing all these organs is reduced by implanting a brain from a pupa or from an adult, or by implanting corpora cardiaca and corpora allata. To reduce the delay, the brain must be transplanted into the decerebrate anterior fragment; it may act by producing prothoracotrophic hormone. Synthetic α-ecdysone can cause the connective between the mesothoracic and metathoracic ganglia to shorten in brainless pupae. The pupal nerve cord has been cultured *in vitro* (2461); it shortens when ecdysone is added to the medium. β-Ecdysone is more effective than α-ecdysone. *In vivo*, shortening coincides with the deposition of new cuticle, the period of physiological activity of ecdysone.

Segments of transplated nerve cord shorten during metamorphosis of the host (2297). Interganglionic connectives which do not normally shorten do so when transplanted. The cause of this added vulnerability is uncertain; perhaps adipohemocytes can invade the perineural connective tissue more easily and destroy the basal lamina in the connectives that are normally unaffected; or perhaps this shortening is inhibited when the connectives lie in their normal positions with all their connections intact. As transplanted connectives shorten, the shortening is unlikely to be due to contractile cells outside the sheath. It is not known which cells of the connective contract; perhaps they all do. Coiling of the axon is not evidence that this process does not contract, but only indicates that it cannot shorten so fast as the rest of the connective.

After axotomy axons in connectives of *Galleria mellonella* degenerate distal to the lesion, and the appearance of the degenerating axon has been described (635, 2963). During metamorphosis some axons have a similar appearance (2964); perhaps they are degenerating. Late in metamorphosis connectives contain many cell processes which might be growing axons. In the connective between the meso- and metathoracic ganglia the number of axons seen in cross-sections increases sevenfold early in metamorphosis. It is probable that the great shortening of this connective and the considerable coiling of axons within it cause many of the axons to be cut and counted more than once in a cross section. Glial cells hardly react to axons which degenerate within connectives after injury; they change in a striking manner during metamorphosis. Cells lying peripherally in the connective develop prominent cisternae of endoplasmic reticulum. This is consistent with active synthesis during the formation of a new basal lamina. If the dense bodies and myeloid bodies seen in these cells result from phagocytosis or from autolysis, it is a pattern of reaction that differs from that which follows axotomy. The pattern of wrapping axons in glial cells is also different at the end of metamorphosis than at its start. It is not clear whether this is a result of axonal death or growth or of primary remodelling of the neuroglia, or a result of both of these processes occurring in concurrent harmony, but independently.

Changes in the ganglia have not been studied in detail, but are likely to be extensive.

If the neurons which come to innervate some muscles of lepidoptera are removed, these muscles fail to develop (1623, 2141, 2142, 2143). The entire central nervous system has been removed from silkworm pupae (3134); these pupae develop into wholly flaccid adult moths, which are in other respects normal. Fully differentiated muscles which normally survive metamorphosis regress during this process if they are denervated in the larva (899, 900, 2400, 2401). It has been proposed that the muscles regress when efferent nerve activity ceases (1822). There is, however, evidence for considerable regression or structural disarray of these muscles at a time when they are still responding electrically to natural stimuli (1822). Loss of excitation does not therefore necessarily precede the regression.

Plasticity of the mature brain

Slow anatomical and biochemical changes occur in the brains of adult rats kept in different environments. The rats also behave differently (198, 464, 690, 695, 1156, 1644, 1645, 1646, 1788, 1789, 2493–2502).

Rats experiencing an enriched environment are housed in groups of ten or twelve. The cage is of wire mesh in a well lit room, and contains ‘toys’ such as ladders, wheels, boxes and platforms. The rats may also spend half an hour daily exploring a large open cage with partitions that can be varied, and half an hour receiving training by conditioning. Not all these features are necessary to evoke the characteristic changes. In contrast, rats experiencing an impoverished environment are housed singly in cages with translucent walls; they have no toys. The room is quiet and dimly lit. After thirty to eighty days in these different environments the brains are examined.

Published photographs show the limits of the regions of brain most studied (1653). The cerebral cortex of a rat exposed to an enriched environment is thicker, and a slice of constant thickness weighs more (2499). This increase is greatest (6%) in the occipital region, and least (2%) in the somesthetic region (2493). Layers II and III are affected most, layer I not at all (692). The enriched brain does not contain more neurons; they are more widely separated. The dendrites of pyramidal cells of layers II, IV and V and the stellate cells of layers II and IV branch more; the numbers of dendrites of the 4th and 5th orders are increased, but those of the 1st and 2nd are not (1378, 3024). The density of spines upon the basal dendrites is greater, but is elsewhere unchanged. Taken together, these observations suggest that the total number of spines is greater for dendrites of each order, and that each neuron receives a greater total number of boutons ending upon spines. Electron microscopy of layer III of the occipital region shows a smaller number of boutons, which are of greater diameter (2083, 2502). The counts of boutons are not necessarily incompatible with those of spines. When the functional state of a neuron changes, synapses ending upon different parts of the cell behave differently, increasing in some parts and decreasing in others (563, 564). Some boutons may also terminate upon the shaft of a dendrite. The medial hippocampus is also thicker in rats reared in an enriched environment (3040).

There are more glial cells in the cortex of an animal that has experienced an enriched environment (690, 693). Their type has not yet been established, and their source is uncertain. Such rats studied 7 days after receiving tritiated thymidine have more labelled cells in the optic radiation but not in the cortex (68). These cells might be migrating to the cortex from a presumed site of origin in the periventricular zone beneath the ependyma (Chapter 1). If so, it suggests that the glia of the cortex cannot under these circumstances divide to increase their number. The further movement, differentiation and fate of these cells is not known. It seems probable that the increased number of glia and the possibly increased rate of turnover are related to similar changes observed in discrete groups of neurons when they are stimulated physiologically (3067), or alternatively to changes seen under circumstances of known synaptic remodelling (3068). The visual cortex of rats which have been exposed to an enriched environment have fewer capillaries per unit volume; this is a result of the increased thickness of the cortex. The capillaries are larger (692). There are no values reported for the cortical blood flow or tissue pO_2.

The acetylcholinesterase and cholinesterase activities of the occipital cortex are greater in animals which have experienced the enriched environment (2499). The functional significance of these changes is uncertain. Acetylcholinesterase is by no means confined to cholinergic synapses; cholinesterase is present in capillaries and some glial cells.

Rats which have experienced an enriched environment have less exploratory activity (1908). Their greater capacity to solve problems (1279) is not due to this diminished activity (2749). They perform no better than their impoverished littermates in simple problems involving light-dark discrimination, but much better in more difficult problems (1653). This would be expected if the nervous system is responding with better discrimination to sensory information of which it has had greater experience. Both impoverished and enriched rats have received experience of light and dark. Related studies have shown that if a rat experiences stimuli of relevance to a future problem, it becomes predisposed to react to appropriate cues. It is very probable that this predisposition is caused by changes resembling those found in the visual cortex of the kitten when its visual experience is closely controlled (945) (p. 178). Conversely early experience of insoluble problems impairs later learning; this is consistent with the great care that must be taken to re-educate visually a kitten that has had a limited visual experience; the steps of advance must each be small and great encouragement is needed. Even so, the improvement is small. Rats exposed to an enriched environment forget a learned avoidance response more rapidly than do rats kept in an impoverished state (2219).

In the initial studies, the environment was enriched in all the ways described above. Not all are necessary. Rats need not be in such an environment for more than two hours each day. A total period of 30 days is sufficient (1653). Formal conditioning of behaviour is not needed. Rats need not be housed communally; a single rat in an enriched environment will show changes, provided it can be induced to explore its exciting surroundings adequately, and play with its toys (1969, 2495). Communal housing by itself without access to toys does not produce cortical changes. Although the changes are most pronounced in the occipital cortex, the experience does not have to be visual. Blind rats exposed to an enriched environment respond (1653). Cortical changes are produced by blinding, but the alterations are very different and do not interfere with the assessment of the influence of the environment. The cortex responds in rats of all ages after weaning (2439); it is most susceptible in rats about 55 days old (2147). Although it has been shown that the cortical changes following enrichment can be produced at any age, the extent to which the changes can be reversed if such a rat is returned to an impoverished environment is uncertain.

Mice and gerbils are similarly susceptible (1295, 1296, 1297, 2494). Strains differ in the degree of their response.

The cerebral responses to an enriched environment are not due to 'stress' or to an altered rate of maturation (2440, 2498). They do not depend upon hormonal events mediated through the pituitary (691, 1149, 1150, 2496, 2498).

Plasticity of experimental epilepsy

Epilepsy can in some situations be considered as a learned response. The nature of the plastic changes that cause continuing interictal after-discharges or overt seizures in animals with experimental epilepsy is unknown. The primary effect of the large number of different agents which can produce an epileptic focus in a given part of the brain of a given species is unlikely to be identical; such agents include cooling, irritants, antibiotics, hydrolytic enzymes, and a wide range of drugs which act upon neurotransmission or as antimetabolites (863, 1097, 2050, 2051, 2340, 2796, 3058). If a primary focus is produced by one of these methods, its rate of development and irradiation cannot be predicted with confidence. The rate of development of secondary foci, and their features are similarly variable (2050). Secondary foci often emerge in a region of brain that receives a projection from the primary focus; such a region receives a pattern of stimulation that is abnormal. After a period of weeks or months the secondary focus can become autonomous. It will then continue to discharge after the primary focus has been removed. Although the term 'mirror focus' has been used to describe the secondary focus formed by the projection to a symmetrical site in the opposite hemisphere, secondary foci can also be formed through pathways other than the corpus callosum, and in subcortical nuclei (1173, 2051, 2343, 3028). A secondary focus is not found opposite a primary cortical lesion in a slab of cortex, isolated except for its callosal projection (1089, 2050). This might indicate that afferent connections other than those derived directly from the primary focus are required. Alternatively, the failure might result from axotomy of the pyramidal cells, which might alter the synaptic connections received by these cells (Chapter 7).

The continuation of an abnormal pattern of discharge in the secondary focus after the primary focus has been removed indicates that the brain has 'learned' to behave in this manner. It is possible but not certain that the biochemical, anatomical and physiological changes of such learning occur primarily within the secondary focus. It is not clear that the abnormal discharge of a secondary focus can be considered to be a learned *response*; if it is, the afferent pattern of discharge that characteristically produces it is not always apparent. It is far from clear whether abnormal patterns of discharge which are not obviously triggered by a defined stimulus should be viewed as arising spontaneously from the basic biochemical or structural abnormality, or as being generated by intrinsic and autonomous patterns of activity which can be perpetuated in such an abnormal zone. A focus secondary to one produced in the manner described above is unsatisfactory to study as a model of plasticity, because the altered pattern of stimulation to which it is being subjected is not known, and may vary considerably in rhythm, intensity, duration and frequency.

A method that permits closer control of these features is the production of an epileptic focus by the process of 'kindling' (1093, 1094). An electrode is implanted in the cortex, and the brain is stimulated for several seconds once daily, using a current that is too small to produce an afterdischarge, a fit or other behavioural change. After several days this intensity of current causes a local after-discharge, associated with increased activity in the fibres passing from the stimulated zone to other parts of the brain; later, behavioural changes and fits also occur. The latency of the response and the rate of its development differ from one part of brain to another. The enhanced susceptibility to afterdischarge persists after the period of daily stimulation has ended; some change has occurred in the brain which is either irreversible or only slowly reversed. A very small lesion made at the site of stimulation abolishes the enhanced susceptibility. In different studies (3046), repeated subthreshold stimulation has also been found to produce behavioural changes. Secondary foci can develop, and the complexity of the afterdischarge increases as these appear (2361–2364). The reduced excitability which follows an afterdischarge or a fit, postictal depression, is different in a number of respects (1308, 1309, 1551, 2956). Such depression seldom lasts more than 36 hours and usually much less; enhanced susceptibility persists for weeks without reinforcement. The depression must be preceded by an afterdischarge or seizure; the enhanced susceptibility progresses in the early stages without producing an afterdischarge at all.

When penicillin is applied locally to the brain, seizure discharges develop rapidly. Both recurrent excitation and inhibition are enhanced (118, 696, 697, 698, 1940, 2338–2341). Penicillin is known to diminish the efficacy of some inhibitory synapses (604, 634, 746). Many nerve cells must be affected before a local afterdischarge can be elicited. It is likely that the minimum number depends upon the internal connections of the part of the brain affected, and upon the density and nature of its projections to other parts (1912).

Afterdischarges can be produced in isolated slabs of neocortex perfused *in vivo*. These develop in response to electrical stimulation or drugs. The gross architecture of the dendrites of the pyramidal cells is not changed by such isolation; the severed axons develop many collateral branches which ramify within the isolated cortex. The nature of the boutons formed, if any, by these collaterals has not been established. The development of afterdischarge in an isolated slab of cortex can be suppressed by frequent electrical stimulation; the intensity of stimulation required is less than that required to produce an afterdischarge; this process might be related to the 'kindling' described above.

The ability of neuroglia of an epileptic focus to buffer changes in the extracellular concentration of potassium is not reduced, but might be enhanced (1092). During ictal discharges the resting potential of glial cells falls (696). As these act as potassium electrodes (Chapter 1), the extracellular concentration of potassium ions is rising. After each discharge the glial cells repolarize. If the discharges are frequently repeated the resting potential of glial cells sometimes does not return to a normal value. It seems certain that this failure is a consequence of the persisting increase in the concentration of extracellular potassium ions rather than a primary defect of the neuroglial cells which causes the seizure. Glial cells within an epileptic focus are commonly large. While this hypertrophy of the astrocytes is undoubtedly often a direct consequence of the pathological process which caused the focus to develop, such hypertrophy might also be a reaction to the abnormal neural discharge. Astrocytes hypertrophy around active nerve cells.

Experimental epilepsy clearly raises many problems related to plasticity of the nervous system, even though it is itself an abnormal manner of excitation. The division of its possible primary cause into two alternatives 'changes in cells' and 'changes in circuits' clearly reflects similar controversy in adjacent fields, for example, that of learning (p. 159). It is also clear that speculation surpasses data in both fields, and that it is almost certainly false to consider these as alternatives; more probably, they are complementary. It is possible, for example that the development of giant postsynaptic potentials within either a primary or a secondary focus is due to heterosynaptic facilitation of activity that is frequently concurrent (p. 199). The mechanism of such facilitation is not known, but it almost certainly results from, and leads to, metabolic responses in the postsynaptic cell.

As in related fields, it is necessary and difficult to distinguish slow changes that result from particular patterns of stimulation from those related more generally to use, or in this case also from injury. Reported changes in protein and nucleic acid metabolism are as uninformative in this context as in others.

Homosynaptic plasticity

In homosynaptic plasticity the altered synaptic behaviour is by definition limited to the synapse affected by the agent or state investigated. Several changes take place in a synapse following use.

Posttetanic potentiation (761, 769, 772, 774, 878, 3077) is closely related to posttetanic hyperpolarization of the affected presynaptic terminals. Hyperpolarization increases the size of the action potential and so augments the release of transmitter from the nerve terminal (1219, 1419, 1809).

Fatigue follows prolonged stimulation. The number of synaptic vesicles within a neuromuscular junction (1510, 1511) and in preganglionic sympathetic nerve terminals (978, 1245, 2353, 2354) is then less than normal; the circumference of the terminal increases but its area does not (2353, 2354). The area of membrane lost by the synaptic vesicles is similar to that gained by the plasma membrane. If a neuromuscular junction is depolarized *in vitro* by 20 mM potassium chloride, fewer vesicles are found in the terminal, both adjacent to, and remote from the plasma membrane (1418). Synapses examined by freeze-etching (2277, 2278) are changed by general anaesthesia (2842), and the changes might be due to a reduced release of neurotransmitter (3078). The synaptic membrane is 'lifted' less, and fewer stomata are seen, which very probably are formed by vesicles communicating with the surface. Although the size of the presynaptic grid is different in different synapses, activity has not yet been shown to alter it; nor have definite changes in coated vesicles been seen (1131, 1529, 2514). If a nerve terminal is bathed in a medium that contains an exogenous protein marker, the number of vesicles labelled increases after stimulation; it still remains remarkably low, however (1383).

The formation of new vesicles is not tightly coupled to the synthesis of neurotransmitter. If acetylcholine synthesis is reduced by hemicholinium, the sympathetic junction becomes fatigued much more rapidly, but the rate of depletion of vesicles is not greatly reduced (2353). Fatigue also causes striking changes in the mitochondria of presynaptic terminals (1510, 1511, 2214); these are reversible. They may reflect either the increased rate of oxidative phosphorylation, or sequestration of considerable quantities of calcium which has entered the presynaptic terminal during repeated depolarization.

Some poisons also release transmitter from nerve terminals and reduce the number of synaptic vesicles. The venom of the black widow spider (1833, 2201), β-bungarotoxin (489), and the venom of the South American rattlesnake (331) act in this way. The depletion of acetylcholine at the neuromuscular junction follows a period of greatly augmented discharge of acetylcholine, indicated by the increased frequency of miniature end-plate potentials. This action of the black widow spider is not prevented by removing calcium from the extracellular fluid, or by raising the concentration of magnesium.

Heterosynaptic facilitation

Interaction between the synapses of different nerves can be demonstrated in the abdominal ganglion of *Aplysia* (1531, 1532, 2898). A weak stimulus applied to one pathway produces an EPSP; this is the 'test' stimulus. It precedes by about 300 ms a 'conditioning' stimulus, which is applied to a different nerve and causes the postsynaptic nerve cell to discharge. On repetition, the conditioning stimulus may facilitate the EPSP produced by the test stimulus, so that it becomes larger. This occurs only in some cells of the ganglion; it is not identical in all the cells in which it is found. In unidentified nerve cells facilitation affects only those pathways in which the test stimulus was paired with the conditioning stimulus; EPSPs resulting from stimulation of other pathways are not facilitated if these have not been paired with the conditioning stimulus. A less specific form of facilitation is found in the right upper quadrant giant cell; test stimuli are given at different times to two different afferent pathways, and only one is coupled to a conditioning stimulus. Both the paired and the unpaired test stimulus are facilitated. In both situations facilitation is greater when a stronger conditioning stimulus is given. The facilitation once produced lasts up to 40 min. The capacity of a ganglion to show heterosynaptic facilitation is susceptible to fatigue and habituation.

This facilitation can only be due to interaction between presynaptic terminals, or between postsynaptic changes of potential induced transsynaptically. Neither alteration of the passive membrane properties of the postsynaptic cell, nor the generation of action potentials within it by direct electrical stimulation produces facilitation. One pathway in which facilitation can be demonstrated is monosynaptic, judged by electrophysiological criteria; the facilitation cannot therefore result from interaction between interneurons common to each path. It is likely that the mechanism is one of facilitation and not of reduced inhibition; most inhibitory synapses in the abdominal ganglion are cholinergic, and are blocked by *d*-tubocurarine. Heterosynaptic facilitation is found in the presence of this drug. The mechanism of heterosynaptic facilitation is uncertain, and the primary interaction could be taking place either between the presynaptic terminals themselves, or in the postsynaptic cell. Interaction between presynaptic terminals would resemble the process of presynaptic inhibition in some respects, and the prolonged effect could be due to persistence of the neurotransmitter or to prolonged biochemical consequences of its biochemical action. Alternatively biochemical changes initiated in the postsynaptic cell as a result of transsynaptic stimulation (Chapter 5) might interact to produce such electrophysiological facilitation.

7 RESPONSES TO INJURY

The manner of reaction of the cells of brain to injury illustrates aspects of their interaction, which are not otherwise obvious.

Degeneration and regeneration of axons

Vertebrates

In vertebrates and in many, but not all, invertebrates an axon dies after it has been separated from the cell body. The sequence of degenerative change occurring in the isolated axon is called Wallerian degeneration (1514). The rate of degeneration varies according to species, age, temperature and fibre size. Not all the fibres of a given size present within one severed nerve degenerate at the same rate (1016, 1514, 2938).

Polymerized microtubules and neurofilaments disappear within a few hours after unmyelinated or small myelinated axons have been divided, and the cytoplasm contains a flocculent deposit of increased electron density (1389, 1740, 2102, 2103, 2104, 2164, 3009). It is probable that this flocculent deposit contains depolymerized tubulin and neurofilament protein (Chapter 2), which has been precipitated either *in vivo* or by fixation. This deposition seems to be less in large axons. It is likely that this early change results from calcium entering the axoplasm, so that its concentration rises; the concentration of calcium within an axon is normally very low, about 0.3 μM in the giant axon of a squid (124). Microtubules do not polymerize in the presence of a high concentration of calcium ions (3101), and tubulin is precipitated under these conditions (3103, 3143). An excised nerve incubated *in vitro* degenerates more slowly when kept in a calcium-free solution, or in a medium containing a chelating agent, and more rapidly than usual in a solution containing a concentration of calcium that is greater than normal (2589, 2591, 2593). Degeneration takes place much more rapidly in the presence of substances which limit oxidative phosphorylation, but only if calcium ions are present in the medium; it therefore appears that the degenerating axon sequesters or extrudes a significant fraction of the calcium which enters (2592). Calcium could enter either at the site of injury or through the membrane of the distal stump of the axon. If the time course of an injury potential indicates the rate at which a membrane seals the end of a severed axon, the half-time of the process is about 90 min; within this period calcium ions could presumably enter and cause the disruption of the microtubules and neurofilaments, a process that leads to the aggregation of organelles. No information is available about the composition or permeability of this newly formed membrane. Calcium is needed for the sealing of the membrane of erythrocytes. It seems to be unnecessary to postulate that calcium ions enter the axon more readily through the uninjured membrane of the separated axon. The possibility could be studied by measuring the rate of degeneration *in vitro* when only the severed end is bathed in a calcium-free or in a calcium-rich medium, while the rest of the axon is bathed in a physiological solution. The actions of cobalt or manganese, or of tetrodotoxin could be studied; these agents block the entry of calcium through calcium channels and through sodium channels respectively. The acceleration of degeneration by stimulation (8, 550) might result from an additional inflow of calcium, this time entering through the ionophores of the uninjured membrane (124).

Various organelles pass along axons (Chapter 2). Much of this traffic is closely associated with microtubules. Some axoplasmic transport can still continue for several hours after axotomy. This has been shown in three ways. First, a neuromuscular junction continues to release acetylcholine after the motor nerve has been divided; it does so for longer if the length of the distal stump of nerve is greater (1997, 3008). Second, in isolated segments of nerve, noradrenaline and acetylcholinesterase accumulate at the ends of the isolated segment (613, 614, 1840). This accumulation, which is more marked at the distal than at the proximal end of an isolated segment, is not due to local synthesis, because the total quantity of noradrenaline, and the total activity of acetylcholinesterase of the whole isolated segment does not alter. Third, an axon in tissue culture can continue to extend distal processes for several hours after isolation from the nerve cell body (1778, 2443).

After the axon has been injured, the axoplasm immediately adjacent becomes granular (3235). Organelles accumulate near the severed end of the isolated axon. The pattern of accumulation is fairly constant, and the end of the degenerating axon has been classified into four zones (Fig. 7.1). The first, next to the injury, initially contains granular and disrupted cytoplasm, but later is clear and watery. This zone is usually about 100 μm to 500 μm long. In the second zone organelles accumulate: particularly mitochondria, lysosomes and fragments of endoplasmic reticulum (328, 714, 1740, 3086, 3235). Vesicles resembling synaptic vesicles are seldom seen (1536). Within this zone the mitochondria are not usually arranged with their long axes parallel to the long axis of the axon, but more variably. The mitochondria are fatter and shorter than normal, and their cisternae are fragmented. It is possible that these changes result from increased sequestration of calcium ions as the concentration in the axoplasm increases. Oxidative phosphorylation would diminish as their fragmentation progresses, and lysosomal enzymes might be activated by the rising concentration of hydrogen ions. The third zone is also between 100 μm and 500 μm long; mitochondria, lysosomes and fragments of endoplasmic reticulum accumulate beneath the axolemma, leaving empty the central core of the axon, except for neurofilaments. The organelles are less tightly packed than in the second segment and are better ordered. The mitochondria lie parallel to the long axis of the axon, and in longitudinal sections of

Figure 7.1 Diagram of the distribution of organelles at the proximal end of the distal stump of a divided axon to show the four zones. (1). The zone of clear cytoplasm next to the injury. (2). The zone of accumulated organelles with random orientation. (3). The zone of accumulated mitochondria which are longitudinally oriented. (4). The zone within which organelles accumulate at nodes of Ranvier.

the axon they often seem to lie in trains. The fourth zone is between 5 mm and 10 mm long. In this, mitochondria and lysosomes accumulate near nodes of Ranvier. The remainder of the isolated segment of axon has no accumulation of organelles, and may have fewer than normal. Most evidence indicates that this concentration of mitochondria results from redistribution rather than from synthesis. It is therefore unnecessary to postulate that they arise from the adaxonal cytoplasm of the Schwann cell (2716). The source of the lysosomes is less clear; they may be derived from the fragments of endoplasmic reticulum. The formation of lysosome-like structures from fragments of endoplasmic reticulum has been described in axon growth cones. Hydrolytic enzymes can be demonstrated more readily in an injured axon than in a normal one (18, 372, 386, 1384, 1868), and the activity of protease increases.

This appearance of the end of the injured axon is probably due to the entry of calcium ions. The calcium disrupts the microtubules necessary for the movement of organelles. The concentration is likely to be greatest at the severed end and by the adjacent nodes of Ranvier. Organelles are therefore arrested at these sites.

Organelles also accumulate at the proximal end of a segment of axon isolated between two lesions. These are either organelles which are passing centripetally along the axon, and are uninfluenced in this passage by the injury, or they are induced to move in this direction by the injury. There is some evidence for each of these views. Organelles can be seen passing in both directions along an axon of a cultured neuron, and bidirectional transport has been repeatedly demonstrated *in vivo*. The movement of organelles is influenced by alterations in the ionic composition of the external medium, which alter the injury potential, and it has been suggested that some movement of organelles occurs by electrophoresis. It is necessary, however that the influence of these alterations upon the integrity and activity of the microtubular transporting system should be studied. It seems very improbable that movement by electrophoresis is a significant part of the story, as the time course of the injury potential is totally different from the time course of the accumulation of organelles.

It seems probable that the distal axon dies after axotomy because many of its mitochondria accumulate at the injured end, while others continue to pass distally. The centre is therefore left deprived.

Oxidative phosphorylation would then fail, sodium and water would enter, and the sequestration of calcium would be disordered. A proximodistal sequence of degeneration would be expected (1514). The axon of a peripheral nerve can degenerate and disappear within 48 hours (3009). Three factors may contribute to its removal: hydrolysis by enzymes activated within the axon, phagocytosis by the adaxonal cytoplasm of reacting Schwann cells, and phagocytosis by invading macrophages derived from blood monocytes (2180). Unmyelinated axons are phagocytosed at least in part by Schwann cells (328, 1470).

The Schwann cells of myelinated nerve fibres alter within minutes of nerve transection. When isolated living fibres are gently teased apart, the Schwann cells, the nodes of Ranvier and the Schmidt–Lantermann clefts can be seen (Fig. 7.2). These clefts widen and narrow as cytoplasm enters and leaves them (2710). As they seldom open more than once every eight hours this movement cannot be identical with pulsations reported in these cells in culture (829). Myelin around the processes of cultured dorsal root ganglion cells also moves. Within minutes of axotomy the Schwann cells retract from the nodes, so that these become wider (3136). This occurs within several millimetres of the site of injury but not more remotely. This distance is about the same as that over which nodal accumulations of organelles are found within the axon, and might be the distance over which the injury potential influences the flow of ions through the node. Rather later the Schmidt–Lantermann clefts

Figure 7.2 Diagram of the distal end of the proximal stump of an axon, divided several hours previously, to show the retraction of myelin at the nodes and the greater and more prolonged opening of the Schmitt-Lantermann clefts. *Top*: a normal axon. *Bottom*: after axotomy.

become more mobile, larger and more frequently seen (3087, 3136). Much of the Schwann cell cytoplasm becomes interposed between the layers of the myelin sheath and between the myelin sheath and the axon; the volume of the adaxonal layer increases considerably. The widened Schmidt–Lantermann clefts split the myelin sheath into segments which become ovoids of myelin with a poor preservation of the lamellar structure. These ovoids are ultimately absorbed into the Schwann cell cytoplasm. During this degeneration of myelin cholesterol is converted into cholesterol ester. Cholesterol which is laid down early during myelination is esterified late during Wallerian degeneration; that which is laid down late is esterified early. This suggests that the myelin that is laid down early is removed late (2699). It is probable that this indicates that those axons in a nerve which gain their myelin sheaths early lose them late; it seems less probable that within a given myelin sheath, myelin laid early is degraded late. This latter interpretation would require too great a biochemical stability (Chapter 3). The disappearance of degenerating myelin is not due to phagocytosis, but is rather a process in which the Schwann cells readjust the area and composition of their own plasma membranes. Very little myelin still persists two weeks after a peripheral nerve has been injured (1853); this differs from the situation in the central nervous system (p. 220). The cholesterol of myelin, and perhaps other constituents of myelin also, is stored within the Schwann cell, for it is used again when the cells lay down new myelin sheaths around regenerating fibres (1280, 1281, 2414, 2417). This makes most unlikely the hypothesis that the Schwann cells which myelinate regenerated axons are formed from mesenchymal cells (1819, 1820). The Schwann cells of unmyelinated nerve fibres also react (328, 1470). There is no substantial evidence yet obtained from vertebrates that a Schwann cell sustains a peripheral axon either normally, or after axotomy.

Macrophages derived from blood monocytes invade a degenerating peripheral nerve after axotomy. The capillaries become more permeable. It is not certain how these changes are brought about. The Schwann cell reaction cannot be due to an increased concentration of plasma proteins within the nerve, as the reaction occurs *in vitro* also, in a medium of reasonably constant composition.

After an axon has been divided, the reaction is not confined to the distal severed part. The proximal stump also reacts. It becomes greatly distended with material that is not synthesized locally (1654), but is derived from the cell body. The end of the proximal stump can be divided into zones similar to those described above (Fig. 7.3). In the proximal stump, however, unlike the isolated segment of a nerve, many vesicles accumulate which resemble synaptic vesicles. Noradrenaline accumulates here in noradrenergic axons, and acetylcholine in cholinergic axons. The myelin sheath surrounding the distended proximal stump has fewer lamellae than normal (966). The total area of myelin membrane around the distended axon does not alter (Fig. 7.4). This change of geometry of the myelin sheath probably results from slippage, a process in which the myelin lamellae can slide past one another.

Between 1 and 15 days after axotomy sprouts are formed at the proximal stump (44, 328, 754, 1182, 2408, 3218, 3235). These do not arise from the tip but from a short distance behind it, in the zone of accumulated organelles. Although there is little information about sprouting *in vivo*, it has been studied in some detail in cultured growing nerve cells. As the growth of axons *in vitro* resembles closely the way in which other cells extend processes and move (1446,

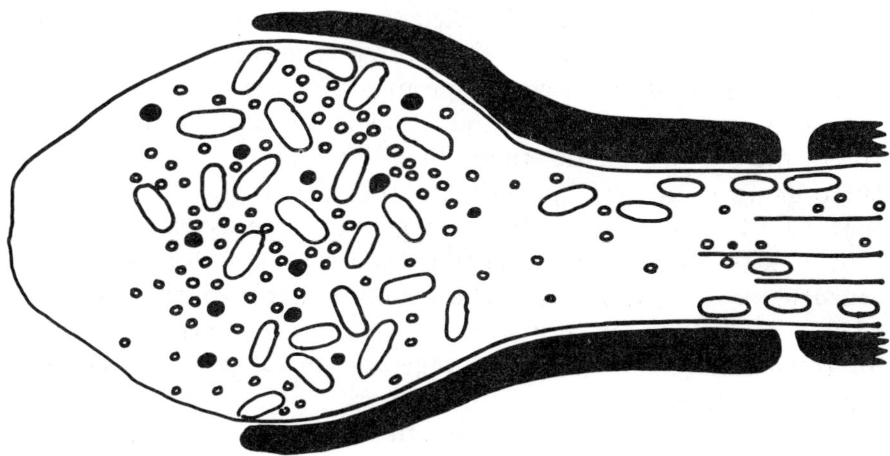

Figure 7.3 Diagram of the distal end of the proximal stump of an axon to show the swelling of the axon and the accumulated organelles. In addition to mitochondria and lysosomes many synaptic vesicles are present. The zones of accumulation resemble those at the proximal end of the distal stump.

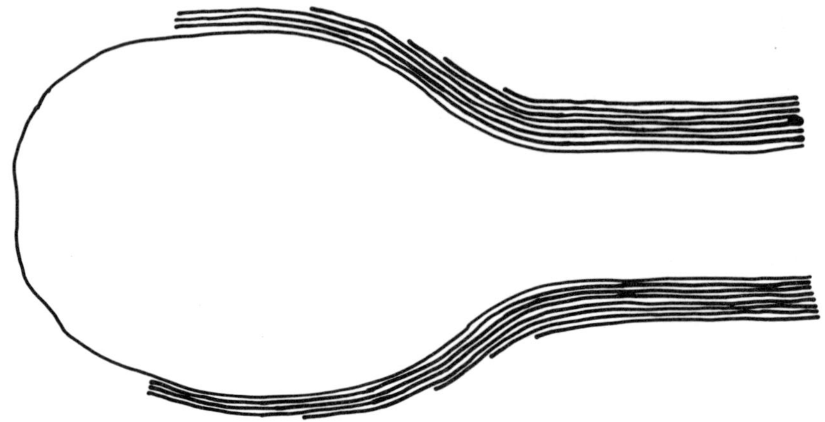

Figure 7.4 Diagram of the myelin sheath around the swollen axon in the zone of accumulation of organelles. The number of myelin lamellae is reduced around the swollen axon. The total area of the myelin sheath is unaltered.

1447), it is likely that this mechanism is responsible also for the extension of sprouts *in vivo.*

Dissociated sympathetic ganglion cells extend processes in culture; their growth cones contain mitochondria, tubules of endoplasmic reticulum, and a variety of vesicles: coated vesicles, dense core vesicles, arrays of elongated vesicles often arranged in columns, and lysosomes (385, 386). When the sprout is first formed many of these organelles come from the accumulation in the axon stump (3088, 3089); the swelling of the stump diminishes as this occurs (Fig. 7.5). Microfilaments of actin lie beneath the membrane of the growth cone (3201, 3202). In culture the growth cone extends in direct contact with the supporting medium, and is seldom in contact with neuroglia or other cells. Neuroglia are not essential for the extension of processes or for the formation of synapses; they may be essential for regeneration, that is for the organization of the cells and processes into a tissue.

The surface of a growth cone has filipodia or a ruffling membrane (2088). A model of the growth cone has been proposed to explain the extension of filipodia at the front of the growth cone, their passage backwards across its surface, and their reabsorption at its back (324, 325, 326) (Fig. 7.6). It is suggested that the inner surface of the plasma membrane contains oriented myosin; oriented myosin

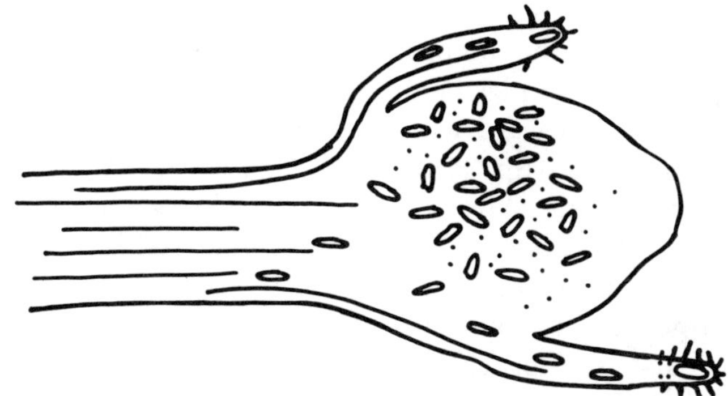

Figure 7.5 Emergence of sprouts from the injured end of an axon. They arise behind the tip of the axon, usually just proximal to the accumulated organelles. Each sprout ends in at least one growth cone.

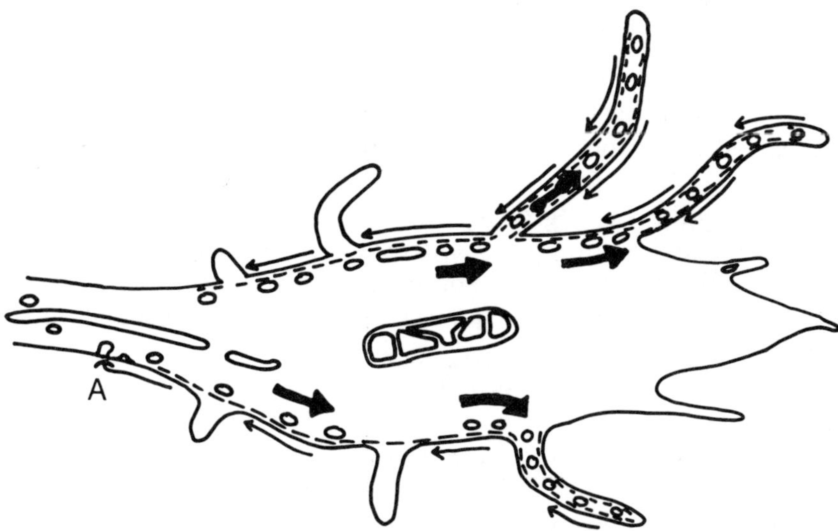

Figure 7.6 Mode of movement of membrane at a growth cone (324, 325). Vesicles are formed at 'A' by pinocytosis. They pass distally within the growth cone, as shown by the broad arrows. Filaments of actin lie between the vesicles and the inner surface of the plasma membrane. These vesicles fuse with the plasma membrane at the extending tips of the process. The plasma membrane itself moves backwards, as indicated by the thin arrows. If the axon is extending, fresh membrane is added to the system from tubules within the axon and from vesicles passing down the axon. The backward movement of the membrane would make the growth cone pass forwards if the membrane adheres to the substratum.

of opposite polarity is present on the outer surface of the vesicles. With such polarity the myosin can react with the actin filaments which lie between the plasma membrane and the vesicles, so that the vesicles are driven forward within the growth cone, along the lumen of a filipodium to its tip. Here the vesicle is everted and fuses with the plasma membrane. After eversion the myosin that was upon the outer surface of the vesicle is now lining the freshly formed plasma membrane, and is oriented in the same direction as the myosin of the membrane; hence the next vesicle can pass to the tip of the extended filipodium, and the process can be repeated. At the same time membrane is being removed from the back of the growth cone by pinocytosis as vesicles are formed. The site at which membrane is pinocytosed is determined by the polarity of the myosin. Such a site can be represented as a region of radial divergence of myosin, while a site of membrane formation can be considered as a region of myosin convergence. As membrane is added to the front of the growth cone and removed at the back, the overall direction of the movement of the membrane is backwards. This is compatible with the observation that particles on the surface of a growth cone (324) or ruffling membrane (1464) pass backwards. The constant position of a branching point of a growing axon also indicates that the new membrane required for gross extension of the axon is deposited distally (326). Additional membrane probably passes down the axon in the form of vesicles, and joins trains of vesicles 'cycling' in the growth cone.

Actin and tropomyosin (887, 888) have been identified in brain in considerable amounts. The microfilaments beneath the plasma membrane of the growth cone have been shown histochemically to be actin (3200, 3201), but the myosin and tropomyosin have not yet been localized in this way. It is not yet known how calcium is released and sequestered within a growth cone; it might enter mitochondria, or the vesicles rich in cAMP and phosphodiesterase which pass along the axon (330). The regulation of sprouting at the cut end of an axon is not understood. The growth of sprouts *in vivo* can be influenced by charged dyes (1369, 1370, 1371); this study by light microscopy done some time ago deserves to be repeated and extended both *in vivo* and *in vitro* using electron microscopy and electrophysiology to assess regeneration. It is possible that in this situation the drugs act in a way similar to the action of similar drugs upon other cells to influence the rate of pinocytosis. Such an effect could change the rate of regeneration by altering the rate at which the membranous vesicles 'cycle'. The suggestion that nerve growth factor acts directly by increasing the activity of a growth cone is based more upon a failure to find alternative modes of action than upon direct evidence (3202).

The growth cone adheres more firmly than the perikaryon or axon to the substrate in tissue culture. *In vivo* the growing nerve sprouts extend upon the surface of Schwann cells. It is not known if a growth cone adheres differently to normal and to reacting Schwann cells. The reacting cells extend fine processes and form complex surfaces along which the axon extends (3137). The Schwann cell is an epithelial cell and the similarity between the response of the Schwann cells as a tissue to axotomy and the response of other epithelia to injury is close. In both situations there is division and migration of the epithelial cells, division of the fibroblasts, and invasion by macrophages; in both circumstances the epithelial cells can migrate as a sheet to bridge a gap (1057); in both, new basal lamina is formed between epithelial cell and fibroblast, and is deposited at the surface of the former. Apparent differences of behaviour between Schwann cells on the one hand and epithelial cells on the other, result from the geometrical differences rather than from fundamental dissimilarities. As in other epithelia considerable amounts of collagen may be deposited if healing is delayed, or if an extensive gap is to be bridged. This collagen can impede the extension of the axon and of processes of Schwann cells. If the synthesis of collagen is reduced by giving rats *cis*-hydroxyproline, the regenerated nerve contains more myelin (2302); this might indicate that the regeneration of the axon has been less hindered, and that the diameter of the axon is therefore greater. The difference in myelination is then secondary. On the other hand, as the Schwann cell is the cell that both forms myelin and precipitates the collagenous precursors of the basal lamina a more direct influence cannot be excluded. There can be no doubt that in the peripheral nervous system the architecture of the regenerating nerve is determined by the interaction between mesenchyme and epithelium (Chapter 1).

A number of estimates have been made of the centrifugal flow of material along a regenerating axon. The composition of axoplasm changes after axotomy (617), and this appears to be associated with a considerable change in the metabolic direction of the cell as a whole (p. 238). Observed changes in the rate of proximodistal movement are likely to reflect variation in the nature of the substances transported just as much as alteration in the rate of movement of given materials (427, 428, 979, 1656). Centripetal transport cannot be demonstrated with exogenous protein early after axotomy, but reappears about the time that neuromuscular contact is re-established (1658). It is then greater than normal for a period which coincides with early maturation, re-extension of the dendrites and restoration of presynaptic terminals to the axotomised nerve cell.

Several sprouts pass peripherally from the stump of an injured axon, but the number of sprouts diminishes later (327). It is not known how this reduction is initiated or achieved, but it is usually assumed to coincide with the success of one sprout in making contact with an end organ. Sprouting is regulated regionally in the nerve cell; unsuccessful sprouts which originate from the severed stump of the axon are withdrawn at the same time as new collateral sprouts are extended from the region of the successful terminal to innervate adjacent muscle fibres or other areas of postsynaptic surface and so form the compact motor unit seen after regeneration. A nerve cell can pass material preferentially along one of its processes, or along one branch of a process, in a dorsal root ganglion cell, for example, (79), in noradrenergic neurons (616), and in neurosecretory cells (1314). The successful sprout increases in diameter as the others are withdrawn. This maturation is not simply caused by the successful branch ceasing to grow longer, and so 'filling out' from the material which continues to pass into it; such maturation is not seen when the extension of regenerating axons is stopped in other ways (44, 45, 1392). It seems only to follow the recognition of the postsynaptic surface. The release of neurotransmitter is concerned in this recognition, but the evidence is indirect. First, if the release of acetylcholine from a nerve terminal is prevented with botulinum toxin, axonal sprouts emerge from the terminal and pass across the surface of the muscle fibres without being arrested, despite the occurrence of 'denervation' hypersensitivity and an expansion of the area of the cholinoceptive surface (740, 741). A regenerating axon similarly is unable to establish contact with a muscle in the presence of botulinum toxin. Second, if the action of released acetylcholine upon the postsynaptic cell is prevented, specific adherence and connections fail to develop. α-Bungarotoxin applied to cells of the optic tectum prevents ingrowing nerve fibres from making effective junctions; the growing axons pass the cells upon which they normally end without being arrested. It would therefore appear that a growth cone responds to interaction between the neurotransmitter that it normally releases and the postsynaptic responding cell. This can be reconciled with the failure to demonstrate neurotransmitter within a regenerating axon: if it is assumed that the quantity needed is small, and below the threshold that can be detected by the methods used, or if the neurotransmitter is being released by another cell in the immediate vicinity. The former seems perhaps the more likely, but the release of neurotransmitter in an atypical manner by Schwann cells is found at the denervated neuromuscular junction of some species, and neuroglial cells are capable of concentrating, synthesizing, and releasing some neurotransmitters. The responding postsynaptic membrane presumably increases the rate of the forward reaction of adherence, or slows the backward reaction of release (p. 175).

The conversion of a growth cone into a secreting synapse is likely to be a simple process. The model in which 'cycling' of the membrane occurs is very similar to the model of the synapse. In both, membrane of vesicles fuses with the most distal part of the axon, flows laterally, and is then taken up again as vesicles. The lipid constituents of the membrane flow past the fixed points of the synapse which stabilize the junction. The vesicles of the established synapse release more neurotransmitter, but this is likely to a difference of degree if the release of transmitter is a necessary part of the process of reaction between the growing terminal and the cell with which it makes contact. Both in the growth cone and in the synapse it seems almost certain that actomyosin is concerned in moving the vesicles.

As the regenerated axon increases in diameter, so the myelin sheath develops. Over a wide range of axon size, there is a linear relationship between the circumference of the axon and the number of lamellae of compact myelin surrounding it (969, 970). It is not known how this precise matching is achieved. After axotomy, the diameter of the axon immediately proximal to the site of injury swells rapidly. The total area of the myelin lamellae does not change, and so the number of lamellae decreases. Under these circumstances there is no increase in the amount of myelin formed. Even if the circumference of a constricted axon remains greater than normal for 30 days under conditions in which there is no accumulation of organelles, the area of the myelin sheath does not increase (967). When axons regenerate and mature the myelin sheath becomes thicker, but frequently it is observed that the myelin sheath fails to regain the area that would be appropriate for an axon of that diameter (567, 1392, 2553, 2612, 2613). The causal relationships between end-organ contact, axonal maturation and perikaryal metabolic change on the one hand, and myelination on the other, have not been worked out. It is interesting to note that if axoplasmic flow is disrupted by drugs (Chapter 2), the axons distal to the site of disruption remain myelinated; this suggests that the maintenance of the myelin sheath by the Schwann cell does not depend upon continuing centrifugal axoplasmic transport.

Arthropods

The fate of the axon isolated by injury is more varied in arthropods. Many degenerate rapidly (283, 744, 858, 1200, 1202, 1315, 1471, 1472, 1473). Others survive much longer. Two examples which differ from each other in several respects are crayfish motor axons, and some fibres of the connectives of locusts and mantids (300, 301, 2526).

Crayfish

The distal axon of the nerve to the opener muscle of the crayfish can survive for more than one year after separation from the nerve cell body (110, 254, 1410, 1411, 1577, 2134) (Fig. 7.7). The transected motor axons to the abdominal flexors usually degenerate rather more rapidly (1409); nevertheless the axons to the flexors can survive for 45 days. Miniature end-plate potentials persist for at least 60 days after axotomy in the opener muscle; the nerve remains excitable and the muscle does not atrophy. After this period the nerve conducts action potentials less well, and it is more easily fatigued, axonal conduction fails earlier than the release of neurotransmitter, and miniature end-plate potentials can still be detected when the nerve is no longer excitable. Electron microscopy of the distal stump of the axon, which was severed in the meropodite 6 months earlier, reveals no evidence of degeneration.

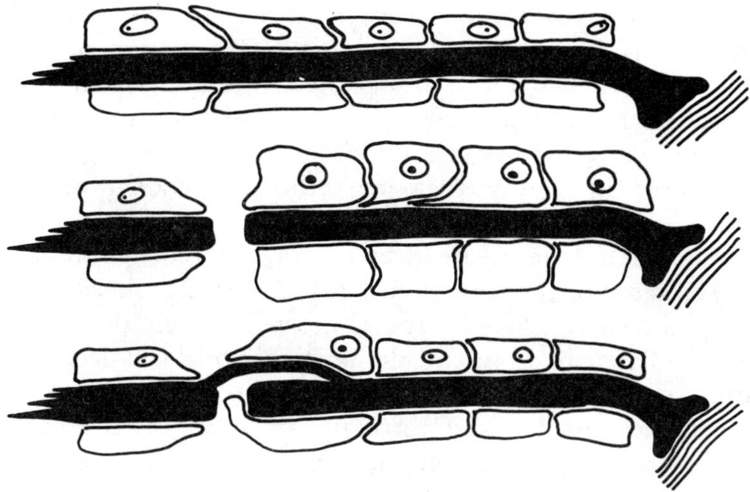

Figure 7.7 A motor nerve of an arthropod is surrounded by glial cells (top). When the nerve is divided, the distal isolated stump of the axon survives and the surrounding glial cells hypertrophy (middle). Sprouts are formed by the proximal stump; these establish contact with the surviving distal stump either by fusion or through electrotonic junctions (bottom).

Within 12 hours of injury the end of the axon is sealed with an intact membrane, and the end of the axon is swollen; it is swollen with organelles, including mitochondria, multivesicular bodies, dense bodies, large clear vacuoles, small vesicles and glycogen (2134). The duration of the injury potential in these nerves is not known, and the period not established for which calcium can gain easy access. It is also not known if the crayfish axon has an exceptional capacity to sequester or to expel calcium. It would be interesting to know if the longer survival of the axon is associated with an initially much smaller increase in the concentration of the axoplasmic calcium. The severed axon is not sustained by contact with muscle, because isolated segments of nerve divided both centrally and distally also survive. It would be interesting to know if such an isolated segment of axon would survive for a long time in culture, or if transplanted to another part of the crayfish. Within 48 hours of axotomy the glial cells around the distal axon hypertrophy. Mitochondria and smooth and rough endoplasmic reticulum accumulate within the adaxonal region of the cell. Although it is possible that these glial cells support the axon, this has not been proved; neither the cause of this glial change nor its consequences are known. The nature of any such possible 'support' has not been suggested. Alternatively the adaxonal accumulation might represent a process of reaction to degeneration that becomes arrested because the process of axonal degeneration is itself arrested in a way that might be independent of the glial reaction.

Giant axons in the connectives of crayfish remain excitable for 230 days after axonal transection (1501, 3148); they become thinner and conduct more slowly. Transmission between the atrophying isolated medial giant fibre and the postjunctional cell decreases. The proximal segment of the axon which retains contact with the cell body becomes swollen and tortuous. Its commisural electrotonic junction increases in efficiency, possibly because the swelling of the fibre reduces the intra-axonal longitudinal resistance.

The motor axon regenerates readily in the crayfish. Within 4 weeks of injury, axonal sprouts are embedded in the hypertrophied glial cells surrounding the surviving axon. Function is then rapidly restored. It is almost certain that the regenerating sprouts either fuse with the surviving distal axon, or communicate with the surviving axon through large electrotonic junctions (Fig. 7.7). The distal axon is known to be incorporated into the conducting path when it is re-established, because regeneration results in abrupt recovery of function simultaneously at all points along the path of the opener nerve, and not in proximo-distal sequence as would be expected if regeneration occurred only by a slowly growing axon, because only one excitatory junctional potential is found when the distal nerve is stimulated after regeneration, and because the regenerated nerve contains only the normal number of axons (1410).

Cockroach

The isolated axons of the cockroach do not survive for long. Peripheral motor nerves degenerate rapidly (2983); axonal conduction fails after 4 to 6 days, and miniature end-plate potentials can be recorded for a further three days. Giant miniature end-plate potentials do not occur. The axon has degenerated to a considerable extent within 2 weeks, but a few fragments can still be seen after 9 weeks (1200, 1202). The nerves regenerate with considerable precision but not with perfect accuracy (1202, 2236). Immature cockroaches recover the proper use of affected limbs better than adults. Because nerves can regenerate, both ganglia and limbs have been transplanted. Transplanted ganglia can innervate muscles of the limb if the motor nerve is divided (283, 1201, 1203, 1472, 1473), but the leg usually does not function well. It is difficult to study a transplanted ganglion electrophysiologically, as they become surrounded by a fibrous tissue capsule, and because distortion of the ganglion makes it difficult to recognize particular nerve cells. Rhythmical discharges have been recorded within such ganglia. Limbs can also be transplanted (3217). A muscle of a transplanted limb commonly receives the nerve which normally passes to the homologous muscle of the limb which has been removed. This indicates that specificity of regeneration, however it is attained in this species, may not be unique in all respects but is at least in part segmentally repeated.

Locust

Motor axons of the locust can survive rather longer than those of the cockroach. Evoked potentials remain for 9 days and sometimes for 24 days after axotomy, and spontaneous miniature end-plate potentials persist for at least 32 days. The miniature end-plate potentials are abnormal in frequency and amplitude (2979–2982). They are recorded as bursts of high frequency separated by periods of low frequency or of silence. During the bursts of high frequency discharge, giant miniature end-plate potentials occur. Electron microscopy shows the presynaptic vesicles are aggregated at this stage.

It is possible, but still uncertain, that isolated axons in connectives may persist (300, 301). The neck connective between the suboesophageal and the prothoracic ganglia contains several thousand axons, most of which are very small (2526). When this connective is divided, only 2% of axons on each side of the lesion die (Fig. 7.8). The remaining axons show only accumulations of the usual range of organelles, which probably result from transport to a site of arrest, as in the divided axons of vertebrates. Additionally about 2% of the surviving axons contain osmiophilic granules resembling neurosecretory material. If the connective is divided in two places, just behind the suboesophageal ganglion and just in front of the prothoracic ganglion, all the axons in the isolated segment degenerate. Some

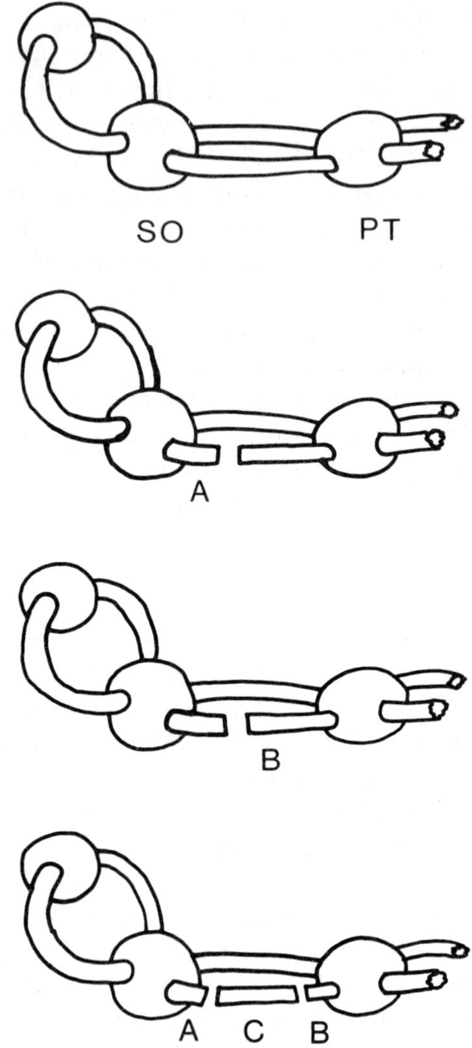

Figure 7.8 The suboesophageal ganglion (SO) and prothoracic ganglion (PT) of the locust are joined by paired connectives (top). If one of these is divided only 2% of the axons degenerate in segment 'A' (second from top). This would not be surprising if 98% of the axons of this connective had their cell bodies on the cranial side of the lesion. But if segment 'B' is examined (third from the top) again only 2% of the axons degenerate. If the connective is divided in two places, all the axons in segment 'C' die.

glial cells also die. The interpretation of these findings rests critically upon the position of the nerve cell bodies. If each axon has only one cell body, then the findings suggest that the presence of intact axons within the connective helps those which are isolated to survive. This interpretation has lead to the suggestion that the neurosecretory material may have a 'trophic' role, being released from intact nerve cells to support the severed axons. On the other hand, if each axon has two cell bodies, one rostral to the lesion and the other caudal to it, then both parts of the axon would be sustained after axotomy. Some giant axons are syncitia; the lateral giant fibre of *Leander* may have a cell body in each abdominal ganglion. Such an axon should survive on both sides of a transection; indeed if so its mode of regeneration would make an interesting study. It might be by fusion, as in the crayfish. It seems a little difficult to believe, however, that 98% of axons in the suboesophageal-prothoracic connective have at least two cell bodies each, one above and one below the lesion. Indeed, it is known that the most medial giant axon of the locust has only one cell body; this is in the terminal ganglion.

The cause of this survival of axons in the locust connective is not clear. No trophic effect has yet been proved and there is no evidence that the 'neurosecretory material' has any role other than its well established physiological one. Nevertheless, it remains an interesting possibility with superficial analogies to the situation during regeneration in *Hydra* (Chapter 8); it is worth further study.

Central nervous system of vertebrates

Wallerian degeneration is very varied in the central nervous system of vertebrates, and the cause of the variability is not understood. The capacity of neuronal processes to regenerate is different in different species, in animals of the same species at different ages, and in different tracts of the same animal. Most nerve tracts of the central nervous system of mammals fail to reconstitute their previous connections after injury. The reasons for this are far from clear, and might differ from one experimental model to another (1190). The models described here have been selected to illustrate different aspects of the problem, and to show the varied pattern of response to injury (505, 748, 1197, 1198).

Optic nerve of mammals

After an optic nerve has been divided most of the axons degenerate, and show many of the features observed in degenerating axons of the peripheral nervous system (551, 552, 1698, 1810), disappearance of microtubules, increased electron density of axoplasm, appearance of a flocculent precipitate, and sometimes also hypertrophy of neurofilaments. Beneath the axolemma of the degenerating optic nerve of the guinea pig many tubules appear (1698). It would be interesting to know whether these communicate with the exterior, and whether they are able to sequester or extrude calcium ions. The axons are removed more slowly from a degenerating optic nerve than from a degenerating peripheral nerve (241, 586, 587, 1868, 3000); it is not clear whether this is a result of a longer latency before degeneration begins, or whether the process of degeneration is slower. Some isolated segments of axon can be seen by light microscopy several months after axotomy; these have been called 'preserved fibres' (414). There is at present no evidence to indicate whether they are living or dead. Each hypothesis seems unlikely; if living, they resemble in some respects the surviving axonal segments of the crayfish; if dead, their resistance to phagocytosis is peculiar, which appears to affect the entire length of the axon between the lesions. It is possible that the axoplasm has been 'fixed' as a result of the massive entry of calcium; or perhaps the isolated segment lacks lysosomes or their precursors, a lack which might prevent an initial and necessary hydrolysis from within; it is unlikely that an external variable would affect the whole length of a given axon, for example, a lack of macrophages. A study of these fibres by electron microscopy might be interesting. Axons of immature animals are removed more rapidly; before myelination removal is very fast, and no axonal debris persists (551, 552, 1849, 1850, 2433).

The cause of slow removal of axoplasm in the mature animal's central nervous system is not precisely known. At least three extra-axonal factors are likely to be important: the greater stability of the surrounding myelin acting as a protective sheath, the smaller likelihood of phagocytosis by the adaxonal cytoplasm of an oligodendrocyte, and the very restricted access of macrophages from the blood to the degenerating tract, except at the site of the injury itself.

Myelin disappears much more slowly from an optic nerve than

from a peripheral nerve after axotomy (1853, 1868, 2433). Often, myelin is still present 250 days after the lesion. Some of this might be newly formed and can surround processes other than axons (243). The manner of removal of myelin is different from that seen in the peripheral nervous system. Most is not taken up again by the cell which formed it; rather, it is phagocytosed by cells intrinsic to the nervous system (2730, 3000, 3003). Although some macrophages penetrate the brain at the site of injury, they cannot invade the whole of the degenerating tract. This failure is probably a cause of the increase in DNA being less in a degenerating tract of the central nervous system than in a peripheral nerve (1868). The intrinsic macrophages enter the blood after ingesting myelin, and can be identified in the spleen. The processes of myelinating oligodendrocytes show a slow reactive change during the degeneration of myelin (552); while this is hardly surprising its significance is unclear. It seems that myelin is removed as an entity from the central nervous system after axotomy and that the removal of no one component precedes that of the others (241).

The optic nerve of mammals does not regenerate. The axon stump swells and a few sprouts may be formed, but these soon cease to grow and are retracted. The axonal stumps which retain contact with the cell body atrophy slowly, and the number of axons in the optic nerve decreases. In some such studies it is not always clear to what extent the blood supply of the retina might have been impaired by the division of the optic nerve.

Terminals of retinogeniculate fibres disappear quite rapidly. Neurofilamentous hypertrophy begins on the second day and is greatest about the eighth day. Small cells seem to act as phagocytes and remove them. The processes of astrocytes become more prominent around the tenth day; the significance of this is uncertain.

Spinal dorsal roots of mammals

Regeneration of the central process of the dorsal root ganglion cell has been studied often. Part of this process lies within the dorsal root where it is surrounded by Schwann cells and part within the central nervous system.

After axotomy, both isolated axoplasm and the related myelin are removed rapidly in the dorsal root (2100, 2103, 2104). The reaction of the Schwann cells is the same as in a peripheral nerve. Within the spinal cord the axon degenerates just as fast, but is removed more slowly (1874, 2058, 2100, 2101). A similar rate of removal is seen if the dorsal columns rather than the dorsal roots are injured (1706), except in the immediate vicinity of the injury, where macrophages gain ready access. Evidence of axonal degeneration is apparent within 3 days but some degenerated axons can still be seen after 30 days. The myelin sheath of these fibres is removed even more slowly. The oligodendrocytes seem to be substantially normal, but the astrocytes hypertrophy. Axons of the dorsal column are included amonst those that degenerate rapidly within the dorsal root; this suggests that the slow rate of removal of an axon within the central nervous system is not due to a property of the axon itself, but to the nature of the cells investing it and gaining access to it during degeneration.

The central process can regenerate within the dorsal root. Here it is closely related to Schwann cells which have reacted as a result of axotomy. Such fibres are readily remyelinated (2101, 2104). Under ill-defined favourable conditions the regenerating fibres can penetrate the central nervous system, grow within it, and form synapses which appear normal (1461, 2098–2101). On other occasions fibres regenerate within the dorsal root, but fail to enter the spinal cord; they may pass along vessels of the pia or double back within the dorsal root (429, 2058). When a ventral root is anastomosed to a dorsal root (151), the fibres grow into the spinal cord and form transmitting synapses. Grafted fibres can be traced within the dorsal horn and central zone of grey matter; none enter the ventral horn. When the grafted root is stimulated, a volley can be obtained from the ventral root fibres of the contralateral side. The latency indicate that a polysynaptic pathway is involved. No evidence indicates the restoration of a monosynaptic link between the grafted fibres and the motor nerve cells. If the grafted root and an adjacent normal dorsal root are both stimulated, the pattern of interaction, judged by recording the evoked ventral root volley, is very variable. Reexamination of this model with intracellular microelectrodes might be interesting.

Several factors might contribute to the variability of the regeneration of the dorsal root fibres. When a dorsal root is divided, it is very easy to injure the radicular artery, or even the posterolateral artery of the spinal cord; the blood supply of the regenerating root and of the dorsal horn may be impaired, perhaps causing the slow regeneration, or an increased synthesis of collagen. The technique of injuring the root might be important; if a nerve is divided and anastomosed, the regeneration might be less successful than if it is crushed. If vascular cuffs or millipore sleeves are used to align the stumps and to restrain misdirected sprouting, these supports might impede growth; a vascular cuff commonly becomes fibrosed, and a millipore sleeve calcified. Penetration of the cord by regenerating axons might be critically dependent upon the state of reacting cells within the cord when the axon attempts to re-enter. If their attempted penetration is delayed or premature, or if the glial response of the cord is retarded or accelerated, or if the rate of collateral sprouting of intact fibres within the cord is great, then entry may be more difficult. If collateral sprouting is an important impediment, it might be expected that regenerative entry of the axons of one root would be facilitated by the division and diversion of adjacent dorsal roots. Although various hormones, especially steroids, have been shown to influence the formation of 'glial scars', their actions in this model are unknown.

Aminergic tracts of the mammalian central nervous system

Neurons which release noradrenaline 5-hydroxytryptamine or dopamine can be identified histologically by fluorescence; cell bodies and axons can both be seen. These fibres can be divided in the diencephalon or in the spinal cord, and the degenerative and regenerative sequence studied.

After axotomy fluorescent amines disappear from the isolated axons within 4 days. The axon degenerates and the terminals of the fibres are removed. The fluorescence of the proximal stump increases considerably during the first day after axotomy as the amine accumulates in vesicles in the proximal stump. This accumulation persists for about 12 days and then decreases (613, 614, 1560). Between 7 days and 10 days after injury fine sprouts are formed near the end of the proximal stump. These extend vigorously through the brain, but fail to reconstitute the lost tract; many fibres pass to adjacent blood vessels and ramify upon their surfaces. Others follow emerging cranial nerves. The axons cannot penetrate easily the necrotic and organizing scar tissue of the injury which severed the tract. Despite the vigour of the sprouting, there is no indication that effective synaptic contact is restored. Nevertheless, the fibres are capable of forming synaptic junctions.

If a piece of iris is implanted into the lesion made in the caudal diencephalon, which severs dopaminergic, noradrenergic and indolaminergic tracts, regenerating fibres ramify around the implant. Initially the graft is penetrated by both noradrenergic and dopaminergic fibres; later nearly all the fibres which contain dopamine disappear, while the noradrenergic fibres ramify extensively upon and within the implant. This considerable innervation persists for at least 6 months, and is not influenced by sympathectomy. Junctions are made, which have dense core vesicles (259, 2817). Nerve growth factor augments the growth of the noradrenergic fibres and the extent to which they invade the implant (255, 2816), whether it is given by systemic or by intracranial injection. When given by intracranial injection, nerve growth factor potentiates the growth of sprouts whether it is injected at the site of the lesion or near the cell bodies of the noradrenergic nerve cells in the locus coeruleus. If it is also shown that nerve growth factor fails to act in this way when injected into a part of brain remote from the affected nerve cells, then it is probable that the sensitivity of the noradrenergic neurons to this agent is not confined to one part of the nerve cell. Nerve growth factor is most effective if it is injected in the early phase of regeneration. A mitral valve implanted into the caudal diencephalon becomes similarly extensively invaded by noradrenergic axons; these fibres remain subendothelial, occupying a similar position to noradrenergic axons in the normal mitral valve of the rat. Descending bulbospinal noradrenergic axons also sprout and innervate iris and mitral valve implanted at the site of the lesion (256, 257). Descending axons containing 5-hydroxytryptamine also form sprouts, but these do not innervate these implanted tissues. After the spinal cord has been crushed or divided, aminergic fibres grow rapidly into the necrotic tissue, but wander tortuously. They do not cross the lesion to enter the severed distal part of the spinal cord. The picture is one of vigorous but undirected growth, such as explants of brain achieve in tissue culture.

There is some evidence that a noradrenergic axon which has two fields of ramification may show increased fluorescence in one group of terminals if the axons passing to the other are divided (2285). It is not yet clear whether the increased fluorescence indicates a greater amount of amine in a constant number of terminals, or sprouting. If the amount of noradrenaline contained in a very small nerve fibre increases, the nerve fibre could become visible; such a change might be interpreted as sprouting. The interpretation of the observation is difficult for other reasons also: because each branch of the axon ends in a complicated synaptic bed, because the lesion is partial, and because the two synaptic areas – the cerebellum and the hippocampus – can interact physiologically in many ways. The observation is interesting, in indicating either that diversion of amine takes place under these conditions to the residual branches of the axon, or that distal sprouting can be induced in such a way. Possibly related to this observation is the finding that normal adrenergic terminals of the cervical and upper thoracic spinal cord fluoresce more strongly than normal after the cord has been divided in the mid-thoracic region (256, 257).

5,6-Dihydroxytryptamine is a strongly reducing congener of 5-hydroxytryptamine. If it is injected into the cerebral ventricles it is taken up by central indolaminergic terminals and destroys them and their axons. The characteristic fluorescence of 5-hydroxytryptamine disappears; the concentration of this indolamine falls both in brain and in spinal cord, and the number of sites which can take up 5-hydroxytryptamine decreases (175, 176, 555, 2133). Some superficial cell bodies which contain this neurotransmitter are also killed. Noradrenergic fibres and other cells are not affected except immediately beneath the ependyma, where the concentration of drug is presumably high.

The surviving stumps of the axons become swollen. Between 10 and 17 days after administering the drug, fine fibres extend from these distorted stumps (258); after 3 months they are abundant. Many of these newly formed axons are found where none are normally seen; they may follow other intact nerve tracts, for example the deep surface of the spinal tract of the trigeminal nerve. As would be expected, growth in the spinal cord progresses from the cranial end, because the cell bodies are present in the lower brain stem, but not in the spinal cord. Three months after injection regrowth is well advanced; the density of terminals, judged by fluorescence is more than half normal, and the uptake of labelled 5-hydroxytryptamine is 80% of normal. As the terminals reappear, pharmacologically demonstrable hypersensitivity diminishes (2146). It seems likely therefore that bulbospinal axons regenerate to form functioning synapses in the spinal cord under conditions in which the formation of a large zone of necrosis is avoided, and in which the orientation of many of these bulbospinal fibres is not lost. Even so the majority of fibres pursue a predominantly aberrant and meandering course; many may gain a suitable site to end, but do so without any evidence of having been guided by the other elements of the neuropil.

Neurosecretory cells

The degeneration and regeneration of axons of these nerve cells have been studied in many different preparations (49, 668).

The ferret has a long hypothalamo-hypophyseal tract, which is applied to the ventral aspect of the median eminence (19–23, 624). The tract cannot be divided without causing venous infarction and oedema of the pituitary. Retraction balls are found on both the distal and proximal stumps (624); these contain neurosecretory material. The distal swelling might result from the venous infarction and oedema of the pituitary; if the pressure becomes great within the pituitary fossa, axoplasm might be extruded from the end of the divided axon. Such distal retraction balls are not seen in the rat, possibly because the pituitary does not lie in a rigid bony fossa. The axons degenerate and are phagocytosed by pituicytes; it is not clear if macrophages from the blood take part in this process. Their participation would be expected, both because they can enter the tract at the lesion and because the capillaries of the neural lobe are fenestrated. If regeneration is prevented by a barrier inserted at the site of section, the neural lobe atrophies.

After injury the proximal stump also swells; the swelling contains packed vesicles, mitochondria and tubules; neurosecretory material is abundant. After a few days sprouts grow vigorously; if an impermeable barrier is placed at the site of injury, a large mass is formed; this projects through the median eminence into the infundibular recess of the third ventricle. Initially the mass is not covered by ependyma, and contains few cells. After a few weeks this apparently structureless mass of tangled axons becomes organized into an ectopic infundibular process, containing pituicytes, many capillaries, axons and their terminals and loose connective tissue.

A similar poorly structured mass is also found early after the lesion in animals which have no barrier at the site of the lesion. Two weeks after axotomy some axons can be found passing through the scar. They grow slowly along the tract, but may take one year to reinnervate the pituitary fully. Even then, the swelling above the lesion may not subside completely: perhaps because some fibres fail to penetrate the scar and so remain here. Regenerating axons also invade the pars distalis, which normally does not contain neurosecretory fibres. This aberrant growth may be diffuse, when the fibres ramify widely among the parenchymal cells, or it may be more localized and form a bundle of axons which sometimes end in a small ectopic neural lobe. The response of the supraoptic neurons of the ferret is variable but usually slight, possibly because the long hypothalamo-hypophyseal tract protects the nerve cell bodies from traction when the axons are divided, or possibly because the axons have collateral terminals above the level of the lesion. The shorter the axon stump that remains, the greater is the cell death (625).

In some respects the hypothalamo-hypophyseal tract of the rat behaves similarly after axotomy. This also can regenerate in two ways: either by forming a new and ectopic neural lobe at the end of the proximal stump, or by re-establishing the connection with the normal posterior lobe (669, 1582). The ectopic neural lobe is formed through hyperplasia in the central stump and contains a core of neurosecretory axons and their terminals, surrounded by pituicytes, loose connective tissue and a rich capillary plexus. These neurosecretory fibres do not grow among Schwann cells; it is unlikely that they can grow and terminate among other types of glia than pituicytes (1583). Axons of peripheral nerves cannot regenerate among pituicytes; perhaps other tracts of the central nervous system cannot regenerate among pituicytes. The pituicyte seems to be essential for the formation of functioning terminals, but it is less clear that they are necessary to permit the extension of the neurosecretory axon. The normal axon passes for a considerable distance without contact with pituicytes. When a regenerating axon penetrates the loose connective tissue of the scar, it does not appear to be growing along a column of pituicytes in the manner of a peripheral nerve fibre growing along a column of Schwann cells. The effect of hypophysectomy upon the supraoptic nerve cells of the rat is marked. Most disappear within a few months. Within the first two weeks degenerating cells and axons within the proximal stump can be found (2375, 2376). Later they are few. The few cells which remain ultimately hypertrophy. The fate of the presynaptic boutons which terminate upon supraoptic neurons which die is not known; boutons are occasionally seen upon the surface of a cell that is obviously degenerating.

These unmyelinated fibres can therefore regenerate to an extent at least as great as the aminergic nerve cells of the central nervous system. Not only can they penetrate loose connective tissue and organizing scar tissue, but they can traverse it, and, gaining the other side restore functional connections. The pituicyte is essential for this success. They resemble ependymal cells in many respects, a type of cell closely involved in the regeneration of tracts in the central nervous system of the non-mammalian vertebrates.

Regeneration of the mammalian spinal cord

Other studies have been directed at the healing of the spinal cord as a whole, rather than at regeneration of particular tracts. After transection there is usually a region on either side of the primary lesion, within which the tissue rapidly dies (Fig. 7.9). This region is at least 100 μm thick, and often 1 or 2 mm. This might result directly from the trauma of injury, from damage to blood vessels at the time of injury, or from the considerable release of catecholamines which occurs locally under these conditions. This necrotic region can be removed by phagocytosis only slowly. Two processes then occur simultaneously: the inward migration of fibroblasts from the pia and perhaps from vascular pericytes, and the removal of the dead ends of the stumps, sometimes called the 'preserved segments'. If the ingrowth of fibroblasts and the inward collapse of the pia preponderates, then the site of the lesion is converted to a fibrous strand. If the removal of the preserved segments is slow, they are finally replaced by cystic spaces, separated by thin fibrous sheets and strands. The connective tissue is initially loose, but later becomes collagenous and dense. On either side of this mesenchymal scar is a zone containing fibrous astrocytes. This pia-glial scar is usually considered to be a

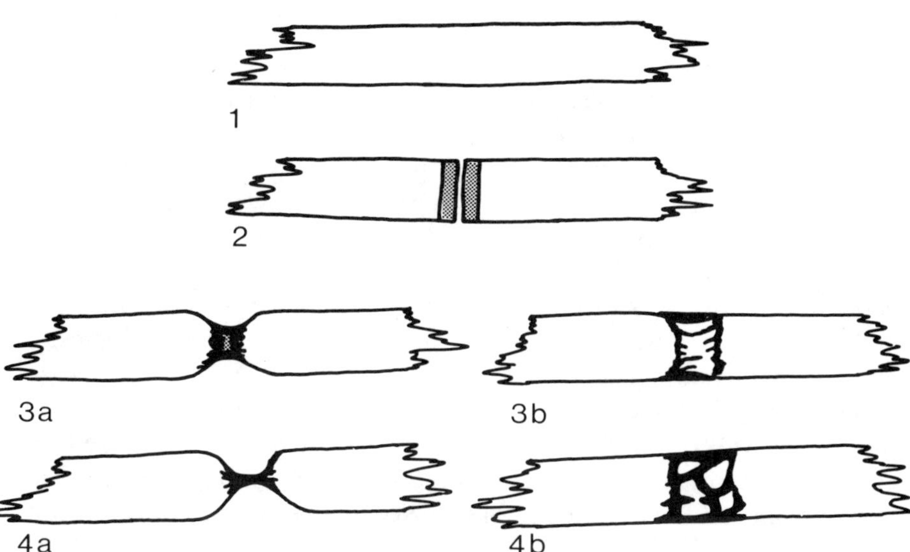

Figure 7.9 When the spinal cord of a mammal is divided, the region on each side dies (stippled zone, 2). These regions are invaded by macrophages and fibroblasts. If these regions are rapidly removed (3a, 4a), the affected segment collapses and the ends of the cord are united by a fibrous cord. If they are removed slowly (3b, 4b), the injured zone becomes surrounded by fibrous tissue and the preserved segments are replaced by cysts.

severe impediment to regeneration. When the scar is first being formed, a few axons can be seen entering the loose connective tissue; these are commonly aligned along the connective tissue strands. They have never been observed to cross the lesion completely. Efforts have been made to reduce the severity of the mesenchymal scarring by giving steroids or ACTH, or by inducing the release of ACTH with pyrogens (1890, 3146). These procedures reduce the scarring considerably, but do not significantly facilitate regeneration in adult animals. The situation is much less clear in immature animals. If the cord of a rat is divided on the 5th or 7th postnatal day, some function of the lower limbs is sometimes found if the scarring is reduced in this way (1890). In these animals it is also possible to trace histologically the tortuous nerve fibres across the lesion; unfortunately electrophysiological studies were not made. It is necessary to distinguish clearly, however between two types of fibre that might cross the defect: the regenerating fibre and the growing fibre. Many fibres of the spinal cord of the rat of this age are still developing; it is very likely that many of these have not been divided by the transection of the cord, but reach the site of the lesion later. Regeneration is not facilitated by implanting fragments of peripheral nerve at the site of the lesion to provide Schwann cells, nor by implanting neural tissue from other sources (358). It is possible that *L*-thyroxine may facilitate regeneration (1271), but no incontrovertible evidence was provided that the initial lesions were complete, and regeneration was not conclusively demonstrated histologically and electrophysiologically.

The size of the cysts formed at the site of the lesion can be reduced if the removal of the 'preserved segments' is facilitated by enzymatic debridement, by bathing the severed ends of the cord in fibrinolysin and DNAase (2244). Neuronal regeneration is not enhanced.

After spinal cord injury, many nerve cells in adjacent segments lose their afferent connections. Their dendrites at first become beaded and may retract (229, 230, 231, 234). After several weeks boutons often reappear on these neurons. None of these is derived from fibres which have regenerated across the defect, and very few come from axons of the long tracts. Most are derived from collateral sprouts of fibres which end within the segment studied or close to it, and many arise from the short axons of interneurons. In the rat, many of these new connections acquired shortly after the injury are later lost.

Attention has been directed almost exclusively at attaining regeneration by facilitated regrowth. The possibility of healing by restoring effective contact with the distal axon has hardly been entertained seriously for many years. This might lead to healing in the manner of the crayfish axon. There is no obvious reason *a priori* why conditions should not be sought that favour such survival; after all, the cytoplasm of the red cell survives for months without a nucleus. Cell fusion by membrane union has been achieved in other cells using viruses, lysolecithin or phospholipid dispersions (2323–2326). Conditions might be found in which growing sprouts of a regenerating axon can re-establish contact with a sustained and living distal axon.

Spinal cord regeneration in other vertebrates

Some axons of the spinal cord of goldfish can traverse a transection (226, 232, 233, 2282). After the cord has been transected, ependymal cells divide and the neural central canal extends from both divided surfaces, usually in the form of closed terminal ventricles; these come together and fuse. The ependymal cells are tanycytes, which have long basal processes passing to expansions resting upon the pia and separated from it only by basal lamina. Regenerating axons then grow between these processes and enter the cord beyond. Astrocytic hypertrophy is slight or absent (242). This is a clear example of the restoration of continuity by the cells of brain which obviously retain their epithelial characteristics (Chapter 1). Axonal sprouts grow proximally and distally. There is no evidence to indicate that ascending and descending sprouts interact. The absence of such interaction is interesting, as the ends of the growth cones are very convex and have little glycocalyx: properties which in other cells facilitate cell fusion. The regenerating fibres form synapses with nerve cells beyond the lesion. Although swimming movements are restored, the pattern of synaptic connection which is re-established differs considerably from the normal pattern. Most of the regenerating axons form synapses within the first available segment beyond the transection; fewer fibres than normal pass to the more remote segments. Those that do so have a diameter that is greater than normal, presumably because the field of their terminal innervation is great. The boutons formed by these fibres lie in an abnormal place (233), upon nerve cells with which they do not normally communicate. Many of the boutons which end upon nerve cells distal to the transection and which appear after the injury come not from the fibres of long descending tracts, but from short intersegmental fibres. Despite this extensive imprecise regeneration swimming movements are restored to the tails by the 30th day after the lesion; such swimming movements are again abolished if the cord is divided again immediately rostral to the first injury. Connections formed by the descending long tracts are therefore necessary.

Regeneration of the goldfish spinal cord can be impeded by interposing a Teflon barrier at the lesion (Fig. 7.10). This can be removed after 30 days. Regeneration does not then follow. Although

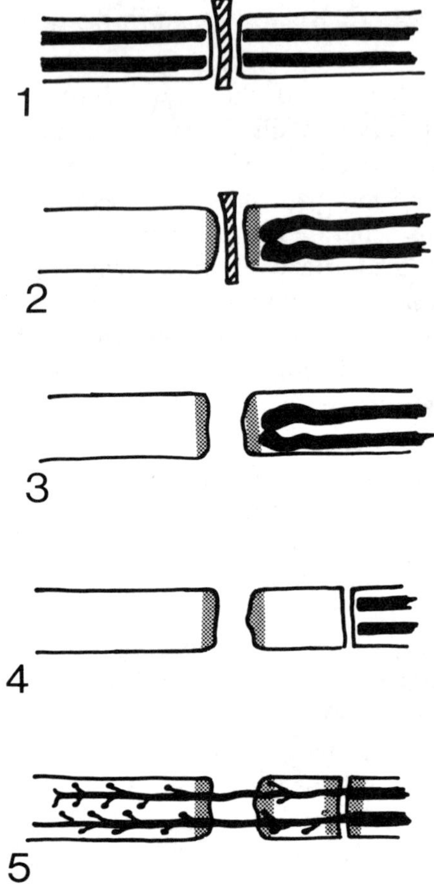

Figure 7.10 When the spinal cord of a goldfish is divided, regeneration can be prevented by the insertion of a mechanical barrier (1). The distal axons degenerate, a glio-mesenchymal scar is formed (stippled zone, 2) and the remaining axon stumps become stabilized by forming axo-axonal junctions. When the barrier is then removed (3) no regeneration follows. If a second division is made on the cranial side of the first at the time that the barrier is removed (4), regrowth of the divided axons occurs through the scar tissue around both incisions (5). The glio-mesenchymal scar does therefore not act as an effective barrier to prevent axonal penetration. Note that the growing axons do not pass to their normal terminations.

a mesenchymal scar is formed at the site of the lesion, this does not itself impede regeneration; if the cord is divided for a second time just proximal to the first lesion, the axons regrow, penetrate the scar tissue and enter the distal cord (227, 228). This indicates that the growth of the axons is itself impeded. Electron microscopy of the fibres in the immediate vicinity of the barrier shows many axo-axonal junctions, and it seems likely that these are stabilizing the axon and inhibiting growth. A situation which is similar in many respects is found in the peripheral nervous system when two nerves are joined end-to-end. In such a collision anastomosis (3073), interpenetration of nerve fibres is very slight.

The response of the spinal cord of amphibia to injury is very variable, according to the species and age. Severed axons of the spinal cord of the adult toad or frog degenerate as rapidly as those of the peripheral nervous system (1515, 2284) and no regeneration occurs.

When the lizard *Anolis carolensis* regenerates its tail after amputation, the spinal cord within the tail also regenerates and contributes to organizing the differentiation of other tissues. As the tail grows the ependymal tube extends in advance of the regenerating axons (790, 2705, 2706, 2707). Between the processes of the ependymal cells which extend to the basal lamina beneath the pia there are spaces filled in the fixed preparation with flocculent material. Later large bundles of unmyelinated axons pass through these spaces. The regenerated cord consists only of ependyma, oriented axons and the connective tissue of regenerated meninges.

The severed spinal cord of the newt behaves similarly. Within two or three days of injury the neural canal is closed. By the sixth day the ependymal layer has many mitoses and the tube is extending caudally. By the twentieth day the regenerated cord is well differentiated; it contains nerve cells as well as fibres (1382). It is uncertain if these cells have been newly formed or if they have migrated from adjacent segments of the cord. Some tissues of the tail only differentiate if the spinal cord regenerates.

Regeneration of the postbrachial region of the spinal cord of the larva of the urodele *Ambystoma maculatum* is similar. After 2 mm of cord have been removed, ependymal cells migrate and close the defect in the central canal (407, 408). Initially the central canal dilates at each cut surface to form a terminal ventricle; these ventricles gradually elongate and fuse. Within the walls of the terminal ventricles are cells of the grey matter and many fibres. When the ependymal vesicles fuse the fibres pass into the cord beyond.

The reaction of the ependymal cells is again evident in larvae of the lamprey *Petromyzon marinus* (1322); they proliferate early. By the second day after injury growth cones enter the blood clot, lying between the severed ends of the cord, and by the seventh day these have traversed the defect. Morphologically and behaviourally regeneration is complete by the twentieth day. The lamprey has no blood vessels within the spinal cord, but a rich vascular plexus lies upon its surface. The supply of metabolites to the cord is probably not much affected by the lesion. Occasionally very considerable ependymal reaction is seen. It may then appear as though the extreme proliferation and the associated enlargement of the central canal are transiently blocking the advance of the fibres.

Studies related to the regeneration of the central nervous system

If a peripheral nerve is implanted into brain, the capacity of its axons to penetrate the brain can be measured; conversely the growth of axons of the central nervous system into degenerated nerves of the peripheral nervous system can be studied (501, 502). Schwann cells grow out from the end of an implanted nerve, usually into a region of haemorrhage and organizing scar. A clear demarcation persists between the Schwann cells and the surrounding brain. Usually mesenchymal cells derived from the pia and from local pericytes lie between the two tissues; two basal laminae are also interposed: one between mesenchyme and Schwann cells and the other between mesenchyme and the basal processes of astrocytes. Injured axons of the peripheral nerve grow out over the surface of the Schwann cells, but do not pass beyond to penetrate the brain and to mingle with fibres of the central nervous system or to form boutons. This finding can be contrasted with the demonstrated ability of ventral root motor axons to grow along dorsal roots and to invade the dorsal horn and form functioning synapses (p. 222). The circumstances differ. When a nerve is implanted bleeding occurs and the brain is damaged. When the nerve is implanted the continuity of the glia-ependymal epithelium is destroyed; on the other hand the ventral root fibres are able to enter the central nervous system along a route which does not involve disruption of the epithelial components of brain.

When a peripheral nerve is implanted into brain after the axons of the nerve have degenerated, injured axons of the central nervous system fail to enter it. This indicates that their failure to enter the peripheral nerve when its axons are still present is not caused by contact with axons of the peripheral nervous system, but rather by a failure either to penetrate the mesenchymal zone bounded by the two basal laminae or to grow upon the surface of Schwann cells.

Central axons of a vagus can grow along the distal degenerating stump of the hypoglossal nerve (742, 1800). The central processes of primary sensory neurons can regenerate in dorsal roots. In both of these situations the nerve grows along the surface of Schwann cells.

Irradiation

Degeneration of the cerebral cortex of the rat follows bombardment with α particles (1901, 1954, 1955, 2483). A laminar lesion is produced between 60 and 350 μm thick and about 4 mm in diameter. Within the lesion there is an intense vascular reaction and many cells die; neurons seem to be more sensitive than neuroglia. Within two weeks the degeneration of the nerve cell processes is virtually complete. Seven weeks after irradiation the vascular reaction has subsided and a dense network of axons is seen at the site of the laminar lesion; most of these pass horizontally so that a conspicuous and abnormal horizontal striation is produced. By the eleventh week dendrites can also be identified in this zone. The growth of processes is luxuriant, but it is not known if working synapses are formed.

The source of the cells which act as macrophages under these conditions is not clear. Although there is some evidence from electron microscopy that they are derived from vascular pericytes (1955), a study using labelled blood macrophages or locally administered tritiated thymidine would be of interest. The vascular pericyte resembles the fibroblast of other tissues and it is far from certain that this cell or its progeny act as macrophages. Astrocytes react; on the first day they contain more glycogen. Later they become fibrillary (1956). Perineural oligodendrocytes change in a number of ways: an initial phase of acute swelling is replaced by the appearance of many dense cytoplasmic inclusions, which probably represent autophagic vacuoles. These cells do not act as phagocytes (1957).

If higher doses of irradiation are used, the destruction is more intense; haemorrhagic necrosis leads to the formation of cystic spaces and mesenchymal scarring. No neural regeneration then follows.

In many respects the response of nerve fibres around a lesion produced by the local application of a cold object resembles the response to irradiation; degeneration and regeneration occur around a central zone of necrosis but do not traverse it (834).

Olfactory nerve cells

Primary olfactory neurons are lost and replaced after injury. If their axons are divided the nerve cell bodies in the epithelium degenerate. They are replaced in the mature frog from basal cells of the epithelium which cannot be distinguished from those giving rise to other types of cell. The new nerve cells send axons which penetrate the olfactory bulb and form new contacts with secondary nerve cells in glomeruli. These olfactory glomeruli appear to be poorly organized for several weeks after injury, but later improve (1132–1135, 2874). This reaction is not confined to amphibia but also has been seen in pigeons.

General comments on regeneration in the central nervous system

It emerges from the studies summarized here that a number of features must be considered in comparing unsuccessful regeneration of the central nervous system of mammals with the successful regeneration found in some other vertebrates and chordates: mesenchymal reaction, neuroglial reaction, the vigour of neuronal sprouting, the distractability of the sprouts formed, and the capacity of the reinnervated nervous system to remodel its connections either functionally or structurally, so that correct final responses may be obtained despite the formation of inappropriate connections.

Mesenchymal scarring can impede regeneration, but it is unlikely that it does so by creating an impenetrable barrier. Collagen does not constitute a surface upon which growth cones cannot be formed and over which they cannot migrate. In tissue culture the processes of nerve cells extend well upon a substrate of collagen. Observations upon fibres growing in loose connective tissue between the two severed ends of the spinal cord show that they can grow along the surface of collagen fibres. It is probably the irregular orientation of these in the scar that leads to the poorly directed extension of the growing axons and to their failure to traverse connective tissue bridges.

Much of the evidence cited indicates that some nerve fibres can grow and regenerate provided that the continuity of the neural epithelium has not been destroyed: after selective chemical destruction of the tract, after irradiation or after cold lesions. If this is so, then the hypertrophy and hyperplasia of astrocytes might be regarded as a factor facilitating regeneration rather than impeding it. In some vertebrates, whether regeneration occurs or not depends upon the balance between the growth of mesenchymal cells and the growth of the epithelial cells. It is these cells and not the nerve cells which remodel the nervous system, and it is into the remodelled tissue that nerve cells later grow. Nerve growth without glial influence can be seen in tissue culture under some circumstances; then, a mass of intercommunicating nerve cells can be formed, but it is not a recognizable regenerate.

The vigour of axonal regeneration almost certainly varies, but is difficult to assess. Certainly, some nerve cells develop retraction balls after injury, and either produce no growth cones, or small sprouts that are rapidly lost or withdrawn. Very little is known of the factors which cause either a severed axon to produce a regenerative sprout or a preterminal fibre to produce a collateral sprout. It is possible, but not proven that such sprouts are formed more readily in distal than in proximal parts of the axon. It must also be considered uncertain whether some axons are unable to produce growth cones, or whether they are prevented from doing so by the environment at the end of the axon or through the influence of other connections that the injured nerve cell receives. Even when effective growth cones are produced, their extension can be poorly directed or undirected; this lack of direction is even found in the mammalian central nervous system when a tract is destroyed chemically without associated damage to the astrocytes. It would be interesting to know if the regeneration of such tracts in fish and in some amphibia would be precisely directed.

Regenerating fibres can be stabilized in two disadvantageous ways: by forming junctions locally with neurons which are inappropriate, and by forming junctions with other regenerating axons. The first is a consequence of the lack of directed growth. It is interesting that the number of synapses newly formed after a lesion is slowly reduced, perhaps as a result of selecting out those which are inappropriate.

In some parts of brain, regeneration must be precise for the restoration of proper behaviour: for example in the retinotectal system of fish and amphibia. In contrast swimming movements reappear in the caudal part of a goldfish after regeneration has taken place that is very largely aberrant. It is not known if the restoration of function is a result of the very small number of axons which have made proper reconnection, or a consequence only of an alteration of the level of excitability so that locally determined reflexes might function with greater efficacy, or to structural and functional remodelling in the neural circuits distal to the lesion so that inappropriate signals can be translated into effective action. Considerable inhibition of inappropriate signals must occur. It would be interesting to measure the turnover of inhibitory neurotransmitters in the regenerating cord, or even much less specific parameters such as the pO_2 or blood flow. It is not even known whether the fibres which are removed later are the inappropriate ones.

Responses of the perikaryon to axotomy

Most interpretations of the response of the nerve cell body to axotomy seek to explain the changes within the context of altered patterns of synthesis of nucleic acids and proteins associated with axonal regeneration (565, 1800, 1802, 2943). Two other features are of great importance: regulated intracellular hydrolysis, and altered afferent innervation.

Boutons

Motor nerve cells

If a hypoglossal nerve of a rat is divided, boutons are shed from the surface of the perikaryon (2857); only boutons with round vesicles are shed, which are almost certainly excitatory (2855) (Figs. 7.11, 7.12, 7.13). Boutons with flat vesicles which are probably inhibitory are not lost. This loss can be detected both by light microscopy and by electron microscopy (597, 597(a), 2857, 3073). If the axons of

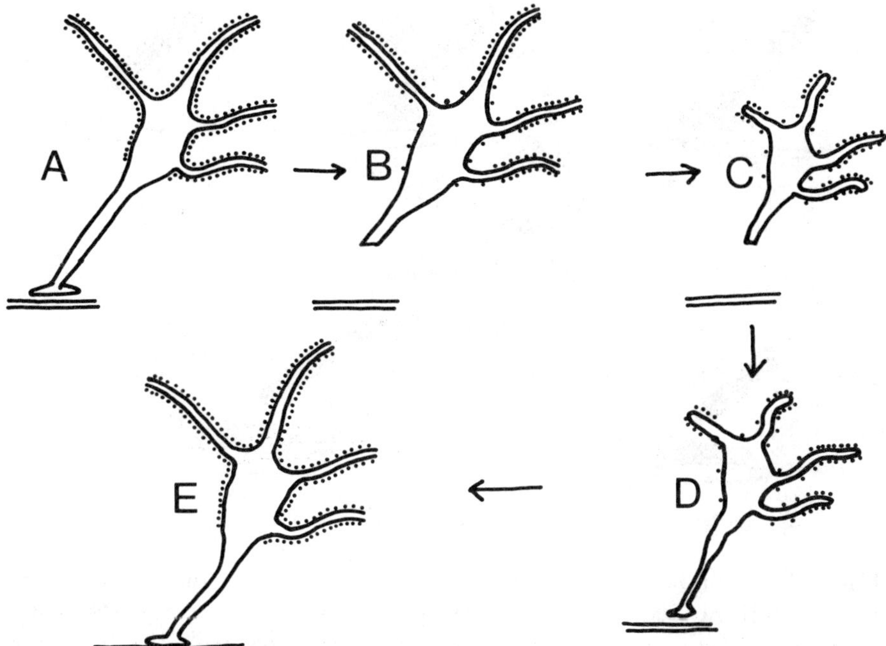

Figure 7.11 Boutons are lost from the surface of an axotomized motor neuron; they are lost earlier from the cell body than from the dendrites (B). The dendrites retract as the dendritic boutons are lost (C), so that the density of boutons on the dendrites does not change. Boutons are restored if nerve-muscle contact is restored (D, E).

the hypoglossal nerve fail to re-establish contact with muscle, the shed boutons are not restored. There is no evidence that the shed boutons are phagocytosed. Although glial cells react around an axotomized neuron (3062, 3072), only exceptionally can intraglial structures be found which might be phagolysosomes. Most axons within a hypoglossal nucleus are probably presynaptic; their number does not alter significantly after the hypoglossal nerve has been divided. This indicates that although many boutons have disappeared, their axons have not been lost; the extent of withdrawal is small. The reappearance of boutons after neuromuscular contact has been restored also makes it likely that boutons withdraw only a short way. Boutons lose their junctional connections and become smaller before they are detached. During the period in which boutons are

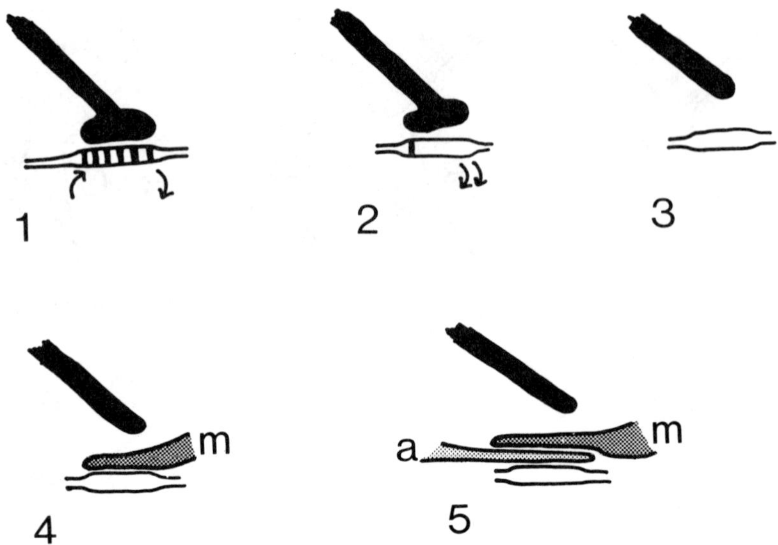

Figure 7.12 Boutons are shed from the surface of an axotomized motor neuron. This is probably due to loss of receptor molecules in the postsynaptic membrane. These are represented as black rectangles in the subsynaptic membrane (1). The arrows indicate that these molecules are being continuously added to the membrane and removed from it. After axotomy, receptor molecules are no longer added, and the rate of removal might be increased (2). This leads to the withdrawal of the bouton (3). The postsynaptic membrane is initially covered by a process of a 'microglial' cell 'm' (4). Later a process of an astrocyte 'a' lies between the microglial cell and the surface of the nerve cell (5). Shortly afterwards the microglial cell disappears.

being lost, more of their vesicles take up horse-radish peroxidase, if this exogenous marker is added to the extracellular fluid (2856). This can be interpreted as increased terminal pinocytosis which leads to the peripheral removal of membrane, so that the end of the process is retracted; it is a growth cone working in reverse.

Normal phasic electrophysiological activity is absent from the proximal stump of the severed phrenic nerve. It is lost about the seventh day and returns after 50 days if regeneration is successful (9). Excitatory postsynaptic potentials (EPSPs) produced in motor neurons monosynaptically by stimulating afferent nerves of muscle are smaller after axotomy (773, 1681, 1682, 1881). Despite this, and

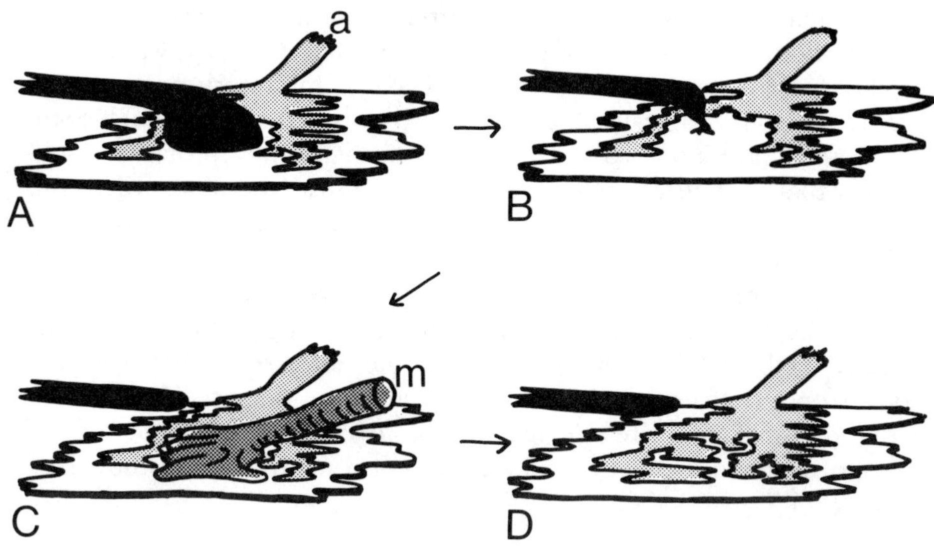

Figure 7.13 A bouton is retracted from the postsynaptic surface (A, B), but does not retract far (C). The bouton is replaced initially by a process of a microglial cell 'm' but later by an extension of the sheath of an astrocyte 'a'.

despite an unaltered membrane resistance and rheobase (773, 1681), an axotomized nerve cell can be excited by fewer afferent nerves than normal (721, 1881); between 10 and 20 afferent nerve fibres must be stimulated to cause an action potential in a normal motor neuron (3078), but after axotomy one sometimes suffices. There is no obvious direct relationship between the amplitude of an EPSP and its capacity to initiate a spike at the initial segment. As such a spike can on occasion be elicited by an EPSP as small as 500 μV this difference is unlikely to be due to a reduced threshold for spike initiation at the initial segment. Rather it is due to amplification of the EPSP by local spike generation in excitable patches of the membrane of the dendrites or nerve cell body. The latency of the response of an axotomized motor neuron to an afferent from muscle is more variable than normal, and usually longer. This is not due to the formation of alternative affective pathways for polysynaptic activation (721), but to the variable threshold of the excitable patches of dendritic or perikaryal membrane which initiate local spikes (773, 1681, 1881).

The altered shape of the EPSP after axotomy suggests that excitatory boutons close to the perikaryon or upon it are lost while those placed more remotely are retained (1682). Inhibitory postsynaptic potentials (IPSPs) caused by Renshaw cells activated through axon collaterals are also prolonged. Electron microscopy, however, indicates that no boutons with oval vesicles are lost from the perikaryon. IPSPs caused by stimulating inhibitory cells of the bulbar reticular formation are not altered; the boutons from these cells lie upon the distal part of the dendrites (1682). This cannot easily be reconciled with the retraction of the dendrites found in axotomized motor nerve cells (2857, 2858). Possibly the differences between the electrophysiological and structural studies are due to differences in the response of different species, or to differences in the rate of response, or to the fact that different motor neurons were studied.

Sympathetic ganglion cells

The sympathetic ganglion cell responds to axotomy in a similar way. Boutons are detached from its surface; their junctional thickenings are lost but they remain clearly identifiable as boutons (1945, 1946). This begins within 24 hours and is well advanced by seven days. By the 40th day after axotomy many of these boutons have regained contact with the ganglion cell; this restoration is probably related to the restoration of contact between the ganglion cell and its postjunctional effector.

Electrophysiological studies of transsynaptic excitation in autonomic ganglia give results which are similar in many respects to those obtained from motor neurons. As one of the transmitter substances of sympathetic ganglia is known to be acetylcholine, these studies can be extended pharmacologically.

Between two and three days after axotomy, the resting membrane potential falls (1441). It is not clear whether this indicates that the cell is more likely to be damaged by penetration with the microelectrode. Sustained depolarization of a normal ganglion cell by current usually leads to a train of spikes. In an axotomized cell, such depolarization often causes only a single, slow and small action potential. The first changes in synaptic transmission are also seen two or three days after axotomy. A failure of conduction is indicated in three related ways: a single maximal preganglionic volley causes a small discharge in the postganglionic nerve, post-tetanic potentiation occurs in the axotomized but not in the normal ganglion, and the relative refractory period of transmission is prolonged (10, 355, 1441). A maximal presynaptic volley usually causes an action potential in virtually all the cells of the ganglion. This is not so after axotomy; a subliminal fringe develops and so post-tetanic potentiation is found. The demonstration of this increasing post-tetanic potentiation is made more difficult by the increasing relative refractory period, which is associated with a great reduction of the slow potentials. Repetitive preganglionic stimulation normally causes a large positive after potential; this is often absent from an axotomized ganglion.

The blood vessels of a sympathetic ganglion can be perfused and the perfusate collected. In the presence of an inhibitor of choline esterase, acetylcholine is present in the perfusate of a stimulated ganglion. After postganglionic axotomy acetylcholine is still released in normal amounts. This indicates that the disorder of synaptic transmission is not caused by failure of the preganglionic terminals to release acetylcholine. It is not known whether the axons of the preganglionic trunk, which have had their boutons disconnected are in fact discharging. It is possible that those preganglionic neurons which have had their terminals removed from the surface of the injured ganglion cells have become silent and behave themselves like 'axotomized' nerve cells. This could be a step in sequential retrograde transneuronal degeneration. As in the normal ganglion, a small quantity of acetylcholine is released from the axotomized ganglion even when the preganglionic fibres are not stimulated (355); this release might be due to spontaneous activity in these fibres or to quantal release. The synthesis of acetylcholine by the axotomized sympathetic ganglion is not reduced (1887), but the capacity of the ganglion to augment this synthesis when the external concentration of potassium ions is raised, is reduced.

Close intra-arterial injection of acetylcholine depolarizes some normal ganglion cells so that they discharge; this does not occur after axotomy (355). Inhibitors of choline esterase do not restore normal conduction or make the ganglion cells respond to intra-arterial injection. *D*-tubocurarine does not restore conduction to an axotomized ganglion, although it might be expected to do so if conduction is blocked by tachyphyllaxis, that is by continuing saturation of the cholinergic receptors.

The evidence therefore indicates that about the time of bouton withdrawal the postganglionic cells lose their sensitivity to acetylcholine. This is not a consequence of the withdrawal of boutons, as division of the presynaptic fibres does not have this effect. It is therefore likely that the loss of sensitivity represents one of the changes in the membrane that leads to bouton loss; this is another example of the probable involvement of the neurotransmitter substances influencing directly or indirectly cell adhesion and so the stability of connections, possibly through the facilitation of the forward reaction and the inhibition of the backward one (p. 175). The measurement of the density of cholinergic receptor site density with α-bungarotoxin would be of interest. No information is available at present concerning the response of the interneurons of sympathetic ganglia to ganglion cell axotomy. It would be possible to measure the receptor site density of the dopaminergic connections by determining the dopamine-dependent adenylate cyclase activity of the ganglion. Information is also lacking about the behaviour of muscarinic receptors in the ganglion, and variations in the activity of regulated guanylate cyclase.

Subsurface cisterns

Boutons are detached because of a primary change in the membrane of the injured cell. Subsurface cisterns are closely applied to the deep surface of the plasma membrane of nerve cells. Fewer are found when boutons are shed (2854). The temporal resolution of the study does not show whether their loss precedes, accompanies or follows the loss of boutons. The reappearance of cisterns precedes the re-establishment of synaptic connections; when the boutons are restored, the cisterns are found only beneath boutons with round vesicles, that is, only beneath boutons of the type that are known to have been shed and restored (Fig. 7.14). On the other hand, only a small proportion of the boutons with round vesicles have cisterns beneath them. It therefore seems likely that the cisterns need to be in contact with the cell membrane only for a short time during re-establishment of synaptic connection. If the axon of the motor

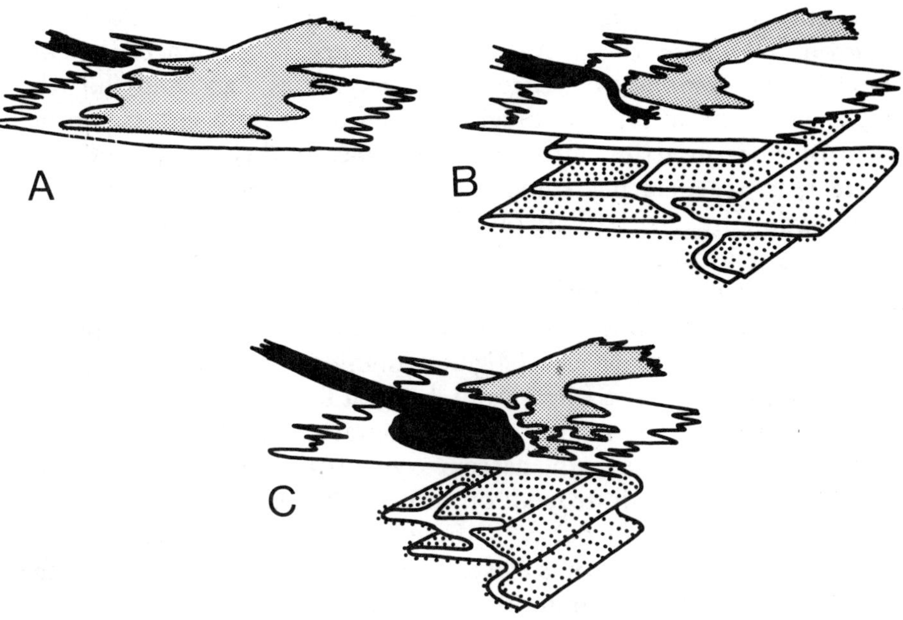

Figure 7.14 (A) shows a presynaptic nerve terminal (black) which has no specialized ending, and which does not make functional contact with the postsynaptic surface (white). When the axon of a motor neuron regains contact with muscle, subsurface cisterns appear (B). A bouton re-establishes contact with the adjacent plasma membrane shortly afterwards, the process of the astrocyte being withdrawn. After the bouton has regained contact (C) the subsurface cistern is withdrawn. It is interesting to note that subsurface cisterns become less common than usual when the boutons are withdrawn. This suggests – but by no means proves – that the factors maintaining normal boutons may be the same as those which restore boutons that have been removed. If so, a continuous but slow insertion of receptors into the membrane might be necessary to maintain the contact with boutons with round vesicles.

neuron is prevented from making contact with muscle that it can innervate, then the boutons are not re-applied to the surface; in this circumstance, the subsurface cisterns do not appear at the expected time. The cisterns are therefore either induced by the returning boutons, or they prepare the membrane to receive them. As the cisterns appear before the boutons are apparent, the latter seems more likely. It seems probable that the subsurface cisterns are responsible for changing the membrane in a way that allows the bouton to adhere to it. As such adherence is influenced by the capacity to respond to neurotransmitters, it is suggested that the subsurface cisterns represent regions of synthesis of receptors and the sites of their insertion into the membrane. If this interpretation is correct, a regenerating sympathetic ganglion cell should develop denervation hypersensitivity at this time if the preganglionic and postganglionic fibres have both been divided. Alternatively the cistern might secrete glycoproteins or glycosyl transferases to promote adherence, or enzymes which modify the glycocalyx and so permit boutons to adhere. N-acetyl-β-glucosaminidase has been sought in subsurface cisterns histochemically, but not found.

If subsurface cisterns are only in contact with the plasma membrane transiently beneath nascent synapses they must either be readily made and destroyed or mobile. There is no evidence at present to distinguish between these alternatives. The outer membrane layer of the cistern is closely applied to the inner aspect of the plasma membrane, no microfilaments have yet been described between them.

Metabolic responses of the perikaryon

After axotomy the synthetic activity of the perikaryon changes both qualitatively and quantitatively. A qualitative change has two aspects: the cessation of one pattern of synthesis and the induction of another.

Intracellular hydrolysis

When the synthetic activity of a secreting cell is curtailed physiologically, for example in the mammotrophic cells of the anterior pituitary when lactation suddenly ceases, then the involution is not only a consequence of the repression of the synthesis of particular proteins and types of RNA. Many intracellular organelles become sequestrated in membrane-bound vacuoles and are destroyed. Primary lysosomes form autophagic vacuoles and multivesicular bodies within which degradation occurs; dense bodies with myelin figures are formed. Endoplasmic reticulum, ribosomes, mitochondria and a number of different kinds of vesicle are found within the vacuoles.

Such a response can be seen clearly within the axotomized nerve cell soon after injury (1948, 2853). In sympathetic ganglion cells it occurs within 24 hours of injury, and coincides with the onset of histologically discernable chromatolysis and with the start of bouton loss. Similar responses have been described in motor neurons (165, 166, 284, 456, 1385). The Golgi apparatus is prominent when this degradation is taking place. Very little is known of the regulation of this response, or whether the organelles degraded are in any way selected for hydrolysis. The severity of this aspect of the reaction to injury is likely to be of great importance in determining whether the nerve cell itself dies or survives. Such death is especially likely to occur in axotomized nerve cells of immature animals, in which it should be considered as a controlled and terminal differentiation of cells which have not retained a competent connection with muscle. The metabolic direction of the motor neuron again changes after the axon has regained contact with muscle; it is not known however, if this qualitative change is also associated with degradation of the synthetic apparatus; if the interpretation of the hydrolytic activity described here is correct, such a late recurrence of the hydrolysis would be expected.

If the synthesis of RNA is inhibited soon after axotomy with actinomycin D, the chromatolytic response can be prevented, delayed or modified (2940, 2941). The suppressed RNA might be responsible for the formation of proteins necessary for this process of degradation.

Among the proteins which may be destroyed in this way are those concerned with the synthesis of neurotransmitter. When a noradrenergic nerve is ligated, noradrenaline accumulates proximal to the constriction (137, 165, 166, 284, 456, 613, 616, 618, 826, 1045, 1046, 1047, 1385). Dopamine-β-hydroxylase also accumulates. Noradrenaline accumulates at a constant rate after axonal ligation for two or three days (137, 310); it is not entirely clear how much of the accumulation is due to local synthesis and how much to transport from elsewhere. If a second ligature is applied four days after the first the accumulation is less (Fig. 7.15). No accumulation is found at all if the second ligature is applied 21 days after the first (Fig. 7.15). This can be interpreted as a decreased synthesis of the enzymes necessary for the formation of neurotransmitter. It is possible that these may also be destroyed. Also among the proteins which might

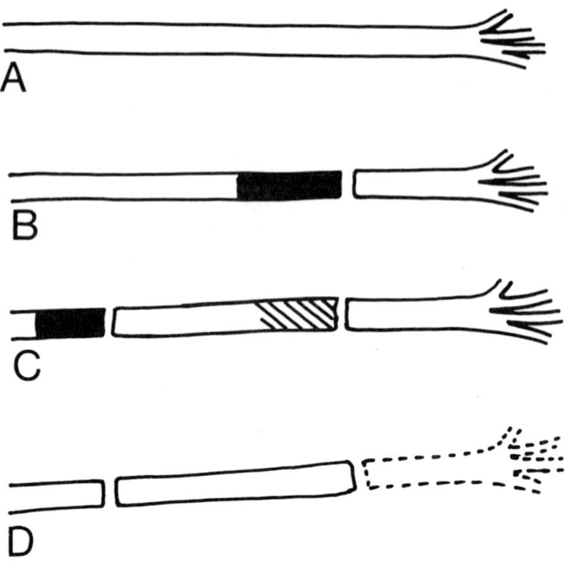

Figure 7.15 When a normal adrenergic nerve is divided (A, B), noradrenaline accumulates proximal to the division (B). If a second division is made 2 days later, proximal to the first, noradrenaline accumulates proximal to the second lesion (C). Alternatively if a second division is made 20 days after the first (D), noradrenaline fails to accumulate proximal to the second lesion.

be destroyed during the early stage of the chromatolytic response are the receptor proteins for the neurotransmitters which act upon the axotomized cell (p. 244). Enzymes which degrade neurotransmitters might also be reduced in this way. The preservation of inhibitory boutons while many excitatory boutons are shed suggests that the process might have some specificity, or that different types of bouton differ fundamentally in the properties regulating their stability.

Nucleic acid synthesis

After axotomy the motor neuron synthesizes more RNA (323, 3062). It does so under circumstances in which the DNA is less repressed (3071). It is possible to dissociate experimentally the response of the nerve cell body to axotomy or to loss of nerve-muscle contact from the response to restoration of this contact

(3073). This separation is obtained by implanting the hypoglossal nerve, severed in the tongue, into an innervated sternomastoid muscle. Under these circumstances the implanted nerve cannot super-innervate the muscle, but the nerve cell body undergoes the sequence of metabolic reactions initiated by axotomy; no reactions due to restoration of nerve-muscle contact can take place. About 70 days later, the spinal accessory nerve is divided; the implanted hypoglossal nerve can then innervate the sternomastoid, and the motor neurons react to this restoration of contact. Hence under these circumstances two episodes of RNA synthesis are found: one due to axotomy, and the other to restoration of contact.

The first episode cannot be interpreted exclusively in terms of the synthesis of the RNA and protein needed for the disconnection of boutons from the cell surface. If a severed hypoglossal nerve is implanted into an innervated sternomastoid, the first episode of synthesis finishes within 30 days and the neuron atrophies; over 90% of the boutons that will be lost by the 150th day have been lost by the 30th day (3073). If such a hypoglossal neuron is divided a second time 70 days after the first division, there is no further measurable loss of boutons; nevertheless, the neuronal RNA content and rate of synthesis of RNA increase. It seems likely that this response is due to the loss of the stabilized axonal tip, or perhaps to the loss of the small amount of axoplasm. By the removal of the stabilized tip the axon has been converted from its 'resting' mode to its 'seeking' state. In a closely related experiment a severed hypoglossal nerve is prevented from growing by 'collision anastomosis'. The proximal stump of the hypoglossal nerve is joined end-to-end to the proximal stump of the median nerve severed at the wrist and drawn into the neck. The hypoglossal nerve does not extend detectably beyond 2 mm into the median nerve. Seventy days later the median nerve is divided proximally near the brachial plexus. The hypoglossal axons then extend along the degenerating median nerve, probably because stabilizing junctions between the terminals of the median axons and those of the hypoglossal nerve have been destroyed. Under these conditions also, the hypoglossal neurons synthesize more RNA. In this experiment also, the boutons are shed from the surface of the hypoglossal nerve cell within 30 days of axotomy, and very few thereafter. The synthesis of RNA cannot therefore be concerned only with the removal of boutons.

The converse experiment, to demonstrate alteration of RNA synthesis when boutons are shed but when axoplasm is not lost is performed by injecting botulinum toxin into the tongue; the dendrites of the motor neurons again retract and boutons are shed (3070, 3073). Although no axoplasm has been lost more RNA is synthesized by the perikaryon. Unfortunately, however, although no axoplasm is lost, it is probable that more is synthesized than normal; the formation of extensive terminal sprouts might require such synthesis.

The measurement of the RNA content of the perikaryon is crude; the estimation of the rate of turnover is also an imprecise parameter. Little is yet known about the types of RNA synthesized under these circumstances, and very little of the kinds of protein formed.

Structural proteins

Silver impregnation studies indicate that considerable changes in the structural proteins of the axotomized nerve cell are probable. Tubulin and actin have been extracted from the hypoglossal nuclei after axotomy. The increases found are difficult to interpret, both because only a part of the nerve cell is studied and because many other kinds of cell are present in these nuclei. Immunofluorescence indicates that the different cells in the nucleus containing the axotomized nerve cells react at different rates, and that the total measured differences are a consequence of many different patterns of alteration. The resolution of immunofluorescence is not adequate to resolve this problem, to show, for example, whether the antibody to actin which can be demonstrated at the periphery of a nerve cell while its boutons are being restored, lies diffusely beneath the plasmalemma of the nerve cell, is confined to subsynaptic cisterns or other discrete regions, or is within nascent presynaptic terminals. Alternatively, the demonstrated actin might be within perineural processes of glial cells. Neurofilament protein has not yet been measured after axotomy.

Enzymes

Activities of enzymes can be measured far more readily than quantities, but are more difficult to interpret, whether they are studied in homogenates or histochemically.

Esterases, including cholinesterase are decreased in the cell bodies of sympathetic ganglion cells within 2 days of axotomy; their activity remains low until about the 50th day (1244). This is entirely consistent with the timing of the detachment and restoration of the boutons, and with the associated electrophysiological changes. The acetylcholinesterase activity of motor neurons also decreases at the time that the boutons are lost, and increases again when they are restored (3064).

The results obtained from measurement of mitochondrial enzymes are varied. The variability has several causes. Mitochondria can pass in either direction along an axon, and they accumulate near the end of a divided axon. Mitochondria are included among the organelles that are contained in multivesicular bodies and dense bodies. It is most unlikely that they can function in this situation, and certain that they are destroyed. The activities of mitochondrial enzymes in the hypoglossal nerve cells of the rat fall early after axotomy, and are later restored (3063). In the hypoglossal nerve cells of the rabbit the activities increase (1233).

Responses of the perineuronal glia

Different kinds of glial cell react characteristically around an axotomized nerve cell (2942, 3065, 3067, 3072). Astrocytes synthesize more RNA and protein both when the boutons are being withdrawn and when they are being restored. They do not respond to metabolic changes in the nerve cell body which are unaccompanied by presynaptic alterations. It seems likely that the astrocytes provide a surface over which boutons can move; or perhaps, by analogy with the role of the radial glial cell during development (Chapter 1) they might guide the processes through the neuropil both during retraction and during restoration. It is not known if the astrocyte guides the moving process or also draws it along; nor is it known if the astrocyte controls the removal or restitution in any way. It seems certain that the first change occurs in the receptive surface of the axotomized nerve cell. If the withdrawn boutons are collateral to boutons that remain, then the regeneration of the withdrawn boutons need not involve any other cell. On the other hand, if all of the boutons of a given afferent neuron are withdrawn, then they cannot obtain information directly from the surface of the injured cell, and their regrowth must involve the carriage of information through other cells. These other cells might be nerve cells or glial.

Oligodendrocytes within the hypoglossal nucleus increase their dry mass and synthesize more RNA and protein only during the period in which the boutons are being reapplied to the surface of the nerve cell. The identity of these cells is uncertain (Chapter 1). If these are the perikarya of cells which myelinate preterminal segments of the axons with boutons which end upon the hypoglossal neurons, then their response might be akin to that of Schwann cells of a peripheral nerve. After the axon of a peripheral nerve has re-established contact with muscle, the axon increases in diameter, and the myelin sheath thickens. This process has been called 'maturation'. It seems possible that the axons of boutons which have been reapplied to the axotomized neuron also undergo a similar process of maturation.

Glial cells which are neither astrocytes nor oligodendrocytes divide around hypoglossal neurons during the phase of bouton withdrawal. These are not derived from vascular pericytes or from blood macrophages. These cells disappear between ten and 30 days after axotomy, that is, after most of the labile boutons have been disconnected. They do not reappear if the hypoglossal nerve is injured 70 days later, provided that the axons have not been allowed to re-establish contact with muscle: that is, provided boutons have not been reapplied to the surface of the injured nerve cell. On the other hand if the motor axon has regenerated and regained contact with muscle, a second lesion leads again to the division of these cells. In both cases the motor nerve cell itself has passed through a phase of increased metabolic activity. Although the most straightforward explanation would indicate that these cells are in some way related directly to the removal of these boutons, this is not necessarily so; for example, the metabolic response of the nerve cell body is likely to be different under conditions in which the boutons are removed from those in which they are not. When neuromuscular contact has not been re-established, the second lesion causes the nerve cell to change from a 'resting' to a 'seeking' mode. When neuromuscular contact has been regained in the interval, the second lesion causes the nerve cell to change from a 'secreting' to a 'seeking' mode (3073). In the former case extensive intracellular activation of lysosomes might not occur; in the latter it does.

The cells which have divided disappear from the nervous system; some can be found in the spleen between ten and 30 days after axotomy; but no quantitative study of their fate has been undertaken. These cells have a variety of names (Chapter 1): microglia, glia of the third type and non-myelinating oligodendrocytes. They are intrinsic cells of the nervous system. It is not known how many there are in the normal hypoglossal nucleus, or whether they have any role except under such circumstances. They do not act as macrophages after axotomy. During the period of bouton withdrawal these cells can be seen lying in the cleft between the postsynaptic membrane and the partially detached bouton (274). It has been suggested that the cell is prising apart the two plasma membranes of the synapse. While this might be so, it seems more likely that the microglial cell occupies this position because it adheres to the membrane. It lies there because it adheres more strongly to the membrane than does the bouton or any other adjacent cell process. After axotomy the microglial cell adheres to the postsynaptic cell; this is further evidence that the membrane which is primarily altered is the postsynaptic membrane. Between ten and 30 days after axotomy processes of astrocytes come to lie between the microglial cell and the perikaryon of the neuron. It seems likely that the astrocyte displaces the microglial cell because at this stage the former adheres more firmly to the nerve cell than the latter. It is attractive to suggest that this indicates a further change in the receptive surface of the neuron; it takes place if the axon fails to re-establish contact with muscle, and coincides with the nerve cell's transition from a 'seeking' to a 'resting' mode (3073). The evidence therefore suggests that the axon can adopt three modes of behaviour, 'seeking', 'secreting' and 'resting', and that each of these modes is associated with a different state of the receptive surface of the nerve cell, indicated respectively by the adherence of microglia, of boutons and of astrocytes, and that the conversion of one mode of behaviour to another is associated with alterations of nucleic acid and protein synthesis.

Responses to axotomy of cells of the central nervous system

The response to axotomy of neurons which have their processes confined to the central nervous system is very varied (1800, 1802). Some nerve cells show chromatolysis, and may later atrophy and die:

for example, large nerve cells of the dentate nucleus and of Clarke's column. Others, such as the neurons of the inferior olive, degenerate very rapidly. Others again seem to show no response in the perikaryon after axotomy; many cells of the cerebral cortex behave in this way, and also pyramidal cells of the hippocampus and Purkinje cells of the cerebellum.

Cells can die in one of two ways; neurons of a given kind can die in either way. In one, the cell becomes electron dense, and many of the organelles become dark and aggregate; in the other, the cells have a clear watery cytoplasm and contain few, poorly preserved organelles. These two types of demise resemble those found in axons separated from their perikarya, and might be similarly caused. If so, 'dark' patterns of degeneration can be ascribed provisionally to a considerable increase in the concentration of cytoplasmic calcium ions; cell death is then due, amongst other things, to the consequent disorder of protein polymerization, to disturbed transport, and to clumping of cytoplasmic organelles. The cytoplasmic concentration of calcium ions may increase because they enter at a lesion close to the nerve cell body, or at a lesion which seals slowly, because the permeability of the cell membrane to calcium ions increases, or because the ability of organelles to sequester them declines. The 'pale' manner of dying can be ascribed provisionally to excess intraneuronal hydrolysis during the phase of increased lysosomal activity. Neither of these two modes explains completely the slow rate of decrease of cell number after axotomy; their number may decrease for several months after axotomy and abortive regeneration. It seems probable that this late slow decline is a specific but indirect consequence of axotomy, mediated through slow retrograde or anterograde transneuronal degeneration or by both (Fig. 7.16). It is possible that an axotomized cell is sustained in some way by those of its afferent connections which are not withdrawn. Different nerve cells of a given type may differ quantitatively in the number of boutons that remain, and those with more may survive longer.

Cells that show chromatolysis and survive may do so either because the axon regenerates, as in some supraoptic nerve cells, or because they are sustained by surviving collaterals of the axon which have not been divided (414). Two models provide evidence that is consistent with this hypothesis: large neurons of Clarke's nucleus and neurons of the lateral mammillary nucleus (981, 1817). Axons of

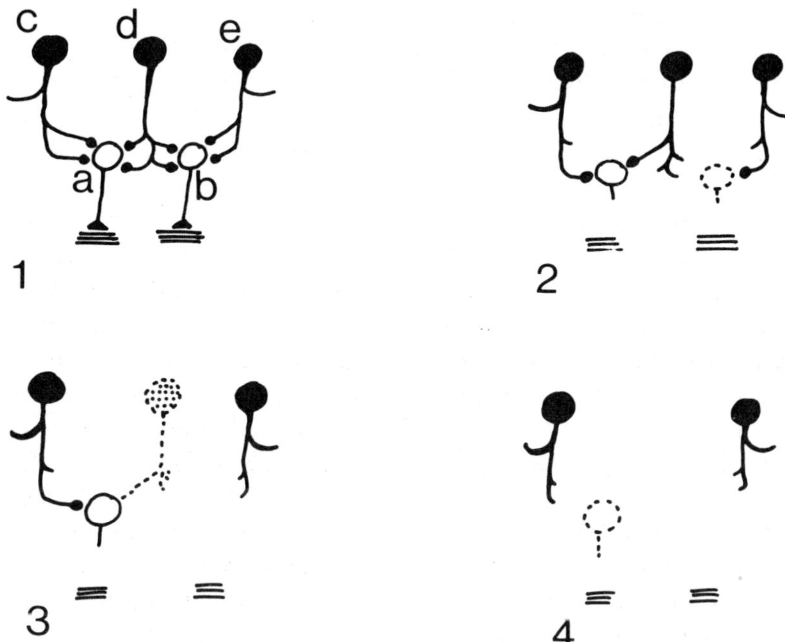

Figure 7.16 After axotomy the injured nerve cells sometimes die; such death can occur long after the lesion. This might be due to a combination of retrograde and anterograde transneuronal degeneration.

Nerve cells 'a' and 'b' (1) are axotomized (2). These normally receive connections from nerve cells 'c', 'd' and 'e'. Cell 'b' might die as an immediate result of the injury (2). Whether it does so or not, many of the boutons of 'c', 'd' and 'e' are withdrawn from the axotomized neurons (2). This withdrawal might cause the death of a cell by retrograde transneuronal degeneration – say cell 'd' (3). The number of boutons remaining on 'a' might then be insufficient to sustain it, so that it dies by anterograde transneuronal degeneration (4). In this situation a combination of early and delayed death of the axotomized cells is seen.

Clarke's column ascend to the cerebellum and give collateral branches in the lower cervical and upper thoracic segments of the spinal cord. If the ascending tract is divided in the lower thoracic region, so that these collateral connections are also lost, more nerve cells of Clarke's column die in the upper lumbar region than if the tract is divided in the upper cervical region, a site that preserves many of the collateral connections. The initial chromatolysis is also more severe with more proximal lesions.

Neurons of the lateral mammillary nucleus of the cat have axons that enter the principal mammillary tract; each axon bifurcates, sending one branch into the mammillothalamic tract and another into the mammillotegmental tract. Some fibres of the mammillothalamic tract again bifurcate and send branches bilaterally to the anterodorsal nucleus of the thalamus. By placing a lesion in one of several tracts, a varying proportion of the efferent connections of the neurons of the lateral mammillary nucleus can be removed. The results show clearly that more neurons die if a greater proportion of the efferent connections is lost. After unilateral destruction of the anterodorsal nucleus of the thalamus no cells die. Lesions of the mammillothalamic tract cause 15% of the cells to die. If the mammillotegmental tract alone is destroyed, all the cells survive. After combined lesions of the mammillothalamic and mammillotegmental tracts, severe retrograde degeneration follows and 60% of the cells disappear. A single lesion of the principal mammillothalamic tract causes degeneration equivalent to the combined lesions. The results of both of these studies are consistent with the hypothesis that cell survival is related to the proportion of efferent connections that remain. The results can be explained equally well, however, by the hypothesis that a greater loss of axoplasm causes a greater degree of cell reaction and cell death. It is known from studies of the reaction of motor neurons that the metabolic response differs quantitatively according to the amount of axoplasm removed; in these cells the number of efferent connections remaining is constant. The results obtained from study of Clarke's column and the lateral mammillary nucleus do not show if each efferent connection is equally able to sustain the perikaryon. It is possible that connections made with particular postsynaptic neurons are more effective in this respect than others.

The apparent failure of other cells of the central nervous system to respond even after a large amount of axoplasm has been lost is interesting; it suggests either that the severed branches of the axon form an insignificant part of the total efferent field of the neuron, or that the metabolic response of the neuron is not determined by the particular branches of the axon removed, or that all axonal connections of the cell are without this property. This would mean that the axonal terminals do not constitute the 'dominant pole' of the neuron; this possibility is entirely consistent with the evidence for heterosynaptic induction in the developing nervous system (1240).

The differentiation of some neurons is regulated through certain afferent connections; especially, the capacity of a neuron to receive some types of afferent fibre is critically dependent upon it having received earlier connections of another kind. This is 'heterosynaptic induction'. Pyramidal cells of the cerebral cortex and Purkinje cells of the cerebellum are known to be influenced in this way. In both dendritic spines develop only after afferent connections have been made; these connections do not themselves terminate upon the spines (1240). The tertiary dendrites of Purkinje cells have spines which receive connections only from parallel fibres, which are the axons of granule cells. Climbing fibres do not form synapses with the spines of tertiary dendrites. The granule cells and other local interneurons can be destroyed selectively in newborn animals by virus infection (1587). The Purkinje cells then receive no contact from parallel fibres; the spines of the dendrites do not receive connections from the climbing fibres, which terminate upon the neuron elsewhere. As the spines still appear on these dendrites they cannot have been induced locally by either type of fibre. They might have been induced remotely by the climbing fibres acting upon another part of the cell. If the climbing fibres are destroyed in a previously normal adult animal a large number of the spines are lost. This suggests that the climbing fibres ending upon one part of the nerve cell facilitate the appearance of spines upon which another kind of afferent terminal ends. The essential experiment, of destroying the climbing fibres during early stages of the development of the cerebellum appears not yet to have been studied. A similar remote induction of spines is found in the somatosensory cortex.

The dominant pole of nerve cells of the central nervous system therefore may or may not be the axon terminal, or it may be a particular type of afferent connection. It seems likely that the division of nerve cells of the central nervous system into those which show chromatolysis and those which do not is really a division into those which have the axon terminal as the dominant pole and those that do not. It might be expected that neurons which make contact with mesenchyme would have as their dominant pole this interface of interaction, which has great importance in other kinds of cell. This seems generally to be so. If the process of a neuron which makes contact with mesenchyme is divided, chromatolysis follows; this is seen in motor nerve cells, the much greater metabolic response of the primary sensory neuron to division of its distal process than to division of its central process, and in the response of neurosecretory cells to axotomy.

The way in which the dominant pole of a nerve cell regulates its biochemical and structural differentiation is not known. There is no evidence that 'trophic factors' must be transferred from one cell to the other (Chapter 8). Although material can be demonstrated ascending the axons of motor neurons this does not provide proof that such material regulates the metabolic direction of the nerve cell. A silastic cuff containing colchicine can block centrifugal axoplasmic flow for seven or more days if it is placed around a nerve, presumably by interfering with the structure or the function of the microtubules (Chapter 2). The axon does not degenerate distal to the cuff. Centripetal flow of an exogenous marker, horse-radish peroxidase, is also arrested. If such a cuff is placed around the hypoglossal nerve, boutons are shed from its surface (597). A cuff containing vinblastine acts similarly. A silastic cuff containing local anaesthetic does not cause a significant loss of boutons. The experiments suggest that the centripetal transport of material in the axon might be important in matching the response of the perikaryon to the state of the axon terminal. Centripetal transport of horse-radish peroxidase cannot be detected for about three weeks after nerve injury; if regeneration occurs, proximal transport then becomes greater than normal for about three weeks before returning to normal (1658). If a severed hypoglossal nerve is implanted into an innervated sterno-mastoid muscle, no horse-radish peroxidase passes proximally. If the spinal accessory nerve is divided 70 days later, so that the previously implanted hypoglossal nerve can establish contact with the altered membrane of the muscle, horse-radish peroxidase can be detected passing proximally within a very few days (597). This suggests that the capacity of the motor nerve cell to ingest material and to pass it proximally depends upon the ability of the terminal to establish contact with the postsynaptic membrane: an ability that depends upon the release of acetylcholine (3073). It is interesting to note in passing that the presence of a silastic cuff impregnated with either colchicine or vinblastine does not lead to the demyelination of the intact distal or proximal axons; this suggests that the maintenance of myelin once formed does not depend upon material transported unidirectionally within the axon.

Transneuronal degeneration

Retrograde transneuronal degeneration is most likely to occur in parts of the brain where there is little divergence of efferent connection; most of the axon terminals are then made by the nerve cells primarily affected by the lesion. Anterograde transneuronal degeneration is most likely to be found where there is little convergence of afferent connections; most of the terminals upon the surface of the neuron then come from one source. Both types of transneuronal degeneration occur more readily in young animals. The severity, rate of degeneration and latency vary considerably between species.

Anterograde

Anterograde transneuronal degeneration is the atrophy or death of nerve cells which receive the terminal projections of axons which have been divided. Favourite models for study include the nerve cells of the lateral geniculate nucleus after the division of the optic nerve (1027, 1028, 1029, 1175, 1686, 1943, 1944, 1947), auditory relay nuclei (2330), olfactory nuclei (1947), somatic sensory relay nuclei (1829) and the inferior olive (2939). The atrophy may be slight, when the cell body is small but dendrites have not been withdrawn (319), or more severe, when the dendrites have also become small (1512, 1818, 1947), or most severe, when the cells die. If the dendrites of a nerve cell receive similar afferents from two different sources, the removal of only one of these causes only the dendrites which have themselves lost their afferent connections to atrophy (1818). Presynaptic terminals of the injured axons are removed; characteristically the specialized postsynaptic membrane persists structurally, sometimes for months (548, 1130, 1440, 1849, 1850, 2290, 2291, 2335, 2763, 2764, 3116, 3117, 3120). The composition of these persisting postsynaptic thickenings is not known. The cause of the atrophy is not known, but possible interpretations have been discussed elsewhere: heterosynaptic induction (p. 257), neurotrophism (Chapter 8) and excitation-metabolism coupling (Chapter 5). Interpretation should not assume that the partially deafferented nerve cell becomes silent or reduces its activity. Several studies indicate that some denervated nerve cells discharge with greater frequency: in the dorsal horn after dividing dorsal roots (1823), in the lateral cuneate nucleus after dividing dorsal roots or the dorsal columns (1600, 1601), and in the nucleus of the spinal tract of the trigeminal nerve after dividing its sensory root (80). These respond similarly; after a period of two to ten days in which the activity is reduced, the activity increases considerably. Bursts of very rapid discharge occur. These can occasionally be influenced by stimulating an adjacent intact nerve root. The rapid discharge is unlikely to be due to the presence of the microelectrode. The rate of high frequency discharge usually does not decrease steadily; rather, it stops abruptly at the end of a burst. It may then begin again and repeat the cycle many times. It is not possible to relate this electrophysiological behaviour precisely to transneuronal degeneration, as the cells from which records are obtained have not been identified by intraneuronal iontophoresis of dye. Nor have sufficient electrophysiological studies been performed at different stages of the response. The cause of the increased discharge is not known; its onset coincides with the time when boutons would be expected to be degenerating and phagocytosed. It might result from generator potentials at the denervated postsynaptic site, similar to those found in denervated muscle; if so, the process is one of neuronal fibrillation. It could also arise from the formation of a pacemaker. Alternatively the bursts of discharge could be considered a result of damage to the postsynaptic membrane when a bouton is engulfed.

If the cause of atrophy is uncertain, the relationship between atrophy and cell death is even less clear. Death is not a necessary consequence of severe atrophy. Cell destruction could be an inevitable result of the loss of a certain proportion of afferent connections, or of the loss of particular connections, perhaps mediated through the entry of calcium, enzymatic hydrolysis, or the cessation of synthesis of essential proteins or other substances. Alternatively, the metabolic state of an atrophied nerve cell may be so fragile that it cannot survive the vicissitudes of later and minor changes in its environment. Quantitative cytochemical studies (1686) do not indicate how atrophy is brought about or why some cells die. It is not clear if the process is closely related to the terminal differentiation which causes controlled cell death during the development of the nervous system.

When some afferent connections degenerate, their place can be taken by collateral sprouts of other axons. If the injured tract is only partially destroyed some of the collaterals may be of the same kind as the terminals lost. Such collaterals might alter the susceptibility of the postsynaptic cell to anterograde transneuronal atrophy. On the other hand, if the injured tract is completely destroyed the collateral connections must come from a different source. It is not known if this collateral bouton substitution results in synapses that are effective both electrophysiologically and trophically, if the latter is indeed a distinct entity (Chapter 8). It is not known if collateral innervation influences the hyperactivity found in many nuclei after denervation.

Species differ in their susceptibility to anterograde transneuronal degeneration. This might result from quantitative differences in the proportion of afferent connections which a neuron pool receives from a given source, from variation in the rate or extent of collateral sprouting and the formation of collateral synapses, or from quantitative differences in the susceptibility of the postsynaptic nerve cell to a constant loss of its afferent connections. It is probable that even within a species these factors vary as a result of genetic polymorphism, and perhaps as a result of the previous and present hormonal, proprioceptive and exteroceptive experiences of the animal.

Anterograde transneuronal degeneration is found in the developing brain. If fibres are divided before they establish contact with postsynaptic cells, these neurons continue to differentiate normally until they reach the stage at which the terminals of the divided fibres would normally have made contact. The nerve cells then either fail to develop further, or die. This suggests that such contact between different neurons is essential for continuing differentiation. Such degeneration may then extend over several synapses, producing degeneration *en cascade*.

Retrograde

Retrograde transneuronal degeneration is the atrophy or cell death which may affect cells which have boutons terminating upon axotomized neurons. The response of optic nerve terminals in the lateral geniculate nucleus after occipital decortication has been investigated (1396, 2394). The corticogeniculate fibres degenerate, and many of the cells of the lateral geniculate die. The rate of disappearance of the optic nerve terminals is very variable in the monkey; six months after injury many optic nerve fibres are swollen and abnormal (1396). The rabbit behaves differently (2394). Within 14 days of cortical ablation, many axodendritic synapses in the lateral geniculate nucleus from the cortical cells have disappeared, and many geniculate cells have died. Four months after injury very few cells remain, but axo-axonal contacts have been formed. They have a characteristic structure. Optic nerve afferents are always the presynaptic component, and an axon with flattened vesicles forms the postjunctional member. The origin of the postsynaptic axon is uncertain. The presynaptic terminals disappear rapidly when the optic nerve is divided.

The withdrawal of some boutons from the perikaryon of an axotomized motor neuron or sympathetic ganglion cell is a lesser example of retrograde transneuronal degeneration.

Retrograde degeneration can occur across more than one synapse; this is seen if the limbic cortex is removed from new born rabbits (272). Retrograde changes are seen across one synapse in the medial mammillary nucleus, and across a second synapse in the ventral tegmental nucleus. The fact that the change can be induced by the first neuron suffering transneuronal influence indicates that loss of axoplasm by the cell inducing the change is not necessary, but that loss of effective contact with the postsynaptic cell suffices to make the cell reject its own afferent connections. This situation has of course affinities with the loss of boutons from the surface of a motor neuron which has had its functional connection with muscle disrupted with botulinum toxin.

The degeneration seen in the cerebellum of the mutant mouse 'staggerer' (Chapter 9) has been interpreted as consecutive retrograde transneuronal degeneration (2765).

Injury and plasticity

The responses of the nervous system to injury provided one method of studying the capacity of the nervous system to alter its pattern of neural connection (1836), or its pattern of synaptic use. Some of the aspects of such plasticity are described in the models discussed below.

Denervation hypersensitivity

Problems of denervation and disuse hypersensitivity have been discussed (425, 426, 816, 817, 923, 2661, 2662, 2952, 2953). Changes which occur in denervated muscle are considered when neurotrophism is discussed (Chapter 8). In such muscle, acetylcholine can evoke a given response with a concentration over 1000 times smaller than normal, and a greater area of the muscle membrane becomes sensitive to the transmitter. Sensitivity to adrenaline and to noradrenaline also increases but to a far smaller extent (306, 419, 949). Denervated muscle does not respond to histamine, 5-hydroxytryptamine, bradykinin or to muscarinic agents. Denervated muscle in some insects fails to show such hypersensitivity to glutamate (2980, 2982).

The nictitating membrane responds differently. Its sensitivity both to acetylcholine and to noradrenaline increases after denervation or disuse; it also becomes responsive to 5-hydroxytryptamine (395, 2594, 2950–2955). The denervated nictitating membrane can accept cholinergic innervation (3005, 3006); when it does so, the enhanced sensitivity to noradrenaline as well as that to acetylcholine disappears.

Living nerve cells in the interatrial septum of the frog can be seen. These are parasympathetic neurons and receive a cholinergic innervation. The boutons on their surfaces are clearly visible. Most of these neurons receive one presynaptic axon, which terminates in about ten boutons; sometimes up to 30 boutons are present, and it is not uncommon for a nerve cell to be innervated by two or three fibres. About 3% of the surface of the perikaryon bears boutons; the rest is free of synapses (679, 1889). The cell body is highly sensitive to acetylcholine when this is applied iontophoretically in the region of the bouton; the rest of the membrane has a lower sensitivity (1262). Examination of these cells has provided the clearest evidence yet available that denervation hypersensitivity can develop in nerve cells (1670). If the vagosympathetic trunks innervating the heart are divided bilaterally, transsynaptic excitation evoked by stimulating the distal stump of the nerve fails after two days. This failure is accompanied by the appearance of new chemosensitive areas. By the fourth day after injury the entire surface has a sensitivity as great as that normally possessed by the subsynaptic site alone, or greater (1670). This enhanced sensitivity persists for at least four weeks. As the possibility that sensitivity to other neurotransmitters might develop on denervation was not studied, the specificity of the response is not known.

Studies of other nerve cells have been less precise and have yielded controversial results. In the central nervous system it is difficult to distinguish the effect of hypersensitivity from effects of altered activity in the remaining synapses, or from other sequelae of denervation, such as collateral sprouting (747, 776, 1479, 2346, 2492, 2535, 2661–2665, 2774, 2810). Even the response of neurons of the sympathetic ganglion is not agreed, although the reasons for disagreement are becoming clearer. If the sensitivity of the ganglion is judged from the response of the nictitating membrane, the contribution of enhanced responses of this muscle must be considered; such enhancement results from disuse hypersensitivity. Furthermore, before the results can be interpreted any effects of partial reinnervation, collateral sprouting of intrinsic connections, sheath cell changes, vascular change and alterations of the extracellular space must be assessed (1141, 1191, 2877, 3025, 3115, 3149). The response of ganglion cells to derivatives of methonium is not only altered quantitatively, but also differs in nature after the preganglionic nerve has been divided (2247).

Although denervation hypersensitivity has been demonstrated in a number of excitable tissues and unequivocally in parasympathetic nerve cells of the frog heart, it cannot yet be assumed that all nerve cells behave in this way; nor is the specificity of any neuronal response known.

Muscle

If a muscle is partially denervated, the remaining axons give branches to reinnervate the deprived muscle fibres at the vacated motor end-plates (1186, 1187, 1369, 1370, 1371). The diameter of these axons increases (778). When the regenerating axons re-enter the muscle, some of these also terminate upon the motor end-plate; such superinnervation can persist.

If a nerve regenerates to a foreign muscle, which on contraction gives a response inappropriate to the original stimulus, then the inappropriate pattern of response persists in mammals. An appropriate pattern of response is not learned. In fish, urodeles and teleosts the situation is different (1919). Some functional or structural remodelling takes place, so that innervation that is inappropriate can result in co-ordinated appropriate responses. The evidence suggests that this plasticity is peripheral, at the neuromuscular junction and not in the central nervous system (1920, 2790).

The restoration of appropriate responses has been studied in the extraocular muscles of teleosts (1920–1925, 2794). The eye rotates when a fish pitches into a tail-up or tail-down position, in a way that tends to keep the eye steady. When the tail of the fish is raised, the upper cornea usually rotates posteriorly, and the lower cornea anteriorly. This rotation is produced by contraction of the superior oblique muscle, innervated by the IVth cranial nerve. When the tail of the fish is lowered, rotation in the opposite direction results from contraction of the inferior oblique, innervated by a branch of the oculomotor nerve. The recti acting alone cannot produce more than 8° of rotation.

If the left IVth nerve is divided and allowed to regenerate, clockwise rotation of the left eye, seen from in front, is lost and then restored. If the IVth nerve is divided and the IIIrd nerve is allowed to innervate the superior oblique, the eye rotates inappropriately when the position of the fish is altered; this pattern of inappropriate response does not alter for as long as the superior oblique remains supplied by the nerve to the inferior oblique alone. This indicates that the fish, like the mammal, does not relearn a new and appropriate pattern of response. If the IVth nerve enters the superior oblique after the muscle has been innervated by the IIIrd, the response again becomes appropriate (1925). The influence of the IIIrd nerve upon the muscle has been stopped. Further experiments have studied the way in which this change is brought about. Electrical recording from the transplanted branch of the IIIrd nerve and from the regenerated IVth nerve show that each is active at the position appropriate for the normal fish. If this pattern of activity found in nerves that had been acutely divided to facilitate recording also occurs in intact nerves, then the activation of the muscle is probably blocked at the nerve terminal. No degenerating terminals are present when the muscle receives again its proper innervation (1922). This cannot, however be accepted as evidence that the terminals remain; withdrawal of boutons from the surface of an axotomized nerve cell takes place without evidence of degeneration or phagocytosis. The process of junctional withdrawal is very different from that of degeneration following axotomy. It is necessary to show that junctions of the IIIrd nerve persist by showing that some of the junctions degenerate when the IIIrd nerve is divided. Some terminals persist after the IVth nerve is divided, but this tells us nothing about their origin, except that they are unlikely to be derived from the IVth nerve. It is also necessary to know if the superior oblique responds to stimulation of the IIIrd nerve when the nerve seems to be causing no longer inappropriate contractions of the muscle.

If the IVth nerve is again divided after it has grown back into the superior oblique and apparently caused the previously implanted IIIrd nerve to stop working, then the fish does not regain the ability to rotate its eye inappropriately. This indicates that the effect of the ingrowth of the IVth nerve upon the previously functioning IIIrd nerve cannot be reversed.

Mammals and teleosts respond similarly in that neither relearns an appropriate movement when the muscle is innervated by the inappropriate nerve alone. They differ, in that teleosts seem to re-establish a co-ordinated appropriate response if muscles are innervated by a mixed population of appropriate and inappropriate terminals (1157, 1920, 1923, 2792, 2794); in mammals only a poorly co-ordinated and often inappropriate result occurs (1641, 2784, 2785, 3109). This difference has been ascribed to the different patterns of neuromuscular innervation. Most mammalian muscle fibres receive only one nerve terminal; even after injury it is uncommon to see a muscle fibre that receives more than two. Multiple innervation is common in urodeles and teleosts (1921). This difference increases the chance that a contact will be made between a muscle fibre and an axon terminal that can elicit an appropriate movement.

Sympathetic nervous system

After the preganglionic trunk has been injured, the distal part of the axon and the terminals degenerate (1440). Postsynaptic specializations persist (2763).

If ganglion cells of the superior cervical sympathetic ganglion are partially denervated, the remaining unmyelinated preganglionic fibres give both terminal sprouts and collateral sprouts. The injured fibres also regenerate and re-establish synapses (2077, 3137, 3138). Unmyelinated fibres of the cervical preganglionic trunk leave the spinal cord of the cat through the ventral roots T_1 to T_5 and perhaps to a much smaller extent to T_7; 90% of the fibres leave through the roots T_1 to T_3. The ventral root of T_1 contains most, perhaps all, of the fibres which make the pupil dilate and none which make the vessels of the ear constrict. Fibres to the nictitating membrane are distributed in most of these roots. The sympathetic fibres in each of the roots can be stimulated separately to identify their connections.

When the sympathetic axons are divided in the ventral roots of T_1 to T_3, 90% of the fibres passing to the superior cervical ganglion degenerate, and all the fibres which make the pupil dilate are lost (1191, 2077). If T_4 and T_5 are stimulated immediately after the section of the roots, the nictitating membrane contracts weakly – as it does on the normal side if stimulation is confined to these roots. After ten to 20 days the nictitating membrane of the injured side becomes more sensitive to intravascular adrenaline; the response of the injured side is greater when the roots T_4 and T_5 are stimulated. The pupil of the injured side dilates slightly, and the dilatation is easily fatigued. By the 28th day after injury, the response of the nictitating membrane to intravascular adrenaline is again normal. Each side again responds equally when either the preganglionic trunk or the postganglionic trunk is stimulated. More acetylcholine is released into the eserinized perfusate from the injured ganglion than from the normal ganglion when the roots T_4 and T_5 are stimulated. The pupil dilates fully when the roots T_4 and T_5 are stimulated.

Histologically many fine axons pass to the dendrites of ganglion cells at this time after partial denervation. These sprouts arise both from the preterminal part of the intact fibres and also from the trunk of the axon (754). Only unmyelinated preganglionic fibres sprout in this way; the myelinated fibres, presumed to be postganglionic do not. The Schwann cells also react, extending many fine processes (3138). Dendrites of the ganglion cells also sprout. Reinnervation of ganglion cells can therefore occur, new terminals can function and their use reverses the apparent hypersensitivity of the nictitating membrane.

If the preganglionic fibres T_1 to T_3 are crushed, they regenerate and re-establish connections with ganglion cells several months after injury, that is after the terminals formed by collateral sprouts have been functioning. After the regenerating fibres have grown back to the ganglion the pupil dilates when T_1 is stimulated and no longer dilates when T_4 and T_5 are stimulated. The vessels of the ear do not constrict when T_1 is stimulated. A normal looking situation has been re-established. There is very little known about how the regenerating fibres regain contact with the ganglion cells, or how they come to restore proper patterns of innervation. Although the correct responses have been restored, it is not known how many of the divided fibres have grown back to their proper sites.

Sympathetic ganglion cells can be innervated by cholinergic axons other than preganglionic sympathetic fibres, by the vagus, or hypoglossal nerve (442, 453, 1183, 1334). Transmission through the ganglion is stopped in such ganglia by ganglion-blocking drugs. The ganglion could only be stimulated through the implanted nerve and not through the preganglionic trunk, indicating that no hidden regeneration had occurred. If the transplanted nerve is divided many degenerating terminals are seen in the ganglion; most of these are axodendritic. After the vagus has been anastomosed both the regulation of pupil size and of the skin temperature have been reported as being normal (453, 1183). This suggests that considerable reorganization of the regulating pathways must take place; it is perhaps surprising that similar reorganization is not found after partial denervation of the ganglion.

The junction between the postganglionic sympathetic axons and the nictitating membrane is adrenergic. If a cholinergic nerve, such as the hypoglossal, is anastomosed to the distal severed end of the postganglionic trunk a cholinergic junction is formed. The muscle contracts when the hypoglossal nerve is stimulated. This response is blocked by a small concentration of atropine, but not by *d*-tubocurarine or dibenamine (3006). The cholinesterase content of the muscle increases (1752). The failure of dibenamine to block the response suggests that the response is not due to ectopic ganglion cells within the postganglionic nerve. The central sensory process of the vagal nodose ganglion can also innervate the smooth muscle of the nictitating membrane; this response is blocked by *d*-tubocucarine but not by atropine (3005).

Hippocampus

In the hippocampus and dentate gyrus the nerve cell bodies lie in a compact layer and the dendrites radiate very largely on one side. Afferent fibres come from different sources in bundles that are widely separated. The boutons of these various fibres terminate upon the dendrites in different layers of the cortex, some of which have sharp boundaries (73–76, 260, 1115, 1116, 1355–1358, 1795, 1861, 1862, 2055, 2335, 2377, 3239–3242). Some of the afferent fibre bundles can be injured individually; in the young animal this either stops the boutons being formed or makes them degenerate; in an adult they degenerate. The response of the remaining fibres that enter the hippocampus through undamaged bundles can then be followed. Such responses can be detected histochemically, or by later studies of the pattern of degeneration that follows a second lesion, applied to another tract.

After perforant fibres have been divided their boutons upon the apical dendrites degenerate. The width of this zone, previously occupied by these fibres decreases; there is a glial reaction which is more marked in older animals (3240). Quantitative studies of the effect upon the dendrites of the pyramidal cells of the hippocampus have not been reported. The depth of the cortex occupied by commisural fibres and by ipsilateral association fibres increases (3239–3242). This is an absolute increase and not just a relative increase resulting from cortical shrinkage. Degeneration studies do not indicate great extension of the zone of septal fibres.

Responses seen in the dentate gyrus are essentially similar. After the fibres from the entorhinal cortex have been divided, the zones occupied by the commisural fibres and by septal fibres increase (1857, 1858, 2838, 3239–3242), expanding into the zone previously occupied by the injured axons and their terminals.

If the commisures are divided, no definite increase is found in the depth of the cortex occupied by the layer of septal fibres (3239, 3240).

Septal nuclei

The septal nuclei of the rat receives fibres from two distinct regions: from the hippocampus and from the medial longitudinal bundle. There are also a number of lesser sources and many intrinsic connections (2109, 2110, 2370, 2371, 2372). The terminals are not arranged in clear layers but can be distinguished. Many terminals derived from the medial longitudinal bundle are noradrenergic; most from the hippocampus are not (1003, 2040). Most boutons from the hippocampus make contact with the spines of dendrites; about half the boutons ending upon spines come from the hippocampus. Virtually none of the afferents from the hippocampus end upon the cell body. On the other hand, over half the boutons derived from the medial forebrain bundle end upon the cell body.

After the fibres from the hippocampus have been divided, their boutons degenerate rapidly (2371, 2372, 2379). The number of healthy boutons which end upon spines is halved within one week, but restored within one month. The rate at which new boutons appear is almost the same as the rate at which degenerating ones are removed (2379). Very few vacated postsynaptic sites are seen at any time. These findings indicate that the unoccupied postsynaptic sites are very rapidly filled by new boutons. Some of the new boutons are certainly derived from the medial forebrain bundle. It is likely that others come from intrinsic connections of the septal nucleus. After the new boutons have been formed, an abnormally high proportion forms junctions with more than one postsynaptic site (2371, 2372). This tendency is not confined to boutons derived exclusively from one source.

Between three and eight days after dividing the fibres from the hippocampus, the noradrenergic terminals of the medial longitudinal bundle appear larger and contain more noradrenaline. After 15 days many fine fibres appear which contain this transmitter (2040); these persist for more than 100 days. If a lesion is made in the medial longitudinal bundle, the fluorescent fibres disappear both in the normal rat and in the rat in which such sprouting has been induced.

Nerve fibres from the hippocampus react similarly when the medial longitudinal bundle is divided; a greater number of boutons derived from the hippocampus then come to end on the cell body (2371).

Leech

The ganglia of the nerve cord of the leech, like those of some other invertebrates contain many large neurons with constant patterns of connection. Thirty out of 350 nerve cells have been identified in each segmental ganglion; some are sensory and others are motor (177, 178, 179, 2125, 2126, 2128, 2845). The ganglia are joined by paired connectives to form a nerve cord. Within these, axons pass between adjacent ganglia and between more remote ganglia. Each connective contains thousands of axons. Fibres do not cross from one of a pair of connectives to the other; if one of a pair of connectives is divided, the disconnected cells lie on the same side of the body.

If a certain sensory cell is stimulated in one ganglion, a characteristic response is evoked in a motor cell of the adjacent rostral ganglion. The cell becomes hyperpolarized after a short latency; this is followed by a much greater depolarization. A response is also evoked in the adjacent caudal ganglion, but it differs in that the response is one of depolarization only (1487).

If one of the paired connectives between adjacent ganglia is divided, the axons regrow and a new connective is formed. The response of the motor cells to stimulation has been studied after regeneration. A disorder of connection is found throughout the length of the leech, both contralateral and ipsilateral to the divided connective. If any pair of segmental ganglia is selected, and the sensory cell is stimulated in the more caudal one, the motor cell responds in the more rostral cell either with hyperpolarization alone, or with a hyperpolarization that greatly exceeds the slower depolarization. If the sensory cell of the more rostral ganglion is stimulated, the motor neuron of the more caudal ganglion responds to depolarizing, as normally. This change has not been explained. It indicates a persisting disorder of synaptic connection. The results indicate that the neurons of the leech can regenerate with a fair accuracy, and that synapses within the nerve cord can change their efficacy.

If a ganglion is isolated by dividing connectives rostrally and caudally, and by cutting the nerve roots, the response of nerve cells to the stimulation of sensory neurons within the same ganglion alters. It is not known if this is due to structural remodelling within the isolated ganglion. This model is in some ways similar to the isolated but perfused slab of cortex. The changes might be due to removal of afferents, to axotomy of many of the cells, or to both of these procedures.

Optic nerve regeneration

The optic nerve of teleosts and of amphibia can regenerate (1033–1042, 1474, 1475, 2786–2794, 2836). Teleosts, urodeles and anurans differ in certain respects.

The optic nerve fibres of the goldfish cross completely at the chiasm and enter the contralateral optic tract. They then enter either the medial or the lateral brachium to reach the optic tectum contralateral to the eye of origin. Regenerating optic nerve fibres all cross the chiasm; they follow the appropriate brachium of the opposite tract to the tectum (109). It therefore appears as though the axons select the proper pathway; the results are, however based upon light microscopy alone, and the route of large impregnated fibres only is known. The regenerating fibres within the tectum pass superficially and then plunge radially to reach the part of the tectum in which they normally terminate; they then divide into their terminal branches. As the ingrowing axons pass superficially they pass across the radial processes of ependymoglial cells. It seems probable both that these processes guide the regenerating axons radially through the neuropil of the tectum, and also indicate to the axons the appropriate place at which they should plunge radially. Such a role is similar to that of radial glial processes during development (Chapter 1). No electrophysiological evidence has been obtained for an initial phase of diffuse regenerative growth within the tectum (1475); in this, the goldfish seems to differ from the newt and the frog. The regenerated fibres produce an essentially normal pattern of projection of the visual field, judged electrophysiologically. The microelectrode does not indicate the precise positions of the terminals, but probably the region of the terminal arborization. The projection of the optic nerve fibres to the tectum is rigidly determined in the mediolateral axis, but not rostrocaudally (1041, 1042). If the lateral half of the optic tectum is removed and the optic nerve is crushed, only those fibres that normally terminate in the medial half of the tectum are accommodated after regeneration has occurred; if the medial half of the tectum is removed only those regenerating fibres that normally end in the lateral half of the tectum form terminal arborizations (1475). The projection is therefore inflexible in the mediolateral axis. If the caudal half of the tectum is removed and the optic nerve is crushed, the whole representation is compressed into the remaining rostral half (1041, 1042, 3212). If the tectum is divided into rostral and caudal halves, and an impervious plate is placed into the incision but neither half is removed, regenerating fibres again form a compressed representation in the rostral half (3213). If the optic nerve is again divided in these animals and the plate is removed, fibres regenerating for a second time can form a normal pattern of projection over the whole of the tectum (3213). The projection can be compressed yet again by dividing the optic nerve again and removing the caudal half of the tectum. If the caudal half of the tectum is removed without dividing the optic nerve, some of the divided fibres which projected to the part removed form arborizations within the caudal part of the remaining rostral half of the tectum. Some of these are distinct from the nerve fibres which originally projected to this region; these fibres appear to have been displaced by the fibres which originally ended in the caudal half of the tectum. Others of these fibres overlap the normal projection. The appearance is therefore one of rigid localization within a mediolateral axis, and of varied termination within the rostrocaudal axis, apparently determined by competition.

The optic nerve also regenerates in urodeles, and the retinotectal projection is restored. The eye can be rotated through 180° when the nerve is sectioned; the extraocular muscles are divided to prevent the rotation being reversed. Under these circumstances optic nerve fibres from a particular part of the retina grow back to the part of the tectum in which they end in the normal animal. In such animals the retinotectal projection is restored, but the representation of the visual field is inappropriate (2786, 2787). This inappropriate representation is not altered by experience. There is therefore no evidence for the remodelling of retinotectal projections as a result of experience. The behavioural responses of the animal are abnormal when vision is restored; optokinetic movements and reactions to visually attractive objects are reversed, and remain reversed. There is therefore no evidence for remodelling of other connections that might have made the behavioural responses appropriate. There is a phase early in regeneration when a small area of the retina projects to a greater area of the tectum than normal (593).

Visual stimulation of one eye of a normal frog gives evoked potentials primarily in the contralateral rectum, and after a further latency of between 20 and 30 ms in the ipsilateral tectum. The contralateral tectum projects to the ipsilateral tectum through the postoptic commisure (1566). The latency indicates that the total pathway between the two tecta is polysynaptic. The ipsilateral projection is such that a given point in the visual field that is seen binocularly, projects through each eye to an identical point in the tectum. This is binocular convergence, similar to that described in the cat (p. 177).

If the optic nerve of a urodele is crushed, it regenerates into the contralateral tectum (2788), where the normal pattern of projection of fibres is restored; electrophysiologically it is possible to map not only the terminal arborizations of optic nerve fibres with reference to the surface co-ordinates of the tectum, but also to identify different kinds of unit discharge at different depths within the tectum (1567). After regeneration, terminal arborizations of fibres having defined properties lie rather more superficially than normal. When the tectum is first reinnervated optic nerve fibres project more diffusely than normal; later the area of tectum from which potentials can be evoked from a given small area of the visual field then decreases. If the regenerated direct retinotectal projection is precise, the indirect ipsilateral projection is unaltered; binocular convergence is therefore unaffected (1038).

An optic nerve can be divided and caused to grow directly into the ipsilateral tectum (2789). The projection of each point of the retina is then correct; that is, the temporal retina of the right eye then projects to the part of the tectum which normally receives the projection from the temporal retina of the left eye. The projection is correct for the retinotectal fibres, but incorrect with respect to the visual field. Inappropriate behavioural responses to visual stimuli are consistent with this anatomical rearrangement. New appropriate responses are not learnt.

If an optic nerve of a urodele is divided and allowed to regenerate, most fibres pass to the contralateral tectum; but a few grow abnormally along the ipsilateral optic tract. This aberrant growth does not occur if the nerve is crushed (1037, 1038). These aberrant fibres project in the same way as fibres deliberately diverted into the ipsilateral tectum. The tectum then receives information from the ipsilateral eye in two ways: through the regenerated nerve to the contralateral tectum and then through the intertectal polysynaptic path, and directly through the aberrant regenerated fibres passing directly to the ipsilateral tectum. These two pathways therefore cause one point in the visual field to project to two different parts of the ipsilateral tectum. The aberrant fibres passing to the ipsilateral tectum excite cells which give axons commencing the intertectal path. The aberrant pathway therefore give a projection which is abnormal to the contralateral tectum. This process gives the double projection to the tecta, which is found after cutting an optic nerve but not after crushing it. No remodelling leads to convergence of the artificially dissociated binocular projections. The experiment demonstrates well two things: that experience does not cause plastic rearrangement of these pathways, and that the mechanisms regulating regeneration have a considerable degree of bilateral symmetry.

Binocular convergence can be adjusted in developing *Xenopus* to abnormalities of retinotectal projection (1040), and the process requires visual experience (1565). This is more likely to be an adjustment of development than an altered pattern of regeneration.

8 NEUROTROPHISM

Neurotrophism is the process by which a nerve cell influences the differentiation of the tissue in which its terminals ramify, and which depends neither upon excitation, nor upon the quantal release of neurotransmitter. The untidiness of this definition lies in its failure to define precisely a 'neurotransmitter'; for example, should ATP as well as acetylcholine be considered to be a neurotransmitter at the neuromuscular junction, and probably at cholinergic synapses also? Should vesiculin, chromogranin A and neurophysin be considered as neurotransmitters? Nevertheless, this is considered to be a more manageable definition than one which considers neurotrophism to be an effect that defies explanation.

Regeneration in Hydra

Hydra behave (336, 337, 2886, 2887). They can move in four different ways; they have complicated feeding habits; they respond to light and to mechanical stimulation; they can adapt to a noxious stimulus. Despite some early expressions of doubt (1318, 2731), the evidence that *Hydra* possess a nervous system is now unequivocal. Nerve cells of characteristic shape can be stained with methylene blue (396, 398, 1757), and with heavy metals (648). Some neurons contain neurosecretory material (398); others have catecholamines or 5-hydroxytryptamine (1763). Three types of nerve cell can be identified either by light (396) or electron (648, 649, 1753) microscopy: sensory, ganglionic and neurosecretory neurons. Nerve cells are most plentiful in the hypostome, in the proximal quarter of the tentacles and in the basal disc (396). Very few are seen in the gastric region. If *Hydra* are immersed in methylene blue for 20 minutes, the neurons die (396, 2544). The animal becomes limp and recovers its repertoire of behaviour only when the nervous system regenerates.

Neurons, like other somatic cells are continuously shed (337), and regenerate from interstitial cells (646, 647). Nerve cells do not divide (646), and all three types differentiate independently (647). If nerve cells are destroyed selectively with methylene blue, regeneration occurs within 24 hours (396, 397). Electrophysiological studies suggest that rhythmic potentials result from nervous activity driven by unstable pacemakers (2221). Electrical silence has not yet been shown to follow specific destruction of the nerve cells, nor has its modification by drugs been studied.

After transection, cells proliferate in the growth zone beneath the hypostome (398). Twice as many nerve cells as normal are found at the cut surface (1771). The amount of stainable neurosecretory material increases for 4 hours and then decreases as it is shed. A regenerate is formed and differentiates. Regeneration is inhibited by low concentrations of physostigmine, hexamethonium, DFP, decamethonium, *d*-tubocurarine, reserpine, xylocaine, ergotamine, amphetamine, methyldopa and chlorpromazine (1771). These drugs also prevent the accumulation of nerve cells at the cut surface. If such an inhibited regenerate is transferred to a normal medium, regeneration and differentiation are usually restored. The action of these drugs might involve the nerve cells in which the appropriate transmitter is found (1763), but it is by no means certain that they do so. These drugs can affect the differentiation of organisms which lack a nervous system. The regeneration of *Hydra* after destruction of the nervous system with methylene blue appears not to have been studied.

Neurosecretory granules have been isolated from *Hydra littoralis* (1754). If these are added to a medium containing a regenerating *Hydra* additional heads are formed. A diffusible substance can be extracted from the hypostome; this causes more heads and tentacles to be formed than normal. Instead of the eight tentacles normally found, 40 may be formed. This effect is not species specific; *Hydra littoralis, Hydra viridis* and *Hydra pirardi* interact. Heterogeneous invertebrate tissue is ineffective (1771). The active material is dialyzable and destroyed by trypsin; it is likely to contain a peptide. It makes interstitial cells divide and directs their differentiation towards becoming somatic cells rather than gametes (397). More material can be extracted from heteropolar than from normal animals, probably because these have more nerve cells (1771). Although not yet proved to be the same, the low molecular weight, head and tentacle-inducing substance (2573, 2574) is probably identical. Most is present in *Hydra* in a bound form. It is almost certainly from large neuronal vesicles (2573). The variation in the amount of this material after *Hydra* is transected does not, however match precisely the variation in the amount of stained neuro-secretory material (1754, 2574). This lack of correspondence between demonstration of neurosecretory material histochemially and the quantity of active principle that can be detected by assay is by no means confined to this animal.

These studies suggest strongly that a material contained in the nervous system of *Hydra* can influence its regeneration.

Amphibian limb

Some amphibia can regenerate a limb after amputation. Immature animals do this more readily than adults (365, 845, 846, 2015, 2017, 2709). The developing opossum can also do so under certain circumstances (2020). Successful regeneration has three phases: wound healing, the formation of a regeneration blastema and the modelling of the regenerating limb. If the stump does not regenerate it atrophies. The age and species of the animal, the site of injury, environmental temperature and the season of the year influence the rate of regeneration, but not the sequence of regenerative changes (365, 846, 2578, 2709, 2712, 2715).

The first phase starts immediately after amputation. The injured surface is covered by a clot. Within one hour epithelial cells start migrating; by 24 hours they have covered the exposed surface. In the first three days epithelial cells accumulate beneath the wound epithelium; most of these cells are derived from differentiated tissues, such as bone and muscle (405, 460, 1272) of the adjacent part of the stump. These cells lose many of their differentiated features but do not divide much at this stage (365). Muscle fibres share in this process; an early feature of this change is the loss of neuromuscular contact (1759). On the second and third days after amputation, axons divided by the injury form sprouts which ramify within the epidermis and the mesenchyme.

The second phase begins about the fourth day and ends between the tenth and the 17th. Epithelial cells near the tip of the stump divide to form an apical cap; no significant cell division occurs elsewhere in the epithelium covering the blastema or the stump (1274). Mesenchymal cells continue to lose their features of differentiation and to migrate into the blastema. They now also divide (365, 943, 1274). In this way is formed a conical blastema. Its mesenchyme lacks features which indicate the varied origin of these cells, or the divergent paths of differentiation that they will later follow (297, 424, 2827, 2828); in this sense they have become dedifferentiated.

In the third stage the blastema becomes angled, flattened and forms rudimentary digits which differentiate progressively. The differentiation of the blastema is not directed from its proximal end, that is, from the stump, but from its distal or epithelial end (845, 2014). Complete differentiation of muscle in the late stage of development depends upon its innervation (1758).

If a normal limb is denervated before amputation, sprouting nerve fibres cannot enter the zone of reaction (2709), and no blastema is formed; the stump then atrophies. The first phase, that of wound healing, is not affected by denervation. The third phase, that of restoration of form, is also not prevented by late denervation, but it is slowed (2611, 2712). A late blastema shows a similar independence of innervation in tissue culture; some differentiation occurs (886, 2826). The second phase, that of the formation and growth of the blastema usually only occurs if nerves invade the blastema (406); exceptions to this are discussed below. If the nerves are divided during the second phase, the synthesis of RNA falls within seven hours and of DNA and protein within 24 hours (729). This decrease may be preceded by a very brief augmentation of synthesis (2711). Within two days of denervation the blastema shrinks (369), a process affecting both mesenchymal and epithelial cells.

A minimum density of innervation is needed for regeneration. This threshold is not constant but depends upon the site of amputation, the amount of associated trauma and the age of the animal. It is very different in different species (2715, 2830). There is further evidence that it is not the number of nerve fibres that is important, but the quantity of axoplasm present in the blastema (2536). These observations need reassessment by electron microscopy, so that the importance of small fibres or of fibres unstained by metal impregnation can be determined. Nerves need not retain their connection with the central nervous system to act in this way (2687). Either motor or sensory nerves or both together facilitate regeneration. Amputation of another limb close to the one being studied slows the rate of regeneration (2970); amputation of a second limb remote from the one examined does not have this effect. This influence is not necessarily mediated neurally.

Hyperinnervation of an amputation site can cause regeneration in adult animals (2704), in which it does not occur normally. It can also cause excessive or bizarre regeneration (2681). Nerve growth factor stimulates regeneration, probably by increasing the number of nerve fibres passing to the amputation site (3097, 3098, 3099); it is only effective in the period in which it is known that the regenerate is sensitive to innervation (3099).

Limbs acquire their dependence upon nerves for successful regeneration. If a limb has developed without nerves it can regenerate after amputation in their continued absence (2317, 2318, 2611, 2913–2918, 3209, 3210). Such a blastema has a fairly normal apical epithelial cap (3210, 3211). A limb which has developed without nerves can acquire a dependence upon them; if it becomes innervated late, and remains innervated for 13 days, an amputation blastema regresses when the nerve supply to the stump is divided (2015, 2016, 2017, 2018). Such an acquired dependance can be lost again after the limb has been denervated for 28 days. A limb which has developed without nerves can therefore gain a reversible dependance upon them for the formation of a regeneration blastema.

A blastema results from the interaction of at least three tissues: skin epithelium, mesenchyme derived from tissues of the stump, and dermis. Their manner of interaction determines whether wound healing alone occurs, or whether regeneration takes place. It appears very likely that it is these interactions that are influenced by the presence of nerve fibres. It also seems likely that regeneration is facilitated when differentiated dermis is prevented from interacting with the epithelium covering the wound. Studies in which skin or mesenchyme are exchanged between aneurogenic and normal limbs show that dependence upon innervation resides primarily in the skin covering the blastema and neither in the mesenchyme of the blastema nor in the skin covering the more proximal part of the stump (1080, 2811). Regeneration only occurs if the amputation stump is covered by skin (2049, 2311, 2484), and only progresses if the dermal mesenchyme dedifferentiates (1060). Epidermis can then interact directly with mesenchyme derived from the stump (2828) to form an apical epidermal cap and a mesenchymal blastema. A basal lamina is formed between the skin and differentiated dermis. Between the apical cap and the mesenchyme of the blastema a basal lamina is very thin or absent. Limbs transplanted remotely grow more readily in regions where the basal lamina is normally thin or where the skin of the recipient site is damaged in a way likely to disrupt the normal relationship of dermis, basal lamina and epidermis (846). Regeneration in an adult animal induced by repeatedly injuring the amputation site may be the result of impeding repeatedly the interaction between epidermis and dermis which leads to wound healing.

The way in which nerves exert their influence upon the regeneration of a limb is not known. Nerve fibres usually penetrate both the dermis and the epidermis of the blastema (2714). A study based upon light microscopy suggests that they need not enter the epidermis (2688); this is difficult to reconcile with the demonstration that the difference between a normal limb and an aneurogenic limb lies in the epidermis. If electron microscopy confirms that nerve fibres can indeed be absent from the epidermis, then it is likely that nerves influence the dermis and prevent it from interacting with the epithelium. Nerves might act in several ways: by making the dermal mesenchyme dedifferentiate so that its cells contribute to the early blastema (1060), by preventing the migration of the dermal mesenchyme beneath the spreading epidermis, by preventing the differentiation of new dermal cells from other mesenchyme under the influence of the covering epithelium, or by impeding one stage of the interaction between dermal mesenchyme and epidermal epithelium. Denervation might facilitate this interaction and so aids wound healing at the expense of regeneration. The manner in which nerves might exert such hypothetical influence is unknown. The process might be related to the observation that regeneration can be facilitated by implanting a bimetallic couple; it is not clear whether the action of this has any specificity or whether it acts just as a chronic irritant impeding wound healing.

Biochemical study of the changes induced by denervation of a blastema have produced results which are entirely consistent with the histology but add nothing to it. Within a few hours of injury the rates of synthesis of RNA, protein and DNA increase; later they decrease. Although cells close to a regenerating nerve bundle incorporate more labelled methionine than the remainder of the blastema (282, 2713), this result cannot be interpreted until it is known if the amino acid has equal access to all cells of the blastema, and if the cells labelled are blastemal or derived either from Schwann cells or from cells of the nerve sheath.

The regression of an established blastema after denervation (369) is prevented by infusing material derived from the innervated and developing blastemata of donors (399, 666). Local injection of extracts of brachial plexus makes a denervated regressing blastema incorporate more labelled amino acid (1735); the effect is transient. A heated extract has no effect. As blastemata of aneurogenic animals have not been used to prepare material for injection, it is not known if the material or materials responsible are derived from nerves, from other tissues, or from both. The regeneration of an amputated tail is accelerated by the injection of extracts of tail blastemata (3085). Regression of a limb blastema can also be delayed by injecting buffered isotonic saline (666), probably as a result of the trauma. Extracts containing ribonucleoprotein or particulate matter are the most effective; both are very heterogeneous (666). Regeneration of a limb of an axolotl is facilitated by injecting extracts of various organs (2317). Fragments of brain, but not of liver or kidney help the limb of an opossum to regenerate (2020). It seems probable that regeneration is facilitated by the administration of tissue extracts in several ways: by the trauma of the initial injection, which is often repeated, by the continuing 'irritation' of the foreign material (whatever that might mean in terms of cellular interaction), by the action upon the cell surface or upon the matrix of enzymes contained within the extracts, by a similar action of enzymes derived from the animals own macrophages or other cells in response to the implantation of material, and possibly by a specific act of facilitation. It is evident that not all of these are likely to be relevent to the 'trophic' function of the nerves.

The nerves which enter a developing blastema are not stable axons, but these too are regenerating. It is interesting that extracts from a normal brachial plexus can aid regeneration, but it is necessary to know if this capacity is greater in axotomized nerves. A regenerating spinal ganglion can facilitate the regeneration of the lens in a newt, whereas a normal one cannot (2329). Regenerating nerves within a blastema contain many vesicles and an increased number of large membrane-bound dense granules (98, 1756). It is attractive but probably wrong to equate these with the neurosecretory vesicles found in a regenerating *Hydra.*

Neurotrophic actions of sensory neurons

It is in principle simple to seek the trophic effects of sensory neurons. As the specialized receptor cell excites the nerve terminal any changes which occur in the cell after axotomy should be due to withdrawal of the trophic influence. Unfortunately, the situation in practice is not so straight forward. Many specialized receptor cells have more than one kind of nerve cell ending upon them; as well as one or more afferent nerves they may receive a specific motor innervation, or sympathetic nerves. Furthermore, afferent nerves commonly branch and serve more than one receptor cell. An action potential passing orthodromically from one receptor cell may then invade other branches antidromically and so excite the terminals in the vicinity of the specialized receptor cells. Such a spread of excitation resembles that seen in the axon reflex.

A denervated receptor cell may atrophy, lose its differentiated features or die. If the cells of the receptor epithelium are renewed by division, the rate of cell division, migration, differentiation or death might be affected. The way in which cells are removed may alter; some organs are invaded by macrophages after denervation.

Taste buds

Taste buds develop in mammals and in the catfish at the time that nerve fibres are first seen to approach the epithelium (856). Lingual taste buds fail to develop in the absence of nerves although the papillae differentiate partially (2747). Epithelial eminences develop which have a connective tissue core; nerve fibres then enter the core and penetrate the basal lamina to gain access to the epithelium. Around the time of nervous invasion the basal cells of the differentiating epithelium contain a large number of dense granules and vesicles (856). Taste buds then differentiate. There is no evidence that material is transferred from nerve to epithelium at this time; the appearance of the granules and vesicles may represent an early stage in the process of differentiation rather than the stage of induction itself. Taste buds do not develop in epithelium removed by transplantation from the influence of the gustatory nerves (857).

Taste buds of amphibia differ in some respects. Their taste buds can develop after the epithelium has been transplanted remotely (2832). It is necessary to assume either that these are formed without the influence of gustatory nerves, or that gustatory nerves can gain access to remotely transplanted epithelium in these animals. The second possibility has not been excluded. It is by no means impossible that gustatory nerve fibres could invade grafts placed in the orbit, and nerves have been seen to invade these grafts, although their origin is not known (2320). It is also not impossible that grafts placed in the liver might be similarly invaded; the liver is supplied by branches of the vagus, a nerve which has been shown to be capable of inducing the differentiation of taste buds in mammals. It is usually assumed that the gustatory fibres of the vagus are responsible; this has not been proved.

The taste buds of mammals and of the catfish degenerate if the gustatory nerves are divided (854, 855, 985, 1184, 1185, 2080, 2409, 3222–3227). This degeneration only takes place if the nerve fibres themselves degenerate. If the sensory nerve is divided distal to the ganglion which contains the cell bodies, then both the axons and the taste buds degenerate; if the nerve is divided proximal to the ganglion, the axons survive and the taste buds do not disappear. Continuing connection with the central nervous system is therefore not essential (1528); limb regeneration shows a similar dependance upon the presence of nerves, and a similar independence of continuing connection with the central nervous system.

Within three to six hours of dividing the distal process of the sensory nerve the terminals of the nerve to the taste bud of the rat contain abnormal dark bodies (855); degeneration is obvious by 12 hours (855, 985, 2080, 2556), and nearly all the nerves have disappeared by 48 hours. On the fourth day the number of taste buds begins to decrease (855, 1184, 2080), and the disappearance continues for six days (1184). The taste buds become smaller within this period and contain many degenerating cells; such cells seem to be extruded through the pore of the taste bud. There is very little inflammatory response or phagocytosis in the rat. The taste buds of the catfish are invaded by leucocytes and disappear more rapidly. Taste buds do not atrophy if the sympathetic innervation is removed by excising the cervical ganglia (715, 1007). The non-gustatory epithelium of the tongue of the rat also becomes thinner after the glossopharyngeal nerve has been divided (1184), an effect that is not found in the cat (1185). After prolonged denervation the mesenchymal core of the lingual papillae also atrophies (1184).

If the denervated tongue is reinnervated by a gustatory nerve, taste buds reappear (1185, 1189, 2148, 2149, 3222–3227); intimate contact between the nerve and the epithelium seems to be necessary for such renewal. The chorda tympani and the glossopharyngeal nerve can each replace the other in inducing taste buds; the vagus can replace either. The hypoglossal nerve cannot induce the reformation of taste buds in denervated epithelium. It has been claimed that the central process of the glossopharyngeal nerve can induce redifferentiation. (3224). Although this is not surprising, as the proximal and distal processes are parts of the same cell, the evidence presented does not fully substantiate the claim. Processes growing from the proximal end of the petrosal ganglion are not necessarily the proximal processes of the cells of the ganglion. If the observation is substantiated, it is of some interest in view of the evidence that different proteins are transported along the central and distal processes of dorsal root ganglion cells (Chapter 2). It would be interesting to see if this is also true for the glossopharyngeal nerve, and if the material transported by each process changed after such altered connection.

Ingrowing gustatory nerve fibres can induce taste buds in areas of lingual epithelium that previously lacked them (3225). More buds are formed if testosterone is administered during the period of nerve regeneration (63, 3223), an effect probably allied to the atrophy of these receptors after castration (2273). It is not known if this action of testosterone is exerted upon the epithelium, upon the regenerating nerve fibre or upon both.

The chorda tympani and the glossopharyngeal nerves contain fibres which differ quantitatively in their response to the various modalities of taste and to temperature change (2148). If the chorda tympani and the glossopharyngeal nerves are divided, and the proximal stump of the glossopharyngeal is anastomosed to the distal stump of the chorda tympani, regeneration occurs and the differentiation of taste buds follows; the sensory responses obtained have features of the epithelium previously supplied by the chorda tympani, and not of the glossopharyngeal nerve now supplying it (2148). Similarly if the proximal stump of the chorda tympani regenerates into the distal stump of the glossopharyngeal nerve, the sensory responses are characteristic of the epithelium and not of the nerve. Hence although innervation by gustatory nerves is necessary for the formation of taste buds in the rat, the spectrum of sensitivity of the receptor remains that of the epithelium and not that of the nerve.

An influence of innervation upon taste buds cannot be doubted; it is due to gustatory nerves and not to sympathetic or motor nerves. It is not known how this action is brought about.

Lateral line organs

The lateral line organs of fish and tadpoles mostly lie within canals on the lateral side of the head and body (122, 3191–3193). In most species the organs are discoid; the receptor cells bearing hairs lie in the centre of the disc and the 'supporting' cells around the periphery. When the fluid within the canal moves, a gelatinous cupula lying upon the hairs is displaced. This movement is detected by the cells bearing hairs. Each hair cell has one kinocilium and several stereocilia (936). Displacement of the stereocilia towards the kinocilium probably causes depolarization and excitation; displacement in the opposite direction causes hyperpolarization and inhibition (935, 936, 3114). Hair cells receive an afferent terminal from the lateral line branch of the vagus and also an efferent terminal from the same source. The efferent fibres, like those of the vestibular apparatus inhibit the discharge in afferent nerves (2457, 2458, 2531, 2532). At the synapse believed to be afferent, vesicles are found within the hair cell close to a synaptic bar, and no vesicles are seen in the nerve terminal. At the synapse believed to be efferent, the terminal contains vesicles and the hair cell has a subsurface cistern (934).

Lateral line organs develop from an epidermal thickening or placode near the vagal ganglion (1268, 2831–2835). Vagal fibres enter this placode before the cells migrate and maintain contact with them during migration. The organs atrophy if this contact is lost. In fish, this atrophy is rapid (122, 351, 688); in amphibia, slow (185, 1507, 2775–2782, 2835, 3191, 3192). If the somatic nerves passing segmentally to the body wall are divided as well as the lateral line nerve, atrophy is accelerated (1507), but still remains slower than that of fish. Fish differ in the apparent interaction between the nerve and the receptor. Lateral line branches of the vagus of the goldfish can make only the epithelium of the canal form organs, whereas the nerve of the catfish makes cells of a wide area of epithelium differentiate (122). It is interesting to compare this with the ability of the canals themselves to regenerate after injury. The canals of the goldfish regenerate readily, but those of the catfish hardly regenerate at all.

Reinnervation of atrophied lateral line organs leads to their restoration. The number of organs increases by the division or budding of those already there. The number of cells in each organ returns to normal; supporting cells are normally restored before hair cells. The cells return to their normal size. Regeneration usually starts close to the site of nerve injury and then passes distally, a process presumed to follow the regeneration of the nerve (351).

At metamorphosis the lateral line organs of the tadpole disappear; this process has two points of interest: first, although the lateral line nerves disappear at about the time that the innervation is lost, the disappearance does not appear to be due to the loss of innervation, and second, the disappearance of the lateral line organs is accompanied by a reduction in the projection of the nerves to the cerebellum; the only innervation of this kind that persists comes from the vestibular organs (1714, 1715).

Despite this close association between presumptive receptors and the nerve during development, and despite the continuing dependence of the organs upon the nerve, the tadpole can form new receptors from those already present in the absence of apparent innervation. If the tail of a tadpole is amputated, it regenerates. The transected lateral line develops a cord of cells near the injured zone; these migrate into the regenerating tail and form new receptors. This still occurs if the nerve to the lateral line is divided before amputating the tail, and is injured repeatedly to prevent regeneration (2775–2782, 2835, 3191). This column of regenerating cells has not been examined by electron microscopy to demonstrate conclusively the absence of innervation. A role for fibres from the somatic segmental nerves has not been excluded.

The saccular maculae in the semicircular canals of mammals resemble lateral line organs; the receptors of the cochlea differ in several important respects (1229, 3114), but have broad similarities. These cells also receive an afferent and an efferent nerve supply. The distal axons of the primary afferent neurons cannot be divided easily to see if the receptor cells of the inner ear atrophy, lose their features of differentiation, or degenerate. The cell bodies of the primary vestibular neurons lie in the internal auditory meatus, and those of the cochlea within the canal of the spiral lamina of the modiolus. Efferent axons to the sensory epithelium of the vestibule transport radioactive protein distally; the radioactivity later appears in the cells of the epithelium (69). While it is possible that this does not represent the degradation of axonal protein and the reutilization of labelled aminoacids, the transfer of a given protein has not been proved. The transfer of material from a nerve cell to another cell does not prove a neurotrophic relationship; a demonstration that such transfer does not occur does not exclude one.

Cutaneous receptors

Encapsulated sensory corpuscles differ from taste buds and from receptors of the lateral line – vestibular – acoustic system; the cells innervated form a complicated buried capsule and do not line a surface; they may not be epithelial. The classification of these receptors has been discussed (447, 2359, 2708).

The Pacinian corpuscle (494, 495, 2239, 2310, 2558) receives one myelinated nerve fibre (449). Two unmyelinated efferent fibres enter through the hilum of the receptor; they remain closely associated with the afferent fibre, but make no junctional contact with it (494, 495, 1826, 2228, 2557). These fibres are noradrenergic (493, 2557). Weak fluorescence of the outer core of the receptor has been ascribed to catecholamines, but might be an artefact (2557). Adrenaline reduces the threshold of the receptor (1749, 1825), and it seems likely that the efferent nerves do so also. As adrenaline acts similarly upon bare terminals this response is not a peculiar feature of this corpuscle. During fetal life a corpuscle develops around a characteristic branched nerve ending (449). It is not known how the terminal is produced. The origin of the cells of the inner core is uncertain; cells of the outer core probably come from the surrounding adventitia. Corpuscles do not develop in the absence of innervation. There is no convincing evidence that new corpuscles develop in the adult. A characteristic sequence of changes follows denervation (495); the first evidence of degeneration is found in the axon within 12 hours. By the second day the cells of the inner core react and these probably ingest the degenerating axon, which disappears by the tenth day. Cells of the outer core survive for at least 120 days. The consequences of reinnervation have not been clearly described.

The Herbst corpuscle (2072, 2084, 2359, 2571) has a central afferent axon (95) surrounded by 60 to 80 lamellae of the core. These lamellar cells have desmosome-like contacts with the nerve terminal (77, 2072). The nuclei of these cells form a characteristic twin row, one on either side of the axon. Nerve fibres pass also between the lamellae of the core; these fibres contain agranular vesicles 45 nm in diameter and dense cored vesicles 80 nm in diameter. These are not branches of the afferent axon and might be autonomic (2434); fluorescence due to catecholamines has not yet been sought. These corpuscles develop only in skin innervated by somatosensory nerves, which retain their inductive capacity after separation from the central nervous system. Autonomic or motor nerves cannot act in this way. Corpuscles develop only in appropriate dermal mesenchyme (2571). After the afferent nerve has been crushed in the adult, the corpuscles atrophy when the nerve degenerates; if reinnervation occurs, the corpuscles become large again (2358). The fate of the Herbst corpuscle which has been denervated for a long time is unknown.

Some touch spots (1456, 1457, 1458) contain tactile cells resembling Merkel discs. Each spot is innervated by a myelinated axon which branches at the base of the receptor. There is no evidence yet for an efferent supply to these receptors. After the afferent nerve has been divided, changes are seen within four days (353); the epidermis becomes thinner, the sac of the tactile cell becomes smaller and the nucleus more central. Between 20 and 30 days after denervation the corpuscle is still recognizable, but tactile cells cannot be seen. If the nerve is crushed, axons may reappear after about 16 days (353). Ten days after this the tactile cells are again evident, with typical large clear sacs. Until the fate of denervated tactile cells is known, it will not be possible to judge whether the cells that appear on reinnervation arise from redifferentiation of the ones or from new cells.

A Meissner corpuscle (443–450, 2708) receives several nerve fibres; each of these can innervate more than one corpuscle. The fibres ramify in planes between the lamellae formed from non-nervous cells. The relation between the nerve fibres and the lamellae therefore differs from that of the Pacinian corpuscle. In some regions the axolemma lies very close to the membrane of the receptor cell; in these regions vesicles are usually found both in the cell and in the axon (450). The receptors develop during fetal life; it is not known if the nerve terminal has special features before the appearance of an identifiable capsule. The origin of cells of the organ is unknown. The number of corpuscles decreases with age, and the number of nerve fibres passing to each corpuscle also decreases (443). It is probable that no new corpuscles develop after birth. The structure of the corpuscle changes with age, especially where it is exposed to trauma. Experiments designed to study the effects of denervation and reinnervation are few. Denervation does not alter the cholinesterase or the alkaline phosphatase activities of the receptor cells, and the corpuscles persist for at least 18 months. The corpuscles do not develop in skin transplanted from an area in which they are not normally found to one in which they are common (2437). As this experiment was performed in adults, this failure may result from an incapacity of nerves to induce receptors in an unsuitable epithelium, or from the inability of this interaction between nerve and any skin to take place in an adult.

Skeletal muscle

Types of fibre

Most muscles contain fibres of different types. These can be distinguished in three ways: by differences in excitation, by differences in excitation-contraction coupling, and by differences in the metabolic pathways preferentially used to obtain energy. Differences in excitation include differences in the structure of the motor end-plates, in the properties of the propagated action potential and in the chemosensitivity of the membrane. Differences in excitation-contraction coupling include differences in the amount and arrangement of the sarcoplasmic reticulum and tubular system, in the myofibrillar or actomyosin ATPase activity, in the structure of myosin, in the interaction of actomyosin and ATP, and in the rate of contraction and relaxation of the muscle. Metabolic differences involve the enzymes of the glycolytic, lipolytic and oxidative pathways, differences which are related to the density of the capillary network and to the muscle's ability to withstand fatigue. Muscle fibres are labile, and can change their excitatory, contractile and metabolic properties.

The types of fibre that a muscle possesses is controlled by its innervation in a number of ways (768, 770, 1188, 1204, 1205). One axon supplies many muscle fibres; these make up the motor unit. For example the anterior tibial muscle of the rat has motor units which each have between 80 and 180 muscle fibres (787). This muscle has about 40 units, and the soleus of the rat about 30 (511). All the muscle fibres of a normal motor unit do not lie together, but are widely dispersed in the muscle (394, 787). With few exceptions (787) the fibres supplied by one axon are histochemically similar (394, 787). As most muscles contain motor units of histochemically different types, a transverse section of a muscle appears as a variegated mosaic of fibres (739, 787, 3205, 3206, 3207). The mechanical responses of different muscle fibres of a single unit has not been measured; it is usually assumed that they are similar. The contractile response of single motor units (394, 511, 1606, 1793, 2174) indicates that units can be divided into a group that is clearly rapidly contracting and a group that is slow. For example, extensor digitorum longus of the rat has about 40 units, all of which are fast, with an average contraction time of 11 ms. In contrast, 27 of 30 motor units of soleus have an average contraction time of 38 ms and the remaining three of 18 ms (511). Motor units which contract slowly and others which contract fast have also been found in the gastrocnemius of the cat (394). Motor units composed of muscles which contract slowly are resistant to fatigue, and have a predominantly oxidative. Other rapidly contracting fibres are easily fatigued; rapidly contracting fibres are also resistant to fatigue and predominantly oxidative. Other rapidly contracting fibres are easily fatigued; these have a high glycolytic and a low oxidative enzyme activity.

Analysis of the response or biochemistry of a whole muscle must be interpreted cautiously because of the different types of fibre they contain. This is especially so if the functional state of the muscle is altered by denervation, stretching, cross-innervation, stimulation, tenotomy or altered use; in these situations the proportions of different types of fibre can alter, and in some of these the number of fibres can change. Even in normal muscle some fibres change: for example through growth, hormonal change or minor degrees of spontaneous injury (103, 104, 105, 106, 146, 2427, 2428, 2584).

Motor end-plates of fast-twitch muscle have a greater mean area than those of slow in some species (518, 519, 1316, 1317, 2198); it is uncertain if the end-plates of all the muscle fibres of one motor unit have similar areas, or if fibres of different sorts within a single muscle differ in this respect. It seems probable that they do. The mean frequency of miniature end-plate potentials is greater in fast-twitch muscles. More detailed analysis might show if this, and the differences in sensitivity to blocking drugs (1496, 1497, 1905, 2226) can be correlated with the end-plate area. Motor nerves to fast-twitch muscle usually discharge at higher peak frequencies than motor nerves to slow muscle. Sensitivity to acetylcholine is confined more closely to the motor end-plate in fast-twitch muscle; end-plates of fast-twitch muscle have a greater activity of cholinesterase than those of slow-twitch muscle. (595). An action potential is propogated more slowly along the membrane of a slow-twitch muscle; a slow-twitch muscle can also continue to conduct more readily when the external concentration of sodium ions is artificially lowered.

The contractile mechanism of muscle has been reviewed a number of times (1445, 1446, 1447, 2027, 2234, 3121). The molar ratios of actin to myosin are equal in muscles having different proportions of various types of fibre (997–1001). Myosin and actomyosin from fast-twitch muscle have a higher ATPase activity (2649, 2806, 2807, 3060), and the actomyosin is superprecipitated more rapidly (999, 2550, 2649, 3204). ATP is hydrolysed more rapidly by fast-twitch fibres (637). A greater proportion of muscle fibres in fast-twitch muscle has a high activity of myofibrillar ATPase (662, 724, 1193, 1194, 1199). The muscle fibres of a single motor unit have a similar ATPase activity.

The ATPase activity of myosin is lost if the light chains (728, 1001, 3090) are removed. Fast-twitch muscles have three light chain myosin components of different molecular weight (1839, 3091). The third light chain component is absent from muscles composed of slow-twitch fibres; the second is slightly heavier in slow-twitch than in fast-twitch fibres; the first has two components of slightly different molecular weight in slow-twitch fibres, but only one in fast-twitch fibres. Fast-twitch oxidative muscle has the same myosin light chains as fast-twitch glycolytic muscle (2545), indicating dissociation between the metabolic preference of the muscle and its determinants of contraction. The light chains of slow-twitch muscle resemble in many respects those of cardiac muscle. The molecular weights of the different light chains are very similar in guinea pigs, sheep, rabbits and chickens (1001, 1839, 3091); recombination studies indicate that some differences between the species are present, however (2839). It is likely that the differences found in the ATPase activity of various kinds of muscle (150, 781), and the differences in the sensitivity of the enzyme to divalent cations (2649) result from these differences in the light chains. The composition of the light chains influences not only the ATPase activity of myosin, but also its combination with F-actin (2839).

The activity of troponin (763, 764, 1558), measured as the specific binding capacity of actomyosin for calcium, is greater in white muscle than in red muscle (998). The absence of a close correlation between activities attributed to troponin A and troponin B suggests that their proportion might differ in different types of muscle fibre (1137, 1269, 1270).

Most fibres in slow-twitch muscle have less sarcoplasmic reticulum (2200) than those in fast-twitch muscle (50, 2584); muscle fibres with exceptionally rapid twitch and relaxation times have a very prominent sarcoplasmic reticulum (2430). Fragmented sarcoplasmic reticulum isolated as microsomes from slow-twitch muscle concentrates calcium more slowly than that isolated from fast-twitch muscle (1916, 2806). The specific activity of glycolytic enzymes associated with the sarcoplasmic reticulum is less in slow-twitch muscle (847). Ultrastructural differences are seen by negative staining between sarcoplasmic reticulum from fast-twitch muscle and that of slow-twitch muscle, and points of similarity are found between the sarcoplasmic reticulum of slow-twitch muscle and that of cardiac muscle; these changes can be related to the prolonged relaxation time of slow-twitch muscle.

The oxidative, glycolytic and lipolytic activities of muscle cannot be correlated simply with the twitch time of muscle. Some muscles with a high oxidative enzyme activity have a slow twitch; others have a fast twitch (1227, 1721, 3207). A far better correlation is found between oxidative enzyme activity and resistance to fatigue (394, 510, 511, 787, 2478, 2584). Muscles with a high oxidative enzyme activity do not necessarily share all other metabolic characteristics; for example, the capacity of chick mitochondria to utilize β-hydroxybutyrate is found only in slow-twitch and not in fast-twitch oxidative muscles (105). Isoenzymes commonly differ between muscle fibres (2570). Aerobic fibres contain many mitochondria, and there is a higher specific activity of succinic dehydrogenase in mitochondria from oxidative than from glycolytic fibres (103–106). It seems that the most useful classification of muscle fibres into types is into three groups: fast and fatiguable, fast and resistant to fatigue, and slow and resistant to fatigue (394, 510, 511, 787, 2584).

Changes following denervation

There is considerable uncertainty whether the biochemical and structural changes found in denervated muscle (3195) result from lack of transsynaptic stimulation, from lack of focal depolarization beneath the neuromuscular junction (miniature end-plate potentials), from the lack of propagated action potentials in the plasma membrane of the muscle fibre, from disuse of muscle, from passive stretching of denervated muscle or from the loss of some other 'trophic' influence of nerve, which might be exerted only at the time of transsynaptic stimulation. It is at least possible that the different components of the response of muscle to denervation might be caused in different ways (2112).

Changes in the membrane

The plasma membrane of an innervated fast muscle, for example extensor digitorum longus of the hind limb of a rat, is sensitive to acetylcholine applied by iontophoresis only at or very near to the neuromuscular junction (56, 1995). The membrane of an innervated slow muscle, for example soleus, is sensitive for a significant distance beyond the margins of the end-plate; this extrajunctional sensitivity is less than that of the junction (56, 1998). After denervation the area of chemosensitivity spreads both in fast and slow muscle (53, 1995). This spread begins within one or two days and is well advanced by six days (849, 1990–1993). The sensitivity of the extrajunctional membrane is then as high as that of the normal junction, and the sensitivity of the junction does not alter (117, 1863). The process begins earlier if only a short length of nerve remains attached to the junction (817, 1841). The development of extrajunctional chemosensitivity is prevented by drugs which reduce or stop the synthesis of RNA or protein (849, 1126). It is not prevented if the synthesis of DNA is stopped by irradiation (1265). Many studies have demonstrated an increase in the synthesis of RNA and protein by muscle after denervation (1590). The proteins necessary have not been identified; more of the proteolipid that might be associated with the acetylcholine receptor is synthesized after denervation (1851, 1852). In such a denervated muscle, the sensitivity of the membrane beneath the neuromuscular junction is not reduced by preventing the synthesis of RNA or protein (849); extrajunctional chemosensitivity once established is also not dependent upon continuing synthesis of protein. Extrajunctional sensitivity develops in muscle kept *in vitro* (905, 906, 2758), and is not prevented by the addition of acetylcholine to the medium (1991). There is no obvious relationship between the frequency of miniature end-plate potentials measured over a short period and the presence or absense of extrajunctional sensitivity in that fibre (59, 1991). Such extended chemosensitivity is found in the muscle fibres of rats when diphtheria toxin or local anaesthetic blocks the passage of action potentials along a mixed nerve passing to the muscle (1863), even though miniature end-plate potentials continue. On the other hand, a local anaesthetic blockade of a rabbit nerve does not produce such sensitivity (2451). Innervated but disused muscle has an extrajunctional membrane that is more sensitive than normal to acetylcholine, but less sensitive than denervated muscle (1498); disused muscle has miniature end-plate potentials of a normal frequency (1498, 2448). Locally injected botulinum toxin prevents the release of acetylcholine from the motor end-plate; the extrajunctional chemosensitivity becomes as great as that of denervated muscle (2907). If colchicine or vinblastine is applied to the external surface of a mixed nerve to impede transport associated with microtubules

(Chapter 2), the extrajunctional sensitivity of the muscle increases (59). In this situation, the frequency and size of the miniature end-plate potentials are normal when measured over a short period. The nerve could conduct action potentials across the region of application of the drug, but no evidence has been presented to show that the muscle is contracting at the normal frequency in the conscious animal. Direct electrical stimulation of muscle reduces considerably the extrajunctional chemosensitivity of denervated muscle (1508, 1863); as this is found after only one hour's stimulation, it cannot be due to a diminished rate of synthesis of protein. Cholinesterase activity which is normally well localized to the motor end-plate and to a lesser extent to the musculotendinous junction, also spreads widely over the surface of the denervated muscle fibre (371, 723, 2960). This spread is not related quantitativily or temporally to the spread of chemosensitivity. The activity of the enzyme at the motor end-plate falls (592). It is unclear if the enhanced action of catecholamines upon denervated muscle is exerted only upon the plasma membrane, or also upon the sarcoplasmic reticulum.

The resistance and capacitance of the membrane of normal muscle can be described as that of two membranes in parallel, with different properties (848). It is likely that one represents the properties of the plasma membrane, and the other the properties of the sarcoplasmic reticulum. The changes found by electrophysiological methods in denervated muscle have not been analyzed in this way, and cannot be ascribed definitely to one set of membranes or the other. The input resistance and length constant of denervated muscle are high (53, 1426, 1729–1731, 2122, 2908); the potassium conductance (1426) and the potassium flux (1263, 1602) fall. Values for the sodium influx vary according to the way in which they are obtained (13, 581). Late exchange, which almost certainly measures influx across both the plasma membrane and the sarcoplasmic reticulum is slowed by denervation (13). The rate constant for sodium efflux is not altered, and the intracellular concentration of sodium is not changed. The resting membrane potential falls by ten to 15 mV (53, 55, 2908). These changes take place at about the same time as the spread of chemosensitivity (2908), but may not coincide precisely with it (53).

There is other evidence which suggests that the channels in the plasma membrane concerned with the passive movement of sodium ions are altered as a result of denervation. The action potential still depends upon the increased sodium conductance of the excited membrane (2422, 2423) and is reduced if the sodium concentration of the bathing medium is lowered. Tetrodotoxin is a drug that blocks passive sodium conductance, probably by reacting with polar groups of cholesterol molecules of the membrane near the sodium channel (420, 421, 3015, 3016, 3018). It prevents the conduction of an action potential in normal muscle; it is less able to do so in denervated muscle (57, 58, 2421, 2422, 2423). The reduced sensitivity of the membrane to tetrodotoxin appears to coincide with the increased membrane resistance, and enhanced chemosensitivity (2421, 2422, 2423), although exact coincidence has been doubted (53). It is not possible to assess at present whether the altered action of tetrodotoxin results from a changed configuration of the wall of the sodium channel, or from altered access of the drug to an unchanged channel. Batrachotoxin, which increases the passive sodium conductance of membranes probably by reacting with sulphydryl groups of membrane proteins near sodium channels (54, 58, 1373) and possibly elsewhere (2097) is able to depolarize both innervated and denervated muscle (57, 58, 1323, 3053, 3054). Tetrodotoxin is able to reverse the action of batrachotoxin only in innervated and not in denervated muscle (57, 3053). These membrane changes do not occur if the synthesis of RNA or of protein is prevented pharmacologically (1126).

At about the same time, denervation also reduces the critical depolarization at which action potentials are generated, and the rate of rise and amplitude of the action potentials (56, 57, 2122, 2422, 2423). These results cannot yet be correlated with changes in composition of the muscle fibre membranes (971, 1118, 1119, 1120).

It appears most reasonable to attempt to explain these changes by the withdrawal of some influence of the nerve terminal upon the muscle (93, 1635, 1636, 1638). Although it is possible that motor nerve terminals might release substances which are taken up by muscle fibres and influence directly the metabolism of the muscle, these have not been demonstrated. The membrane properties of the muscle change when the activity of the muscle is reduced even if the release of acetylcholine in quanta is uninfluenced, and the changes regress in denervated muscle that is stimulated directly (1508, 1863). Two facts, however prevent the conclusion that the membrane alters because it conducts fewer action potentials. First, the changes of denervation persist despite fibrillation potentials (2946) propagated in all directions over the fibre from a pacemaker close to the former neuromuscular junction (196); second, the extrajunctional chemosensitivity appears near a denervated end-plate of a muscle fibre that has a second and normally innervated end-plate (1990). In both circumstances enhanced sensitivity is found despite repeated depolarization of the plasma membrane.

Changes in the rates of synthesis and destruction of the lipid constituents of the membrane of muscle have been found to follow denervation (99, 971, 1118, 1119, 1120, 1285, 2241). Gangliosides increase and alter (556, 1953). Interpretation of these findings will remain difficult until pure preparations of the plasma membrane of muscle are used, and until chemosensitivity and the permeability of the membrane can be expressed in molecular terms. After denervation there is an increased incorporation of ^{3}H-leucine into proteolipid. The increase is particularly great into the fraction which can bind acetylcholine and hexamethonium (1851, 1852) (Chapter 4).

Denervated muscle contains more calcium than normal (1346). The fragmented sarcoplasmic reticulum of muscle concentrates more calcium than normal, but the rate of such accumulation is reduced (352, 1407, 2805). The fragmented sarcoplasmic reticulum derived from innervated muscle accumulates calcium at a rate which is inversely related to the relaxation time of the muscle from which it is derived (882). The increased relaxation time of denervated muscle (770, 1792) is compatible with this. The rate of release of calcium ions from vesicles of sarcoplasmic reticulum is increased by denervation. Caffeine increases this efflux (1313) by acting on the membrane of the sarcoplasmic reticulum (117), possibly by reducing the binding constant of the sites to which this ion is attached (236). The efflux is sufficiently enhanced by caffeine to potentiate the elicited twitch of either normal or denervated muscle (235, 1066, 1210). Caffeine increases the efflux more in denervated than in innervated muscle (1466). Normal mammalian muscle, unlike that of the frog, does not respond to caffeine with a contracture; denervated mammalian muscle does. It seems likely that this difference might result from several factors: preservation or hypertrophy of the sarcoplasmic reticulum (1031, 1032, 1223, 2081, 2614, 2822), the increased calcium content of denervated muscle, and the capacity of caffeine to potentiate the calcium efflux from the sarcoplasmic reticulum of denervated muscle. Caffeine does not penetrate denervated muscle more readily than normal.

Denervated muscle contains less contractile protein than normal muscle (2081, 2240, 2822, 2823). The half-lives of the different proteins have not been measured under these circumstances. The increase in hydrolytic enzymes, and the greater prominence of lysosomes suggests that the rate of destruction might be increased. This might be another example of the initial step of altered differentiation being one of destruction (p. 246). After denervation the myosin ATPase activity of fast-twitch muscle decreases (1206, 2867), and that of slow-twitch muscle increases. Corresponding qualitative changes take place in the light chains of myosin, and in the rates of superprecipitation of actomyosin (760, 761, 762, 999). Actomyosin from a denervated fast muscle is not identical to that from slow muscle. Actomyosin from innervated slow muscle is superprecipitated with greater efficiency (precipitate/mole ATP/mg/min) than that from innervated fast muscle; that is, the efficiency is greater with actomyosin that has a slower rate of superprecipitation (2550). In contrast, superprecipitation from denervated muscle occurs more slowly and less efficiently.

The mechanical response of a denervated fast muscle is not identical with that of a slow muscle. The speed of shortening of muscle is related not only to the force-velocity properties of the contractile proteins, but also to the time course of the active state and the elastic properties of the series mechanical resistance. In denervated muscle the active state is lengthened both by the lower rate of conduction of an action potential in the membrane of the muscle, and by the altered flux of calcium at the sarcoplasmic reticulum. On the other hand, atrophy of the muscle fibres reduces the time necessary for the radial spread of excitation (39, 1104, 2240). The slowing of the contraction and relaxation phases of a muscle twitch after denervation is greater in fast than in slow muscles (770, 1792).

Several factors make the metabolic consequences of denervation difficult to interpret. Within three days of denervation myoblasts are formed from muscle fibres, and these begin to differentiate (273, 1319). Their complete differentiation must await reinnervation. This formation of new fibres and their dependence upon innervation is also seen in innervated injured muscle (2432). Large numbers of myoblasts are seen if a denervated muscle is repeatedly injured. These can be introduced into an empty muscle sheath, and, if reinnervation is permitted, well differentiated muscle fibres are formed (426, 2846). All muscle fibres found in a reinnervated muscle may not have been present before denervation.

A denervated muscle is often stretched by antagonists, or exceptionally by another muscle which retains it innervation, for example, digastric. Such stretching is seen in a denervated hemi-diaphragm (2822, 2823). A stretched muscle changes its contractile proteins, whether it is innervated or not (1209, 1228). This change is produced in innervated muscle by denervating or removing its synergists. The effects of stretching can be avoided in a denervated muscle if the antagonists are removed or also denervated. When an innervated fast-twitch muscle is stretched, it develops a slower twitch, a longer relaxation time, a lower frequency of tetanus fusion and a lower myofibrillar ATPase activity (1208, 2867). An innervated slow-twitch muscle changes less when stretched. A denervated fast-twitch muscle also develops a slower twitch, a longer relaxation time, and a lower myofibrillar ATPase activity. If the denervated muscle is not stretched, the increase in contraction time is not so great (1211). Both the sequelae of denervation and compensatory hypertrophy require critical experimental evaluation in the light of the responses of muscle to stretching (1325, 1770, 3026). In contrast a muscle that is stimulated for short periods either in vivo (779, 780, 781, 1207), or *in vitro* (1208), but that is not subjected to prolonged stretching develops a shorter contraction time and an increased myofibrillar ATPase activity.

A further factor which makes the metabolic consequences of denervation difficult to interpret is the blood flow. After denervation the blood flow of a resting muscle usually increases, at least initially (91).

Following denervation, both glycolytic and aerobic muscles tend to lose their separate metabolic identities (822, 1561, 1562, 1915, 1916). The rate at which they change varies in different muscle fibres and in different species (1548, 1549). Histochemical differences between different fibres within a muscle become less marked (1374, 2479). The reduction of aerobic activity is associated with the fragmentation of mitochondria in some muscle fibres (1996, 2966, 2967, 2968). The concentration of oxidative enzymes beneath the cell membrane decreases (2479); as this decrease is associated with the appearance of fibrillation potentials, it is unlikely to be a result of a decreased frequency of membrane depolarization. Some enzymes probably associated with the sarcoplasmic membrane increase in activity: glyceraldehyde-3-phosphate dehydrogenase and lactic dehydrogenase (847, 1915, 1916, 2966).

The capacity of insulin to increase the uptake of glucose is reduced by denervation (403). The uptake of glucose in the absence of added insulin is not reduced. Denervated muscles store less glycogen and the activity of phosphorylase is low (2967). The reserves of energy are usually greater in a fast-twitch than in a slow-twitch muscle, and a greater proportion of the total energy reserves of a fibre is lost by a fast-twitch muscle (1561, 1562, 1781, 1902). The increased activity of enzymes of the pentose shunt accompanies increased synthesis of RNA (1025); these changes commonly occur together (180) as pentose sugars are required for the synthesis of nucleic acids. The change in the proportions of isoenzymes of lactic dehydrogenase (2966, 2968) result from the altered pO_2 that follows change in blood flow (1108, 1291).

Reinnervation

If a nerve passing to a single muscle is divided and allowed to regenerate, the types of muscle fibre found after several months are the same as in the normal muscle. The distribution of the fibres changes. A section stained histochemically no longer appears as a mosaic (314, 3205, 3206, 3207). Instead, fibres of a closely similar histochemical type lie closely packed together in islands. This appearance probably results from early collateral innervation of adjacent fibres by the same axon; that the adjacent histochemically similar fibres form one motor unit has not yet been demonstrated, but the techniques used to study the distribution of muscle fibres of a motor unit in normal muscle should be applicable. It is likely that a proportion of the muscle fibres studied in reinnervated muscle were not present originally, but have differentiated from myoblasts. The size of the proportion is not known. Provided it is small, the persistence of islands of identical fibres, and the failure of muscle fibres to form the original mosaic form indicates that no resorting of nerve connections takes place after the primary reinnervation. It might be interesting to know if such islands are formed initially, and a mosaic is later established, in those muscles of fish which appear to rearrange their neuromuscular connections to restore their original relationships.

If a nerve that serves several dissimilar muscles is divided, the types of fibre found in any one muscle after regeneration has occurred are no longer characteristic of that muscle alone (1548, 1549). The greater variation presumably results from the invasion of the muscle examined by the axons which originally innervated other muscles which had a different pattern of constituent fibres. In interpreting such studies, the consequences of associated denervation of synergists and antagonists are important (1493). For example, dividing the sciatic nerve denervates both the plantarflexors and the dorsiflexors of the ankle. The gastrocnemius and soleus will not only have been denervated, but also subjected to less passive stretching after denervation than they would have received had they been denervated alone. During reinnervation the process is likely to be complicated by the different rates of reinnervation of the different muscles.

Cross-reinnervation

This kind of study has been extended considerably by cross-union of nerves. A nerve passing to a slow-twitch muscle is implanted into a freshly denervated fast-twitch muscle, and *vice versa*. After flexor hallucis longus or flexor digitorum longus of the cat has been innervated by the nerve to soleus, the myosin ATPase and the actomyosin ATPase activities fall, and the rate of twitch declines (381, 382). The light chains of myosin come to resemble those of soleus. The capacity of the fragmented sarcoplasmic reticulum to accumulate calcium decreases (2028). Innervating soleus with the nerve to flexor hallucis longus or flexor digitorum longus causes little change (382). The maximum rate of shortening is directly related to the actomyosin ATPase activity (140, 141, 142). The responses to excitation show a much more complete slowing of the fast muscle than speeding of the slow (377–383). The myoglobin contents of the muscles do not change significantly on cross-innervation (1894). In soleus the oxidative activity decreases and the glycolytic activity rises; the reverse occurs in flexor digitorum longus (726, 2334, 2480).

In the rat not only can fast muscle be slowed by cross-reinnervation, but slow muscle can also be speeded (509, 510, 512, 1493). Myosin ATPase and actomyosin ATPase change appropriately (141, 383, 1493, 2549). Oxidative and glycolytic activities and the soluble proteins resemble less the pattern seen in normal muscle and more the pattern seen in the muscle normally innervated by the nerve implanted (1195, 2480). Some caution is needed in interpreting the results obtained from the rat. Denervation of the dorsiflexors of the ankle without implanting the peroneal nerve into the soleus can cause the biochemical and mechanical properties of the soleus to change in the direction seen after cross-union. This change, due to decreased stretching, is a result of denervation of the antagonists. The changes seen in soleus after cross-union are usually greater than those resulting from division of the peroneal nerve without implantation (1493).

While the results obtained from study of muscle conclusively demonstrate that the structure of the light chains of myosin and some features of the sarcoplasmic reticulum are changed by innervation, they do not clearly substantiate a trophic role for nerve; nor is there any evidence that the muscles exert a specific trophic influence upon the nerves influencing them (378). Evidence from tissue culture suggests rather that these properties of muscle are determined by the frequency of stimulation, which need not be chemically mediated across the neuromuscular junction, and by the extent to which it is stretched. Differentiation of myoblasts is certainly facilitated by nerves; this facilitation acts at an earlier stage of differentiation than the role of the nervous system in determining the contraction properties of the muscle. Tissue culture studies suggest that this earlier facilitation of differentiation might be produced by sensory as well as by motor nerves (1760, 1761, 2150).

9 GENETICS OF THE NERVOUS SYSTEM

The effects of genetic variation upon the nervous system can be viewed in two ways: from a study of inbred strains reared in a laboratory to reduce pressures of natural selection, and from a study of variation within a natural or controlled population. The former can indicate through the expression of recessive traits the way in which the nervous system is put together and the way in which defined chemical or structural abnormalities can cause characteristic disorders of function. The latter shows the influence of pressures of selection upon the variation already present within the total gene pool of the population; such selection works upon the expression of the genetic variation already present in the total gene pool and does not require further mutation. From the viewpoint of the cell biology of brain, neither approach is yet far advanced.

Inbred strains

The ultimate source of genetic variation is mutation. With few exceptions, dominant traits are not harmful or they would not withstand the pressures of selection; some are advantageous. Most mutations are harmful and can persist in the total genome of the species only if the resulting characteristics are recessive. A striking exception to this general principle is Huntingdon's chorea; this trait can persist because it develops usually only after the breeding period of life has commenced.

The harmful recessive traits must therefore be studied in inbred strains reared and protected in the laboratory. The examples have been selected because of their relevence to other parts of this book, because of the detailed study they have received and because they indicate the range of techniques that can used. Neurological mutants of the mouse have been catalogued (2686), and behavioural mutants of *Drosophila* have been reviewed briefly (217, 219, 1906, 1907).

Mouse: Disorders of myelination

Jimpy

Jimpy is a sex-linked recessive trait. The mouse appears normal at birth, but develops a tremor and unsteadiness on the 12th day, tonic-clonic convulsions about the 18th day, and dies when about 25 days old. By light microscopy no myelin can be seen in the central nervous system, but myelination of the peripheral nerves seems to be normal (1312, 2686). Electron microscopy of the central nervous system shows only occasional fragments of myelin.

Biochemical studies confirm this abnormality, without showing the cause. The first defect is seen when the mouse is seven days old. Thc brain has far less of the constituents of myelin than normal: sulphatides (174, 1312, 2563), cerebrosides (738, 1010, 1375, 2144, 2145), monogalactosyl diglyceride (2144), and sterols (1533, 1534) are all reduced. During normal myelination there is a progressive increase in the proportion of fatty acids that has long chains (1516, 1966, 2151); this increase in the long chain fatty acids of ceramides, cerebrosides and sulphatides is not found in the Jimpy mouse (1516). The capacity of the brain to synthesize myelin is depressed at a stage of development when the absence of myelin is striking (738); the galactosylation of ceramide is markedly reduced (2114, 2115, 2116). Most of the phospholipids which are normally present in myelin are reduced in quantity, whereas those which are not normally found in high concentration in myelin are substantially unaffected (663, 1010, 1375). There is also less proteolipid (738) and 2′3′-cyclic nucleotide-3′-phosphohydrolase (1694, 1695).

The defect of myelination cannot easily be explained as defective action of a single enzyme resulting in a failure to form one essential constituent of myelin, for at the start of myelination the metabolism of cerebrosides and sulphatides is normal (738, 1375), the activities of cerebroside-β-galactosidase and of related enzymes are normal (683), and the rate of turnover of such galactolipid as is synthesized is normal (738). The observations made so far do not explain adequately the lesion either in cytological or in biochemical terms. There is considerable agreement that some myelination begins in this mutant; persisting fragments are seen by electron microscopy, and the early biochemical development resembles that of normal mice (609, 958, 1596, 2144, 2145, 2986). Abnormal degradation of myelin is less well authenticated. There is an absolute decrease in the quantity of some of the protein constituents of myelin, a small accumulation of cholesterol esters (2144), and an appearance of lipid in sudanophilic macrophages (2685). Such changes could result equally from the degeneration of precursors of myelin before their incorporation into the sheath, or from degradation of myelin which has been briefly established in the sheath. Degradation before insertion might result from an abnormal composition of one of the constituents, or from an abnormality of the mechanism responsible for carrying precursors to the sheath and inserting them. No evidence of demyelination has been seen in cultured explants of cerebellum from Jimpy mutants (3168), and the few myelin sheaths that are formed *in vitro* survive for a period exceeding the life span of an affected mouse; myelination is, however, sparse in culture as well as *in vivo*. The failure to form mature compact myelin *in vivo* might result either from a specific defect in oligodendrocytes that are otherwise normal (2144, 2145), from a shortened lifespan of oligodendrocytes in mice with this trait, or from abnormal interaction between the axons of the central nervous system and oligodendrocytes. The studies so far reported do not allow these alternatives to be distinguished; nor is it likely that measurement of the survival of oligodendrocytes with tritiated thymidine will give a clear answer until the life histories of different classes of oligodendrocyte have been clarified further (Chapter 1), and the patterns of response of these cells to demyelination and dysmyelination have been investigated further. Furthermore, the cells showing the most striking pathological change are not necessarily the primary site of expression of the genetic abnormality. The results obtained from tissue culture (3168) show that the deficient myelination does not result from a primary disorder of another organ, for example of the liver, acting at the time of failing myelination. The contribution of studies *in vitro* will be much greater when mixed cultures of cloned cells derived from different strains can be studied; it might then be possible to grow Jimpy oligodendrocytes with normal neurons and *vice versa*. The presence of substances which inhibit or facilitate normal myelination can then be sought in the medium in which cells have grown, with some chance of being able to interpret the results.

The manner in which the disorder of myelination leads to the neurological signs is not clear. While disorders of movement and of muscle tone may be ascribed hypothetically to low rates of conduction in bare axons, the necessary electrophysiological studies have not yet been reported.

Quaking

Quaking is an abnormal recessive trait; it differs in mode of inheritance, life span and pathology from Jimpy (2686). At autopsy about one fifth of the affected mice have pronounced hydrocephalus (663, 1151, 3163); it seems unlikely that this is a secondary communicating hydrocephalus due only to atrophy consequent upon deficient myelination, as it is not a feature of Jimpy in which the defect of myelination is more severe. Although myelin cannot usually be detected by light microscopy, electron microscopy shows that most axons of the central nervous system are surrounded by several loose lamellae of uncompacted myelin (223, 3163). The nodal gaps are exceptionally wide, and this feature can give a misleading appearance of an absence of the myelin sheath if transverse sections only are examined. The cell bodies and processes of oligodendrocytes contain vesicles and some glial cells are vacuolated.

As in the Jimpy trait, the most striking features are probably a consequence of abnormal myelination rather than its cause. During the first ten days of postnatal life the phospholipid, ceramide, cerebroside and sulphatide contents of the brain increase normally. From the 16th to the 20th day the normal rate of increase is not sustained, and instead the values remain fairly constant (1375, 1477, 1478). The brain of the adult mouse contains less ceramide (2114, 2115), cerebroside (304, 1151, 1375, 1478, 2419) and other galactolipids (174, 2419, 2685) than normal. The concentration of some phospholipids is reduced: ethanolamine, triphosphoinositide and phosphatidic acid. Surprisingly, phosphatidyl serine and sphingomyelin are better preserved (663). A relative but not an absolute increase occurs in phosphatidyl choline, diphosphatidyl glycerol and phosphatidyl inositol; these phospholipids are found apart from the myelin fraction of brain, and their relative increase indicates only the diminished amount of myelin. Sphingomyelin, cerebrosides and sulphatides have an abnormally low content of long chain (C_{22} to C_{26}) fatty acids. This is not likely to be due to defective mechanisms of chain elongation because the defect affects different sphingolipids differently (174, 1478, 1596), and because similar defects are not seen in skin and kidney (1596). The myelin of normal ten day old mice is mainly dense; the myelin of Quaking is more poorly compacted and more vesiculated (1146). In Quaking the activities of enzymes that synthesize ketodihydrosphingosine and ceramide are normal, but the capacity to synthesize cerebroside is greatly reduced. The reported low activity of sphingosine galactosyl transferase (2115) may have been a nonspecific measurement of UDP galactose : ceramide galactosyl transferase. Such a depressed galactosylation of ceramide could have several causes: inhibition of enzyme activity, failure to facilitate the activity of the enzyme, repression of enzyme formation, or a defect in the primary structure of the enzyme. It has not yet been purified. Proteolipid is almost absent and the amount of basic protein is reduced (1151).

The immediate cause of defective myelination is not known. It is probable, but not certain, that the cells primarily affected are the oligodendrocytes. Although its formation may be described loosely as 'arrested', the myelin is not identical with that of a normal immature mouse (3163). A primary disorder of cerebroside synthesis seems unlikely. Although its rate of synthesis is only 35% of normal, cerebroside is still made more rapidly than the very low rate of myelin formation demands. A primary disorder of elongation of fatty acid chains is also unlikely unless there exists a mechanism to distribute unevenly long chain fatty acids between different types of lipid. A primary disorder of proteolipid synthesis might impair the incorporation into myelin of cerebroside, which would then be degraded. Alternatively, the oligodendrocyte might be incapable of transporting some necessary myelin constituent from its site of synthesis to the membrane (1050).

Dilute lethal. Wobbler lethal

These are autosomal recessive traits (2686). The adult mice have deficient myelin, but the situation differs from Jimpy and from

Quaking. Homozygotes appear normal until they are nine days old; their lurching gait is then apparent and they become progressively unsteady. Convulsions begin at the age of two weeks and increase in severity and frequency until the animals die around the 20th day. Judged by light microscopy, myelination proceeds normally in a homozygote at the same rate and in the same tracts as in a heterozygous littermate. Between one and five days later demyelination follows (701, 1574). Tracts demyelinate in the same order as they myelinate; for example, myelination and demyelination occur sequentially in the vestibular, spinal and cerebellar systems. By the 15th day the cerebellar tracts contain no detectable normal myelin. In contrast to Jimpy and Quaking, glia of the affected tracts contain products of myelin degeneration.

These mice have not yet been studied intensively. It is probable that the picture that will emerge will be mixed: demonstrating that less myelin is present than normally and showing simultaneous formation and degradation of myelin.

Explants of normal cerebellum only grow well *in vitro* and myelinate if taken from mice less than two days old. At this age homozygotes and heterozygous littermates cannot be distinguished. Explants have therefore been studied from all members of several litters (1238). It is very improbable that all the animals from which explants have been obtained were heterozygotes; the chance of this being so cannot be stated precisely, as the number of animals studied was not stated. In all cultures the myelin begins to appear between the 9th and the 11th day; it is usually extensive by the 14th day, and persists up to the 30th. In no culture does myelin fail to develop and in none does it degenerate prematurely. This suggests that the cause of the degeneration is not an irreversible metabolic defect of brain. It does not necessarily imply that the defect is caused by a primary factor outside brain. For example, the culture medium might be supplying a substrate or other factor that allows a cell with a genetically determined deficiency to function more normally; or the metabolic load upon an oligodendrocyte might be less *in vitro* than *in vivo*; or demyelination might be related in some way to disorder of neural function or contact, for myelination is often correlated with the onset of function (701) and can be augmented in peripheral nerves when the work load of the muscles they supply is increased. On the other hand the blood of these homozygotes might contain a substance that is toxic to oligodendrocytes or to myelin, or it might not contain a factor essential for their normal activity. It seems likely that further study of cultured explants to which the serum of normal, heterozygous or homozygous mice is added will clarify this situation. Parabiotic union of a homozygote and a heterozygote might demonstrate whether stable blood-borne agents play a role *in vivo*.

The myelin that persists *in vitro* might or might not be normal; no analysis has yet been reported.

Mouse: Disorders of cell interaction

Reeler

Reeler is an autosomal recessive trait (2683, 2685). The brain is small and the neuropil is poorly developed. The laminar structure of the neocortex, hippocampus and cerebellum is characteristically disordered (1234, 1235, 1963, 2683). Despite a first impression of histological chaos, disorder is not total. Cells lie within appropriate architectonic areas of the cortex, but the radial organization is abnormal. In normal cortex the layers derived from neurons migrating early lie deeply, while the layers derived from the neurons which migrate late lie superficially. The neurons which migrate late must pass through a well ordered neuropil. Both in the cerebral and cerebellar cortices they move in close association with radial glial fibres (Chapter 1). In Reeler, the radial arrangement is disordered only in those parts of the brain in which such guided migration through neuropil must occur. The disorder is not found in all such regions; the olfactory bulb develops in accordance with this pattern, but is essentially normal in this mutant (452).

All layers of the cerebellum have abnormal features. Purkinje cells lie scattered throughout the depth of the cortex and their dendrites are in considerable, but not total, disarray. The molecular layer is much thinner than the granular layer and many granule cells fail to migrate. The Bergman glia are disordered; their major processes commonly run obliquely instead of radially. In a normal mouse, basket cells and stellate cells lie apart; in the Reeler they are mixed.

The dendrites of Purkinje cells bear spines, but only about 10% of these receive synapses. The spines that lack boutons are invested by the processes of astrocytes. Provided that synaptic contacts have not been previously developed and then lost, this suggests that the formation and maintainance of some spines of Purkinje cells' dendrites do not require presynaptic contact. This is consistent with their probable formation by heterosynaptic induction (p. 257). The structure of these parts of the brain which normally lack a well-defined lamina structure is substantially normal, although some of the pathways of projection are displaced.

The most attractive hypothesis is that the disorder results from abnormal migration in some parts of the brain. Studies *in vitro* indicate that the disorder is intrinsic to the cells themselves. If fragments of cerebellum are explanted from normal mice, between 10 and 20% of well myelinated cultures show a strikingly ordered laminar structure, and a further 20 to 30% of such cultures have easily recognizable features of the cerebellum. The Reeler can be recognized at birth. The cerebellum is small and lacks fissures (3167). The individual cell types of the Reeler cerebellum survive well *in vitro*, but the culture lacks all evidence of characteristic lamination. Cells of the cerebellum can be disaggregated and reaggregated *in vitro*; cells from a normal cerebellum form patterns of association which are characteristic (670). The pattern of reaggregation formed from cells of the cerebellum of Reeler differs from the pattern formed from cells of a heterozygous littermate, and from that of normal mice of other strains (671). The cells of Reeler and of normal mice therefore differ in their social behaviour. It is not yet clear whether the defect is one of the cell surface or of the sequence of change induced by cell contact.

Staggerer

Staggerer is an autosomal recessive trait; it is interesting in that the abnormality may be a failure to form one particular type of synapse. The cerebellum is very small. Granule cells appear and migrate normally but by the 14th postnatal day they degenerate and none can be found on the 33rd postnatal day. Purkinje cells are small and atrophic; the dendrites are small and lack those spines which normally receive boutons from granule cells (2683). The granule cells initially form connections with all the normal types of target cell except Purkinje cells. All other types of synapse seen in the normal cerebellum can be found in the cerebellum of Staggerer. It seems likely that the primary defect lies in the failure of the Purkinje cell to provide a proper postsynaptic surface for the granule cell. The absence of spines cannot be attributed easily to a primary failure of the granule cells to establish contact with the spine, as spines can develop normally without afferent connections in Reeler. The disappearance of the granule cells is almost certainly an example of retrograde transneuronal degeneration (p. 261). The disorder of the Purkinje cell might be one of heterosynaptic induction (p. 257).

Brindled

This X-linked recessive trait is the only example definitely known to affect one transmitter substance. The hemizygote Brindled has a tremor and stiff hind limbs; when held up by the tail, it clasps its hind paws together. It usually dies before weaning. The noradrenaline content of brain is markedly reduced, and that of tyrosine is increased; the dopamine content is normal. The conversion of dopamine to noradrenaline is reduced both in brain and in sympathetic ganglia. It seems likely that the activity of dopamine-β-hydroxylase is low, but it is still uncertain if this is due to an abnormality of enzyme structure or synthesis, or to abnormalities of substances which might impede or facilitate its action.

Muscular dystrophy

This is an autosomal recessive trait. The condition is of interest because it is possible that the muscles are affected indirectly as a result of neural change. If so, study of the mutant is of considerable relevence to problems related to neurotrophism (Chapter 8).

The affected muscles respond both to direct and indirect stimulation. Different kinds of muscle fibre are affected differently. Fibres of soleus of Muscular dystrophy have a specific membrane resistance which is lower than normal; the specific membrane resistance of fibres of extensor digitorum longus is greater than normal (1729, 1730, 1731). The resting membrane potential is low (1264, 1730). The rate of rise of the action potential is less than normal even when the membrane is hyperpolarized; the reversal of polarization is also less than normal. Some early reports suggested that the fibres of dystrophic muscle are functionally denervated, as they respond to direct and not to indirect stimulation (1869). Dystrophic muscle studied *in vitro* can be fatigued very easily. If care is taken to avoid hypoxia very few muscle fibres will in fact be found which fail to respond to stimulation of the nerve; furthermore neither action potentials which are resistant to tetrodotoxin, nor increased sensitivity of the membrane to acetylcholine can be found in the fibres of dystrophic muscle (1264, 1265), although such changes are characteristic of denervated muscle (Chapter 8).

Dystrophic muscle contains less ATP and creatine phosphate than normal muscle, and has more ADP and AMP (864). Mitochondria from livers of dystrophic mice respire at a lower rate than normal; this is due to the abnormally great gradient of hydrogen ion concentration across the mitochondrial membrane, which impedes the efflux of hydrogen ions that is linked to respiration. The passive permeability of the mitochondrial membrane for hydrogen ions is abnormally low (1408). If this defect of respiration is present in nerve and muscle also, it could explain both the low concentration of ATP and creatine phosphate in muscle, and also the ease with which hypoxia blocks conduction *in vitro*. Fragments of plasma membrane have an activity of Na^+-K^+-ATPase that is higher than normal. This might be due to a high intracellular concentration of sodium ions (125), which could contribute to the low resting membrane potential. The phospholipids of dystrophic muscle differ from those of normal muscle, a finding that cannot yet be interpreted in the context of the altered electrophysiological properties of the membrane. There is a defect also in the synthesis of certain series of polyunsaturated fatty acids present in phospholipids (2196). These abnormalities might affect the physical properties of the lipid part of the membrane or the behaviour of intrinsic proteins of the membrane (Chapter 2). Both the reduced content of polyunsaturated fatty acids, and the raised molar ratio of cholesterol to phospholipid could diminish the fluidity of the membrane. If a normal muscle is denervated, the amount of the major sialoglycolipid, GM3, increases; it is not increased in dystrophic muscle (1952, 1953). The rate of turnover of actin is normal, that of myosin is reduced and that of tropomyosin in increased (2808); the nature of the light chains of myosin in fast-twitch muscle has not yet been reported. The rates of turnover of creatine kinase and of adenylate kinase are both increased (1597).

Some evidence now suggests that the muscle of these mice is abnormal in some respects because of its innervation. Muscle obtained from affected mice on the 17th day of fetal life develops *in vitro* in the same way as normal muscle (2328). Mixed cultures have been grown containing spinal cord segments obtained either from fetal normal mice or from fetal dystrophic mice, and muscle obtained either from normal adult mice or from adult mice with muscular dystrophy. When a normal spinal cord innervates either normal muscle or dystrophic muscle, the muscle develops normally. If the spinal cord of a mouse with muscular dystrophy innervates either normal muscle or dystrophic muscle, then both kinds of muscle develop the characteristic changes of dystrophy (1011). A similar experiment has been performed *in vivo*. If a muscle is minced and reintroduced into its fascial compartment the muscle regenerates and differentiates if it is innervated. If muscle from a normal mouse is minced and introduced into the fascial compartment of a mouse with dystrophy, the implanted muscle fails to differentiate normally. Conversely, if dystrophic muscle is minced and inserted into a normal mouse, the muscle differentiates normally (1351, 2542, 2543). Dystrophic muscle responds abnormally to cross-innervation (1729). If a soleus of a normal mouse is denervated and becomes reinnervated by its own nerve, the mean contraction time does not change. If the soleus of a normal mouse is innervated by the nerve to extensor digitorum longus, the contraction time is reduced. If the soleus of a dystrophic mouse is denervated and becomes reinnervated either by its own nerve or by the nerve to extensor digitorum longus, the contraction time of the muscle does not alter. Muscle fibres of a normal soleus have different membrane properties from those of a normal extensor digitorum longus; after cross-innervation both change. The properties also differ in the muscles of the Muscular dystrophy mouse but these do not alter after cross-reinnervation. These experiments are subject to the same potential artefacts as have been described before (p. 299).

Abnormalities of axoplasmic flow have been seen in mice with dystrophy (1618, 2881). It is not known whether this results from the disorder, causes it, or is an unrelated defect; it could, for example, result from an associated disorder of the contractile proteins of nervous tissue, or from the defect of respiration. The number of myelinated nerve fibres passing to tibialis anterior is less in the mutant mouse than in a normal one (1266). It is probable that sensory fibres from muscle are also affected. There is no evidence to suggest that the myelination of the nerves is abnormal as far distally as this; there is defective myelination of nerve fibres in the dorsal root. Motor end-plates of dystrophic muscle fibres are frequently abnormal at a time preceding structural change in the muscle.

Red cells of the Muscular dystrophy mouse have an abnormal shape (2052). No information has yet been published concerning their fragility or biochemistry.

Retinal degeneration

This is an autosomal recessive trait. The mouse appears normal at birth. About the 11th postnatal day, excess rhodopsin accumulates in membranous lamellae lying between the outer segments of the retinal rods, which look normal, and the pigment epithelial cells (720). The proximo-distal passage of photoreceptor discs in the outer segments is slower than normal and they are not engulfed by the pigment epithelial cells when shed (1311). The rate of synthesis of opsin in the retina remains high (1728). It appears likely that the pigment cells also contribute to the accumulation of extracellular lamellar material. These changes are accompanied by a progressive and selective degeneration of the retinal receptor cells. It is not known whether the receptor cell or the pigment cell is primarily affected, or whether the defect is common to both. The cultured retina develops a similar abnormality. Chimeric retinae of mosaic tetraparental mice (2008–2013) show a variable and patchy retinal degeneration (2011). Such mice are formed by combining *in vitro* two 8-cell morulae, one being homozygous for retinal degeneration. It is, however, difficult to establish whether regions of normal tissue influence those parts derived from the mutant unless an independent system of marking the cells is used. This method of analysis is also used to study mutants of *Drosophila* (p. 320). .

Rat: Diabetes insipidus (Brattleboro)

This is probably an autosomal recessive trait controlled by two gene loci. Diabetes insipidus can result from homozygosity at either locus. The daily output of urine is about 70% of body weight, and the water intake rather more. The osmolality of the blood is greater than that of normal rats, and that of the urine is less than that of blood. Little or no vasopressin can be detected in the plasma, even during dehydration. When vasopressin is injected, the output of urine falls, its osmolality rises and the intake of water decreases. The osmolality of urine is not increased by dehydration, by hypertonic saline, or by nicotine (2989, 2990). The response of the kidneys to vasopressin is therefore normal. The picture is one of sustained dehydration from impaired synthesis or release of vasopressin, and the polydipsia is secondary. Neurons of the supraoptic nucleus are hypertrophic with large nuclei and nucleoli; no neurosecretory material can be detected histochemically. Neurons of the paraventricular nucleus are similar, but less extremely affected (2757). The histochemical results therefore indicate that the defect is probably one of synthesis rather than of release of vasopressin. The posterior lobe of the pituitary is also larger and heavier than normal; the degree to which different types of cell in the posterior pituitary contribute to this enlargement is uncertain. The neuronal hypertrophy is likely to be a response to sustained and intense transsynaptic stimulation (Chapter 5). Such stimulation could result from the increased blood osmolality. The rate of discharge of some nerve cells of the supraoptic nucleus increases after hypertonic saline has been injected into the carotid artery (752, 753). An alternative hypothesis is that the hypertrophy results from a failure of recurrent collateral inhibition.

Heterozygotes synthesize less vasopressin than normal rats (2989). The synthesis of oxytocin is unimpaired in both heterozygotes and homozygotes.

Mouse: Audiogenic fits

These provide an example of a genetically determined functional disorder of the nervous system in which no macroscopic, histological or specific biochemical disorder has yet been demonstrated conclusively, and an example of a condition which might depend upon systemic factors.

Susceptibility depends upon a polygenic inheritance (989–992,

1067–1074, 1999). The frequency, severity and threshold of the fits vary diurnally; they vary also with age, sex and endocrine state, and are influenced by diet, temperature, previous experience and drugs (860, 861, 989–992, 1067–1074, 1999, 2303, 2508, 2540, 2541, 3027, 3113). It is not yet agreed whether the amounts of 5-hydroxytryptamine and noradrenaline are less in a strain of DBA mice susceptible to fits (1746, 1747, 1876, 2595). It is likely that measurements of the content and turnover of transmitter substances in small regions of the brain may provide more clear-cut differences. Drugs which alter the brain content of 5-hydroxytryptamine and noradrenaline alter the frequency and severity of audiogenic fits (3027). Fits from other causes, for example, methionine-sulphoxine, are influenced similarly by these drugs.

Glutamate reduced the severity and frequency of these seizures (1070, 1073, 1074). Its manner of action has not been shown. It might lead to the synthesis of more γ-aminobutyric acid. More fits occur in susceptible mice if they are made pyridoxine-deficient (539, 540). Pyridoxine is an essential cofactor for the synthesis of 5-hydroxytryptamine, noradrenaline and γ-aminobutyric acid. This does not indicate that the untreated mice genetically disposed to fits are necessarily deficient either in pyridoxine or in γ-aminobutyric acid.

Parabiotic union between a susceptible mouse and a normal mouse reduces the incidence of fits in the susceptible animal, and does not make the normal animal susceptible. When such joined mice are separated again the susceptible mouse has fits again after a latency of six to eight days. Two joined susceptible mice remain susceptible. This result might indicate that a stable systemic factor can reduce susceptibility (1237). The factor involved is not necessarily an agent absent from the susceptible mouse as a result of its genetic lesion. The animals studied did not show gross 'parabiotic intoxication' (a joint and usually assymmetrical graft versus host reaction) but lesser degrees of such a reaction might have been present. A further control would have been a study of a mild graft versus host reaction induced by standard immunological methods upon the incidence and severity of the fits.

Siamese cat: Visual pathways

In a number of species, albino strains have an abnormal decussation of retinogeniculate fibres at the optic chiasm. This abnormality, and the consequent disordered sequence of visual projection have been studied in the Siamese cat (1174, 1433). Some axons which originate from the temporal half of the retina close to the vertical meridian decussate and pass to the opposite lateral geniculate nucleus, instead of passing to the ipsilateral nucleus. Each lateral geniculate nucleus therefore fails to receive some fibres from the ipsilateral eye, but receives instead some fibres from the contralateral eye. The fibres from the contralateral eye end in that part of the lateral geniculate nucleus which fails to receive fibres from the ipsilateral eye. This type of projection might be expected from the experiments in which the optic nerves of a frog are deliberately uncrossed and innervate the ipsilateral optic tectum (p. 274). Within this part of the lateral geniculate nucleus the aberrant fibres end in an ordered projection upon the cells of the lateral geniculate nucleus. The axons of the postsynaptic cells of the lateral geniculate nucleus pass to the visual cortex of the same side. Normally one hemisphere receives axons which convey information derived only from the contralateral half of the visual field; in these cats the ipsilateral cortex receives such information, but in addition information from the paramedian part of the ipsilateral visual field. The abnormal cortical projection of the ipsilateral field is highly organized. In some cats it forms an ordered projection wedged between areas 17 and 18 at the edge of the area receiving the primary visual input. In other cats, parts of the ipsilateral field project to cells of the cortex very close to cells receiving information from corresponding parts of the contralateral field; this again has affinities with the 'mirror image' type of projection seen in the frog with the deliberately uncrossed chiasm.

The cause of the abnormal decussation is not known. As the aberrant projection is highly ordered, the defect is unlikely to be a defect of discrete cell-to-cell organization. The similarity of the projection to that found in the frog with the surgically uncrossed chiasm makes such a cause unlikely. It is possible that the disorder is one of timing of axonal growth, or of rate of growth, so that fibres fail to arrive at appropriate sites at the correct times. It is also not known whether the ordered abnormalities of cortical projection are also a direct result of the genetic disorder or a secondary plastic response. The varied nature of the cortical projection makes the latter more likely. Binocular vision is disordered over the whole field. No cortical units are found which can be driven in a normal binocular manner; this applies both to units receiving information from parts of the field directly affected by the aberrant projection and to units outside this zone. The absence of binocular convergence is a result of the squint. If a Siamese cat is reared with its eyelids sutured so that it is deprived of experience of patterned vision, units which respond binocularly are found. This effect of the squint of the Siamese cat is similar to the effect of a squint produced experimentally in a previously normal kitten (Chapter 6).

Human disorder

A rational approach to problems of neurological defects in man caused by genetic disorder is impeded by many factors (2331). Relatively few nervous systems have been studied of related patients; in these the lesions often vary considerably in distribution and severity. This might be a pathological correlate of variation in age of onset and in rate of progression of the disorder seen clinically within families. Different families have been classified as having the same disorder even though the patterns of inheritance differ. Similarities and differences between families are difficult to study because of the variability of each family, and because of the relatively limited repertoire of basic pathological response possessed by the nervous system.

Within a family not all similar disorders of the nervous system are necessarily genetic. Within a large closed community a high incidence of striking neurological disorder can be due to social and not to direct genetic influences, for example kuru. Genetic abnormalities may alter the susceptibility of a person to a condition that is not itself genetically determined.

In one respect, however, the study of inherited neurological disease is relatively far advanced. In several conditions biochemical deficiencies have been found which are consistent with the cytological disorder (97, 599, 600, 601, 3141). As in other animals, inherited disorder is not usually confined to the nervous system, and in some disorders the nervous system is only secondarily involved.

Drosophila

Many mutants behave abnormally (217, 218, 219, 837, 838, 1401, 1402, 1403, 1622). Individually these mutants are characteristic. Some, for example have abnormalities of diurnal rhythm, which can be too long, too short, or absent. Another mutant is apparently normal for the first few days of life and then begins to stagger and dies within hours; no gross abnormality of the nervous system precedes staggering, but vacuolation and cavitation then appear. Other mutants have abnormalities of behavior that appear in states of general stress, under anaesthesia, or in response to particular stresses such as abnormal temperature or humidity. At least five mutants have disorders of vision; in two of these the visual pigments are present and the electroretinogram indicates that the response of the primary photoreceptors is normal, but that later events are disordered. Correlated physiological, pharmacological, biochemical and structural studies of such mutants are likely to be very interesting.

Genetic mosaics have been used to indicate the part of the fly affected primarily by the genetic defect. This is not necessarily the part in which functional disorder, histological change or biochemical aberration is first noticed, nor need the genetic defect produce its effect at the time that gross disorder first appears. Individuals are produced in which part of the body is a mutant male, while the rest is a normal female. The dividing line between normal and mutant parts can occur with differing orientation; this is because the orientation of the spindle is random during the first few divisions of the developing egg, because the resulting nuclei remain after sequential division clustered within the blastoderm as they divide, and because the position of the cells in the blastoderm determines the fate of their descendants. This technique will have more power when it is possible to identify particular cells or groups of cells with each genetic endowment; at present only regions of the surface of the animal can be identified in this way.

Songs of flies and crickets

The song pattern of the cricket is produced by a neuronal network which continues to function rhythmically even after it has been isolated from all phasic input (213–215). A polygenic system regulates the song pattern, which, unlike that of birds, contains no component that is learnt. Different strains sing characteristically. The song patterns of hybrids lie within a spectrum extending apparently continuously from the song pattern of one parent to the song pattern of the other. As the song patterns are continuously variable, it is likely that those properties of the nervous system responsible for the song pattern are also continuously variable. It is improbable that such fine and continuous variation of behaviour could result from discrete and specific qualitative differences as crude as the presence or absence of a particular type of cell junction or synapse. More probably the variation results from quantitative differences in neuronal properties or connections, or from differences in synaptic efficacy.

Drosophila of different strains woo with different serenades. The song can be analyzed into a number of variables; some of these variables are inherited independently (837).

Isogeneic animals

Both genetic and environmental constraints influence the development of the nervous system. The relative importance of 'nature' and 'nurture' have been argued for many years. Central to a profitable debate of this issue is the effect of experience upon animals of known genetic identity. This demands an initial study of the nervous system of isogeneic animals to see if they are in fact structurally, chemically, physiologically and behaviorally identical. Animals which produce young by parthenogenesis are very suitable for studying some aspects of these problems. Preliminary study suggests that the nervous systems of such isogeneic animals are very similar but not identical. This is true of *Daphnia*. Although the pathways to the brain from the compound eye are very similar, the precise patterning of boutons upon the processes of the postsynaptic cells is different in animals that are isogeneic. The extent to which this variability is due to random variation in patterns or rates of growth, or due to the different experience of different animals is not known, but could in principle be determined.

Polyploid and hapliod animals

Newts are normally diploid. Polyploid newts have larger, but fewer cells. This affects the nervous system as well as other tissues.

Polyploid newts are less adept at solving problems than their diploid colleagues (3007). Haploid animals have more cells. It is not clear if this affects the nervous system as well as other organs, and it is not known whether the behaviour of the animal is altered.

Inheritance of behavioural traits

Inheritance of behavioural traits has now been demonstrated so often that the question is now not one of whether it occurs, but of how it is brought about (349, 350, 1293–1297, 1354, 1813). Inbred strains of mice differ in maze performance, in the use of preferred sensory modalities in learning a maze, in learning to discriminate, in spontaneous alternation, in open field behaviour, aggression and dominance, in sexual behaviour and in vocalization (350, 568, 569, 977, 988, 1295, 1444, 2188, 2640, 2650, 2769, 2987, 3122). Other species show a similar variation. In some circumstances genetic variation of behaviour is apparent only after they have been exposed to an appropriate environment (1295); this is occasionally self-evident; for example, a mouse with a great genetically determined capacity to run mazes will not demonstrate this particular skill in the absence of a maze. The influence of the environment can, however, be more subtle, especially because of the great importance of early experience not only in learning, but also in determining the later selection of cues (Chapter 6). In some circumstances genetic variation of behaviour is only apparent in mice after they have been exposed to an enriched environment. Carefully correlated bio-chemical, physiological, structural, behavioural and pharmacological measurements need to be made upon strains of mice carefully chosen for their clear and elemental differences of behaviour. Such analysis cannot be confined to the nervous system alone, because of the certain importance of hormones. Analysis cannot be limited to adult animals alone, for the major difference might lie in variation of response to early experience. It is difficult to dissociate genetic influences from environmental factors. The environment acts from the time of conception.

Studies of physical correlates of behavioural variation which have so far been reported are many, but are unco-ordinated and difficult to interpret. For example, two strains of mice which differ in their performance both in a shuttle box and in learning mazes (303) have different concentrations of certain proteins in membranes of synaptosomes (1181). This difference might be either in the membrane of the bouton or in the postsynaptic membrane, or in both. The difference could result from differences between the strains in the proportion of different types of synapse that are present in the brain, or from variation in the amount of certain proteins in the synapses of one given sort; or both factors might contribute. It is impossible to interpret the results until much more is known about the properties of synapses in these strains; furthermore, it remains to be established whether the biochemical differences between the synaptosomes are related to the dissimilarities of behaviour, and if so, whether they are caused by it, or cause it.

There is much anecdotal evidence for the inheritance of behavioural traits in man, and most studies of this difficult field indicate that it occurs. It is a field in which factual evidence is difficult to determine and in which many of the subjects and even some of the investigators are not disinterested in the outcome of the study. It is a field not without political pressures. It seems highly improbable in view of the evidence from other mammals, and in view of the many other similarities between the brains of man and other mammals, that an individual man, or a strain of men, should not have behavioural traits which are determined by the interaction of the genetic endowment with the environment.

Most mutations are harmful; most persist in a population only if the resulting traits are recessive. Animals of an inbred strain are genetically very similar, and are homozygous at many loci. If mice of each of two inbred strains are assumed to be completely homozygous, the first generation of hybrid mice resulting from mating between these two strains will be heterozygous at those loci for which the strains possess different alleles and homozygous at the remainder (2742). The greater number of recessive traits in inbred strains is related to the better performance of hybrids in many formal tests of behaviour (152, 364, 545, 1129, 2596, 2742, 3154, 3155). Such contrived demonstrations in the laboratory of 'hybrid vigour' do not necessarily indicate a superior fitness of the hybrid to withstand pressures of selection in a natural environment and in an evolutionary context. Some types of superiority in laboratory tests may confer natural advantage: a greater ability to obtain food, to care for the young, to escape from a dangerous environment, or to avoid noxious stimuli. Nevertheless heterosis is only an inference drawn from such tests.

Variation in a natural population

In a natural population the behaviour of individual members of a strain differs. This is due in part to polygenic variation which may be continuous or discontinuous, and in part to the influence of the environment interacting with the individual genotype. Variation within a species increases the chance of its survival greatly, and, in a natural environment may be necessary. The variation is of at least two types. First, a range of behaviour expressed by different animals of a given species increases the advantage of the species in selection, just as a range of surface colouring does so. The range of such variation in a population is a balance between two factors: the possession by each animal of the species of all the most suitable traits for survival in the particular and immediate environment, and the capacity to surmount by natural selection the vicissitudes of periodic changes of the environment. The point of balance shown by a species at any one time is itself a consequence of the pressures of natural selection, preserving in the total genome of the species both a favourable position for the species in the present environment and a capacity for subgroups to withstand different pressures of selection resulting from the appearance of different environments. Hence a capacity to vary is itself a trait that increases the ability of the species to survive. Second, for social animals it is the group of animals rather than the individual animal that possesses advantages or disadvantages in natural selection. In animals with organized societies selection operates upon the group as well as upon the individual. A society is composed of individual animals which behave differently; for example, a certain proportion may be dominant and others submissive. A certain range of aggression within a population is likely to produce the most suitable degree of social coherence. A range of behaviour, rather than one given type of behaviour, gives in this way also advantage in selection. Pressures of selection discriminating between different patterns of behaviour are not selecting only for the genetic basis of the behaviour, but rather for the interaction between the genetic and environmental factors which have come to mould the behavioural repertoire of the strain.

It is not only in the direction of genes influencing the response to an environment that genetic and environmental factors interact. The reverse also occurs; environmental factors can influence quantitatively the composition of the overall genome of the species. For example, this occurs through mating preferences, which in at least one species depend upon the early experience of the animal.

REFERENCES

REFERENCES

Readers are requested to note that there are no references for the following numbers: 108, 317, 458, 627, 628, 1692, 1801, 1803, 2718 and 3022.

1. ABDELKADEL, A. B. & MAZLIAK, P. (1970). Echanges de lipides entre mitochondries, microsomes et surngeant cytoplasmique de cellules de pomme de terre ou de chou-fleur. *Europ. J. Biochem.*, **15**, 250–262.

1a ABDEL-LATIF, A. A., YAU, S. J. & SMITH, J. P. (1974). Effect of neurotransmitters on phospholipid metabolism in rat cerebral cortex slices – cellular and subcellular distribution. *J. Neurochem.*, **22**, 383–393.

2. ABE, T., HAGA, T. & KUROKAWA, M. (1974). Retrograde axoplasmic transport; its continuation as anterograde transport. *FEBS Lett.*, **47**, 272.
3. ABERCROMBIE, M. (1967). Contact inhibition: the phenomenon and its biological implications. *Natl Cancer Inst. Monographs*, **26**, 240–277.
4. ABERCROMBIE, M. & AMBROSE, E. J. (1962). The surface properties of cancer cells: a review. *Cancer Res.*, **22**, 525–548.
5. ABERCROMBIE, M., HEAVSMAN, J. E. M. & PEGRUM, S. M. (1970). The locomotion of fibroblasts in culture. *Expl Cell Res.*, **59**, 393–398.
6. ABOOD, L. G. & ABUL-HAJ, S. K. (1952). Histochemistry and characterisation of hyaluronic acid in axons of peripheral nerves. *J. Neurochem.*, **1**, 119–125.
7. ABRAHAMSON, E. W. & OSTROY, S. E. (1967). The photochemical and macromolecular aspects of vision. *Progr. Biophys. molec. Biol.*, **17**, 179–215.
8. ABRAMS, J. & GERARD, R. W. (1933). The influence of activity on the survival of isolated nerve. *Am. J. Physiol.*, **104**, 590–593.
9. ACHESON, G. H., LEE, E. S. & MORISON, R. S. (1942). A deficiency in the phrenic respiratory discharges parallel to retrograde degeneration. *J. Neurophysiol.*, **5**, 269–273.
10. ACHESON, G. & REMOLINA, J. (1955). The temporal course of the effects of postganglionic axotomy on the inferior mesenteric ganglion of the cat. *J. Physiol.*, **127**, 602–616.

11. ADAIR, L. B., WILSON, J. E. & GLASSMAN, E. (1968). Brain function and macromolecules. IV. Uridine incorporation into polysomes of mouse brain during different behavioural experiences. *Proc. natn. Acad. Sci. U.S.A.,* **61**, 917–922.
12. ADAIR, L. B., WILSON, J. E., ZEMP, J. W. & GLASSMAN, E. (1968). Brain function and macromolecules. III. Uridine incorporation into polysomes of mouse brain during short-term avoidance conditioning. *Proc. natn. Acad. Sci. U.S.A.,* **61**, 606–613.
13. ADAMIC, S. (1968). Sodium influx into denervated rat diaphragm muscle fibres. *Biochim. biophys. Acta,* **163**, 137–140.
14. ADAMS, C. W. M. (1965). *Neurohistochemistry.* Elsevier; Amsterdam, London, NY.
15. ADAMS, C. W. M. & DAVISON, A. N. (1959). The occurrence of esterified cholesterol in the developing nervous system. *J. Neurochem.,* **4**, 282–289.
16. ADAMS, C. W. M. & DAVISON, A. N. (1965). The myelin sheath. *In* Adams, C. W. M. *Neurohistochemistry.* Elsevier; Amsterdam, London, NY, pp. 332–400.
17. ADAMS, C. W. M., DAVISON, A. N. & GREGSON, N. A. (1963). Enzyme inactivity of myelin; histochemical and biochemical evidence. *J. Neurochem.,* **10**, 383–395.
18. ADAMS, C. W. M. & TUQAN, N. A. (1961). Histochemistry of myelin. II. Proteins, lipid-protein dissociation and proteinase activity in Wallerian degeneration. *J. Neurochem.,* **6**, 334–341.
19. ADAMS, J. H., DANIEL, P. M. & PRICHARD, M. M. L. (1968). Regrowth of nerve fibres in the hypophysis: regeneration of a tract of the central nervous system. *J. Physiol.,* **198**, 4P.
20. ADAMS, J. H., DANIEL, P. M. & PRICHARD, M. M. L. (1969). Degeneration and regeneration of hypothalamic nerve fibres in the neurohypophysis after pituitary stalk section in the ferret. *J. comp. Neurol.,* **135**, 121–144.
21. ADAMS, J. H., DANIEL, P. M. & PRICHARD, M. M. L. (1969). Degeneration and regeneration of the hypothalamo-hypophyseal tract after pituitary stalk section in the ferret. *J. comp. Neurol.,* **142**, 109–124.
22. ADAMS, J. H., DANIEL, P. M. & PRICHARD, M. M. L. (1969). The blood supply of the pituitary gland of the ferret with special reference to infarction after stalk section. *J. Anat.,* **104**, 209–225.
23. ADAMS, J. H., DANIEL, P. M. & PRICHARD, M. M. L. (1971). Changes in the hypothalamus associated with regeneration of the hypothalamo-hypophyseal tract after pituitary stalk section in the ferret. *J. comp. Neurol.,* **142**, 109–124.
24. ADELMAN, W. J. (1971). *Biophysics and physiology of excitable membranes.* Van Nostrand Reinhold Co.; NY.
25. ADELSTEIN, R. S., POLLARD, T. D. & KUEHL, W. M. (1971). Isolation and characterization of myosin and two myosin fragments from human blood platelets. *Proc. natn. Acad. Sci. U.S.A.,* **68**, 2703.

26. ADEY, W. R. (1971). Cortical excitability changes following interaction of calcium with topical polyanions and glutamic acid. *Anat. Rec.*, **169**, 265–266.
27. ADEY, W. R. (1971). Evidence for cerebral membrane effects of calcium derived from direct current gradient, impedance, and intracellular records. *Expl Neurol.*, **30**, 78–102.
28. ADEY, W. R. (1974). The influence of impressed electrical fields at EEG frequencies on brain and behaviour. *In* Altshuler, H. & Burch, N. R. *Behaviour and brain activity*. Plenum Press; NY.
29. ADEY, W. R., BYSTROM, B. G., COSTIN, A., KADO, R. T. & TARBY, T. J. (1969). Divalent cations in cerebral impedance and cell membrane morphology. *Expl Neurol.*, **23**, 29–50.
30. ADEY, W. R., KADO, R. T. & DIDIO, J. (1962). Impedance measurements in brain tissue of animals using microvolt signals. *Expl Neurol.*, **5**, 47–66.
31. ADEY, W. R., KADO, R. T., DIDIO, J. & SCHINDLER, W. J. (1963). Impedance changes in cerebral tissue accompanying a learned discrimination performance in the cat. *Expl Neurol.*, **7**, 259–281.
32. ADEY, W. R., KADO, R. T., McILWAIN, J. T. & WALTER, D. O. (1966). The role of neuronal elements in regional cerebral impedance changes in alerting, orientating and discriminative responses. *Expl Neurol.*, **15**, 490–510.
33. ADEY, W. R. & ROVNERT, E. N. (1969). Cortical neuronal and silent cell membrane potentials after excess calcium ions. *Physiologist*, **12**, 156.
34. ADLARD, B. P. F. & DOBBING, J. (1971). Phosphofructokinase and fumarate hydrolase in developing rat brain. *J. Neurochem.*, **18**, 1299–1303.
35. ADLARD, B. P. F. & DOBBING, J. (1971). Vulnerability of developing brain. III. Development of four enzymes in the brain of normal and undernourished rats. *Brain Res.*, **28**, 97–107.
36. ADLARD, B. P. F. & DOBBING, J. (1971). Elevated acetylcholinesterase activity in adult rat brain after undernutrition in early life. *Brain Res.*, **30**, 198–199.
37. ADLARD, B. P. F. & DOBBING, J. (1972). Vulnerability of developing brain. 8. Regional acetylcholinesterase activity in the brains of adult rats undernourished in early life. *Br. J. Nutr.*, **28**, 139–143.
38. ADLARD, B. P. F., DOBBING, J. & SMART, J. L. (1970). Undernutrition and the development of certain enzymes in rat brain. *Biochem. J.*, **119**, 46P.
39. ADRIAN, R. H., COSTANTIN, L. L. & PEACHEY, L. D. (1969). Radial spread of contraction in frog muscle fibres. *J. Physiol.*, **204**, 231–257.
40. AGHAJANIAN, G. K. & BLOOM, F. E. (1967). Electron-microscopic localization of tritiated norepinephrine in rat brain; effect of drugs. *J. Pharmac. exp. Ther.*, **156**, 407–416.
41. AGRANOFF, B. W., DAVIS, R. E. & BRINK, J. J. (1965). Memory fixation in goldfish. *Proc. natn. Acad. Sci. U.S.A.*, **54**, 788–793.

42. AGRANOFF, B. W., DAVIS, R. E. & BRINK, J. J. (1966). Chemical studies on memory fixation in goldfish. *Brain Res.*, **1**, 303–309.
43. AGRANOFF, B., DAVIS, R. E., CASOVA, L. & LIM, R. (1967). Actinomycin D blocks formation of memory of shock avoidance in goldfish. *Science, N.Y.*, **158**, 1600–1601.
44. AGUAYO, A. J., PEYRONNARD, J. M. & BRAY, G. M. (1973). A quantitative ultrastructural study of regeneration from isolated proximal stumps of transected unmyelinated nerves. *J. Neuropath. exp. Neurol.*, **32**, 256.
45. AITKEN, J. T. (1949). The effects of peripheral connections on the maturation of regenerating nerve fibres. *J. Anat.*, **83**, 32–43.
46. AKERS, C. K. & PARSONS, D. F. (1970). X-ray diffraction of myelin membrane. I. Optimal conditions for obtaining unmodified small angle diffraction data from frog sciatic nerve. *Biophys. J.*, **10**, 101–115.
47. AKERS, C. K. & PARSONS, D. F. (1970). X-ray diffraction of myelin membrane. II. Determination of the phase angles of the frog sciatic nerve by heavy atom labeling and calibration of the electron density distribution of the membrane. *Biophys. J.*, **10**, 116–136.
48. AKIYAMA, M. & SAKAGAMI, T. (1969). Exchange of mitochondrial lecithin and cephalin with those in rat liver microsomes. *Biochim. biophys. Acta*, **187**, 105–112.
49. AKMAYEV, I. G. (1969). Morphological aspects of the hypothalamo-hypophyseal system. I. Fibres terminating in the neurohypophysis of mammals. *Z. Zellforsch. mikrosk. Anat.*, **96**, 609–624.
50. AL-AMOOD, W. S. & POPE, R. (1972). A comparison of the structural features of muscle fibres from a fast and slow-twitch muscle of the pelvic limb of the cat. *J. Anat.*, **113**, 49–60.
51. ALBUQUERQUE, E. X., BARNARD, E. A., CHIU, T. H., LAPA, A. J., DOLLY, J. O., JANSSON, S. E., DALY, J. & WITKOP, B. (1973). Acetylcholine receptor and ion conductance modulator sites at the murine neuromuscular junction: evidence from specific toxin reactions. *Proc. natn. Acad. Sci. U.S.A.*, **70**, 949–953.
52. ALBUQUERQUE, E. X., BARNARD, E. A., PORTER, C. W. & WARNICK, J. E. (1974). The density of acetylcholine receptors and their sensitivity in the postsynaptic membrane of muscle endplates. *Proc. natn. Acad. Sci. U.S.A.*, **71**, 4640.
53. ALBUQUERQUE, E. X. & McISAAC, R. J. (1970). Fast and slow mammalian muscles after denervation. *Expl Neurol.*, **26**, 183–202.
54. ALBUQUERQUE, E. X., SASA, M., AVNER, B. & DALY, J. (1971). A possible site of action of batrachotoxin. *Nature, Lond. New Biol.*, **234**, 93–95.
55. ALBUQUERQUE, E. X., SCHUH, F. T. & KAUFFMAN, F. C. (1971). Early membrane depolarization of the fast mammalian muscle after denervation. *Pflügers Arch. Europ. J. Physiol.*, **328**, 123–137.
56. ALBUQUERQUE, E. X. & THESLEFF, S. (1968). A comparative study of membrane properties of innervated and chronically denervated fast

and slow skeletal muscles of the rat. *Acta physiol. scand.*, **73**, 471–480.

57. ALBUQUERQUE, E. X. & WARNICK, J. E. (1972). The pharmacology of batrachotoxin. IV. Interaction with tetrodotoxin on innervated and chronically denervated rat skeletal muscle. *J. Pharmac. exp. Ther.*, **180**, 683–697.
58. ALBUQUERQUE, E. X., WARNICK, J. E. & SANSONE, F. M. (1971). The pharmacology of batrachotoxin. II. Effects on electrical properties of the mammalian nerve and skeletal muscle membranes. *J. Pharmac. exp. Ther.*, **176**, 511–528.
59. ALBUQUERQUE, E. X., WARNICK, J. E., TASSE, J. R. & SANSONE, F. M. (1972). Effects of vinblastine and colchicine on neural regulation of fast and slow skeletal muscles of the rat. *Expl Neurol.*, **37**, 607–634.
60. ALESCIO, T. & COLOMBO-PIPERNO, E. (1967). A quantitative assessment of mesenchymal contribution to epithelial growth rate in mouse embryonic lung development *in vitro. J. Embryol. exp. Morph.*, **17**, 213–227.
61. ALESCIO, T. & DANI, A. M. (1971). The influence of mesenchyme on the epithelial glycogen and budding activity in mouse embryonic lung developing *in vitro. J. Embryol. exp. Morph.*, **25**, 131–140.
62. ALESCIO, T. & di MICHELE, M. (1968). Relationship of epithelial growth to mitotic rate in mouse embryonic lung developing *in vitro. J. Embryol. exp. Morph.*, **19**, 227–237.
63. ALLARA, E. (1952). Sull'influenza esercitata dagli ormoni sessuali sulla stuttura delle formazioni gustative di mus rattus albinus. *Riv. Biol.*, **44**, 209–229.
64. ALLEN, D. W. & ZAMECNIK, P. C. (1962). The effect of puromycin on rabbit reticulocyte ribosomes. *Biochim. biophys. Acta*, **55**, 865–874.
65. ALLEN, E. (1912). The cessation of mitosis in the central nervous system of the albino rat. *J. comp. Neurol.*, **22**, 547–568.
66. ALLEN, R. D., COOLEDGE, J. W. & HALL, P. J. (1960). Streaming in cytoplasm dissociated from the giant amoeba *Chaos chaos. Nature, Lond.*, **187**, 896–899.
67. ALLISON, A. C. (1973). The role of microfilaments and microtubules in cell movement, endocytosis and exocytosis. *In Locomotion of tissue cells.* CIBA Foundation Symposium. Elsevier; Amsterdam. pp. 109–143.
68. ALTMAN, J. & DAS, G. D. (1964). Autoradiographic examination of the effects of enriched environment on the rate of glial multiplication in the adult rat brain. *Nature, Lond.*, **204**, 1161–1163.
69. ALVAREZ, J. & PÜSCHEL, M. (1972). Transfer of material from efferent axons to sensory epithelium in the goldfish vestibular system. *Brain Res.*, **37**, 265–278.
70. AMOORE, J. E. (1965). Psychophysics of odour. *Cold Spring Harb. Symp. quant. Biol.*, **30**, 623–637.
71. AMOORE, J. E. (1967). Specific anosmia: a clue to the olfactory code. *Nature, Lond.*, **214**, 1095–1098.
72. AMOORE, J. E., PALMIERI, G., WANKE, E. & BLUM, M. S. (1969).

Ant alarm pheromone activity: correlation with molecular shape by scanning computer. *Science, N.Y.*, **165**, 1266.

73. ANDERSEN, P., BLISS, T. V. P. & SKREDE, K. K. (1971). Lamellar organization of hippocampal excitatory pathways. *Expl Brain Res.*, **13**, 222–238.
74. ANDERSEN, P., HOLMQVIST, B. & VOORHOEVE, P. E. (1966). Excitatory synapses on hippocampal apical dendrites activated by entorhinal stimulation. *Acta physiol. scand.*, **66**, 461–472.
75. ANDERSEN, P., HOLMQVIST, B. & VOORHOEVE, P. E. (1966). Entorhinal activation of dentate granule cells. *Acta physiol. scand.*, **66**, 448–460.
76. ANDERSEN, P. & LØMO, T. (1966). Mode of activation of hippocampal pyramidal cells by excitatory synapses on dendrites. *Expl Brain Res.*, **2**, 247–260.
77. ANDERSEN, A. E. & NAFSTAD, P. H. J. (1968). An electron microscopic investigation of the sensory organs in the hard palate region of the hen (Gallus domesticus). *Z. Zellforsch. mikrosk. Anat.*, **91**, 391–401.
78. ANDERSON, K.-E., EDSTRÖM, A. & HANSON, M. (1972). Heavy water reversibly inhibits fast axonal transport of protein in frog sciatic nerves. *Brain Res.*, **43**, 299–302.
79. ANDERSON, L. E. & McCLURE, W. O. (1973). Differential transport of protein in axons: comparison between the sciatic nerve and dorsal columns of cats. *Proc. natn. Acad. Sci. U.S.A.*, **70**, 1521–1525.
80. ANDERSON, L. S., BLACK, R. G., ABRAHAM, J. & WARD, A. A. (1971). Neuronal hyperactivity in experimental trigeminal deafferentation. *J. Neurosurg.*, **35**, 444–452.
81. ANDERSON, W. A., WEISSMAN, A. & ELLIS, R. A. (1966). A comparative study of microtubules in some vertebrate and invertebrate cells. *Z. Zellforsch. mikrosk. Anat.*, **71**, 1–13.
82. ANDREOLI, T. E. (1966). Formation and properties of thin lipid membranes from sheep red cell lipids. *Science, N.Y.*, **154**, 417.
83. ANDREOLI, T. E., BANGHAM, J. A. & TOSTESON, D. C. (1967). Formation and properties of thin lipid membranes from HK and LK sheep red cell lipids. *J. gen. Physiol.*, **50**, 1729–1749.
84. ANDREOLI, T. E., TIEFENBERG, M. & TOSTESON, D. C. (1967). The effect of valinomycin on the ionic permeability of thin lipid membranes. *J. gen. Physiol.*, **50**, 2527.
85. ANDREOLI, T. E. & TOSTESON, D. C. (1971). The effect of valinomycin on the electrical properties of solutions of red cell lipid in *n*-decane. *J. gen. Physiol.*, **57**, 526–538.
86. ANDREOLI, T. E. & TROUTMAN, S. L. (1971). An analysis of unstirred layers in series with 'tight' and 'porous' lipid bilayer membranes. *J. gen. Physiol.*, **57**, 464–477.
87. ANDREOLI, T. E. & WATKINS, M. L. (1973). Chloride transport in porous bilayer membranes. *J. gen. Physiol.*, **61**, 809–830.

88. ANDRY, D. K. & LUTTGES, M. W. (1972). Memory traces: experimental separation by cycloheximide and electroconvulsive shock. *Science, N.Y.*, **178**, 518–520.
89. ANGEL, C. & BURKETT, M. L. (1971). Effect of hydrocortisone and cycloheximide on blood-brain barrier function in the rat. *Dis. nerv. Syst.*, **32**, 53–58.
90. ANON. (1974). A new role for the glial cell? *Nature, Lond.*, **251**, 100.
91. ANREP, G. V., CERQUA, S. & SAMAAN, A. (1934). The effect of muscular contraction upon the blood flow in the skeletal muscle, in the diaphragm and in the small intestine. *Proc. R. Soc. B*, **114**, 245–257.
92. APPEL, S. H. & BORNSTEIN, M. B. (1963). Factors responsible for in vitro demyelination present in sera from animals with experimental 'allergic' encephalomyelitis. *J. Neuropath. exp. Neurol.*, **22**, 324.
93. APPELTAUER, G. S. & KORR, I. M. (1975). Axonal delivery of soluble, insoluble and electrophoretic fractions of neuronal proteins to muscle. *Expl Neurol.*, **46**, 132–146.
94. ARGIZ, C. A. G., PASQUINI, J. M., KAPLUN, B. & GOMEZ, C. J. (1967). Hormonal regulation of brain development. II. Effect of neonatal thyroidectomy on succinate dehydrogenase and other enzymes in developing cerebral cortex and cerebellum of the rat. *Brain Res.*, **6**, 635–646.
95. ARMSTRONG, J. & QUILLIAM, T. A. (1961). Can the Herbst corpuscle aid studies of peripheral receptor mechanisms? *J. Physiol.*, **161**, 4–5P.
96. ARONSON, J. F. (1965). The use of fluorescein-labeled heavy meromyosin for the cytological demonstration of actin. *J. Cell Biol.*, **26**, 293–298.
97. ARONSON, S. M. & VOLK, B. W. (1962). *Cerebral sphingolipidoses: a symposium on Tay Sachs disease and allied disorders.* Acad. Press; N.Y.
98. van ARSDALL, G. B. & LENTZ, T. L. (1968). Neurons: secretory activity during limb regeneration and induction in the newt. *Science, N.Y.*, **162**, 1296–1298.
99. ARTOM, C. (1941). Phospholipid metabolism in denervated muscles. *J. biol. Chem.*, **139**, 953–961.
100. ARTYUKHINA, N. I. (1967). Some data on the gliosynaptic relationships in the brain cortex of animals (rats). *Arkh. Anat.*, **52**, 38–45.
101. ARVY, L., FONTAINE, M. & GABE, M. (1955). Modifications histologiques de l'organe sous-commisural au cours du cycle evolutif de *Salmo salar* L. *Archs. Anat. microsc. Morph. exp.*, **44**, 313–322.
102. ASADA, Y. & BENNET, M. V. L. (1971). Experimental alteration of coupling resistance at an electrotonic synapse. *J. Cell Biol.*, **49**, 159–172.
103. ASHMORE, C. R. & DOERR, L. (1971). Postnatal development of fiber types in normal and dystrophic skeletal muscle of the chick. *Expl Neurol.*, **30**, 431–466.
104. ASHMORE, C. R. & DOERR, L. (1971). Comparative aspects of muscle fibre types in different species. *Expl Neurol.*, **31**, 408–418.
105. ASHMORE, C. R., TOMKINS, G. & DOERR, L. (1972). Postnatal

development of muscle fiber types in domestic animals. *J. Anim. Sci.*, **34**, 37–41.

106. ASHMORE, C. R., TOMKINS, G. & DOERR, L. (1972). Comparative aspects of mitochondria isolated from αW, αR and βR muscle fibres of the chick. *Expl Neurol.*, **35**, 413–420.
107. ATKINSON, A. J. & WEISS, M. F. (1969). Kinetics of blood-cerebrospinal fluid glucose transfer in the normal dog. *Am. J. Physiol.*, **216**, 1120–1126.
109. ATTARDI, D. G. & SPERRY, R. W. (1963). Preferential selection of central pathways by regenerating optic fibres. *Expl Neurol.*, **7**, 46–64.
110. ATWOOD, H. L. & GOVIND, C. K. (1973). Ultrastructure of nerve terminals and muscle fibres in denervated crayfish muscle. *Z. Zellforsch. mikrosk. Anat.*, **146**, 155–165.
111. AUBERT, X., CHANCE, B. & KEYNES, R. D. (1964). Optical studies of biochemical events in the electrical organ of *electrophorus. Proc. R. Soc. B*, **160**, 211–245.
112. AUCLAIR, W. & SIEGEL, B. W. (1966). Cilia regeneration in the sea urchin embryo; evidence for a pool of ciliary protein. *Science, N.Y.*, **154**, 913.
113. AUERBACH, A. A. & BENNET, M. V. L. (1969). A rectifying synapse in the central nervous system of a vertebrate. *J. gen. Physiol.*, **53**, 211–237.
114. AUTILIO, L. A., APPEL, S. H., PETTIS, P. & GAMBETTI, P. (1968). Biochemical studies of synapses *in vitro*. I. Protein synthesis. *Biochemistry, N.Y.*, **7**, 2615–2622.
115. AUTILIO, L. A., NORTON, W. T. & TERRY, R. D. (1964). The preparation and some properties of purified myelin from the central nervous system. *J. Neurochem.*, **11**, 17–27.
116. AUTILIO-GAMBETTI, L., GAMBETTI, P. & SHAFER, B. (1973). RNA and axonal flow. Biochemical and autoradiographic study in the rabbit optic system. *Brain Res.*, **53**, 387–398.
117. AXELSSON, J. & THESLEFF, S. (1959). A study of supersensitivity in denervated mammalian skeletal muscle. *J. Physiol.*, **147**, 178–193.
118. AYALA, G. F., DICHTER, M., GUMNIT, R. J., MATSUMOTO, H. & SPENCER, W. A. (1973). Genesis of epileptic interictal spikes. New knowledge of cortical feedback systems suggest a neurophysiological explanation of brief paroxysms. *Brain Res.*, **52**, 1–13.
119. AZUMA, M., AZUMA, K. & KITO, Y. (1973). Circular dichroism of visual pigment analogues containing 3-dehydroretinal and 5,6,-epoxy-3-dehydroretinal as the chromophore. *Biochim. biophys. Acta*, **295**, 520–527.
120. AZZI, A., CHANCE, B., RADDA, G. K. & LEE, C. P. (1969). A fluorescent probe of energy-dependent structure changes in fragmented membranes. *Proc. natn. Acad. Sci. U.S.A.*, **62**, 612–619.
121. BABAKOV, A. V., ERMISHKIN, L. N. & LIBERMAN, E. A. (1966). Influence of electric field on the capacity of phospholipid membranes. *Nature, Lond.*, **210**, 953–955.

122. BAILEY, S. W. (1937). An experimental study of the origin of lateral line structures in embryonic and adult teleosts. *J. exp. Zool.*, **76**, 187–233.
123. BAKER, P. F., HODGKIN, A. L. & SHAW, T. J. (1961). Replacement of the protoplasm of a giant nerve fibre with artificial solutions. *Nature, Lond.*, **190**, 885–887.
124. BAKER, P. F., HODGKIN, A. L. & RIDGWAY, E. B. (1971). Depolarization and calcium entry in squid giant axons. *J. Physiol.*, **218**, 709–755.
125. BAKER, R., BLAND, W. H. & HART, P. (1958). Concentration of K and Na in skeletal muscle of mice with a hereditary myopathy (Dystrophia muscularis). *Am. J. Physiol.*, **193**, 530–533.
126. BAKER, R. & LLINAS, R. (1971). Electrotonic coupling between neurones in the rat mesencephalic nucleus. *J. Physiol.*, **212**, 45–63.
127. BALAZS, R., BROOKSBANK, B. W. L., DAVISON, A. N., EAYRS, J. T. & WILSON, D. A. (1969). The effect of neonatal thyroidectomy on myelination in the rat brain. *Brain Res.*, **15**, 219–232.
128. BALAZS, R. & COTTERELL, M. (1972). Effect of hormonal state on cell number and functional maturation of the brain. *Nature, Lond.*, **236**, 348–350.
129. BALAZS, R., KOVACS, S., TEICHGRÄBER, P., COCKS, W. A. & EAYRS, J. T. (1968). Biochemical effects of thyroid deficiency on the developing brain. *J. Neurochem.*, **15**, 1335–1349.
130. BALAZS, E. A. & LAURENT, T. L. (1951). Viscosity function of hyaluronic acid as a polyelectrolyte. *J. Polymer Sci.*, **6**, 665–668.
131. BANGHAM, A. D. (1968). Membrane models with phospholipids. *Prog. Biophys. molec. Biol.*, **18**, 29–95.
132. BANGHAM, A. D. (1972). Lipid layers and biomembranes. *A. Rev. Biochem.*, **41**, 753–776.
133. BANGHAM, A. D., de GIER, J. & GREVILLE, G. D. (1967). Interaction of phosphatidyl serine monolayers with metal ions. *Chem. Phys. Lipids*, **1**, 225–246.
134. BANGHAM, A. D. & HORNE, R. W. (1964). Negative staining of phospholipids and their structural modification by surface-active agents as observed in the electron microscope. *J. molec. Biol.*, **8**, 660–668.
135. BANGHAM, A. D. & DAVISON, A. N. (1969). Enzyme deficiency and composition of myelin and subcellular fractions in the developing rat brain. *Biochem. J.*, **115**, 1051–1062.
136. BANIK, N. L. & DAVISON, A. N. (1971). Exchange of sterols between myelin and other membranes of developing rat brain. *Biochem. J.*, **122**, 751–758.
137. BANKS, P., MANGNALL, D. & MAYOR, D. (1969). The redistribution of cytochrome oxidase, noradrenaline and adenosine triphosphatase in adrenergic nerves constricted at two points. *J. Physiol.*, **200**, 745–762.
138. BANKS, P., MAYOR, D., MITCHELL, M. & TOMLINSON, D. (1971). Studies on the translocation of noradrenaline-containing vesicles in postsynaptic sympathetic nerves *in vitro*. Inhibition of movement by

colchicine and vinblastine, and evidence for the involvement of axonal microtubules. *J. Physiol.*, **216**, 625–639.

139. BÄR, T., & WOLFF, J. R. (1972). The formation of capillary basement membranes during internal revascularization of the rat's cerebral cortex. *Z. Zellforsch. mikrosk. Anat.*, **133**, 231.
140. BARANY, M. (1967). ATPase activity of myosin correlated with speed of muscle shortening. *J. gen. Physiol.*, **50** (suppl. 2), 197–218.
141. BARANY, M. & CLOSE, R. I. (1971). The transformation of myosin in cross-innervated rat muscles. *J. Physiol.*, **213**, 455–474.
142. BARANY, M., CONOVER, T. E., SCHLISELFELD, L. H., GAETJENS, E. & GOFFART, M. (1967). Relation of properties of isolated myosin to those of intact muscles of the cat and sloth. *Europ. J. Biochem.*, **1**, 156–164.
143. BARBERA, A. J., MARCHASE, R. B. & ROTH, S. (1973). Adhesive recognition and retinotectal specificity. *Proc. natn. Acad. Sci. U.S.A.*, **70**, 2482.
144. BARDACH, J., FUJUYA, M. & HOLL, A. (1967). Investigation of external chemoreceptors of fishes. *In* Hayashi, T. *Olfaction and Taste, II.* Pergamon; NY.
145. BARFORT, P., ARQUILLA, E. R. & VOGELHUST, P. O. (1968). Resistance changes in lipid bilayers: immunological applications. *Science, N.Y.*, **160**, 1119.
146. BARKER, D. & IP, M. C. (1966). Sprouting and degeneration of mammalian motor axons in normal and de-afferented skeletal muscle. *Proc. R. Soc. B*, **163**, 538–554.
147. BARLOW, H. B. & PETTIGREW, J. D. (1971). Lack of specificity of neurones in the visual cortex of young kittens. *J. Physiol.*, **218**, 98–100P.
148. BARLOW, R. M., AGOSTINO, A. N. & CANCILLA, P. A. (1967). A morphological and histochemical study of the subcommisural organ of young and old sheep. *Z. Zellforsch. mikrosk. Anat.*, **77**, 299–315.
149. BARNARD, E. A., WIECKOWSKI, J. & CHIU, T. H. (1971). Cholinergic receptor molecules and cholinesterase molecules at mouse skeletal junctions. *Nature, Lond.*, **234**, 207.
150. BARNARD, J., EDGERTON, V. R., FURUKAWA, T. & PETER, J. B. (1971). Histochemical, biochemical and contractile properties of red, white and intermediate fibres. *Am. J. Physiol.*, **220**, 410–414.
151. BARNES, C. D., & WORRAL, N. (1968). Reinnervation of spinal cord by cholinergic neurones. *J. Neurophysiol.*, **31**, 689–694.
152. BARNETT, S. A. & SCOTT, S. G. (1964). Behavioural 'vigour' in inbred and hybrid mice. *Anim. Behav.*, **12**, 325–337.
153. BARONDES, S. H. (1964). Studies with an RNA polymerase from brain. *J. Neurochem.*, **11**, 663–669.
154. BARONDES, S. H. (1967). Axoplasmic transport. Report of a NRP work session. *Neurosci. Res. Prog. Bull.*, **5**, 307–419.
155. BARONDES, S. H. & COHEN, H. D. (1966). Puromycin effect on successive phases of memory storage. *Science, N.Y.*, **151**, 594–595.

156. BARONDES, S. H. & COHEN, H. D. (1967). Delayed and sustained effect of acetoxycycloheximide on memory in mice. *Proc. natn. Acad. Sci. U.S.A.*, **58**, 157–164.
157. BARONDES, S. H. & COHEN, H. D. (1967). Comparative effects of cycloheximide and puromycin on cerebral protein synthesis and consolidation of memory in mice. *Brain Res.*, **4**, 44–51.
158. BARONDES, S. H. & COHEN, H. D. (1968). Arousal and the conversion of 'short term' to 'long term' memory. *Proc. natn. Acad. Sci. U.S.A.*, **61**, 923–929.
159. BARONDES, S. H. & JARVIK, M. E. (1964). The influence of Actinomycin-D on brain RNA synthesis and on memory. *J. Neurochem.*, **11**, 187–195.
160. BARRACLOUGH, C. A. (1958). Induction of sterility in the female rat by single injections of testosterone propionate. *Anat. Rec.*, **130**, 267.
161. BARRACLOUGH, C. A. (1966). Modification in the CNS regulation of reproduction after exposure of prepubertal rats to steroid hormones. *Recent Prog. Horm. Res.*, **22**, 503–539.
162. BARRACLOUGH, C. A. & GORSKI, R. A. (1961). Evidence that the hypothalamus is responsible for androgen-induced sterility in the female rat. *Endocrinology*, **68**, 68–79.
163. BARRACLOUGH, C. A. & LEATHAM, J. H. (1954). Infertility induced in mice by a single injection of testosterone propionate. *Proc. Soc. exp. Biol. Med.*, **85**, 673–674.
164. BARRON, K. D., KOEPPEN, A. H. W. & ORDINARIO, A. T. (1968). Aberrant nerve fibres in the spinal cord. *J. Neuropath. exp. Neurol.*, **27**, 112–113.
165. BARRON, K. D. & TUNCBAY, T. O. (1962). Phosphatase activities of the cuneate nuclei after brachial plexectomy. *Archs. Neurol. Psychiat., Chicago*, **7**, 203–218.
166. BARRON, K. D. & TUNCBAY, T. O. (1964). Phosphatase histochemistry of feline cervical spinal cord after brachial plexectomy. Hydrolysis of β-glycerophosphate, thiamine pyrophosphate and nucleotide diphosphate. *J. Neuropath. exp. Neurol.*, **23**, 368–386.
167. BARTELS, E., WASSERMAN, N. H. & ERLANGER, B. F. (1971). Photochromic activation of the acetylcholine receptor. *Proc. natn. Acad. Sci. U.S.A.*, **68**, 1820–1823.
168. BASKIN, F., MASIARZ, F. R. & AGRANOFF, B. (1972). Effect of various stresses on the incorporation of ^{3}H-orotic acid into goldfish brain RNA. *Brain Res.*, **39**, 151–162.
169. BASS, L. & MOORE, W. J. (1968). A model of nervous excitation based on the Wien dissociation effect. *In* Rich, A. and Davidson, N. *Structural chemistry and molecular biology.* Freeman; San Francisco, California, pp. 356–369.
170. BASS, N. H. & HESS, H. H. (1969). A comparison of cerebrosides proteolipid proteins and cholesterol as indicators of myelin in the architecture of rat cerebrum. *J. Neurochem.*, **16**, 731–750.

171. BASS, N. H., NETSKY, M. G. & YOUNG, E. (1969). Microchemical studies of postnatal development in rat cerebrum. I. Migration and differentiation of cells. *Neurology,* **19**, 405–414.
172. BATESON, P. P. G., HORN, G. & ROSE, S. P. R. (1969). Effects of imprinting procedure on regional incorporation of tritiated lysine into protein of chick brain. *Nature, Lond.,* **223**, 534–535.
173. BATESON, P. P. G., HORN, G. & ROSE, S. P. R. (1972). Effects of early experience on regional incorporation of precursors into RNA and protein in chicken brain. *Brain Res.,* **39**, 449.
174. BAUMANN, N. A., JAQUE, C. M., POLLET, S. A. & HARPIN, M. L. (1968). Fatty acid and lipid composition of the brain of a myelin deficient mutant, the 'Quaking' mouse. *Europ. J. Biochem.,* **4**, 340–344.
175. BAUMGARTEN, H. G., BJÖRKLUND, A., HOLSTEIN, A. F. & NOBIN, A. (1972). Chemical degeneration of indolamine axons in rat brain by 5,6,-dihydroxytryptamine. *Acta physiol. scand.*, Suppl., **373**, 1–15.
176. BAUMGARTEN, H. G., BJÖRKLUND, A., HOLSTEIN, A. F. & NOBIN, A. (1972). Chemical degeneration of indolamine axons in rat brain by 5,6,-dihydroxytryptamine, an ultrastructural study. *Z. Zellforsch. mikrosk. Anat.,* **129**, 256–271.
177. BAYLOR, D. A. & NICHOLLS, J. G. (1969). Changes in extracellular potassium concentration produced by neuronal activity in the central nervous system of the leech. *J. Physiol.,* **203**, 555–569.
178. BAYLOR, D. A. & NICHOLLS, J. G. (1969). After-effects of nerve impulses on signalling in the central nervous system of the leech. *J. Physiol.,* **203**, 571–589.
179. BAYLOR, D. A. & NICHOLLS, J. G. (1969). Chemical and electrical synaptic connexions between cutaneous mechanoreceptor neurones in the central nervous system of the leech. *J. Physiol.,* **203**, 591–609.
180. BEACONSFIELD, P. & READING, H. W. (1964). Pathways of glucose metabolism and nucleic acid synthesis. *Nature, Lond.,* **202**, 464.
181. BEAN, R. C. (1972). Multiple conductance states in single channels of variable resistance lipid bilayer membranes. *J. mem. Biol.,* **7**, 15–28.
182. BEAN, R. C., SHEPHERD, W. C., CHAN, H. & EICHNER, J. T. (1969). Discrete conductance fluctuations in lipid bilayer protein membranes. *J. gen. Physiol.,* **53**, 741–757.
183. BEAR, D. M., SASAKI, H. & ERVIN, F. R. (1971). Sequential change in receptive fields of striate neurones in dark adapted cats. *Expl Brain Res.,* **13**, 256–272.
184. BECKER, N. H., HIRANO, A. & ZIMMERMAN, H. M. (1968). Observations on the distribution of exogenous peroxidase in the rat cerebrum. *J. Neuropath. exp. Neurol.,* **27**, 439–452.
185. BEDELL, S. G. (1939). The lateral-line organs of living amphibian larvae with special reference to orange coloured granules of the sensory cells. *J. comp. Neurol.,* **70**, 231–248.
186. BEETS, M. G. J. (1970). The molecular parameters of olfactory response. *Pharmac. Rev.,* **22**, 1–34.

187. BEHNKE, O. & FORER, A. (1969). Evidence for four classes of microtubules in individual cells. *J. Cell. Sci.*, **2**, 169–192.
188. BEHNKE, O., FORER, A. & EMMERSEN, J. (1971). Actin in sperm tails and mitotic spindles. *Nature, Lond.*, **234**, 208–210.
189. BEHNSEN, G. (1927). Über der Farbstoffspeicherung in Zentralnervensystem der weissen Maus in verschiedenen Älterizständen. *Z. Zellforsch. mikrosk. Anat.*, **4**, 515–572.
190. BEIDLER, L. M. (1954). A theory of taste stimulation. *J. gen. Physiol.*, **38**, 133–139.
191. BEIDLER, L. M. (1965). Comparison of gustatory receptors, olfactory receptors and free nerve endings. *Cold Spring Harb. Symp. quant. Biol.*, **30**, 191–200.
192. BEIDLER, L. M. (1971). Taste receptor stimulation with salts and acids. *In* Beidler, L. M. *Taste. Handbook of sensory physiology IV*, (2), Springer Verlag; pp. 200–220.
193. BEIDLER, L. M., NEJAD, M. S., SMALLMAN, R. L. & TATEDA, H. (1960). Rat taste cell proliferation. *Fedn Proc. Fedn Am. Socs exp. Biol.*, **19**, 302.
194. BEIDLER, L. M. & SMALLMAN, R. L. (1965). Renewal of cells within taste buds. *J. Cell Biol.*, **27**, 263–272.
195. BEIDLER, L. M. & TUCKER, D. (1955). Response of nasal epithelium to odor stimulation. *Science, N.Y.*, **122**, 76–77.
196. BELMAR, & EYZAGUIRRE, C. (1966). Pacemaker site of fibrillation potentials in denervated mammalian muscle. *J. Neurophysiol.*, **29**, 425–441.
197. BENDA, P., LIGHTBODY, J., SATO, G., LEVINE, L. & SWEET, W. (1968). Differentiated rat glial cell strain in tissue culture. *Science, N.Y.*, **161**, 370–371.
198. BENNET, E. L., DIAMOND, M. C., KRECH, D. & ROSENZWEIG, M. R. (1964). Chemical and anatomical plasticity of brain. *Science, N.Y.*, **146**, 610–619.
199. BENNETT, G. S. & EDELMAN, G. M. (1969). Amino acid incorporation into rat brain protein during spreading cortical depression. *Science, N.Y.*, **163**, 393–395.
200. BENNETT, M. V. L. (1966). Physiology of electrotonic junctions. *Ann. N.Y. Acad. Sci.*, **137**, 509–539.
201. BENNETT, M. V. L. (1969). Electrical impedance of brain surfaces. *Brain Res.*, **15**, 584–590.
202. BENNETT, M. V. L. (1969). Function of electrotonic junctions in embryonic and adult tissues. *Fedn Proc. Fedn Am. Socs exp. Biol.*, **32**, 65–75.
203. BENNETT, M. V. L. (1973). Permeability and structure of electrotonic junctions and intercellular movement of tracers. *In* Kater, S. D. and Nicholson, C., *Intracellular staining in neurobiology*. Elsevier; Amsterdam, pp. 115–133.
204. BENNETT, M. V. L. (1974). Flexibility and rigidity in electrotonically

coupled systems. *In* Bennett M. V. L., *Synaptic transmission and neuronal interaction.* Raven Press; N.Y., pp. 153–178.

205. BENNETT, M. V. L. (1974). Gap junction ultrastructure. *In* Katchalsky, A. K., Rowland, V. and Blumenthal, R., *Dynamic patterns of brain cell assemblies. Neurosci. Res. Prog. Bull.*, **12**, 90–94.
206. BENNETT, M. V. L. (1974). *Synaptic transmission and neuronal interaction.* Raven Press; N.Y.
207. BENNETT, M. V. L., ALJURE, E., NAKAJIMA, Y. & PAPPAS, G. D. (1963). Electrotonic junctions between teleost spinal neurons: electrophysiology and ultrastructure. *Science, N.Y.*, **141**, 262–264.
208. BENNETT, M. V. L., DUNHAM, F. B. & PAPPAS, G. D. (1967). Ion fluxes through a tight junction. *J. gen. Physiol.*, **50**, 1094.
209. BENNETT, M. V. L., GIMINEZ, M., NAKAJIMA, Y. & PAPPAS, G. D. (1964). Spinal and medullary nuclei controlling electric organ in the eel *Electrophorus. Biol. Bull. mar. biol. Lab. Woods Hole,* **127**, 362.
210. BENNETT, M. V. L. & TRINKAUS, J. P. (1970). Electrical coupling between embryonic cells by way of extracellular space and specialised junctions. *J. Cell Biol.*, **44**, 592–610.
211. BENSCH, K. G. & MALAWISTA, S. E. (1969). Microtubular crystals in mammalian cells. *J. Cell Biol.*, **40**, 95–107.
212. BENSCH, K. G., MARANTZ, R., WISNIEWSKI, H. & SHELANSKI, M. (1969). Induction in vitro of microtubular crystals by Vinca alkaloids. *Science, N.Y.*, **165**, 495.
213. BENSON, A. A. (1966). Lipid protein interaction. *J. Am. Oil Chem. Soc.*, **43**, 265.
214. BENTLEY, D. R. (1969). Intracellular activity in cricket neurones during the generation of behaviour patterns. *J. Insect Physiol.*, **15**, 677–699.
215. BENTLEY, D. R. (1969). Intracellular activity in cricket neurones during generation of song patterns. *Z. vergl. Physiol.*, **62**, 267–283.
216. BENTLEY, D. R. & HOY, R. R. (1972). Genetic control of the neuronal network generating cricket (*Teleogryllus gryllus*) song patterns. *Anim. Behav.*, **20**, 478–492.
217. BENTON, J. W., MOSER, H. W., DODGE, P. R. & CARR, S. (1966). Modification of the schedule of myelination in the rat by early nutritional deprivation. *Pediatrics,* **38**, 801.
218. BENZER, S. (1967). Behavioural mutants of *Drosophila* isolated by countercurrent distribution. *Proc. natn. Acad. Sci. U.S.A.*, **58**, 1112–1119.
219. BENZER, S. (1971). From gene to behaviour. *J. Am. med. Ass.*, **218**, 1015–1022.
220. BENZER, T. I. & RAFTERY, M. A. (1972). Partial characterization of a tetrodotoxin-binding component from nerve membrane. *Proc. natn. Acad. Sci. U.S.A.*, **69**, 3634–3637.
221. BENZER, T. I. & RAFTERY, M. A. (1973). Solubilization and partial characterization of the tetrodotoxin binding component from nerve axons. *Biochem. biophys. Res. Commun.*, **51**, 939–944.

222. BERG, D. K., KELLY, R. B., SARGENT, P. B., WILLIAMS, P. & HALL, Z. (1972). Binding of α-bungarotoxin to acetylcholine receptors in mammalian muscle. *Proc. natn. Acad. Sci. U.S.A.*, **69**, 147–159.
223. BERGER, B. (1971). Quelques aspects ultrastructuraux de la substance blanche chez souris quaking. *Brain Res.*, **25**, 35.
224. BERING, E. A. & SATO, O. (1963). Hydrocephalus; changes in formation and absorption of cerebrospinal fluid within the cerebral ventricles. *J. Neurosurg.*, **20**, 1050–1063.
225. BERL, S., PUSZKIN, S. & NICKLAS, W. J. (1972). Actomyosin-like protein in brain. *Science, N.Y.*, **179**, 441–445.
226. BERNSTEIN, J. J. (1964). Relation of spinal cord regeneration to age in adult goldfish. *Expl Neurol.*, **9**, 161–174.
227. BERNSTEIN, J. J. & BERNSTEIN, M. E. (1967). Effect of glial-ependymal scar and teflon arrest on the regenerative capacity of goldfish spinal cord. *Expl Neurol.*, **19**, 25–32.
228. BERNSTEIN, J. J. & BERNSTEIN, M. E. (1968). A mechanism of abortive regeneration in the goldfish spinal cord. *Anat. Rec.*, **160**, 315–316.
229. BERNSTEIN, J. J. & BERNSTEIN, M. E. (1971). Axonal regeneration and formation of synapses proximal to the site of lesion following hemisection of the rat spinal cord. *Expl Neurol.*, **30**, 336–351.
230. BERNSTEIN, J. J. & BERNSTEIN, M. E. (1973). Neuronal alteration and reinnervation following axonal regeneration and sprouting in mammalian spinal cord. *Brain Behav. Evol.*, **8**, 135–161.
231. BERNSTEIN, M. E. & BERNSTEIN, J. J. (1973). Regeneration of axons and synaptic complex formation rostral to the site of hemisection in the spinal cord of the monkey. *Int. J. Neurosci.*, **5**, 15.
232. BERNSTEIN, J. J. & GELDERD, J. B. (1970). Regeneration of the long spinal tracts in the goldfish. *Brain Res.*, **20**, 33–38.
233. BERNSTEIN, J. J. & GELDERD, J. B. (1973). Synaptic reorganization following regeneration of goldfish spinal cord. *Expl Neurol.*, **41**, 402–410.
234. BERNSTEIN, J. J., GELDERD, J. B. & BERNSTEIN, M. E. (1974). Alteration of neuronal synaptic complement during regeneration and axonal sprouting of rat spinal cord. *Expl Neurol.*, **44**, 470–482.
235. BHOOLA, K. D., EVANS, R. H. & SMITH, J. W. (1972). Catecholamine induced contractures in denervated muscle. *Brit. J. Pharmac. Chemother.*, **46**, 531–532P.
236. BIANCHI, C. P. (1961). The effect of caffeine on radiocalcium movement in frog sartorius. *J. gen. Physiol.*, **44**, 845–858.
237. BIDDER, T. G. (1970). Hexose translocation across the blood-brain interface: configurational aspects. *J. Neurochem.*, **15**, 867–874.
238. BIEL, W. C. (1939). The effects of early inanition on a development schedule for the albino rat. *J. comp. Psychol.*, **28**, 1–15.
239. BIGNAMI, A. & DAHL, D. (1973). Differentiation of astrocytes in the cerebellar cortex and pyramidal tract of the newborn rat. An immuno-

fluorescent study with antibodies to a protein specific to astrocytes. *Brain Res.*, **49**, 393–396.

240. BIGNAMI, A. & DAHL, D. (1974). Astrocyte-specific protein and radial glia in the cerebral cortex of newborn rat. *Nature, Lond.*, **252**, 55.
241. BIGNAMI, A. & ENG, L. F. (1973). Biochemical studies of myelin in Wallerian degeneration of rat optic nerve. *J. Neurochem.*, **20**, 165–173.
242. BIGNAMI, A., FORNO, L. & DAHL, D. (1974). The neuroglial response to injury following spinal cord transection in the goldfish. *Expl Neurol.*, **44**, 60–70.
243. BIGNAMI, A. & RALSTON, H. J. (1968). Myelination of fibrillary astroglial processes in long-term Wallerian degeneration. The possible relation to 'status marmoratus'. *Brain Res.*, **11**, 710–713.
244. BIKLE, D., TILNEY, L. G. & PORTER, K. R. (1966). Microtubules and pigment granule migration in the melanophores of *Fundulus heteroclitus* L. *Protoplasma*, **61**, 322–328.
245. BILLENSTEIN, D. & LEVEQUE, T. F. (1955). The reorganization of the neurohypophyseal stalk following hypophysectomy in the rat. *Endocrinology*, **56**, 704–717.
246. BIRD, M. M. & JAMES, D. W. (1973). The development of synapses *in vitro*: between previously dissociated chick spinal cord neurones. *Z. Zellforsch. mikrosk. Anat.*, **140**, 203–216.
247. BIRGE, W. J. (1960). Tissue interactions associated with the differentiation of ependymal stem cells. *Anat. Rec.*, **137**, 340.
248. BIRZIS, L. & CARREGAL, E. J. A. (1966). Electrical impedance and vasomotor changes accompanying synaptic changes in sympathetic ganglia. *Expl Neurol.*, **15**, 1–17.
249. BISCHOFF, A. & MOOR, H. (1967). The ultrastructure of the 'difference factor' in the myelin. *Z. Zellforsch. mikrosk. Anat.*, **81**, 571–580.
250. BISCHOFF, A. & MOOR, H. (1967). Ultrastructural differences between the myelin sheaths of peripheral nerve fibres and CNS white matter. *Z. Zellforsch. mikrosk. Anat.*, **81**, 303–310.
251. BITENSKY, M. W., GORMAN, R. E. & MILLER, W. H. (1971). Adenyl cyclase as a link between photon capture and changes in membrane permeability of frog photoreceptors. *Proc. natn. Acad. Sci. U.S.A.*, **68**, 561–562.
252. BITENSKY, M. W., GORMAN, R. E. & MILLER, W. H. (1972). Digitonin effects on photoreceptor adenylate cyclase. *Science, N.Y.*, **175**, 1363–1364.
253. BITO, L. Z. & DAVSON, H. (1966). Local variation in cerebrospinal fluid composition and its relationship to the composition of the extracellular fluid of the cortex. *Expl Neurol.*, **14**, 264–280.
254. BITTNER, G. D. (1973). Degeneration and regeneration in crustacean neuromuscular systems. *Am. Zool.*, **13**, 379–409.
255. BJERRE, B., BJÖRKLUND, A. & STENEVI, U. (1973). Stimulation of growth of new axonal sprouts from lesioned monoamine neurones in adult rat brain by nerve growth factor. *Brain Res.*, **60**, 161.

256. BJÖRKLUND, A., KATZMAN, R., STENEVI, U. & WEST, K. A. (1971). Development and growth of axonal sprouts from noradrenaline and 5-hydroxytryptamine neurones in the rat spinal cord. *Brain Res.,* **31**, 1–20.
257. BJÖRKLUND, A., KATZMAN, R., STENEVI, U. & WEST, K. A. (1971). Development and growth of axonal sprouts from noradrenaline and 5-hydroxytryptamine neurones in the rat spinal cord. *Brain Res.,* **31**, 21–33.
258. BJÖRKLUND, A., NOBIN, A. & STENEVI, U. (1973). Regeneration of central serotonin neurones after axonal degeneration induced by 5,6,-dihydroxytryptamine. *Brain Res.,* **50**, 214–220.
259. BJÖRKLUND, A. & STENEVI, U. (1971). Growth of central catecholamine neurones into smooth muscle grafts in the rat mesencephalon. *Brain Res.,* **31**, 1–20.
260. BLACKSTAD, T. W. (1956). Commisural connections of the hippocampal region in the rat, with special reference to their mode of termination. *J. comp. Neurol.,* **105**, 417–437.
261. BLAKEMORE, C. & COOPER, G. F. (1970). Development of the brain depends on the visual environment. *Nature, Lond.,* **228**, 477–478.
262. BLAKEMORE, C. & MITCHELL, D. E. (1973). Environmental modification of the visual cortex and the neural basis of learning and memory. *Nature, Lond.,* **241**, 467–468.
263. BLAKEMORE, C. & TOBIN, E. A. (1972). Lateral inhibition between orientation detectors in the cat's visual cortex. *Expl Brain Res.,* **15**, 439–440.
264. BLAISIE, J. K. (1972). The location of photopigment molecules in the cross-section of frog retinal receptor disc membranes. *Biophys. J.,* **12**, 191–204.
265. BLAISIE, J. K. (1972). Net electric charge on photopigment molecules and frog retinal receptor disc membrane structure. *Biophys. J.,* **12**, 205–213.
266. BLAISIE, J. K., DEWEY, M. M., BLAUROCK, A. E. & WORTHINGTON, C. R. (1965). Electron microscope and low angle diffraction studies on outer segment membranes from the retina of the frog. *J. molec. Biol.,* **14**, 143–152.
267. BLAISIE, J. K. & WORTHINGTON, C. R. (1969). Planar liquid-like arrangement of photopigment molecules in frog retinal receptor disc membranes. *J. molec. Biol.,* **39**, 417–439.
268. BLAISIE, J. K., WORTHINGTON, C. R. & DEWEY, M. M. (1969). Molecular localization of frog retinal receptor photopigment by electron microscopy and low-angle X-ray diffraction. *J. molec. Biol.,* **39**, 407–416.
269. BLAUROCK, A. E. (1967). *X-ray diffraction studies of the myelin sheath of nerve.* Thesis. Univ. of Michigan.
270. BLAUROCK, A. E. (1971). Structure of the nerve myelin membrane: proof of the low resolution profile. *J. molec. Biol.,* **56**, 35–52.

271. BLAUROCK, A. E. & WILKINS, M. H. F. (1969). Structure of frog photoreceptor membranes. *Nature, Lond.*, **223**, 906–909.
272. BLEIER, R. (1969). Retrograde transsynaptic cellular degeneration in mammillary and ventral tegmental nuclei following limbic decortication in rabbits of various ages. *Brain Res.*, **15**, 365–393.
273. BLINTIFF, S. & WALKER, B. E. (1960). Radioautographic study of skeletal muscle regeneration. *Am. J. Anat.*, **106**, 233–245.
274. BLINZINGER, K. & KREUTZBERG, G. (1968). Displacement of synaptic boutons from regenerating motoneurones by microglial cells. *Z. Zellforsch. mikrosk. Anat.*, **85**, 145–157.
275. BLOCK, J. B. & ESSMAN, W. B. (1965). Growth hormone administration during pregnancy: a behavioural difference in offspring rats. *Nature, Lond.*, **205**, 1136–1137.
276. BLOMSTRAND, C. & HAMBURGER, A. (1969). Protein turnover in cell-enriched fractions from rabbit brain. *J. Neurochem.*, **16**, 1401–1407.
277. BLOOM, F. E. & AGHAJANIAN, G. K. (1968). Fine structural and cytochemical analysis of the staining synaptic junctions with phosphotungstic acid. *J. Ultrastruct. Res.*, **22**, 361–375.
278. BLOOM, F. E. & IVERSEN, L. L. (1971). Localizing ^{3}H-GABA in nerve terminals of rat cerebral cortex by electron microscopic autoradiography. *Nature, Lond.*, **229**, 628–630.
279. BLUMBERG, B. S., OSTER, G. & MEYER, K. (1955). Changes in the physical characteristics of the hyaluronate of ground substance with alterations in sodium chloride concentration. *J. clin. Invest.*, **34**, 1454–1461.
280. BLUNT, M. J., WENDELL-SMITH, C. P., PAISLEY, P. B. & BALDWIN, F. (1967). Oxidative enzyme activity in macroglia and enzymes of cat optic nerve. *J. Anat.*, **101**, 13–26.
281. BOCCI, V. (1964). Enzyme and metabolic properties of isolated neurones. *Nature, Lond.*, **212**, 826–827.
282. BODEMER, C. & EVERETT, R. (1959). Localization of newly synthesized proteins in regenerating newt limbs as determined by radioautographic localization of injected methionine S 35. *Devl Biol.*, **1**, 327–340.
283. BODENSTEIN, D. (1957). Studies on nerve regeneration in *Periplaneta americana*. *J. exp. Zool.*, **136**, 89–116.
284. BODIAN, D. & MELLORS, R. C. (1945). The regenerative cycle of motoneurones, with special reference to phosphatase activity. *J. exp. Med.*, **81**, 469–488.
285. BOHDANECKA, M., BOHDANECKA, Z. & JARVIK, M. E. (1967). Amnesic effects of small bilateral brain punctures in the mouse. *Science, N.Y.*, **157**, 334–336.
286. BONDAREFF, W. (1965). The extracellular compartment of the cerebral cortex. *Anat. Rec.*, **152**, 119–128.
287. BONDAREFF, W. An intercellular substance in rat cerebral cortex: submicroscopic distribution of ruthenium red. *Anat. Rec.*, **157**, 527–536.

288. BONDAREFF, W. & SJÖSTRAND, J. (1969). Cytochemistry of synaptosomes. *Expl Neurol.*, **24**, 450–458.
289. BONDY, S. C. (1971). Axonal transport of macromolecules I. Protein migration in the central nervous system. *Expl Brain Res.*, **13**, 127–134.
290. BORISY, G. G. & OLMSTED, J. B. (1972). Nucleated assembly of microtubules in porcine brain extracts. *Science, N.Y.*, **177**, 1196–1197.
291. BORISY, G. G., OLMSTED, J. B. & KLUGMAN, R. A. (1972). *In vitro* aggregation of cytoplasmic microtubule subunits. *Proc. natn. Acad. Sci. U.S.A.*, **69**, 2890–2894.
292. BORISY, G. G., OLMSTED, J. B., MARCUM, J. M. & ALLEN, C. (1974). Microtubule assembly *in vitro. Fedn Proc. Fedn Am. Socs exp. Biol.*, **33**, 167–174.
293. BORISY, G. G. & TAYLOR, E. W. (1967). The mechanism of action of colchicine. Binding of colchicine-^{3}H to cellular protein. *J. Cell Biol.*, **34**, 525–534.
294. BORISY, G. G. & TAYLOR, E. W. (1967). The mechanism of action of colchicine. Colchicine binding to sea urchin eggs and the mitotic apparatus. *J. Cell Biol.*, **34**, 535–539.
295. BORNSTEIN, M. B. & MODEL, P. G. (1972). Development of synapses and myelin in culture of dissociated embryonic mouse spinal cord, medulla and cerebrum. *Brain Res.*, **37**, 287–293.
296. BOSMAN, H. B. (1972). Acetylcholine receptor I. Identification and biochemical characteristics of a cholinergic receptor of guinea pig cerebral cortex. *J. biol. Chem.*, **247**, 130–145.
297. De BOTH, N. J. (1965). Enhancement of the self-differentiation capacity of the early limb blastema by various experimental procedures. *In* Kiortis, V. and Trampusch, H. A. L., *Regeneration in animals.* North Holland; Amsterdam, pp. 420–426.
298. BOTTS, J., CASHIN, A. & SCHMIDT, L. (1966). Computation of metal binding in bi-metal-bichelate systems. *Biochemistry, N.Y.*, **5**, 1360–1364.
299. BOUCEK, R. J. & ALVAREZ, T. R. (1970). 5-hydroxytryptamine; a cytospecific growth stimulator of cultured fibroblasts. *Science, N.Y.*, **167**, 898–899.
300. BOULTON, P. S. (1969). Degeneration and regeneration in the insect central nervous system. I. *Z. Zellforsch. mikrosk. Anat.*, **101**, 98–118.
301. BOULTON, P. S. & ROWELL, C. H. F. (1969). Degeneration and regeneration in the insect nervous system. II. *Z. Zellforsch. mikrosk. Anat.*, **101**, 119–134.
302. BOURGEOIS, J. P., RYTER, A., MENEZ, A., FROMAGEOT, P., BOQUET, P. & CHANGEUX, J. P. (1972). Localization of the cholinergic receptor protein in *electrophorus* electroplax by high resolution autoradiography. *FEBS Lett.*, **25**, 127–133.
303. BOVET, D., BOVET-NITTI, F. & OLIVERIO, A. (1969). Genetic aspects of learning and memory in mice. *Science, N.Y.*, **163**, 139.
304. BOWEN, D. M. & RADIN, N. S. (1969). Hydrolase activities in brain of neurological mutants: cerebroside galactosidase, nitrophenyl galactoside

hydrolase, nitrophenyl glucoside hydrolase and sulphatase. *J. Neurochem.*, **16**, 457–463.

305. BOWEN, F. P. (1968). Immunologic reactions after cortical lesions in rabbits. *Archs. Neurol. Psychiat., Chicago*, **19**, 398–402.
306. BOWMAN, W. C. & ZAIMIS, E. (1961). Actions of triethylcholine on neuromuscular transmission. *J. Physiol.*, **157**, 20P.
307. BOWNDS, D. (1967). Site of attachment of retinal in rhodopsin. *Nature, Lond.*, **216**, 1178–1184.
308. BOWNDS, D., DAWES, J., MILLER, J. & STAHLMAN, M. (1972). Phosphorylation of frog photoreceptor membranes induced by light. *Nature, Lond., New Biol.*, **237**, 125–127.
309. BOWNDS, D. & GAIDE-HUGUENIN, A. C. (1970). Rhodopsin content of frog photoreceptor outer segments. *Nature, Lond.*, **225**, 870–872.
310. BOYLE, F. C. & GILLESPIE, J. S. (1970). Accumulation and loss of noradrenaline central to a constriction on adrenergic nerves. *Europ. J. Pharmac.*, **12**, 77–84.
311. BRACHO, H. & ORKAND, R. K. (1972). Neurone-glia interaction: dependence on temperature. *Brain Res.*, **36**, 416–419.
312. BRADBURY, M. W. B. & DAVSON, H. (1965). The transport of potassium between blood, cerebrospinal fluid and brain. *J. Physiol.*, **181**, 151–174.
313. BRADLEY, W. G., MURCHISON, D. & DAY, M. J. (1971). The range of velocities of axoplasmic flow. A new approach, and its application to mice with genetically inherited spinal muscular atrophy. *Brain Res.*, **35**, 185–197.
314. BRADLEY, W. G. & PAPAPETROPOULOS, T. A. (1972). Repeated denervation and reinnervation of skeletal muscle. *Nature, Lond.*, **236**, 401–402.
315. BRADY, U. E., TUMLINSON, J. H., BROWNLEE, R. G. & SILVERSTEIN, R. M. (1971). Sex stimulant and attractant in the Indian meal moth and in the Almond moth. *Science, N.Y.*, **171**, 802–804.
316. BRAGG, P. D. & HOU, C. (1972). Organization of proteins in the native and reformed outer membrane of E. coli. *Biochim. biophys. Acta*, **274**, 478–488.
318. BRAND, L. & GOHLKE, J. R. (1972). Fluorescent probes for structure. *A. Rev. Biochem.*, **41**, 843–868.
319. BRANDES, J. S. (1971). Dendritic branching patterns in lateral geniculate nucleus following deafferentation. *Expl Neurol.*, **31**, 444–450.
320. BRANTON, D. (1966). Fracture faces of frozen membranes. *Proc. natn. Acad. Sci. U.S.A.*, **55**, 1048–1056.
321. BRANTON, D. (1967). Fracture faces of frozen myelin. *Expl Cell Res.*, **45**, 203–207.
322. BRANTON, D. & PARK, R. B. (1967). Subunits in chloroplast lamellae. *J. Ultrastruct. Res.*, **19**, 283–303.
323. BRÅTTGARD, S. O., EDSTRÖM, J. E. & HYDEN, H. (1958). The

productive capacity of the neurone in retrograde reaction. *Expl Cell Res.*, Suppl., **5**, 185–200.
324. BRAY, D. (1970). Surface movements during the growth of single explanted neurons. *Proc. natn. Acad. Sci. U.S.A.*, **65**, 905–910.
325. BRAY, D. (1973). Model for membrane movements in the neural growth cone. *Nature, Lond.*, **244**, 93–96.
326. BRAY, D. (1973). Branching patterns of individual sympathetic neurons in culture. *J. Cell Biol.*, **56**, 702–712.
327. BRAY, G. M. & AGUAYO, A. J. (1974). Regeneration of peripheral unmyelinated nerves. Fate of the axonal sprouts which develop after injury. *J. Anat.*, **117**, 517–529.
328. BRAY, G. M., PEYRONNARD, J. M. & AGUAYO, A. J. (1972). Reactions of unmyelinated nerve fibres to injury. An ultrastructural study. *Brain Res.*, **42**, 297–309.
329. BRAY, J. J. & AUSTIN, L. (1968). Flow of protein and ribonucleic acid in peripheral nerve. *J. Neurochem.*, **15**, 731–740.
330. BRAY, J. J., KON, C. M. & BRECKENRIDGE, B. M. (1971). Adenyl cyclase, cyclic nucleotide phosphodiesterase and axoplasmic flow. *Brain Res.*, **26**, 285–394.
331. BRAZIL, O. V. & EXCELL, B. J. (1970). Action of crotoxin and crotactin from the venom of *crotalus terrificus* (South American rattlesnake) on the frog neuromuscular junction. *J. Physiol.*, **212**, 34P.
332. BRECKENRIDGE, B. M. (1972). Cyclic AMP and neurotransmitters. *Res. Publs. Ass. Res. nerv. ment. Dis.*, **50**, 216–228.
333. BRECKENRIDGE, B. M., BURN, J. H. & MATCHINSKY, F. M. (1967). Theophylline, epinephrine and neostigmine facilitation of neuromuscular transmission. *Proc. natn. Acad. Sci. U.S.A.*, **57**, 1893–1897.
334. BRETSCHER, M. S. (1971). A major protein which spans the human erythrocyte membrane. *J. molec. Biol.*, **59**, 351–357.
335. BRETSCHER, M. S. (1972). Asymmetrical lipid bilayer structure for biological membranes. *Nature, Lond., New Biol.*, **236**, 11–12.
336. BRIEN, P. (1961). Etude d'*Hydra pirardi. Bull. biol. Fr. Belg.*, **95**, 301–360.
337. BRIEN, P. & RENIERS-DECOEN, M. (1949). La croissance, la blastogenèse, l'orogenèse chez *Hydra fuscia* (Pallas). *Bull. biol. Fr. Belg.*, **82**, 293–386.
338. BRIERLEY, G. P., FLEISCHMAN, D., HUGHES, S. D., HUNTER, G. R. & McCONNELL, G. R. (1968). On the permeability of isolated retinal outer segment fragments. *Biochim. biophys. Acta*, **163**, 117–120.
339. BRIGHTMAN, M. W. (1965). The distribution within the brain of ferritin injected into cerebrospinal fluid compartments. *J. Cell Biol.*, **26**, 99–123.
340. BRIGHTMAN, M. W. (1966). Pinocytosis in the brains of dead rats. *J. Anat.*, **154**, 322.
341. BRIGHTMAN, M. W. & PALAY, S. L. (1963). The fine structure of the ependyma in the brain of the rat. *J. Cell Biol.*, **19**, 415–440.

342. BRIGHTMAN, H. W. & REESE, T. S. (1967). Astrocytic and ependymal junctions in the mouse brain. *J. Cell Biol.*, **35**, 16–17A.
343. BRIGHTMAN, M. W. & REESE, T. S. (1969). The role of intercellular junctions in the movement of substances within the brain. *J. Neuropath. exp. Neurol.*, **28**, 167P.
344. BRIGHTMAN, M. W. & REESE, T. S. (1969). Junctions between intimately apposed cell membranes in the vertebrate brain. *J. Cell Biol.*, **40**, 648–677.
345. BRIGHTMAN, M. W., REESE, T. S. & FEDER, N. (1970). Assessment with the electron microscope of the permeability to peroxidase of cerebral endothelium and epithelium in mice and sharks. *In* Crone, C. and Thompson, A. M., *Capillary permeability.* Munksgaard; Copenhagen, pp. 468–476.
346. BRIGHTMAN, M. W., REESE, T. S., OLSSON, Y. & KLATZO, I. (1970). Morphological restriction to the passage of exogenous protein within shark brain. *J. Neuropath. exp. Neurol.*, **29**, 123–124.
347. BRIMBLE, M. J. & WALLIS, D. S. (1974). The role of muscarinic receptors in synaptic transmission and its modulation in the rabbit superior cervical ganglion. *Europ. J. Pharmac.*, **29**, 117–132.
348. BRINK, J. J., DAVIS, R. E. & AGRANOFF, B. W. (1966). Effects of puromycin, acetoxycycloheximide and actinomycin D on protein synthesis in goldfish brain. *J. Neurochem.*, **13**, 889–896.
349. BROADHURST, P. L., FULKER, D. W. & WILCOCK, J. (1974). Behavioural genetics. *A. Rev. Psychol.*, **25**, 389–416.
350. BROADHURST, P. L. & JINKS, J. L. (1966). Stability and change in the inheritance of behaviour in rats; a further analysis of statistics from a diallel cross. *Proc. R. Soc.*, **165**, 450–472.
351. BROCKLEBANK, M. C. (1925). Degeneration and regeneration of lateral-line organs in *Ameiurus nebulosis. J. exp. Zool.*, **42**, 293–305.
352. BRODY, I. A. (1966). Relaxing factor in denervated muscle: a possible explanation for fibrillation. *Am. J. Physiol.*, **211**, 1277–1280.
353. BROWN, A. G. & IGGO, A. (1962). The structure and function of cutaneous 'touch corpuscles' after nerve crush. *J. Physiol.*, **165**, 28–29P.
354. BROWN, D. D. & AFIFI, A. K. (1965). Histological and ablation studies on the relation of the subcommisural organ and rostral midbrain to sodium and water metabolism. *Anat. Rec.*, **153**, 255–264.
355. BROWN, G. L. & PASCOE, J. E. (1954). The effect of degenerative section of ganglionic axons on transmission through the ganglion. *J. Physiol.*, **123**, 565–573.
356. BROWN, J. E. & BLINKS, J. R. (1974). Changes in intracellular free calcium concentration during illumination of invertebrate photoreceptors. Detection with aquorin. *J. gen. Physiol.*, **64**, 643.
357. BROWN, J. H. & MAKMAN, M. H. (1972). Stimulation by dopamine of adenylate cyclase in retinal homogenates and of adenosine 3′:5′-cyclic monophosphate formation in intact retina. *Proc. natn. Acad. Sci. U.S.A.*, **69**, 539–543.

358. BROWN, J. O. & McCOUCH, G. P. (1947). Abortive regeneration of the transected spinal cord. *J. comp. Neurol.*, **87**, 131–137.
359. BROWN, P. K. (1971). Visual pigments. *Biophys. Soc. Abstr.*, **11**, 248A.
360. BROWN, P. K. (1972). Rhodopsin rotates in the visual receptor membrane. *Nature, Lond., New Biol.*, **236**, 25–38.
361. BROWN, P. K. & WALD, G. (1963). Visual pigments in human and monkey retinas. *Nature, Lond.*, **200**, 37.
362. BROWN-GRANT, K. & SHERWOOD, M. R. (1971). The early androgen syndrome in the guinea pig. *J. Endocr.*, **49**, 277–291.
363. BRUCKDORFER, K. R., DEMEL, R. A., de GIER, J. & van DEENEN, L. L. M. (1969). The effect of partial replacements of membrane cholesterol by other steroids on the osmotic fragility and glycerol permeability of erythrocytes. *Biochim. biophys. Acta,* **183**, 334.
364. BRUELL, J. H. (1964). Inheritance of behavioural and physiological characteristics of mice and the problem of heterosis. *Am. Zool.*, **4**, 125–138.
365. BRUNST, V. V. (1961). Some problems of regeneration. *Q. Rev. Biol.*, **36**, 178–206.
366. BRYAN, J. (1974). Biochemical properties of microtubules. *Fedn Proc. Fedn Am. Socs exp. Biol.*, **33**, 152–157.
367. BRYAN, J. & WILSON, L. (1971). Are cytoplasmic microtubules heteropolymers? *Proc. natn. Acad. Sci. U.S.A.*, **68**, 1762–1766.
368. BRYANS, W. A. (1959). Mitotic activity in the brain of the adult rat. *Anat. Rec.*, **133**, 65–71.
369. BRYANT, S. V., FYFE, D. & SINGER, M. (1971). The effects of denervation on the ultrastructure of young limb regenerates in the newt *Triturus. Devl Biol.*, **24**, 577–595.
370. BRYANT, S. V. & SINGER, M. (1969). Movements in the myelin sheaths of peripheral nerve fibres. *Anat. Rec.*, **169**, 345–346.
371. BRZIN, M. & MAJCEN-TKACEV, Z. (1963). Cholinesterase in denervated end plates and muscle fibres. *J. Cell Biol.*, **19**, 349–358.
372. BUBIS, J. J. & WOLMAN, M. (1965). Hydrolytic enzymes in Wallerian degeneration. *Israel J. med. Sci.*, **1**, 410–414.
373. BUCHANEN, A. R. & ROBERTS, J. E. (1948). Relative lack of myelin in optic tracts as a result of underfeeding in young albino rat. *Proc. Soc. exp. Biol. Med.*, **69**, 101–104.
374. BULKIN, B. J. (1972). Raman spectroscopic study of human erythrocyte membranes. *Biochim. biophys. Acta,* **274**, 649–651.
375. BULKIN, B. J. & KRISHNAMACHARI, N. (1970). Infrared spectroscopic measurements of phosphatidyl ethanolamine – water liquid crystals. *Biochim. biophys. Acta,* **211**, 592–594.
376. BULKIN, B. J. & KRISHNAMACHARI, N. (1972). Infrared and Raman spectroscopy of phospholipid-water mixtures. *J. Am. chem. Soc.*, **94**, 1109.
377. BULLER, A. J. (1970). The neural control of the contractile mechanism in skeletal muscle. *Endeavour,* **29**, 107–111.

378. BULLER, A. J., ECCLES, J. C. & ECCLES, R. M. (1960). Interaction between motoneurones and muscles in respect of the characteristic speeds of their responses. *J. Physiol.*, **150**, 417–439.
379. BULLER, A. J. & LEWIS, D. M. (1963). Factors affecting the differentiation of mammalian fast and slow muscle fibres. *In* Gutmann, E. and Hnik, P. (1963) *Proceedings of a symposium on the effect of use and disuse on neuromuscular function.* Czechoslovak Acad. Sci.; Prague, pp. 149–159.
380. BULLER, A. J. & LEWIS, D. M. (1965). Further observations on mammalian cross-innervated skeletal muscle. *J. Physiol.*, **178**, 343–358.
381. BULLER, A. J. & MOMMAERTS, W. F. H. M. (1969). Myofibrillar ATPase as a determining factor for contraction velocity, and its changes upon experimental cross innervation. *J. Physiol.*, **201**, 46–47P.
382. BULLER, A. J., MOMMAERTS, W. F. H. M. & SERAYDARIAN, K. (1969). Enzymic properties of myosin in fast and slow twitch muscles of the cat following cross innervation. *J. Physiol.*, **205**, 581–597.
383. BULLER, A. J., MOMMAERTS, W. F. H. M. & SERAYDARIAN, K. (1971). Neural control of myofibrillar ATPase activity in rat skeletal muscle. *Nature, Lond., New Biol.*, **233**, 31–32.
384. BULLOCK, T. H. & HORRIDGE, G. A. (1965). *Structure and function in the nervous system of invertebrates.* Freeman; San Francisco.
385. BUNGE, M. (1973). Uptake of peroxidase by growth cones of cultured neurones. *Anat. Rec.*, **175**, 280.
386. BUNGE, M. B. (1973). Fine structure of nerve fibres and growth cones of isolated sympathetic neurones in culture. *J. Cell Biol.*, **56**, 713–735.
387. BUNGE, M. B., BUNGE, R. P. & PETERSON, R. P. (1967). The onset of synapse formation in spinal cord cultures, as studied by electron microscopy. *Brain Res.*, **6**, 728–749.
388. BUNGE, R. P. (1968). Glial cells and the central myelin sheath. *Physiol. Rev.*, **48**, 197–251.
389. BUNGE, R. P. & BUNGE, M. B. (1964). Electron microscopic observations on glial cells in tissue culture. *Anat. Rec.*, **148**, 266.
390. BUNGE, R. P., BUNGE, M. B. & PETERSON, E. R. (1965). An electron microscope study of cultured rat spinal cord. *J. Cell Biol.*, **24**, 163–192.
391. BUNT, A. H., LUND, R. D. & LUND, J. S. (1974). Retrograde axonal transport of horseradish peroxidase by ganglion cells of the albino rat retina. *Brain Res.*, **73**, 215–228.
392. BURGER, M. M. (1969). A difference in the architecture of the surface membrane of normal and virally transformed cells. *Proc. natn. Acad. Sci. U.S.A.*, **62**, 994–1001.
393. BURGER, M. M. & NOONAN, K. D. (1970). Restoration of normal growth by covering of agglutinin sites on tumour cell surface. *Nature, Lond.*, **228**, 512–515.
394. BURKE, R. E., LEVINE, D. N., ZAJAX, F. E., TSAIRIS, P. & ENGEL,

W. K. (1971). Mammalian motor units: physiological and histochemical correlation in three types in cat gastrocnemius. *Science, N.Y.*, **174**, 709–712.

395. BURN, J. H., LEACH, E. H., RAND, M. J. & THOMPSON, J. W. (1959). Peripheral effects of nicotine and acetylcholine resembling those of sympathetic stimulation. *J. Physiol.*, **148**, 332–352.
396. BURNETT, A. L. & DIEHL, N. (1964). The nervous system of *Hydra*. I. Types, distribution, and origin of nerve elements. *J. exp. Zool.*, **157**, 217–226.
397. BURNETT, A. L., & DIEHL, N. A. (1964). The nervous system of *Hydra*. III. The initiation of sexuality with special reference to the nervous system. *J. exp. Zool.*, **157**, 237–250.
398. BURNETT, A. L., DIEHL, N. & DIEHL, F. (1964). The nervous system of *Hydra*. II. Control of growth and regeneration by neurosecretory cells. *J. exp. Zool.*, **157**, 227–236.
399. BURNETT, A. L., KARY, C. E. & LAGORIO, A. M. (1971). Induction of growth in newt and frog limbs after perfusion with extracts from newt blastemas. *Nature, Lond.*, **234**, 98–99.
400. BURNS, R. B. & POLLARD, T. D. (1974). A dynein-like protein from brain. *FEBS Lett.*, **40**, 274–280.
401. BURNSTOCK, G. (1972). Purinergic nerves. *Pharmac. Rev.*, **24**, 509–581.
402. BURNSTOCK, G., CAMPBELL, G., SATCHELL, D. & SMYTHE, A. (1970). Evidence that adenosine triphosphate or a related nucleotide is the transmitter substance released by non-adrenergic inhibitory nerves in the gut. *Br. J. Pharmac.*, **40**, 668–688.
403. BUSE, M. G. & BUSE, J. (1959). Glucose uptake and response to insulin of the isolated rat diaphragm. *Diabetes*, **8**, 218–225.
404. BUTENANDT, A. & HECKER, E. (1961). Lure substance of *Bombyx mori*. *Angew. Chem.*, **73**, 349.
405. BUTLER, E. G. & O'BRIEN, J. P. (1942). Effects of localized X-irradiation on regeneration of the urodele limb. *Anat. Rec.*, **84**, 407–413.
406. BUTLER, E. & SCHOTTE, O. (1949). Effects of delayed denervation on regenerative activity in limbs of urodele larvae. *J. exp. Zool.*, **112**, 361–392.
407. BUTLER, E. G. & WARD, M. (1965). Reconstitution of the spinal cord following ablation in urodele larvae. *J. exp. Zool.*, **160**, 47–66.
408. BUTLER, E. G. & WARD, M. (1967). Reconstitution of the spinal cord after ablation in adult *Triturus*. *Devl Biol.*, **15**, 464–486.
409. BUTLER, K., DUGAS, H., SCHNEIDER, H. & SMITH, I. (1970). Cation-induced organization changes in a lipid bilayer model membrane. *Biochem. biophys. Res. Comm.*, **40**, 770.
410. BUZNIKOV, G. A., KOST, A. N., KUCHEROVA, N. F., MNDZHOYAN, A. L., SUVOROV, N. N. & BERDYSHEVA, L. V. (1970). The role of neurohumors in early embryogenesis. III. Pharmacological analysis of the

role of neurohumors in cleavage divisions. *J. Embryol. exp. Morph.*, **23**, 549–569.

411. BYERS, B. & PORTER, K. R. (1964). Orientated microtubules in elongating cells of the developing lens rudiment after induction. *Proc. natn. Acad. Sci. U.S.A.*, **52**, 1091–1099.
412. CABAK, V. & NAJDANVIC, R. (1965). Effects of undernutrition in early life on physical and mental development. *Archs. Dis. Child.*, **40**, 532–534.
413. CADENHEAD, D. A. & KATTI, S. S. (1971). A comparative study of cholesterol and a cholesterol-like ESR probe in pure and mixed macromolecular films. *Biochim. biophys. Acta*, **241**, 709–712.
414. CAJAL, S. R. y (1928). *Degeneration and regeneration of the nervous system. II.* Translated and edited by May, R. M. Oxford Univ. Press.
415. CARALESU, F. R. & HENRY, J. L. (1971). Sex difference in the number of sympathetic neurons in the spinal cord of the cat. *Science, N.Y.*, **173**, 343–344.
416. CALISSANO, P. & BANGHAM, A. D. (1971). Effect of two brain specific proteins (S-100 and 14-3-2) on cation diffusion across artificial lipid membranes. *Biochem. biophys. Res. Commun.*, **43**, 504–509.
417. CALVET, M. C. (1974). Patterns of spontaneous electrical activity in tissue cultures of mammalian cerebral cortex *vs* cerebellum. *Brain Res.*, 69, 281–296.
418. CALVIN, M., WANG, H. H., ENTING, G., GILL., FERRUT, P., HARPOLD, M. A. & KLEIN, M. P. (1969). Biradical spin labeling for nerve membranes. *Proc. natn. Acad. Sci. U.S.A.*, **63**, 1–8.
419. CAMBRIDGE, G. W. (1961). Comparison of the properties of denervated rat diaphragm muscle and an invertebrate smooth muscle. *J. Physiol.*, **158**, 25P.
420. CAMEJO, G. & VILLEGAS, R. (1969). Tetrodotoxin interaction with cholesterol. *Biochim. biophys. Acta*, **173**, 351–353.
421. CAMEJO, G., VILLEGAS, G. M., BARNOLA, F. V. & VILLEGAS, R. (1969). Characterization of two different membrane fractions isolated from the first stellar nerves of the squid *Dosedicus gigas. Biochim. biophys. Acta*, **193**, 247–259.
422. CAMERON, I. R. & KLEEMAN, C. R. (1970). The effect of acute hyperkalaemia on the blood-c.s.f. potential difference. *J. Physiol.*, **207**, 68–69P.
423. CAMMERMEYER, J. (1966). Morphologic distinctions between oligodendrocytes and microglia cells in the rabbit cerebral cortex. *Am. J. Anat.*, **118**, 227–248.
424. LeCAMP, M. (1947). Les tissus de régénération en culture *in vitro. C. r. hebd. Séanc. Acad. Sci., Paris*, **224**, 674–675.
425. CANNON, W. B. & ROSENBLUETH, A. (1949). *The supersensitivity of denervated structures.* Macmillan; N.Y.
426. CANNON, W. W. (1939). A law of denervation. *Am. J. med. Sci.*, **198**, 737–750.

426a CARLSON, B. M. (1968). Regeneration of completely excised gastrocnemius muscle in the frog and rat from minced muscle fragments. *J. Morph.*, **125**, 447–471.

427. CARLSSON, C. A., BOLANDER, P. & SJÖSTRAND, J. (1971). Changes in axonal transport during regeneration of feline ventral roots. *J. neurol. Sci.*, **14**, 75–93.

428. CARLSSON, C. A., BOLANDER, P. & SJÖSTRAND, J. (1971). Biochemical and histochemical changes in the feline gastrocnemius muscle during regeneration of ventral roots. *J. neurol. Sci.*, **14**, 95–105.

429. CARLSSON, C. A. & THULIN, C. A. (1967). Regeneration of feline dorsal roots. *Experentia*, **23**, 125–126.

430. CARNAY, L. D. & BARRY, W. H. (1969). Turbidity, birefringence and fluorescence changes in skeletal muscle coincident with the action potential. *Science, N.Y.*, **165**, 608–609.

431. CARNEGIE, P. R. (1971). Aminoacid sequence of the encephalitogenic basic protein from human myelin. *Biochem. J.*, **123**, 57–67.

432. CARNEGIE, P. R., KEMP, B. E., DUNKLEY, P. R. & MURRAY, A. W. (1973). Phosphorylation of myelin basic protein by an adenosine 3′:5′-cyclic monophosphate-dependent protein kinase. *Biochem. J.*, **135**, 569–572.

433. CARTAUD, J., BENEDETTI, E. L., KASAI, M. & CHANGEUX, J. P. (1971). *In vitro* excitation of purified membrane fragments by cholinergic agonists. IV. Ultrastructure at high resolution of the ACh E-rich and ATPase-rich microsomes. *J. mem. Biol.*, **6**, 81–88.

434. CARTER, C. S. & MARR, J. N. (1970). Olfactory imprinting and age variables in the guinea pig. *Anim. Behav.*, **18**, 238–244.

435. CARTOUZOU, G., ATTALI, J. C. & LISSITZKY, S. (1968). Acides ribonucleiques messagers de la glande thyroide. I. RNA à marquage rapide des noyaux et des polysomes. *Europ. J. Biochem.*, **4**, 41–54.

436. CARVALHO, A. P., SANUI, H. & PACE, N. (1963). Calcium and magnesium binding properties of cell membrane materials. *J. cell comp. Physiol.*, **62**, 311–317.

437. CASPAR, D. L. D. & KIRSCHNER, D. A. (1971). Myelin membrane structure at 10 Å resolution. *Nature, Lond., New Biol.*, **231**, 46–52.

438. CASS, A. & FINKELSTEIN, A. (1967). Water permeability of thin lipid membranes. *J. gen. Physiol.*, **50**, 1765–1784.

439. CASS, A., FINKELSTEIN, A. & KRESPI, V. (1970). The ion permeability induced in thin lipid membranes by the polyene antibiotics nystatin and amphotericin B. *J. gen. Physiol.*, **56**, 100–124.

440. CASTELLUCCI, V. & GOLDRING, S. (1970). Contribution to steady potential shifts of slow depolarization by cells presumed to be glia. *Electroenceph. clin. Neurophysiol.*, **28**, 109–118.

441. del CASTILLO, J., RODRIGUEZ, A. & ROMERO, C. H. (1967). Pharmacological studies on an artificial transmitter-receptor system. *Ann. N.Y. Acad. Sci.*, **144**, 803–815.

442. de CASTRO, F. (1934). Note sur la régeneration fonctionnelle hétéro-

genique dans les anastomoses des nerfs pneumogastrique et hypoglosse avec le sympathique cervical. *Trab. Lab. Invest. biol. Univ. Madr.*, **29**, 397–416.

443. CAUNA, N. (1956). Nerve supply and nerve endings in Meissner's corpuscles. *Am. J. Anat.*, **99**, 315–350.
444. CAUNA, N. (1958). Structure of digital touch corpuscles. *Acta Anat.*, **32**, 1–23.
445. CAUNA, N. (1959). The mode of termination of the sensory nerves and its significance. *J. comp. Neurol.*, **113**, 169–199.
446. CAUNA, N. (1965). The effects of aging on the receptor organs of the human dermis. *Adv. Biol. Skin*, **6**, 63–96.
447. CAUNA, N. (1966). Fine structure of the receptor organs and its probable functional significance. *In* de Reuck, A. V. S. and Knight, J. *Touch, heat and pain.* CIBA Fdn Symp. Churchill; London. pp. 117–127.
448. CAUNA, N. & MANNAN, G. (1958). Structure of human digital corpuscles and its functional significance. *J. Anat.*, **92**, 1–20.
449. CAUNA, N. & MANNAN, G. (1959). Development and postnatal changes of digital Pacinian corpuscles in the human hand. *J. Anat.*, **93**, 271–286.
450. CAUNA, N. & ROSS, S. L. (1960). The fine structure of Meissner's touch corpuscles of human fingers. *J. Cell Biol.*, **8**, 467–482.
451. CAVANAGH, J. B. (1970). The proliferation of astrocytes around a needle wound in the rat brain. *J. Anat.*, **106**, 471–487.
452. CAVINESS, V. S. & SIDMAN, R. L. (1972). Olfactory structures of the forebrain in the reeler mutant mouse. *J. comp. Neurol.*, **145**, 85–104.
453. CECCARELLI, B., CLEMENTI, F. & MANTEGAZZA, P. (1971). Synaptic transmission in the superior cervical ganglion of the cat after reinnervation by vagus fibres. *J. Physiol.*, **216**, 87–98.
454. CEDAR, H., KANDEL, E. R. & SCHWARTZ, J. H. (1972). Cyclic adenosine monophosphate in the nervous system of *Aplysia californica.* I. Increased synthesis in response to synaptic stimulation. *J. gen. Physiol.*, **60**, 558–569.
455. CEDAR, H. & SCHWARTZ, J. H. (1972). Cyclic adenosine monophosphate in the nervous system of *Aplysia californica.* II. Effect of serotonin and dopamine. *J. gen. Physiol.*, **60**, 570–587.
456. CERF, J. A. & CHACKO, L. W. (1958). Retrograde reaction in motoneuron dendrites following ventral root section in the frog. *J. comp. Neurol.*, **109**, 205–216.
457. CHACKO, G. K., GOLDMAN, D. E., MALHOTRA, H. C. & DEWEY, M. M. (1974). Isolation and characterization of plasma membrane fractions from garfish olfactory nerve. *J. Cell Biol.*, **62**, 831–843.
459. CHALCROFT, J. P. & BULLIVANT, S. (1970). An interpretation of liver cell membrane and junction structure based on observation of freeze-fracture replicas of both sides of the fracture. *J. Cell Biol.*, **47**, 49–60.
460. CHALKLEY, D. T. (1954). A quantitative histological analysis of forelimb regeneration in *Triturus viridescens. J. Morph.*, **94**, 21–71.

461. CHANDLER, W. K. & MEVES, H. (1965). Voltage clamp experiments on internally perfused giant axons. *J. Physiol.*, **180**, 788–820.
462. CHANG, C. C. & LEE, C. Y. (1963). Isolation of neurotoxins from the venom of Bungarus multicinctus and their modes of neuromuscular blocking action. *Archs. int. Pharmacodyn. Thér.*, **144**, 241–257.
463. CHANG, H. C., HSIEH, W. M., LI, T. H. & LIM, R. K. S. (1939). Studies on tissue acetyl choline. VI. The liberation of acetyl choline from nerve trunks during stimulation. *Chin. J. Physiol.*, **14**, 19–29.
464. CHANG, S. Y. (1969). *Changes produced in rat brain by environmental complexity and drug injection.* Doctoral dissertation. Univ. California; Berkeley.
465. CHANGEUX, J. P. (1966). Responses of acetylcholinesterase from *Torpedo marmorata* to salts and curarizing drugs. *Molec. Pharmac.*, **2**, 369–392.
466. CHANGEUX, J. P., KASAI, M. & LEE, C. Y. (1970). Use of a snake venom toxin to characterize the cholinergic receptor protein. *Proc. natn. Acad. Sci. U.S.A.*, **67**, 1241–1247.
467. CHANGEUX, J. P., MEUNIER, J. C. & HUCHET, M. (1971). Studies on the cholinergic receptor protein of *Electrophorus electricus*. I. An assay *in vitro* for the cholinergic receptor site and solubilization of the receptor protein from electric tissue. *Molec. Pharmac.*, **7**, 538–553.
468. CHANGEUX, J. P., MEUNIER, J. C. & HUCHET, M. (1971). Studies on the cholinergic receptor protein of *Electrophorus electricus*. *Molec. Pharmac.*, **7**, 554–573.
469. CHANGEUX, J. P., PODLESKI, T. R. & WOFSY, L. (1967). Affinity labeling of the acetyl choline receptor. *Proç. natn. Acad. Sci. U.S.A.*, **58**, 2063–2070.
470. CHANGEUX, J. P. & THIERY, J. (1967). On the mode of action of colicins: a model of regulation at the membrane level. *J. theor. Biol.*, **17**, 315–318.
471. CHANGEUX, J. P. & THIERY, J. (1968). On the excitability and cooperativity of biological membranes. *In* Järnefelt, J., *Regulatory functions of biological membranes.* Elsevier; Amsterdam.
472. CHANY, C., GRESSER, I., VENDRLY, C. & ROBBE-FOSSAT, F. (1966). Persistant polioviral infection of the intact amniotic membrane. II. Existence of a mechanical barrier to viral infection. *Proc. Soc. exp. Biol. Med.*, **123**, 960–968.
473. CHAPMAN, D. (1965). *Structure of lipids.* Methuen; London.
474. CHAPMAN, D. (1968). Recent physical studies of phospholipids and natural membranes. *In* Chapman, D., *Biological membranes, physical fact and function.* Acad. Press; N.Y. pp. 125–202.
475. CHAPMAN, D. (1968). *Biological membranes.* Acad. Press; N.Y.
476. CHAPMAN, D. (1970). The chemical and physical characteristics of biological membranes. *In* Bittar, E. *Membranes and ion transport.* Wiley Interscience; London. pp. 24–63.
477. CHAPMAN, D. (1972). The role of fatty acids in myelin and other

important brain structures. *In Lipids, malnutrition and the developing brain.* CIBA Fdn Symp. Churchill; London. pp. 31–57.

478. CHAPMAN, D. (1972). Nuclear magnetic resonance spectroscopic studies of biological membranes. *Ann. N.Y. Acad. Sci.,* **195**, 179–206.
479. CHAPMAN, D., CHERRY, R. J., FINER, E. G., HAUSER, H., PHILLIPS, M. C., SHIPLEY, G. G. & McMULLEN, A. I. (1969). Physical studies of phospholipid/alamethicin interactions. *Nature, Lond.,* **224**, 692.
480. CHAPMAN, D. & DODD, G. M. (1971). Physicochemical probes of membrane structure. *In* Rothfield, L. I., *Structure and function of biological membranes.* Acad. Press; N.Y. pp. 13–81.
481. CHAPMAN, D. & McLAUGHLIN, K. A. (1967). Oriented water in sciatic nerve of rabbit. *Nature, Lond.,* **215**, 391–393.
482. CHARNOCK, J. S., COOK, D. A., & CASEY, R. (1973). The role of cations and other factors on the apparent energy of activation of Na^+ and K^+ ATPase. *Archs. Biochem. Biophys.,* **147**, 323–329.
483. CHARNOCK, J. S., DOTY, D. M. & RUSSELL, J. S. (1973). The effect of temperature on the activity of (Na^+ +K^+)-ATPase. *Archs. Biochem. Biophys.,* **142**, 633–637.
484. CHASE, H. P., DORSEY, J. & McKHANN, G. M. (1967). The effect of malnutrition on the synthesis of a myelin lipid. *Paediatrics,* **40**, 551–559.
485. CHASE, H. P., LINDSLEY, W. F. B. & O'BRIEN, D. (1969). Undernutrition and cerebellar development. *Nature, Lond.,* **221**, 554–555.
486. CHASIN, M., MAMRAK, F. & SAMANIEGO, S. G. (1974). Preparation and properties of a cell-free, hormonally responsive adenylate cyclase from guinea pig brain. *J. Neurochem.,* **22**, 1031–1038.
487. CHASIN, M., MAMRAK, F., SAMANIEGO, S. G. & HESS, S. M. (1973). Characteristics of the catecholamine and histamine receptor sites mediating accumulation of cyclic adenosine 3′5′-monophosphate in guinea pig brain. *J. Neurochem.,* **21**, 1415–1427.
488. CHASIN, M., RIVKIN, I., MAMRAK, F., SAMANIEGO, S. G. & HESS, S. M. (1971). α- And β-adrenergic receptors as mediators of accumulation of cyclic adenosine 3′5′-monophosphate in specific areas of guinea pig brain. *J. biol. Chem.,* **246**, 3037–3041.
489. CHEN, I. & LEE, C. Y. (1970). Ultrastructural changes in the motor nerve terminals caused by *beta*-bungarotoxin. *Virchows Arch. Abt. Zellpathol.,* **6**, 318.
490. CHEN, J. S. & LEVI-MONTALCINI, R. (1969). Axonal outgrowth and cell migration *in vitro* from nervous system of cockroach embryos. *Science, N.Y.,* **166**, 631–632.
491. CHERKIN, A. (1969). Kinetics of memory consolidation role of amnesic treatment parameters. *Proc. natn. Acad. Sci. U.S.A.,* **63**, 1094–1101.
492. CHERRY, R. J., CHAPMAN, D. & GRAHAM, D. E. (1972). Studies of the conductance changes induced in bimolecular lipid membranes by alamethicin. *J. mem. Biol.,* **7**, 325–344.
493. CHOUCHKOV, H. N. (1968). Histochemical demonstration of primary

catecholamines in Pacinian corpuscles of the cat. *Experentia,* **24**, 826–827.

494. CHOUCHKOV, H. N. (1971). Ultrastructure of Pacinian corpuscles in men and cats. *Z. mikrosk.-anat. Forsch.*, **83**, 17–32.
495. CHOUCHKOV, H. N. (1971). Ultrastructure of Pacinian corpuscles after section of nerve fibres. *Z. mikrosk.-anat. Forsch.*, **83**, 33–46.
496. CHOW, K. L. & STEWART, D. L. (1972). Reversal of structural and functional effects of long-term visual deprivation in cats. *Expl Neurol.*, **34**, 409–433.
497. CICERO, T. J., COWAN, W. M. & MOORE, B. W. (1970). Changes in the concentration of the two brain specific proteins S-100 and 14-3-2 during the development of the avian optic tectum. *Brain Res.*, **24**, 1–10.
498. CLARK, D. G., WOLCOTT, R. G. & RAFTERY, M. A. (1972). Partial characterization of an α-bungarotoxin-binding component of electroplax membranes. *Biochem. biophys. Res. Commun.*, **48**, 1061–1067.
499. CLARK, R. B. & PERKINS, J. P. (1971). Increase of cyclic AMP in cultured human glial cells mediated by norepinephrine and histamine. *Pharmacologist,* **13**, 256.
500. CLARK, R. B. & PERKINS, J. P. (1971). Regulation of adenosine 3′5′-cyclic monophosphate concentration in cultured human astrocytoma cells by catecholamines and histamine. *Proc. natn. Acad. Sci. U.S.A.*, **68**, 2757–2760.
501. CLARK, W. E. LeG. (1942). The problem of neuronal regeneration in the central nervous system. I. The influence of spinal ganglia and nerve fragments grafted in the brain. *J. Anat.*, **77**, 20–48.
502. CLARK, W. E. LeG. (1943). The problem of neuronal regeneration in the central nervous system. II. The insertion of peripheral nerve stumps into the brain. *J. Anat.*, **77**, 251–259.
503. CLAUSEN, J. & HANSEN, A. (1963). Acid mucopolysaccharides of human brain: identification by means of infra-red analysis. *J. Neurochem.*, **10**, 165–168.
504. CLELAND, R. L. (1968). Comparison of polyelectrolyte behaviour of hyaluronate with that of carboxymethyl cellulose. *Biopolymers,* **6**, 1519–1529.
505. CLEMENTE, C. D. (1964). Regeneration in the vertebrate central nervous system. *Int. Rev. Neurobiol.*, **6**, 257–301.
506. CLENDINNIN, B. G. & EAYRS, J. T. (1961). The anatomical and physiological effects of prenatally administered somatotropin on cerebral development in rats. *J. Endocr.*, **22**, 183–193.
507. CLONEY, R. A. (1972). Cytoplasmic filaments and morphogenesis: effects of cytochalasin B on contractile epidermal cells. *Z. Zellforsch. mikrosk. Anat.*, **132**, 167–192.
508. CLOS, J. & LeGRAND, J. (1970). Influence de la déficience thyroidienne et de la sous-alimentation sur la croissance et la myélinisation des fibres nerveuses du nerf sciatique chez le jeune rat blanc. Etude au microscope éléctronique. *Brain Res.*, **22**, 285–297.

509. CLOSE, R. (1965). Effects of cross union of motor nerves to fast and slow skeletal muscles. *Nature, Lond.*, **206**, 831–832.
510. CLOSE, R. (1965). The relation between intrinsic speed of shortening and duration of the active state of muscle. *J. Physiol.*, **180**, 542–559.
511. CLOSE, R. (1967). Properties of motor units in fast and slow skeletal muscles of the rat. *J. Physiol.*, **193**, 45–52.
512. CLOSE, R. (1969). Dynamic properties of fast and slow skeletal muscles of the rat after nerve cross-union. *J. Physiol.*, **204**, 331–347.
513. CLOWES, A. W., CHERRY, R. J. & CHAPMAN, D. (1971). Physical properties of lecithin-cerebroside bilayers. *Biochim. biophys. Acta*, **249**, 301–317.
514. CLOWES, A. W., CHERRY, R. J. & CHAPMAN, D. (1972). Physical effects of tetanus toxin on model membranes containing ganglioside. *J. molec. Biol.*, **67**, 49–57.
515. COATES, P. W. (1973). Supraependymal cells: light and transmission electron microscope demonstration. *Brain Res.*, **57**, 502.
516. COCKS, J. A., BALAZS, R. & EAYRS, J. T. (1969). The effect of thyroid hormone on the biochemical maturation of the rat brain. *Biochem. J.*, **111**, 18P.
517. COENEN, A. M. L. & VENDRIK, A. J. H. (1972). Determination of the transfer ratio of cat's geniculate neurons through quasi-intracellular recordings and the relation with the level of alertness. *Expl Brain Res.*, **14**, 227–242.
518. COËRS, C. (1967). Structure and organization of the myoneural junction. *Int. Rev. Cytol.*, **22**, 239–267.
519. COËRS, C. & WOOLF, A. L. (1959). *The innervation of muscle.* Blackwell Sci. Publ., Oxford. pp. 24–30.
520. COGGESHALL, R. E. (1967). A light and electron microscope study of the abdominal ganglion of *Aplysia californica. J. Neurophysiol.*, **30**, 1263–1287.
521. COGGESHALL, R. E. & FAWCETT, D. W. (1964). The fine structure of the central nervous system of the leech, *Hirudo medicinalis. J. Neurophysiol.*, **27**, 229–289.
522. COHEN, H. D. & BARONDES, S. H. (1967). Puromycin effect on memory may be due to occult seizures. *Science, N.Y.*, **157**, 333–334.
523. COHEN, H. D., ERVIN, F. & BARONDES, S. H. (1966). Puromycin and cycloheximide: different effects on hippocampal electrical activity. *Science, N.Y.*, **154**, 1557.
524. COHEN, J. B., WEBER, M., HUCHET, M. & CHANGEUX, J. P. (1972). Purification from Torpedo marmorata electric tissue of membrane fragments particularly rich in cholinergic receptor protein. *FEBS Lett.*, **26**, 43–47.
525. COHEN, L. B. (1973). Changes in neuron structure during action potential propagation and synaptic transmission. *Physiol. Rev.*, **53**, 373–418.
526. COHEN, L. B., HILLE, B. & KEYNES, R. D. (1969). Light scattering and

birefringence changes during activity in the electric organ of *Electrophorus electricus. J. Physiol.,* **203**, 489–509.
527. COHEN, L. B., HILLE, B. & KEYNES, R. D. (1970). Changes in axon birefringence during the action potential. *J. Physiol.,* **211**, 495–515.
528. COHEN, L. B., HILLE, B., KEYNES, R. D., LANDOWNE, D. & ROJAS, E. (1971). Analysis of the potential-dependent changes in optical retardation in the squid giant axon. *J. Physiol.,* **218**, 205–237.
529. COHEN, L. B. & KEYNES, R. D. (1971). Changes in light scattering associated with the action potential in crab nerves. *J. Physiol.,* **212**, 259–275.
530. COHEN, L. B., KEYNES, R. D. & HILLE, B. (1968). Light scattering and birefringence changes during nerve activity. *Nature, Lond.,* **218**, 438–441.
531. COHEN, L. B., KEYNES, R. D. & LANDOWNE, D. (1972). Changes in light scattering that accompany the action potential in squid giant axons: potential-dependent components. *J. Physiol.,* **224**, 701–725.
532. COHEN, L. B. & LANDOWNE, D. (1971). Optical studies of action potentials. *In* Adelman, W. J. *Biophysics and physiology of excitable membranes.* Van Nostrand Reinhold; N.Y. pp. 247–263.
533. COHEN, L. B., LANDOWNE, D., SHRIVASTAV, B. B. & RITCHIE, M. J. (1970). Changes in fluorescence of squid axons during activity. *Biol. Bull. mar. biol. Lab., Woods Hole,* **139**, 418.
534. COHEN, L. B., SALTZBERG, B. M., DAVILA, H. V., ROSS, W. N., LANDOWNE, D., WAGGONER, A. S. & WANG, C. H. (1974). Changes in axon fluorescence during activity: molecular probes of membrane potential. *J. mem. Biol.,* **19**, 1–12.
535. COHEN, M. S., HAGIWARA, S. & ZOTTERMAN, Y. (1955). The response spectrum of taste fibres in the cat: a single fibre analysis. *Acta physiol. scand.,* **33**, 316–332.
536. COHEN, M. W. (1970). The contribution by glial cells to surface recordings from the optic nerve of an amphibian. *J. Physiol.,* **210**, 565–580.
537. COHEN, S. & FISCHBACH, G. (1971). *Soc. Neurosci. First Meeting Abstr.,* P 162.
538. COLE, K. S. & CURTIS, H. J. (1939). Electrical impedance of the squid giant axon during activity. *J. gen. Physiol.,* **22**, 649–670.
539. COLEMAN, D. L. (1960). Phenylalanine hydroxylase activity in dilute and nondilute strains of mice. *Archs. Biochem. Biophys.,* **91**, 300–306.
540. COLEMAN, D. L. & SCHLESINGER, K. (1965). Effects of pyridoxine deficiency on audiogenic seizure activity in inbred mice. *Proc. Soc. exp. Biol. Med.,* **119**, 264–266.
541. COLEMAN, M. S., PFINGST, B., WILSON, J. E. & GLASSMAN, E. (1971). Brain function and macromolecules. VIII. Uridine incorporation into brain polysomes of hypophysectomized rats and ovariectomized mice during avoidance conditioning. *Brain Res.,* **26**, 349–360.
542. COLEMAN, M. S., WILSON, J. E. & GLASSMAN, E. (1971). Incorpor-

ation of uridine into polysomes of mouse brain during extinction. *Nature, Lond.*, **229**, 54–55.

543. COLEMAN, R., FINEAN, J. B. & THOMPSON, J. E. (1969). Structural and functional modifications induced in muscle microsomes by trypsin. *Biochim. biophys. Acta,* **173**, 51–61.
544. COLLINS, R. L., KIRKPATRICK, J. B. & McDOUGAL, D. (1970). Some regional pathologic and metabolic consequences in mouse brain of pyrithiamine induced thiamine deficiency. *J. Neuropath. exp. Neurol.,* **29**, 57–69.
545. COLLINS, R. L. (1964). Inheritance of avoidance conditioning in mice. A diallel study. *Science, N.Y.,* **143**, 1188–1190.
546. COLONNIER, M. (1968). Synaptic patterns on different cell types in different laminae of the cat visual cortex. An electron microscope study. *Brain Res.,* **9**, 268–287.
547. CONGER, A. D. & WELLS, M. A. (1969). Radiation and aging effects on taste structure and function. *Radiat. Res.,* **37**, 31–49.
548. CONRADI, S. (1969). Observations on the ultrastructure of the axon hillock and initial axon segment of lumbosacral motoneurones in the cat. *Acta physiol. scand.* Suppl. 332, 65–84.
549. CONTI, F. & TASAKI, I. (1970). Changes in extrinsic fluorescence in squid axons during voltage clamp. *Science, N.Y.,* **169**, 1322–1324.
550. COOK, D. D. & GERARD, R. W. (1931). The effect of stimulation on the degeneration of a severed peripheral nerve. *Am. J. Physiol.,* **97**, 412–425.
551. COOK, R. D., GHETTI, B. & WISNIEWSKI, H. M. (1974). The pattern of Wallerian degeneration in the optic nerve of newborn kittens: an ultrastructural study. *Brain Res.,* **75**, 261.
552. COOK, R. D. & WISNIEWSKI, H. M. (1973). The role of oligodendroglia and astroglia in Wallerian degeneration of the optic nerve. *Brain Res.,* **61**, 191–206.
553. COOMBS, J. S., ECCLES, J. C. & FATT, P. (1955). The electrical properties of the motoneurone membrane. *J. Physiol.,* **130**, 291–325.
554. COOPER, R. A. & JANDL, J. H. (1968). Bile salts and cholesterol in the pathogenesis of target cells in obstructive jaundice. *J. clin. Invest.,* **47**, 809–822.
555. COSTA, E., DALY, J., LeFEVRE, H., MEEK, J., REVUELTA, A., SPANO, F. & STRADA, R. (1972). Serotonin and catecholamine concentration in brain of rats injected intracerebrally with 5,6,-dihydroxytryptamine. *Brain Res.,* **44**, 304–308.
556. COTRUFO, R. & APPEL, S. H. (1973). Effects of denervation on glycoproteins of rabbit gastrocnemius and soleus muscle. *Expl Neurol.,* **39**, 58–69.
557. COTTERELL, M., BALAZS, R. & JOHNSON, A. L. (1972). Effects of corticosteroids on the biochemical maturation of rat brain: postnatal cell formation. *J. Neurochem.,* **19**, 2151–2167.
558. COWAN, W. M., GOTTLIEB, D. I., HENDRICKSON, A. E., PROCE, J.

L. & WOOLSEY, T. A. (1972). The autoradiographic demonstration of axonal connections in the central nervous system. *Brain Res.*, **37**, 21–51.

559. COWLEY, J. J. & GRIESEL, R. D. (1963). The development of second generation low protein rats. *J. genet. Psychol.*, **103**, 233–242.
560. COWLEY, J. J. & WISE, D. R. (1972). Some effects of mouse urine on neonatal growth and reproduction. *Anim. Behav.*, **20**, 499–506.
561. COX, R. P., KRAUSS, M. R., BALIS, M. E. & DANCIS, J. (1970). Evidence for transfer of enzyme product as the basis of metabolic cooperation between tissue culture fibroblasts of Lesch – Nyhan disease and normal cells. *Proc. natn. Acad. Sci. U.S.A.*, **67**, 1573.
562. COX, R. P., KRAUSS, M. R., BALIS, M. E. & DANCIS, J. (1972). Communication between normal and enzyme-deficient cells in tissue culture. *Expl Cell Res.*, **74**, 251–268.
563. CRAGG, B. G. (1967). Changes in visual cortex on first exposure of rats to light: effect on synaptic dimensions. *Nature, Lond.*, **215**, 251–253.
564. CRAGG, B. G. (1968). Are there structural alterations in synapses related to functioning? *Proc. R. Soc. B*, **171**, 319–323.
565. CRAGG, B. G. (1970). What is the signal for chromatolysis? *Brain Res.*, **23**, 1–22.
566. CRAGG, B. G. (1975). The density of synapses and neurones in normal, mentally defective and aging human brains. *Brain*, **98**, 81–90.
567. CRAGG, B. G. & THOMAS, P. K. (1964). The conduction velocity of regenerated peripheral nerve fibres. *J. Physiol.*, **171**, 164–175.
568. CRAIG, J. V. & BARUTH, R. A. (1965). Inbreeding and social dominance ability in chickens. *Anim. Behav.*, **13**, 109–113.
569. CRAIG, J. V., ORTMAN, L. L. & GUHL, A. M. (1965). Genetic selection for social dominance ability in chickens. *Anim. Behav.*, **13**, 114–131.
570. CRAIN, S. M. (1956). Resting and action potentials of cultured chick embryo spinal ganglion cells. *J. comp. Neurol.*, **104**, 285.
571. CRAIN, S. M. (1966). Development of 'organotypic' bioelectric activities in central nervous tissues during maturation in culture. *Int. Rev. Neurobiol.*, **9**, 1–43.
572. CRAIN, S. M. (1969). Electrical activity of brain tissue developing in culture. *In* Jasper, H. H., Ward, A. A. & Pope, A., *Basic mechanisms of the epilepsies.* Little, Brown & Co.; Boston. pp. 506–516.
573. CRAIN, S. M., ALFEI, L. & PETERSON, E. R. (1970). Neuromuscular transmission in cultures of adult human and rodent skeletal muscle after innervation *in vitro* by fetal rodent spinal cord. *J. Neurobiol.*, **1**, 471–489.
574. CRAIN, S. M. & BORNSTEIN, M. B. (1964). Bioelectric activity of neonatal mouse cerebral cortex during growth and differentiation in tissue culture. *Expl Neurol.*, **10**, 425.
575. CRAIN, S. M. & BORNSTEIN, M. B. (1972). Organotypic bioelectric activity in cultured reaggregates of dissociated rodent brain cells. *Science, N.Y.*, **176**, 182–184.
576. CRAIN, S. M. & PETERSON, E. R. (1964). Complex bioelectric activity

in organized tissue cultures of spinal cord. (Human, rat, chick). *J. cell. comp. Physiol.*, **64**, 1–14.

577. CRAIN, S. M. & PETERSON, E. R. (1967). Onset and development of functional interneuronal connections in explants of rat spinal cord – ganglia during maturation in culture. *Brain Res.*, **6**, 750–762.
578. CRAIN, S. M., PETERSON, E. R. & BORNSTEIN, M. B. (1968). Formation of functional interneuronal connections between explants of various mammalian central nervous tissues during development in vitro. *In* Wolstenholme, G. E. W. & O'Connor, M. (1968). *Growth of nervous system.* Churchill; London. pp. 13–31.
579. CRAIN, S. M. & POLLACK, E. D. (1972). Restorative effects of cyclic AMP on complex bioelectric activities after acute Ca^{++} deprivation in cultured CNS tissues. *J. Cell Biol.*, **55**, 52a.
580. CRAIN, S. M. & POLLACK, E. D. (1973). Restorative effects of cyclic AMP on complex bioelectric activities of cultured fetal rodent CNS tissues after acute Ca^{++} deprivation. *J. Neurobiol.*, **4**, 321.
581. CREESE, R., EL-SHAFIE, A. L. & VRBOVA, G. (1968). Sodium movements in denervated muscle and the effects of antimycin A. *J. Physiol.*, **197**, 279–294.
582. CREIGHTON, M. O. & TREVITHICK, J. R. (1974). Effect of cyclic AMP, caffeine and theophylline on differentiation of lens epithelial cells. *Nature, Lond.*, **249**, 767–768.
583. CREMER, J. E., JOHNSTON, P. V., ROOTS, B. I. & TREVOR, A. J. (1968). Heterogeneity of brain fraction containing neuronal and glial cells. *J. Neurochem.*, **15**, 1361–1370.
584. CREUTZFELDT, O. D., FROMM, G. H. & KAPP, H. (1962). Influence of transcortical d.c. currents on cortical neuronal activity. *Expl Neurol.*, **5**, 436–452.
585. CREUTZFELDT, O. & ITO, M. (1968). Functional synaptic organization of primary visual cortex neurones in the cat. *Expl Brain Res.*, **6**, 324–352.
586. van CREVEL, H. & VERHAART, W. J. C. (1963). The rate of secondary degeneration in the central nervous system. I. The pyramidal tract of the cat. *J. Anat.*, **97**, 429–449.
587. van CREVEL, H. & VERHAART, W. J. C. (1963). The rate of secondary degeneration in the central nervous system. II. *J. Anat.*, **97**, 451–464.
588. CROFT, C. B. & TARIN, D. (1970). Ultrastructural studies of wound healing in mouse skin. I. Epithelial behaviour. *J. Anat.*, **106**, 63–77.
589. CRONE, C. (1965). The permeability of brain capillaries to non-electrolytes. *Acta physiol. scand.*, **64**, 407–417.
590. CRONE, C. (1965). Facilitated transfer of glucose from blood into brain tissue. *J. Physiol.*, **181**, 103–113.
591. CRONE, C. & THOMPSON, A. M. (1970). Permeability of brain capillaries. *In* Crone, C. & Lassen, N. A. *Capillary permeability.* Munksgaard; Copenhagen. pp. 447–453.
592. CRONE, H. D. & FREEMAN, S. E. (1972). The acetylcholinesterase activity of the denervated rat diaphragm. *J. Neurochem.*, **19**, 1207–1208.

593. CRONLY-DILLON, J. R. (1968). Pattern of retinotectal connections after retinal regeneration. *J. Neurophysiol.*, **31**, 410–418.
594. CROOKS, R. F. & McCLURE, W. O. (1972). The effect of cytochalasin B on fast axoplasmic transport. *Brain Res.*, **45**, 643–646.
595. CSILLIK, B. (1965). *Functional structure of the postsynaptic membrane in the myoneural junction.* Budapest. Hung. Acad. Sci., pp. 50–53.
596. CUATRECASUS, P., TELL, G. P. & SICA, V. (1974). Noradrenaline binding and the search for catecholamine receptors. *Nature, Lond.*, **247**, 92–97.
597. CULL, R. (1974). Role of nerve-muscle contact in maintaining synaptic connections. *Expl Brain Res.*, **20**, 307–310.
597a CULL, R. E. (1975). Effect of sensory nerve division on the afferent synapses of axotomized neurones. *Expl Brain Res.*, **22**, 421–425.
598. CUMINGS, J. N. (1961). Soluble cerebral proteins in normal and edematous brain. *J. clin. Path.*, **14**, 289–298.
599. CUMINGS, J. N. (1962). Some lipid diseases of brain. *Proc. R. Soc. Med.*, **58**, 21–28.
600. CUMINGS, J. N. (1968). Genetically determined neurological diseases of children. *In* Cumings, J. N. & Kremer, M., *Biochemical aspects of neurological disorder.* Blackwells; Oxford. pp. 268–288.
601. CUMINGS, J. N. (1971). Inborn errors of metabolism in neurology (Wilson's disease, Refsum's disease and Lipoidosis). *Proc. R. Soc. Med.*, **64**, 313–322.
602. CUMMINS, J. & HYDEN, H. (1962). ATP levels and ATPases in neurons, glia and neuronal membranes of the vestibular nucleus. *Biochim. biophys. Acta,* **60**, 271–283.
603. CURTIS, A. S. G. (1971). Cell adhesion. *Prog. Biophys. molec. Biol.*, **27**, 315–386.
604. CURTIS, D. R., GAME, C. J. A., JOHNSTON, G. A. R., McCULLOCH, R. M. & McLACHLIN, R. M. (1972). Convulsive action of penicillin. *Brain Res.*, **43**, 242–245.
605. CURTIS, D. R. & ECCLES, R. M. (1958). The excitation of Renshaw cells by pharmacological agents applied electrophoretically. *J. Physiol.*, **141**, 435–445.
606. CURTIS, D. R. & ECCLES, R. M. (1968). The effect of diffusional barrier upon the pharmacology of cells within the central nervous system. *J. Physiol.*, **141**, 446–463.
607. CURTIS, D. R. & ECCLES, J. C. (1959). The time course of excitatory and inhibitory synaptic action. *J. Physiol.*, **145**, 529–546.
608. CURTIS, D. R. & FELIX, D. (1971). The effect of bicuculline upon synaptic inhibition in the cerebral and cerebellar cortices of the cat. *Brain Res.*, **34**, 301–321.
609. CUZNER, M. L., DAVISON, A. N. & GREGSON, N. A. (1965). Chemical and metabolic studies of rat myelin of the central nervous system. *Ann. N.Y. Acad. Sci.*, **122**, 86.
610. DAEMEN, F. J. M., deGRIP, W. J. & JANSEN, P. A. (1972). Biochemical

aspects of the visual process. XX. The molecular weight of rhodopsin. *Biochim. biophys. Acta,* **271**, 419–428.

611. DAEMEN, F. J. M., JANSEN, P. A. & BONTING, S. L. (1971). Biochemical aspects of the visual process. XIV. The binding site of retinaldehyde in rhodopsin studied with model aldimines. *Archs. Biochem. Biophys.,* **145**, 300–309.
612. DAHLQUIST, F. W. & RAFTERY, M. A. (1968). A nuclear magnetic resonance study of associated equilibria and enzyme-bound environments of N-acetyl-D-glucosamine anomers and lysozyme. *Biochemistry, N.Y.,* **7**, 3269.
613. DAHLSTRÖM, A. (1965). Observations on the accumulations of noradrenaline in the proximal and distal parts of peripheral adrenergic nerves after compression. *J. Anat.,* **99**, 677–689.
614. DAHLSTRÖM, A. (1967). The transport of noradrenaline between two simultaneously performed ligations of the sciatic nerves of rat and cat. *Acta physiol. scand.,* **69**, 158–166.
615. DAHLSTRÖM, A. (1968). The effect of colchicine on transport of amine storage granules in sympathetic nerves of rat. *Europ. J. Pharmacol.,* **5**, 111–113.
616. DAHLSTRÖM, A. (1969). Synthesis, transport and lifespan of amine storage granules in sympathetic adrenergic neurones. *In* Barondes, S. H., Cellular dynamics of the neuron. *Symp. int. Soc. Cell Biol.,* **8**, 153–174.
617. DAHLSTRÖM, A. (1971). Axoplasmic transport (with particular respect to adrenergic neurons). *Phil. Trans. R. Soc. Ser. B,* **261**, 325–328.
618. DAHLSTRÖM, A. & HÄGGENDAL, J. (1966). Studies on the transport and life-span of amine storage granules in a peripheral adrenergic neuron system. *Acta physiol. scand.,* **67**, 278–288.
619. DALAL, K. B. & EINSTEIN, E. R. (1969). Biochemical maturation of the central nervous system. I. Lipid changes. *Brain Res.,* **16**, 441–451.
620. DALTON, M. M., HOMMES, O. R. & LEBLOND, C. P. (1968). Correlation of glial proliferation with age in the mouse brain. *J. comp. Neurol.,* **134**, 397–400.
621. DALY, J. W. (1975). Cyclic adenosine 3′5′monophosphate role in the physiology and pharmacology of the central nervous system. *Biochem. Pharmac.,* **24**, 159.
622. DAMERON, F. (1961). Influence de divers mesenchymes sur la différenciation de l'épithelium pulmonaire de l'embryon de poulet en culture *in vitro. C.R. hebd. Séanc. Acad. Sci. Paris,* **252**, 3879–3881.
623. DANCIS, J., COX, R. P., BERMAN, P. H., JANSEN, V. & BALIS, M. E. (1969). Cell population density and phenotypic expression of tissue culture fibroblasts from heterozygotes of Lesch-Nyhan's disease (inosinate pyrophosphorylase deficiency). *Biochem. Genet.,* **3**, 609–615.
624. DANIEL, P. M. & PRICHARD, M. M. L. (1966). Distal retraction balls in the neurohypophysis after transection of the pituitary stalk. *J. comp. Neurol.,* **127**, 321–333.
625. DANIEL, P. M. & PRICHARD, M. M. L. (1970). Regeneration of

hypothalamic nerve fibres after hypophysectomy in the goat. *Acta Endocr.*, **64**, 696–704.

626. DAS, G. D. (1972). Influences of the pia mater on the precursors of nerve cells. *Z. Anat. EntwGesch.*, **138**, 227.
629. DAS, G. D. & ALTMAN, J. (1972). Studies on the transplantation of developing neural tissue in the mammalian brain. I. Transplantation of cerebellar slabs into the cerebellum of neonate rats. *Brain Res.*, **38**, 233–249.
630. DASTOLI, F. R. (1968). Bitter-sensitive protein from porcine taste buds. *Nature, Lond.*, **218**, 884–885.
631. DASTOLI, F. R., LOPIEKES, D. V. & PRICE, S. (1968). A sweet-sensitive protein from bovine taste buds. Purification and partial characterization. *Biochemistry, N.Y.*, **7**, 1160–1164.
632. DASTOLI, F. R. & PRICE, S. (1966). Sweet-sensitive protein from bovine taste buds: isolation and assay. *Science, N.Y.*, **154**, 905–907.
633. DAVENPORT, V. D. (1949). Relation between brain and plasma electrolytes and electroshock seizure thresholds in adrenalectomized rats. *Am. J. Physiol.*, **156**, 322–327.
634. DAVIDOFF, R. A. (1972). Penicillin and presynaptic inhibition in the amphibian spinal cord. *Brain Res.*, **36**, 218–222.
635. DAVIDOFF, R. A., GRAHAM, L. T., SHANK, R. P., WERMAN, R. & APRISON, M. H. (1967). Changes in aminoacid concentrations associated with loss of spinal interneurones. *J. Neurochem.*, **14**, 1025–1031.
636. DAVIDSON, J. M. & LEVINE, S. (1972). Endocrine regulation of behaviour. *A. Rev. Physiol.*, **34**, 375–408.
637. DAVIES, R. E., GOLDSPINK, G. & LARSON, R. E. (1970). ATP utilization by fast and slow muscles during the development and maintenance of isometric tension. *J. Physiol.*, **206**, 28–29P.
638. DAVIES, J. T. (1962). The mechanism of olfaction. *Symp. Soc. exp. Biol.*, **16**, 170–179.
639. DAVIES, J. T. (1965). A theory of the quality of odours. *J. theor. Biol.*, **8**, 1–7.
640. DAVIES, J. T. (1970). Recent developments in the 'penetration and puncturing' theory of odour. *In* Wolstenholme, G. E. W. & Knight, J. (1970). *Taste and smell in vertebrates.* CIBA Fdn Symp. Churchill; London. pp. 265–281.
641. DAVIES, J. T. & TAYLOR, F. H. (1954). A model system for the olfactory membrane. *Nature, Lond.*, **174**, 693–694.
642. DAVIES, J. T. & WIGGILL, J. B. (1959). Diffusion across the oil–water interface. *Proc. R. Soc. A*, **255**, 277.
643. DAVILLA, H. V., SALZBERG, B. M. & COHEN, L. B. (1973). A large change in axon fluorescence that provides a promising method for measuring membrane potential. *Nature, Lond.*, **241**, 159–160.
644. DAVIS, D. G. & INESI, G. (1971). Proton nuclear magnetic resonance studies of sarcoplasmic reticulum membranes. Correlation of the tempera-

ture-dependent Ca^{++} efflux with reversible structural transition. *Biochim. biophys. Acta,* **241**, 1–8.

645. DAVIS, D. G. & INESI, G. (1972). Phosphorus and proton nuclear magnetic resonance studies in sarcoplasmic reticulum membranes and lipids. A comparison of phosphate and proton group mobilities in membranes and lipid bilayers. *Biochim. biophys. Acta,* **282**, 180–186.
646. DAVIS, L. E. (1969). Differentiation of neurosecretory cells in *Hydra. J. Cell Sci.,* **5**, 699–726.
647. DAVIS, L. E. (1971). Differentiation of ganglionic cells in *Hydra. J. exp. Zool.,* **176**, 107–128.
648. DAVIS, L. E., BURNETT, A. L. & HAYNES, J. F. (1968). Histological and ultrastructural study of the muscular and nervous system in *Hydra.* II. Nervous system. *J. exp. Zool.,* **167**, 295–332.
649. DAVIS, L. E. & BURSZTAJN, S. (1973). Histological and ultrastructural studies of the basal disc of *hydra.* II. Nerve cells and other epithelial cells. *Z. Zellforsch. mikrosk. Anat.,* **139**, 29–45.
650. DAVIS, R. E. & AGRANOFF, B. W. (1966). Stages of memory formation in goldfish: evidence for an environmental trigger. *Proc. natn. Acad. Sci. U.S.A.,* **55**, 555–559.
651. DAVIS, W. J. (1970). Motoneurone morphology and synaptic contacts: determination by intracellular dye injection. *Science, N.Y.,* **168**, 1358–1360.
652. DAVISON, A. N. (1969). Biochemistry and the myelin sheath. *Sci. Basis Med. A. Rev.,* Athlone Press; London. pp. 220–235.
653. DAVISON, A. N. (1970). Lipid metabolism in nervous tissue. *In* Florkin, M. & Stotz, E. H., *Comprehensive Biochemistry,* **18**, 293–329.
654. DAVISON, A. N. (1972). Biosynthesis of the myelin sheath. *In Lipids, malnutrition and the developing brain.* CIBA Fdn Symp. Elsevier, North Holland; Amsterdam.
655. DAVISON, A. N. & DOBBING, J. (1966). Myelination as a vulnerable period in brain development. *Br. med. Bull.,* **22**, 40–44.
656. DAVISON, A. N. & WAJDA, M. (1962). Cerebral lipids in multiple sclerosis. *J. Neurochem.,* **9**, 427–432.
657. DAVISON, P. E. (1970). Microtubules and neurofilaments: possible implications in axoplasmic transport. *Adv. Biochem. Psychopharmac.,* **2**, 168.
658. DAVISON, P. F. & TAYLOR, E. W. (1960). Physical–chemical studies of squid nerve axoplasm with special reference to the axon fibrous protein. *J. gen. Physiol.,* **43**, 801–823.
659. DAVOREN, P. R. & SUTHERLAND, E. W. (1963). The cellular location of adenyl cyclase in the pigeon erythrocyte. *J. biol. Chem.,* **238**, 3016–3023.
660. DAVSON, H. (1967). *Physiology of the cerebrospinal fluid.* Churchill; London.
661. DAVSON, H. & DANIELLI, J. F. (1943). *The permeability of natural membranes.* Cambridge University Press; Cambridge.

662. DAWSON, D. M. & ROMANUL, F. C. A. (1964). Enzymes in muscle. *Archs. Neurol.*, **11**, 355–368.

663. DAWSON, R. M. C. & CLARKE, N. (1971). Cerebral phospholipids in quaking mice. *J. Neurochem.*, **18**, 1313–1316.

664. DEAMER, D. W. & BRANTON, D. (1967). Fracture planes in an ice bilayer model membrane system. *Science, N.Y.*, **158**, 655–657.

665. DEAMER, D. W., LEONARD, R., TARDIEU, A. & BRANTON, D. (1970). Lamellar and hexagonal lipid phases as visualized by freeze etching. *Biochim. biophys. Acta*, **219**, 47.

666. DECK, J. D. & FUTCH, C. B. (1969). The effects of infused materials upon the regeneration of newt limbs. I. Blastemal extracts in denervated limb stumps. *Devl Biol.*, **20**, 332–348.

667. DELLMAN, H.-D. (1965). Age variation in the structure of the subcommisural organ of the dog. *Anat. Rec.*, **151**, 449.

668. DELLMAN, H.-D. (1973). Degeneration and regeneration of neurosecretory system. *Int. Rev. Cytol.*, **36**, 216–315.

669. DELLMAN, H.-D., STOEKKEL, M. E., PORTE, A. & STUTINSKY, F. (1974). Ultrastructure of the neurohypophyseal glial cells following stalk transection in the rat. *Experentia*, **30**, 1120.

670. DELONG, G. R. (1970). Histogenesis of fetal mouse isocortex and hippocampus in reaggregating cell cultures. *Devl Biol.*, **22**, 563–583.

671. DELONG, G. R. & SIDMAN, R. L. (1970). Alignment defect of reaggregating cells in cultures of developing brains of reeler mutant mice. *Devl Biol.*, **22**, 584–600.

672. DENBURG, J. L. (1972). An axon plasma membrane preparation from the walking legs of the lobster *Homarus americanus*. *Biochim. biophys. Acta*, **282**, 453–458.

673. DENBURG, J. L. (1973). Interaction between ions and the axon plasma membrane: effects of cations and anions on the axonal cholinergic binding macromolecule of lobster nerves. *J. mem. Biol.*, **11**, 47–56.

674. DENBURG, J. L., ELDEFRAWI, M. E. & O'BRIEN, R. D. (1972). Macromolecules from lobster axon membranes that bind cholinergic ligands and local anaesthetics. *Proc. natn. Acad. Sci. U.S.A.*, **69**, 177–181.

675. DENENBERG, V. H., HUDGENS, G. A. & ZARROW, M. X. (1966). Mice reared with rats: effects of mother on adult behaviour patterns. *Psychol. Rep.*, **18**, 451–456.

676. DENENBERG, V. H., ROSENBERG, K. M., PASCHKE, R., HESS, J. L., ZARROW, M. X. & LEVINE, S. (1968). Plasma corticosteroid values as a function of cross-species fostering and species differences. *Endocrinology*, **83**, 900–903.

677. DENENBERG, V. H., ROSENBERG, K. M., PASCHKE, R. & ZARROW, M. X. (1969). Mice reared with rat aunts: effects upon plasma corticosterone and open field activity. *Nature, N.Y.*, **221**, 73–74.

678. DENNIS, M. J. & GERSCHENFELD, H. M. (1969). Some physiological properties of identified mammalian neuroglial cells. *J. Physiol.*, **203**, 211–222.

679. DENNIS, M. J., HARRIS, A. J. & KUFFLER, S. W. (1971). Synaptic transmission and its duplication by focally applied acetyl choline in parasympathetic neurons in the heart of the frog. *Proc. R. Soc. B,* **177**, 509–539.

680. DENTON, E. J. (1954). On the orientation of molecules in the visual rods of Salamandra maculosa. *J. Physiol.,* **124**, 17–18P.

681. DENTON, E. J. (1959). The contributions of the oriented photosensitive and other molecules to the absorption of the whole retina. *Proc. R. Soc. B,* **150**, 78–94.

682. DERVICHIAN, D. G. (1964). The physical chemistry of the phospholipids. *Progr. Biophys. molec. Biol.,* **14**, 265–342.

683. DESHMUKH, D. S., INOUE, T. & PIERINGER, R. A. (1971). The association of the galactosyl diglycerides of brain with myelination. II. The inability of the myelin-deficient mutant, Jimpy mouse, to synthesize galactosyl diglycerides effectively. *J. biol. Chem.,* **246**, 5695–5699.

684. DETTBARN, W.-D. (1967). The acetyl choline system in peripheral nerve. *Ann. N.Y. Acad. Sci.,* **144**, 483–503.

685. DETTBARN, W.-D. & DAVIS, F. A. (1962). Effect of acetylcholine on the electrical activity of somatic nerves of the lobster. *Science, N.Y.,* **136**, 716.

686. DETTBARN, W.-D. & ROSENBERG, P. (1966). Effect of ions on the efflux of acetylcholine from peripheral nerve. *J. gen. Physiol.,* **50**, 447–560.

687. DEVAUX, P. & McCONNELL, H. M. (1973). Equality of the rates of lateral diffusion of phosphatidyl ethanolamine and phosphatidyl choline spin labels in rabbit sarcoplasmic reticulum. *Ann. N.Y. Acad. Sci.,* **222**, 489–498.

688. DEVILLERS, C. (1948). La genèse des organes sensoriels latéraux de la truite (Salmo fario – S. iridens). *C. R. hebd. Séanc. Acad. Sci. Paris,* **226**, 354–356.

689. DEWEY, M. M., BLASIE, J. K., DAVIS, P. K. & BARR, L. (1969). Localization of rhodopsin antibody in the retina of the frog. *J. molec. Biol.,* **39**, 395–405.

690. DIAMOND, M. C. (1967). Extensive cortical depth measurements and neuron size increases in the cortex of environmentally enriched rats. *J. comp. Neurol.,* **131**, 357–364.

691. DIAMOND, M. C., JOHNSON, R. E., INGHAM, C. & STONE, B. (1968). Lack of direct effect of hypophysectomy and growth hormone on postnatal brain morphology. *Expl Neurol.,* **23**, 51–57.

692. DIAMOND, M. C., KRECH, D. & ROSENZWEIG, M. R. (1964). The effects of an enriched environment on the histology of rat cerebral cortex. *J. comp. Neurol.,* **123**, 111–119.

693. DIAMOND, M. C., LAW, F., RHODES, H., LINDNER, B., ROSENZWEIG, M. R., KRECH, D. & BENNETT, E. L. (1966). Increases in cortical depth and glial numbers in rats subjected to enriched environment. *J. comp. Neurol.,* **128**, 117–125.

694. DIAMOND, M. C., LINDNER, B. & RAYMOND, A. (1967). Extensive cortical depth measurements and neurone size increases in the cortex of environmentally enriched rats. *J. comp. Neurol.*, **131**, 357–364.

695. DIAMOND, M. C., ROSENZWEIG, M. R., BENNETT, E. L., LINDNER, B. & LYON, L. (1972). Effects of environmental enrichment and impoverishment on rat cerebral cortex. *J. Neurobiol.*, **3**, 47–64.

696. DICHTER, M. A., HERMAN, C. J. & SELZER, M. (1972). Silent cells during interictal discharges and seizures in hippocampal penicillin foci. Evidence for the role of extracellular K^+ in the transition from the interictal state to the seizure. *Brain Res.*, **48**, 173–183.

697. DICHTER, M. A. & SPENCER, W. A. (1969). Penicillin induced interictal discharges from cat hippocampus. I. Characteristics and topographical features. *J. Neurobiol.*, **32**, 649–662.

698. DICHTER, M. A. & SPENCER, W. A. (1969). Penicillin induced interictal discharges from the cat hippocampus. II. Mechanisms underlying origin and restriction. *J. Neurophysiol.*, **32**, 663–687.

699. DICKERSON, J. W. T. & DOBBING, J. (1967). Prenatal and postnatal growth and development of the central nervous system of the pig. *Proc. R. Soc. B*, **166**, 384–395.

700. DICKERSON, J. W. T., DOBBING, J. & McCANCE, R. A. (1967). The effect of undernutrition on the postnatal development of the brain and cord in pigs. *Proc. R. Soc. B*, **166**, 396–414.

701. DICKIE, M. M., SCHNEIDER, J. & HARMAN, P. J. (1952). A juvenile 'wobbler lethal' in the house mouse. *J. Hered.*, **43**, 283–286.

702. DIERICKX, K. (1962). The dendrites of the preoptic neurosecretory nucleus of Rana temporaria and the osmoreceptor. *Archs. int. Pharmacodyn. Thér.*, **140**, 708–725.

703. DINGMAN, C. W. & PEACOCK, A. C. (1968). Analytical studies on nuclear ribonucleic acid using polyacrylamide gel electrophoresis. *Biochemistry, N.Y.*, **7**, 659–668.

704. DINGMAN, W. & SPORN, M. B. (1963). The incorporation of 8-azaguanine into rat brain RNA and its effect on maze learning by the rat: an enquiry into the biochemical basis of memory. *J. psychiat. Res.*, **1**, 1–11.

705. DISMUKES, K. & DALY, J. W. (1974). Norepinephrine-sensitive systems generating adenosine 3′5′-monophosphate: increased responses in cerebral cortical slices from reserpine-treated rats. *Molec. Pharmac.*, **10**, 933–940.

706. DOBBING, J. (1964). The influence of early nutrition on the development and myelination of the brain. *Proc. R. Soc. B*, **159**, 503–509.

707. DOBBING, J. (1972). Vulnerable periods of brain development. *In Lipids, malnutrition and the developing brain.* CIBA Fdn Symp., Elsevier, North Holland; Amsterdam.

708. DOBBING, J., HOPEWELL, J. W. & LYNCH, A. (1971). Cerebral edema in developing brain. I. Normal water and cation content in developing rat brain and postmortem changes. *Expl Neurol.*, **32**, 439–447.

709. DOBBING, J. & WIDDOWSON, E. M. (1965). The effect of undernutri-

tion and subsequent rehabilitation on myelination of rat brain as measured by its composition. *Brain,* **88**, 357–366.

710. DOBLER, M., DUNITZ, J. D. & KRAJEWSKI, J. (1969). Structure of the K^+ complex with enniatin B, a macrocyclic antibiotic with K^+ transport properties. *J. molec. Biol.,* **42**, 603–606.
711. DODGE, F. A. & RAHAMIMOFF, R. (1967). Cooperative action of calcium ions in transmitter release at the neuromuscular junction. *J. Physiol.,* **193**, 419–432.
712. DODGE, J. T., MITCHELL, E. & HANAHAN, D. J. (1963). The preparation and chemical characteristics of hemoglobin-free ghosts of human erythrocytes. *Archs. Biochem. Biophys.,* **100**, 119–130.
713. DODSON, J. W. (1967). The differentiation of epidermis. I. The interrelationship of epidermis and dermis in embryonic chicken skin. *J. Embryol. exp. Morph.,* **17**, 83–105.
714. DONAT, J. R. & WISNIEWSKI, H. M. (1973). The spatiotemporal pattern of Wallerian degeneration in mammalian peripheral nerves. *Brain Res.,* **53**, 41–53.
715. DONEGANI, G. & GABELLA, G. (1967). Effeto della sezione intracranica del nervo glossa-faringeo sui corpuscoli gustativi del coniglio. *Boll. Soc. ital. Biol. sper.,* **43**, 1165–1167.
716. DORFMAN, A. & PEI-LEE HO (1970). Synthesis of acid mucopolysaccharides by glial tumour cells in tissue culture. *Proc. natn. Acad. Sci. U.S.A.,* **66**, 495–499.
717. DOUGLAS, W. W. (1968). Stimulus–secretion coupling: the concept and clues from chromaffin and other cells. *Br. J. Pharmac.,* **34**, 451–474.
718. DOUGLAS, W. W. (1974). Involvement of calcium in exocytosis and the exocytosis–vesiculation sequence. *Biochem. Soc. Symp.,* **39**, 1–28.
719. DOWLING, J. E. & GIBBONS, I. R. (1962). The fine structure of the pigment epithelium in the albino rat. *J. Cell Biol.,* **14**, 459–474.
720. DOWLING, J. E. & SIDMAN, R. L. (1962). Inherited retinal dystrophy in the rat. *J. Cell Biol.,* **14**, 73–89.
721. DOWNMAN, C. B. B., ECCLES, J. C. & McINTYRE, A. K. (1953). Functional changes in chromatolyzed motoneurones. *J. comp. Neurol.,* **98**, 9–36.
722. DRABIKOWSKI, W., SARZALA, M. G., WRONISZEWSKA, A., LAGWINSKA, E. & DRZEWIECKA, B. (1972). Role of cholesterol in the Ca^{++} uptake and ATPase activity of fragmented sarcoplasmic reticulum. *Biochim. biophys. Acta,* **274**, 158–170.
723. DRACHMAN, D. B. (1972). Neurotrophic regulation of muscle cholinesterase: effects of botulinum toxin and denervation. *J. Physiol.,* **226**, 619–627.
724. DRACHMAN, D. B. & JOHNSTON, D. M. (1973). Development of a mammalian fast muscle: dynamic and biochemical properties correlated. *J. Physiol.,* **234**, 29–42.
725. DRÄGER, U. C. (1974). Autoradiography of tritiated proline and fucose

transported transneuronally from the eye to the visual cortex in pigmented and albino mice. *Brain Res.*, **82**, 284.

726. DRAHOTA, Z. & GUTMANN, E. (1963). Long-term regulatory influences of the nervous system on some metabolic differences in muscles of different function. *In* Gutmann, E. & Hnik, P. *Proceedings of a symposium on the effect of use and disuse of neuromuscular function.* Czechoslovak. Acad. Sci.; Prague. pp. 143–148.
727. DRATZ, E. A., GAW, J. E., SCHWARTZ, S. & CHING, W.-M. (1972). Molecular organization of photoreceptor membranes of rod outer segments. *Nature, Lond., New Biol.*, **237**, 99–102.
728. DREIZEN, P. & GERSHAM, L. C. (1970). Relationship of structure to function in myosin. II. Salt denaturation and recombination experiments. *Biochemistry, N.Y.*, **9**, 1688.
729. DRESDEN, M. H. (1969). Denervation effects on newt limb regeneration: DNA, RNA and protein synthesis. *Devl Biol.*, **19**, 311–320.
730. DREYER, W. J., PAPERMASTER, D. S. & KÜHN, H. (1972). On the absence of ubiquitous structural protein subunits in biological membranes. *Ann. N.Y. Acad. Sci.*, **195**, 61–74.
731. DREYFUS, H., REBEL, G., EDEL-HARTH, S. & URBAN, P. F. (1974). The effect of light stimulation on phospholipids of calf rod outer segments. *Brain Res.*, **66**, 139–146.
732. DROPP, J. J. & SODETZ, F. J. (1971). Autoradiographic study of neurons and neuroglia in autonomic ganglia of behaviourally stressed rats. *Brain Res.*, **33**, 419–430.
733. DROUIN, H. & THE, R. (1969). The effect of reducing extracellular pH on the membrane currents of the Ranvier nodes. *Pflügers. Arch. ges. Physiol.*, **313**, 80.
734. DROZ, B. (1969). Protein metabolism in nerve cells. *Int. Rev. Cytol.*, **25**, 363–390.
735. DROZ, B. (1973). Renewal of synaptic proteins. *Brain Res.*, **62**, 383–394.
736. DROZ, B., KOENIG, H. L. & GIAMBERARDINO, L. D. (1973). Axonal migration of protein and glycoprotein to nerve endings. I. Radioautographic analysis of the renewal of protein in nerve endings of chicken ciliary ganglion after injection of ^{3}H-lysine. *Brain Res.*, **60**, 93–127.
737. DROZ, B. & LEBLOND, C. P. (1968). Axonal migration of proteins in the central nervous system and peripheral nerves as shown by radioautography. *J. comp. Neurol.*, **121**, 325–345.
738. DRUSE, M. J. & HOGAN, E. L. (1972). Metabolism *in vivo* of brain galactolipids: the jimpy mutant. *J. Neurochem.*, **19**, 2435–2441.
739. DUBOWITZ, V. & PEARSE, A. G. E. (1960). A comparative histochemical study of oxidative enzyme and phosphorylase activity in skeletal muscle. *Histochemie*, **2**, 105–117.
740. DUCHEN, L. W. (1971). An electron microscopic study of the changes induced by botulinum toxin in the motor end-plates of slow and fast skeletal muscle fibres of the mouse. *J. neurol. Sci.*, **14**, 47–60.

741. DUCHEN, L. W., STOLKIN, C. & TONGE, D. A. (1972). Light and electron microscopic changes in slow and fast skeletal muscle fibres and their motor end plates in the mouse after local injection of tetanus toxin. *J. Physiol.*, **222**, 136–137P.
742. DUCHEN, L. W., STOLKIN, C. & TONGE, D. A. (1972). Changes in neuromuscular transmission in slow and fast muscles of the mouse after local injection of tetanus toxin. *J. Physiol.*, **222**, 147–148P.
743. DUDAI, Y., SILMAN, I. SHINITZKY, M. & BLUMBERG, S. (1972). Purification by affinity chromatography of the molecular forms of acetylcholinesterase present in fresh electric organ tissue of electric eel. *Proc. natn. Acad. Sci. U.S.A.*, **69**, 2400–2403.
744. DUDEL, J. & KUFFLER, S. W. (1962). Presynaptic inhibition at the crayfish neuromuscular junction. *J. Physiol.*, **155**, 543–562.
745. DUFFY, P. E. & MENEFEE, M. (1965). Electron microscopic observations of neurosecretory granules, nerve and glial fibers and blood vessels in the median eminence of the rabbit. *Am. J. Anat.*, **117**, 251–286.
746. van DUIJN, H., SCHWARTZKROIN, R. A. & PRINCE, D. A. (1973). Action of penicillin on inhibitory processes in the cat's cortex. *Brain Res.*, **53**, 470–476.
747. DUNCAN, J., RUTLEDGE, L. & DOMINO, E. (1968). Acetylcholinesterase activity in partially isolated cerebral cortex after prolonged intermittent stimulation. *Expl Neurol.*, **20**, 268.
748. DUNKERLEY, G. B. & DUNCAN, D. (1969). A light and electron microscopic study of the normal and degenerating corticospinal tract in the rat. *J. comp. Neurol.*, **137**, 155–184.
749. DUNN, A. (1971). Brain protein synthesis after electroshock. *Brain Res.*, **35**, 254–259.
750. DUNN, G. A. (1971). Mutual contact inhibition of extension of chick sensory nerve fibres *in vitro*. *J. comp. Neurol.*, **143**, 491–508.
751. DUPONT, Y., COHEN, J. B. & CHANGEUX, J.-P. (1974). X-ray diffraction study of membrane fragments rich in acetyl choline receptor protein prepared from the electric organ of *Torpedo marmorata*. *FEBS Lett.*, **40**, 130–133.
752. DYBALL, R. E. J. (1971). Oxytocin and ADH secretion in relation to electrical activity in antidromically identified supraoptic and paraventricular units. *J. Physiol.*, **214**, 245–256.
753. DYBALL, R. E. J. (1973). Single unit activity in the supraoptic nucleus of Brattleboro rats. *J. Physiol.*, **231**, 39–40P.
754. DYCK, P. J. & HOPKINS, A. P. (1972). Electron microscopic observations on degeneration and regeneration of unmyelinated fibres. *Brain*, **95**, 223–234.
755. EAYERS, J. T. (1955). The cerebral cortex of normal and hypothyroid rats. *Acta anat.*, **25**, 160–183.
756. EAYERS, J. T. (1960). Influences of the thyroid on the central nervous system. *Br. med. Bull.*, **16**, 122–127.

757. EAYERS, J. T. (1966). Thyroid and central nervous development. *Sci. Basis Med. A. Rev.*, Athlone Press; London. pp. 317–339.
758. EAYERS, J. T. & GOODHAND, B. (1959). Postnatal development of the cerebral cortex in the rat. *J. Anat.*, **93**, 385–402.
759. EAYERS, J. T. & HORN, G. (1955). The development of cerebral cortex in hypothroid and starved rats. *Anat. Rec.*, **121**, 53–61.
760. EBASHI, S. (1961). Calcium binding activity of vesicular relaxing factor. *J. Biochem.*, **50**, 236–244.
761. EBASHI, S. (1963). Third component participating in the superprecipitation of 'natural actomyosin'. *Nature, Lond.*, **200**, 1010.
762. EBASHI, S. & EBASHI, F. (1964). A new protein component participating in the superprecipitation of myosin B. *J. Biochem., Tokyo,* **55**, 604–613.
763. EBASHI, S., EBASHI, F. & KODAMA, A. (1967). Troponin as the Ca^{++}-receptive protein in the contractile system. *J. Biochem., Tokyo,* **62**, 137–138.
764. EBASHI, S. & ENDO, M. (1968). Calcium and muscular contraction. *In* Butler, J. A. V. & Noble, D. *Prog. Biophys. molec. Biol.*, **18**, 123–183.
765. EBERT, J. D. (1965). *Interacting systems in development.* Holt Rinehart and Winston.
766. ECCLES, J. C. (1957). *The physiology of nerve cells.* Johns Hopkins Press; Baltimore.
767. ECCLES, J. C. (1961). The mechanism of synaptic transmission. *Ergebn. Physiol.*, **51**, 299–430.
768. ECCLES, J. C. (1963). Interrelationship between the nerve and muscle cell. *In* Gutmann, E. & Hnik, P., *Proceedings of the symposium on the effect of use and disuse on neuromuscular function.* Czechoslovak. Acad. Sci.; Prague. pp. 19–28.
769. ECCLES, J. C. (1964). *The physiology of synapses.* Springer Verlag; Berlin.
770. ECCLES, J. C., ECCLES, R. M. & KOZAK, W. (1962). Further investigations on the influence of motoneurons on the speed of muscle contraction. *J. Physiol.*, **163**, 324–339.
771. ECCLES, J. C., FATT, P. & KOKETSU, K. (1954). Cholinergic and inhibitory synapses in a pathway from motor axon collaterals to motorneurons. *J. Physiol.*, **126**, 524–562.
772. ECCLES, J. C. & KRNJEVIC, K. (1959). Presynaptic changes associated with post-tetanic potentiation in the spinal cord. *J. Physiol.*, **149**, 274–287.
773. ECCLES, J. C., LIBET, B. & YOUNG, R. R. (1958). The behaviour of chromatolyzed motoneurones studied by intracellular recording. *J. Physiol.*, **143**, 11–40.
774. ECCLES, J. C. & RALL, W. (1951). Effects induced in a monosynaptic reflex path by its activation. *J. Neurophysiol.*, **14**, 353–376.
775. ECCLESTON, D., ASHCROFT, G. W., MOIR, A. T. B., PARKER-RHODES, A., LUTZ, W. & O'MAHONEY, D. P. (1968). A comparison of

5-hydroxyindoles in various regions of dog brain and cerebrospinal fluid. *J. Neurochem.*, **15**, 447–457.

776. ECHLIN, F. A. & McDONALD, J. (1954). The supersensitivity of chronically isolated cerebral cortex as a mechanism in focal cortical epilepsy. *Trans. Am. neurol. Ass.*, **79**, 75.
777. ECKERT, R. (1963). Electrical interaction of paired ganglion cells in the leech. *J. gen. Physiol.*, **46**, 573–588.
778. EDDS, M. V. (1950). Hypertrophy of nerve fibers to functionally overloaded muscles. *J. comp. Neurol.*, **93**, 290–295.
779. EDGERTON, V. R., BARNARD, R. J., PETER, J. B., GILLESPIE, C. A. & SIMPSON, D. R. (1972). Overloaded skeletal muscles of a nonhuman primate (Galago Senegalensis). *Expl Neurol.*, **37**, 322–339.
780. EDGERTON, V. R., GERSCHMAN, L. & CARRON, R. (1969). Histochemical changes in rat skeletal muscle after exercise. *Expl Neurol.*, **24**, 110–123.
781. EDGERTON, V. R. & SIMPSON, D. R. (1969). The intermediate muscle fibre of rats and guinea pigs. *J. Histochem. Cytochem.*, **17**, 828–838.
782. EDGINGTON, T. I. & DALESSIO, D. J. (1970). The assessment by immunofluorescence methods of humoral antimyelin antibodies in man. *J. Immun.*, **105**, 248–255.
783. EDSTRÖM, A. (1964). The ribonucleic acid in the Mauthner neuron of the goldfish. *J. Neurochem.*, **11**, 309–314.
784. EDSTRÖM, A., HANSSON, H.-A. & NORSTRÖM, A. (1973). Inhibition of axonal transport *in vitro* in frog sciatic nerves by chlorpromazine and Lidocaine. A biochemical and ultrastructural study. *Z. Zellforsch. mikrosk. Anat.*, **143**, 53–69.
785. EDSTRÖM, A. & MATTISON, H. (1972). Fast axonal transport *in vitro* in sciatic system of the frog. *J. Neurochem.*, **19**, 205–221.
786. EDSTRÖM, A. & SJOSTRAND, J. (1969). Protein synthesis in the isolated Mauthner nerve fibre of the goldfish. *J. Neurochem.*, **12**, 343–355.
787. EDSTRÖM, L., & KUGELBERG, E. (1968). Histochemical composition, distribution of fibers and fatiguability of single motor units. *J. Neurol. Neurosurg. Psychiat.*, **31**, 424–433.
788. EDWARDS, C., NAHORSKI, S. R. & ROGERS, K. J. (1974). *In vivo* changes of cerebral cyclic adenosine 3′5′monophosphate induced by biogenic amines: association with phosphorylase activation. *J. Neurochem.*, **22**, 565–572.
789. EDWARDS, D. A. & BURGE, K. G. (1971). Early androgen treatment and male and female sexual behaviour in mice. *Horm. Behav.*, **2**, 49–58.
790. EGAR, M., SIMPSON, S. B. & SINGER, M. (1970). The growth and differentiation of the regenerating spinal cord of the lizard *Anolis carolinensis*. *J. Morph.*, **131**, 131–152.
791. EHRENPREISS, S. (1960). Isolation and identification of an acetylcholine receptor protein from electric tissue. *Biochim. biophys. Acta*, **44**, 561–577.

792. EHRENPREISS, S. (1962). Immunohistochemical localization of drug-binding protein in tissues of the electric eel. *Nature, Lond.*, **194**, 586–587.
793. ENRENPREISS, S. (1964). Acetyl choline and nerve activity. *Nature, Lond.*, **201**, 887–893.
794. EHRENPREISS, S. (1965). An approach to the molecular basis of nerve activity *J. cell. comp. Physiol.*, **66**, Suppl. 159–164.
795. EHRENPREISS, S. (1967). Possible nature of the cholinergic receptor. *Ann. N.Y. Acad. Sci.*, **144**, 720–734.
796. EHRENPREISS, S., FLEISCH, J. H. & MITTAG, T. W. (1969). Approaches to the molecular nature of pharmacological receptors. *Pharmac. Rev.*, **21**, 131–181.
797. EHRENSTEIN, G. (1971). Excitability in lipid bilayer preparations. *In* Adelman, W. J. *Biophysics and physiology of excitable membranes.* Van Nostrand Reinhold Co.; NY. pp. 463–476.
798. EHRENSTEIN, G., LECAR, H. & NOSSAL, R. (1970). The nature of the negative resistance in bimolecular lipid membranes containing excitability-inducing material. *J. gen. Physiol.*, **55**, 119–133.
799. EICHBERG, J. & HESS, H. H. (1967). The lipid composition of frog retinal rod outer segments. *Experentia*, **23**, 993–994.
800. EICHELBERGER, L. & RICHTER, R. B. (1944). Water, nitrogen and electrolyte concentration in brain. *J. biol. Chem.*, **154**, 21.
801. EIGEN, M. & WINKLER, R. (1970). Alkali-ion carriers: dynamics and selectivity. *In* Schmitt, F. O. *The neurosciences second study programme.* Rockefeller; N.Y. pp. 685–706.
802. EISENMAN, G. (1973). *Membranes. 2. Lipid bilayers and antibiotics.* Marcel Dekker Inc.; NY.
803. ELAM, J. S. & AGRANOFF, B. W. (1971). Rapid transport of protein in the optic system of the goldfish. *J. Neurochem.*, **18**, 375–387.
804. ELAM, J. S., GOLDBERG, J. M., RADIN, N. S. & AGRANOFF, B. W. (1970). Rapid axonal transport of sulfated mucopolysaccharide proteins. *Science, N.Y.*, **170**, 458–459.
805. ELDEFRAWI, M. E., BRITTEN, A. G. & ELDEFRAWI, A. T. (1971). Acetylcholine binding to Torpedo electroplax: relationship to acetylcholine receptors. *Science, N.Y.*, **173**, 338–340.
806. ELDEFRAWI, M. E., ELDEFRAWI, A. T., GILMOUR, L. P. & O'BRIEN, R. D. (1971). Multiple affinities for binding of cholinergic ligands to a particulate fraction of Torpedo electroplax. *Molec. Pharmac.*, **7**, 420–428.
807. ELDEFRAWI, M. E., ELDEFRAWI, A. T. & O'BRIEN, R. D. (1971). Binding of five cholinergic ligands to housefly brain and Torpedo electroplax: relationship to acetylcholine receptors. *Molec. Pharmac.*, **7**, 104–110.
808. ELDEFRAWI, M. E., ELDEFRAWI, A. T. & O'BRIEN, R. D. (1971). Binding sites for cholinergic ligands in a particulate fraction of *Electrophorus* electroplax. *Proc. natn. Acad. Sci. U.S.A.*, **68**, 1047–1050.

809. ELDEFRAWI, M. E. & O'BRIEN, R. D. (1971). Autoinhibition of acetyl choline binding to *Torpedo* electroplax; a possible molecular mechanism for desensitization. *Proc. natn. Acad. Sci. U.S.A.*, **68**, 2006–2007.
810. ELETR, S. & INESI, G. (1972). Phospholipid orientation in sarcoplasmic membranes; spin label ESR and proton NMR studies. *Biochim. biophys. Acta*, **282**, 174–179.
811. ELETR, S. & KEITH, A. D. (1972). Spin-label studies of dynamics of lipid alkyl chains in biological membranes: role of unsaturated sites. *Proc. natn. Acad. Sci. U.S.A.*, **69**, 1353–1357.
812. ELFVIN, L. G. (1963). The ultrastructure of the superior cervical sympathetic ganglion of the cat. I. The structure of the ganglion cell as studied by serial sections. *J. Ultrastruct. Res.*, **8**, 403–440.
813. ELLWORTHY, P. H. (1961). The adsorption of water vapour by lecithin and lysolecithin, and the hydration of lysolecithin micelles. *J. Chem. Soc.*, 5385–5389.
814. EMBREE, L. J., HESS, H. H. & SHEIN, H. M. (1971). Sodium-potassium-ATPase activity of normal and virally transformed hamster astroglia grown subcutaneously. *Brain Res.*, **27**, 422–425.
815. EMBREE, L. J., SHEIN, H. M., BENDA, P. & HESS, H. H. (1969). ATPase activity in viral and chemically induced astrocytomas. *J. Neuropath. exp. Neurol.*, **28**, 155–156.
816. EMMELIN, N. (1961). Supersensitivity following 'pharmacological denervation'. *Pharmac. Rev.*, **13**, 17–37.
817. EMMELIN, N. & MALM, L. (1965). Development of supersensitivity as dependent on the length of degenerating fibres. *Q. Jl. exp. Physiol.*, **50**, 142–145.
818. ENDO, A. & ROTHFIELD, L. (1969). Studies of a phospholipid-requiring bacterial enzyme. I. Purification and properties of uridine diphosphate galactose: lipopolysaccharide α-3-galactosyl transferase. *Biochemistry, N.Y.*, **8**, 3500–3507.
819. ENESCO, M. & LEBLOND, C. P. (1962). Increase in cell number as a factor in the growth of the organs of the young male rat. *J. Embryol. exp. Morph.*, **10**, 530–562.
820. ENG, L. F., CHAO, F. C., GERSTL, B., PRATT, D. & TAVASTSJERNA, M. G. (1968). The maturation of human white matter myelin. Fractionation of the myelin membrane proteins. *Biochemistry, N.Y.*, **7**, 4455–4465.
821. ENG, L. F., VANDERHAEGEN, J. J., BIGNAMI, A. & GERSTL, B. (1971). An acidic protein isolated from fibrous astrocytes. *Brain Res.*, **28**, 351–354.
822. ENGEL, W. K., BROOKE, M. H. & NELSON, P. G. (1966). Histochemical studies of denervated or tenotomized cat muscle illustrating difficulties relating experimental animal conditions to human neuromuscular disease. *Ann. N.Y. Acad. Sci.*, **138**, 160–185.
823. ENGELMAN, D. M. (1970). X-ray diffraction studies of phase transitions in the membrane of Mycoplasma laidlawii. *J. molec. Biol.*, **47**, 115–117.

824. ENNIS, H. L. & LUBIN, M. (1964). Cycloheximide: aspects of inhibition of protein synthesis in mammalian cells. *Science, N.Y.*, **146**, 1474–1476.

825. ENTINGH, D. & DAMSTRA-ENTINGH, T. (1974). Avoidance training and incorporation of ^{3}H-uridine into RNA, UMP, and UDP sugars in mouse brain. *Pharmac. biochem. Behav.*, **2**, 579.

826. ERÄNKO, O. & HÄRKÖNEN, M. (1965). Effect of axon division on the distribution of noradrenaline and acetylcholinesterase in sympathetic neurons of the rat. *Acta physiol. scand.*, **63**, 411–412.

827. ERICKSON, R. P., DOETSCH, G. S. & MARSHALL, D. A. (1965). The gustatory neural response function. *J. gen. Physiol.*, **49**, 247–263.

828. ERMISCH, A., STERBA, G., HARTMANN, G. & FREYER, K. (1968). Autoradiographische Untersuchungen über das Wachstum des Reissnerschen Fadens von *Cyprinus carpio* (L). *Z. Zellforsch. mikrosk. Anat.*, **91**, 220–235.

829. ERNYEI, S. & YOUNG, M. R. (1966). Pulsatile and myelin-forming activities of Schwann cells *in vitro*. *J. Physiol.* **183**, 469–480.

830. ESCHANDIA, E. L. & PIEZZI, R. S. (1968). Microtubules in the nerve fibers of the toad Bufo arenarum Hensel. *J. Cell Biol.*, **39**, 491–497.

831. ESFAHANI, M., LIMBRICK, A. R., KNUTTON, S. OKA, T. & WAKIL, S. T. (1971). The molecular organization of lipids in the membrane of *Escherichia coli*: phase transitions. *Proc. natn. Acad. Sci. U.S.A.*, **68**, 3180–3184.

832. ESKOV, A. I. (1961). The efferent control of receptors (on the example of the chemoreceptors of the tongue). *Bull. exp. Biol. Med.*, **51**, 257–262.

833. Van ESSEN, D. & KELLY, J. (1973). Correlation of cell shape and function in the visual cortex of the cat. *Nature, Lond.*, **241**, 403–405.

834. ESTABLE-PUIG, J. F. & ESTABLE-PUIG, R. F. (1972). Paralesional reparative remyelination after chronic local cold injury of the cerebral cortex. *Expl Neurol.*, **35**, 239–253.

835. ETEROVIC, V. A. & BENNETT, E. L. (1974). Nicotinic cholinergic receptor in brain detected by binding of α-[^{3}H]-bungarotoxin. *Biochim. biophys. Acta,* **362**, 346.

836. EVANS, R. & ALEXANDER, P. (1970). Cooperation of immune lymphoid cells with macrophages in tumour immunity. *Nature, Lond.*, **228**, 620–662.

837. EWING, A. W. (1969). The genetic basis of sound production in *Drosophila pseudo obscura* and *D. persimilis*. *Anim. Behav.*, **17**, 555–560.

838. EWING, A. W. & MANNING, A. (1967). The evolution and genetics of insect behaviour. *A. Rev. Ent.*, **12**, 471–494.

839. EYLAR, E. H. (1970). Amino acid sequence of the basic protein of the myelin membrane. *Proc. natn. Acad. Sci. U.S.A.*, **67**, 1425–1431.

840. EYLAR, E. H. (1972). The structure and immunological properties of the basic proteins of myelin. *Ann. N.Y. Acad. Sci.*, **195**, 481–491.

841. EYLAR, E. H., BROSTOFF, S., HASHIM, G., CACCAM, J. & BURNETT, P. (1971). Basic A1 protein of the myelin membrane. *J. biol. Chem.*, **246**, 5770–5784.
842. EYLAR, E. H., BROSTOFF, S., JACKSON, J. & CARTER, H. (1972). Allergic encephalitis in monkeys induced by a peptide formed from the A1 protein. *Proc. natn. Acad. Sci. U.S.A.*, **69**, 617–619.
843. EYLAR, E. H., CACCAM, J., JACKSON, J., WESTALL, F. & ROBINSON, A. (1970). Experimental allergic encephalomyelitis: synthesis of disease-inducing site of the basic protein. *Science, N.Y.*, **168**, 1220–1223.
844. EYLAR, E. H. & THOMPSON, M. (1969). Allergic encephalomyelitis: the physicochemical properties of the basic protein encephalitogen from bovine spinal cord. *Archs. Biochem. Biophys.*, **129**, 468–479.
845. FABER, J. (1965). Autonomous morphogenetic activities of the amphibian regeneration blastema. *In* Kiortis, V. & Trampusch, H. A. L. *Regeneration in animals and related problems.* North Holland; Amsterdam. pp. 404–419.
846. FABER, J. (1971). Vertebrate limb ontogeny and limb regeneration: Morphogenetic parallels. *Adv. Morphogen.*, **9**, 127–147.
847. FAHIMI, H. D. & KARNOVSKY, M. J. (1966). Cytochemical localization of two glycolytic dehydrogenases in white skeletal muscle. *J. Cell Biol.*, **29**, 113–128.
848. FALK, G. & FATT, P. (1964). Linear electrical properties of striated muscle fibres observed with intracellular electrodes. *Proc. R. Soc. B*, **160**, 69–123.
849. FAMBROUGH, D. M. (1970). Acetylcholine sensitivity of muscle fibre membranes: mechanism of regulation by motoneurons. *Science, N.Y.*, **168**, 372–373.
850. FAMBROUGH, D. M., DRACHMAN, D. B. & SATYAMURTI, S. (1973). Neuromuscular junction in myasthenia gravis: decreased acetylcholine receptors. *Science, N.Y.*, **182**, 293.
851. FAMBROUGH, D. M. & HARTZELL, H. C. (1972). Acetylcholine receptors: number and distribution at neuromuscular junctions in rat diaphragm. *Science, N.Y.*, **176**, 189–191.
852. FAMBROUGH, D. M., HARTZELL, H. C., POWELL, J. A., RASH, J. E. & JOSEPH, N. (1974). On the differentiation and organization of the surface membrane of a postsynaptic cell – the skeletal muscle fibre. *In* Bennett, M. V. L., *Synaptic transmission and neuronal interaction.* Raven Press; N.Y. pp. 285–313.
853. FARBMAN, A. I. (1965). Electron microscope study of the developing taste bud in rat fungiform papilla. *Devl Biol.*, **11**, 110–135.
854. FARBMAN, A. I. (1967). Degenerative changes in denervated taste buds. *Anat. Rec.*, **157**, 242–243.
855. FARBMAN, A. I. (1969). Fine structure of degenerating taste buds after denervation. *J. Embryol. exp. Morph.*, **22**, 55–68.
856. FARBMAN, A. I. (1971). Development of the taste bud. *In* Beidler, L.

M., *Taste. Handbook of sensory physiology IV*, (2), Springer Verlag: Berlin, N.Y. pp. 51–62.

857. FARBMAN, A. I. & ZEIGNER, M. (1968). Differentiation of fetal tongue grafts in the anterior chambers of the eye. *Anat. Rec.*, **160**, 347.
858. FARLEY, R. D. & MILBURN, N. S. (1969). Structure and function of the giant fibre systems in the cockroach *Periplaneta americana. J. Insect Physiol.*, **15**, 457–476.
859. FARO, M. D. (1968). Lactic dehydrogenase activity in neuron-free spinal cord tissue obtained by a new method. *Expl Neurol.*, **20**, 233–244.
860. FARRIS, E. J. & YEAKEL, E. H. (1942). The effect of age upon susceptibility to audiogenic seizures in albino rats. *J. comp. Psychol.*, **33**, 249–251.
861. FARRIS, E. J. & YEAKEL, E. H. (1942). Sex and increasing age as factors in frequency of audiogenic seizures in albino rats. *J. comp. Psychol.*, **34**, 75–78.
862. FARROW, J. T. & O'BRIEN, R. D. (1973). Binding of atropine and muscarone to rat brain fractions and its relation to the acetylcholine receptor. *Molec. Pharmac.*, **9**, 33–40.
863. FAUGIER, S. & WILLOWS, A. O. D. (1973). Behavioural and nerve cell membrane effects of an epileptic agent (metrazol) in a mollusk. *Brain Res.*, **52**, 243–260.
864. FEDESOLVA, M., TOFFLER, I., MOORHOUSE, J. A. & DHALLA, N. S. (1970). High energy phosphate stores in cardiac and skeletal muscle of genetically dystrophic hamsters. *Fedn Proc. Fedn Am. Socs exp. Biol.*, **29**, 713.
865. FEINSTEIN, M. B. & FELSENFELD, H. (1971). The detection of ionophorous antibiotic–cation complexes in water with fluorescent probes. *Proc. natn. Acad. Sci. U.S.A.*, **68**, 2037–2041.
866. FEIT, H., SLUSAREK, L. & SHELANSKI, M. L. (1971). Heterogeneity of tubulin subunits. *Proc. natn. Acad. Sci. U.S.A.*, **69**, 2028–2031.
867. FELDMAN, J. D., GAZE, R. M. & KEATING, M. J. (1971). The effect on intertectal neuronal connections of rearing *Xenopus laevis* in total darkness. *J. Physiol.*, **212**, 44P.
868. FELTZ, A. & MALLART, A. (1971). Ionic permeability changes induced by some cholinergic agonists on normal and denervated frog muscle. *J. Physiol.*, **218**, 101–116.
869. FENSTERMACHER, J. D., RALL, D. P., PATLAK, C. S. & LEVIN, A. (1970). Ventriculocisternal perfusion as a technique for analysis of brain capillary permeability and intracellular transport. *In* Crone, C. & Lassen, N. *Capillary permeability.* Munksgaard; Copenhagen. pp. 483–490.
870. FERCHMIN, P. A., ETEROVIC, V. A. & CAPUTTO, R. (1970). Studies of brain weight and RNA content after short periods of exposure to environmental complexity. *Brain Res.*, **20**, 29–57.
871. FERNANDEZ, H. L., BURTON, P. R. & SAMSON, F. E. (1971). Axoplasmic transport in the crayfish nerve cord. *J. Cell Biol.*, **51**, 176–197.

872. FERNANDEZ, H. L., HUNEEUS, F. C. & DAVISON, P. F. (1970). Studies of the mechanism of axoplasmic transport in the crayfish cord. *J. Neurobiol.*, **1**, 395–409.
873. FERNANDEZ, H. L. & SAMSON, F. E. (1973). Axoplasmic transport: differential inhibition by cytochalasin B. *J. Neurobiol.*, **4**, 201.
874. FERNANDEZ-MORAN, H. (1962). Cell membrane ultrastructure. Low temperature electron microscopy and X-ray diffraction studies of lipoprotein components in lamellar systems. *Circulation*, **26**, 1039–1065.
875. FERNANDEZ-MORAN, H. (1962). Cell membrane ultrastructure: low temperature electron microscopy and X-ray diffraction studies of lipoprotein components in lamellar systems. *In* Ultrastructure and metabolism of the nervous system. *Res. Publs. Ass. Res. nerv. ment. Dis.*, **40**, 235–267.
876. FERNANDEZ-MORAN, H. (1972). Electron microscopy: a glimpse into the future. *Ann. N.Y. Acad. Sci.*, **195**, 376–389.
877. FERRENDELLI, J. A., KINSCHERF, D. A. & CHANG, M. M. (1973). Regulation of levels of guanosine 3′5′monophosphate in the central nervous system: effects of depolarizing agents. *Molec. Pharmac.*, **9**, 445–454.
878. FESSARD, A. & TAUC, L. (1958). Effets de répétition sur l'amplitude des potentiels postsynaptiques d'un soma neuronique. *J. Physiol., Paris*, **50**, 277–281.
879. FEWSTER, M. E. & MEAD, J. F. (1968). Lipid composition of glial cells isolated from bovine white matter. *J. Neurochem.*, **15**, 1041–1052.
880. FEWSTER, M. E. & MEAD, J. F. (1968). Fatty acid and fatty aldehyde composition of glial cell lipids isolated from bovine white matter. *J. Neurochem.*, **15**, 1303–1312.
881. FEWSTER, M. E., SCHEIBEL, A. B. & MEAD, J. F. (1967). The preparation of isolated glial cells from rat and bovine white matter. *Brain Res.*, **6**, 401–408.
882. FIEHN, W. & PETER, J. B. (1971). Properties of the fragmented sarcoplasmic reticulum from fast twitch and slow twitch muscles. *J. clin. Invest.*, **50**, 570–573.
883. FIELD, E. J., RAINE, C. S. & HUGHES, D. (1968). Failure to induce myelin sheath formation around artificial fibres with a note on the toxicity of polyester fibres for nervous tissue *in vitro*. *J. neurol. Sci.*, **8**, 129–141.
884. FIELD, P. M. (1972). A quantitative ultrastructural analysis of amygdaloid fibres in the preoptic area and the ventromedial hypothalamic nucleus. *Expl Brain Res.*, **14**, 527–538.
885. FIFKOVA, E. (1967). The effect of unilateral visual deprivation on optic centres. *Brain Res.*, **6**, 763–766.
886. FIMIAN, W.·J. (1959). The *in vitro* cultivation of amphibian blastema tissue. *J. exp. Zool.*, **140**, 125–144.
887. FINE, R. E., BLITZ, A. L., HITCHCOCK, S. E. & KAMINER, B. (1973). Tropomyosin in brain and growing neurones. *Nature, Lond., New Biol.*, **245**, 182–186.

888. FINE, R. E. & BRAY, D. (1971). Actin in growing nerve cells. *Nature, Lond., New Biol.*, **234**, 115–118.
889. FINEAN, J. B. (1960). Electron microscope and X-ray diffraction studies of the effects of dehydration on the structure of nerve myelin. *J. biophys. biochem. Cytol.*, **8**, 31–37.
890. FINEAN, J. B. (1962). The nature and stability of the plasma membrane. *Circulation*, **26**, 1151–1162.
891. FINEAN, J. B. (1965). Molecular parameters in the nerve myelin sheath. *Ann. N.Y. Acad. Sci.*, **122**, 51–56.
892. FINEAN, J. B. (1966). The molecular organization of cell membranes. *Prog. Biophys. molec. Biol.*, **16**, 143–170.
893. FINEAN, J. B. (1969). Biophysical contributions to membrane structure. *Q. Rev. Biophys.*, **2**, 1–19.
894. FINEAN, J. B. & BURGE, R. E. (1963). The determination of the Fourier transform of the myelin layer from a study of the swelling phenomenon. *J. molec. Biol.*, **7**, 672–682.
895. FINEAN, J. B., COLEMAN, R., GREEN, W. G. & LIMBRICK, A. R. (1966). Low angle X-ray diffraction and electron-microscopic studies of isolated cell membranes. *J. Cell Sci.*, **1**, 287–296.
896. FINEAN, J. B., COLEMAN, R., KNUTTON, S., LIMBRICK, A. R. & THOMPSON, J. E. (1968). Structural studies of cell membrane preparations. *J. gen. Physiol.*, **51**, 19–25.
897. FINK, B. R., KENNEDY, R. D., HENDRICKSON, A. E. & MEDDAUGH, M. E. (1971). Lidocaine inhibition of rapid axonal transport. *Anesthesiol.*, **36**, 422–432.
898. FINKELSTEIN, A. & CASS, A. (1967). Water permeability and the composition of a bilayer membrane. *J. gen. Physiol.*, **52**, 145 S.
899. FINLAYSON, L. H. (1956). Normal and induced degeneration of abdominal muscles during metamorphosis in the lepidoptera. *Q. Jl. microsc. Sci.*, **97**, 215–233.
900. FINLAYSON, L. H. (1960). A comparative study of the effects of denervation on the abdominal muscles of saturniid moths during pupation. *J. Insect Physiol.*, **5**, 108–119.
901. FISCHBACH, G. D. (1970). Synaptic potentials recorded in cell cultures of nerve and muscle. *Science, N.Y.*, **189**, 1331–1333.
902. FISCHBACH, G. D. (1972). Synapse formation between dissociated nerve and muscle cells in low density cell cultures. *Devl Biol.*, **28**, 407–429.
903. FISCHBACH, G. D. & COHEN, S. A. (1973). The distribution of acetylcholine sensitivity over uninnervated and innervated muscle fibres grown in cell culture. *Devl Biol.*, **31**, 147–162.
904. FISCHBACH, G. D., HEMKART, M. P., COHEN, S. A., BREUER, A. C., WHYSNER, J. & NEAL, F. M. (1974). Studies on the development of neuromuscular junctions in cell culture. *In* Bennett, M. V. L., *Synaptic transmission and neuronal interaction.* Raven Press; N.Y. pp. 259–284.

905. FISCHBACH, G. & ROBBINS, N. (1969). Changes in contractile properties of disused soleus muscles. *J. Physiol.*, **201**, 305–320.
906. FISCHBACH, G. D. & ROBBINS, N. (1971). Effect of chronic disuse of rat soleus neuromuscular junction on postsynaptic membrane. *J. Neurophysiol.*, **34**, 562–569.
907. FISCHER, S., CELLINO, M., GOMEZ, X. & CANESSA-FISCHER, M. (1969). Isolation of a plasmalemma-rich fraction from squid axon. *Biophys. J.*, **9**, 176A.
908. FISCHER, S., CELLINO, M., ZAMBRANO, F., ZAMPIGHI, G., TELLEZ-NAGEL, M., MARCUS, D. & CANESSA-FISCHER, M. (1970). The molecular organization of nerve membranes. I. Isolation and characterization of plasma membranes from the retinal axons of the squid. *Archs. Biochem. Biophys.*, **138**, 1–15.
909. FISH, I. & WINNICK, M. (1969). Effect of malnutrition on regional growth of the developing rat brain. *Expl Neurol.*, **25**, 534–540.
910. FISHBEIN, W., McGAUGH, J. L. & SWARZ, J. R. (1971). Retrograde amnesia: electroconvulsive shock effects after termination of rapid eye movement sleep deprivation. *Science, N.Y.*, **172**, 80–82.
911. FISHMAN, I. Y. (1957). Single fibre gustatory impulses in rat and hamster. *J. cell. comp. Physiol.*, **49**, 319–334.
912. FISHMAN, M. A., PRENSKY, A. L. & DODGE, P. R. (1969). Low content of cerebral lipids in infants suffering from malnutrition. *Nature, Lond.*, **221**, 552–553.
913. FISZER, S. & de ROBERTIS, E. (1968). Subcellular distribution and possible nature of the α-adrenergic receptor in the CNS. *Life Sci.*, **7**, 1093–1103.
914. FISZER, S. & de ROBERTIS, E. (1969). Subcellular distribution and chemical nature of the receptor for 5-hydroxytryptamine in the central nervous system. *J. Neurochem.*, **16**, 1201–1209.
915. FISZER, S. & de ROBERTIS, E. (1969). Subcellular distribution and possible nature of the binding for ^{14}C-dibenamine and ^{14}C-propranolol. *Life Sci.*, **8**, 1247–1262.
916. FISZER, S. & de ROBERTIS, E. (1972). Isolation of a proteolipid from spleen capsule binding (±)^{3}H-norepinephrine. *Biochim. biophys. Acta*, **266**, 246–254.
917. FJERDINGSTAD, E. J. (1971). *Chemical transfer of learned information.* North Holland; Amsterdam.
918. FLANGAS, A. L. & BOWMAN, R. E. (1970). Differential metabolism of RNA in neuronal-enriched and glial-enriched fractions of rat cerebrum. *J. Neurochem.*, **17**, 1237–1245.
919. FLEISCHER, S., FLEISCHER, B. & STOECKENIUS, W. (1967). Fine structure of lipid-depleted mitochondria. *J. Cell Biol.*, **32**, 193–208.
920. FLEISCHHAUER, K. (1957). Untersuchungen am Ependym des Zwischen- und Mittelhirn der Landschildröte (Testudo graeca). *Z. Zellforsch. mikrosk. Anat.*, **46**, 729–769.
921. FLEISCHHAUER, K. (1961). Regional differences in the structure of the

ependyma and subependymal layers of the cerebral ventricles of the cat. *In* Kety, S. & Elkes, Z., *Regional Neurochemistry.* Pergamon Press; London. pp. 279–283.

922. FLEISCHHAUER, K. & PETROVICKY, P. (1968). Über den Bau der Wanderungen des Aquaeductus cerebriund des IV Ventrikels der Katze. *Z. Zellforsch. mikrosk. Anat.,* **88**, 113–125.
923. FLEMING, W. W., McPHILLIPS, J. J. & WESTFALL, D. P. (1973). Postjunctional supersensitivity and subsensitivity of excitable tissues to drugs. *Ergebn. Physiol.,* **68**, 56–119.
924. FLERKO, B., MESS, B. & ILLEI-DONHOFFER, A. C. (1969). On the mechanism of androgen sterilization. *Neuroendocrinology,* **4**, 164–169.
925. FLEXNER, L. B. & FLEXNER, J. B. (1966). Effect of acetoxycycloheximide and of an acetoxycycloheximide-puromycin mixture on cerebral protein synthesis and memory in mice. *Proc. natn. Acad. Sci. U.S.A.,* **55**, 369–376.
926. FLEXNER, J. B. & FLEXNER, L. B. (1967). Restoration of expression of memory lost after treatment with puromycin. *Proc. natn. Acad. Sci. U.S.A.,* **57**, 1651–1654.
927. FLEXNER, J. B. & FLEXNER, L. B. (1967). Adrenalectomy and the suppression of memory by puromycin. *Proc. natn. Acad. Sci. U.S.A.,* **66**, 48–52.
928. FLEXNER, J. B. & FLEXNER, L. B. (1971). Pituitary peptides and the suppression of memory by puromycin. *Proc. natn. Acad. Sci. U.S.A.,* **68**, 2519–2521.
929. FLEXNER, L. B., FLEXNER, J. B. & ROBERTS, R. B. (1966). Stages of memory in mice treated with acetoxycycloheximide before or immediately after learning. *Proc. natn. Acad. Sci. U.S.A.,* **56**, 730–735.
930. FLEXNER, L. B., FLEXNER, J. B. & ROBERTS, R. (1967). Memory in mice analyzed with antibiotics. *Science, N.Y.,* **155**, 1377.
931. FLEXNER, J. B., FLEXNER, L. B. & STEELAR, E. (1963). Memory in mice as affected by intracerebral puromycin. *Science, N.Y.,* **141**, 57–59.
932. FLEXNER, J. B., FLEXNER, L. B., STELLAR, E. & HABA, G. de la & ROBERTS, R. B. (1962). Inhibition of protein synthesis in brain and learning and memory following puromycin. *J. Neurochem.,* **9**, 595–605.
933. FLEXNER, L. B., GAMBETTI, P., FLEXNER, J. B. & ROBERTS, R. B. (1971). Studies on memory: distribution of peptidyl-puromycin in subcellular fractions of mouse brain. *Proc. natn. Acad. Sci. U.S.A.,* **68**, 26–28.
934. FLOCK, Å. (1965). Electron microscopic and electrophysiological studies on the lateral line organ. *Acta otolaryngol.,* Suppl. **199**, 1–22.
935. FLOCK, Å. (1965). Transducing mechanisms in the lateral line canal organ receptors. *Cold Spring Harb. Symp. quant. Biol.,* **30**, 133–145.
936. FLOCK, Å. & WERSÄLL, J. (1962). A study of the orientation of the sensory hairs of the receptor cells in the lateral line organs of fish, with special reference to the function of the receptors. *J. Cell Biol.,* **15**, 19–30.

937. FOLCH-PI, J. (1955). Composition of the brain in relation to maturation. *In* Waelsch, H., *Biochemistry of the developing nervous system.* Acad. Press; N.Y. pp. 121–136.
938. FOLCH-PI, J. (1967). Composition of brain proteolipid. *Prog. Brain Res.*, **29**, 1–28.
939. FOLCH-PI, J. & STOFFYN, P. J. (1972). Proteolipids from membrane systems. *Ann. N.Y. Acad. Sci.*, **195**, 86–107.
940. FÖLDVARI, I. P. & PALKOVITS, M. (1964). Effects of sodium and potassium restriction on the functional morphology of the subcommisural organ. *Nature, Lond.*, **202**, 905–906.
941. FOLCH, J., LEES, M. & SLOANE-STANLEY, G. H. (1956). The role of acidic lipids in the electrolyte balance of the nervous system of mammals. *In* Richter, D., *Metabolism of the nervous system.* Pergamon; London.
942. FOLEY, J. O. (1945). The sensory and motor axons of the chorda tympani. *Proc. Soc. exp. Biol.*, **60**, 262–267.
943. FORET, J. E. (1971). Regeneration of larval urodele limb containing homoplastic transplants. *J. exp. Zool.*, **175**, 297–322.
944. FORGAYS, D. G. & FORGAYS, J. W. (1952). The nature of the effect of free-environment experience in the rat. *J. comp. Physiol. Psychol.*, **45**, 322–328.
945. FORGUS, R. H. (1954). The effect of early perceptual learning on the behavioural organization of adult rats. *J. comp. Physiol. Psychol.*, **47**, 331–336.
946. FORMAN, D. S. & LEDEEN, R. W. (1972). Axonal transport of gangliosides in the goldfish optic nerve. *Science, N.Y.*, **177**, 630–632.
947. FORMAN, D. S., McEWEN, B. S. & GRAFSTEIN, B. (1971). Rapid transport of radioactivity in goldfish optic nerve following injections of labeled glucosamine. *Brain Res.*, **28**, 119–130.
948. FORN, J., TAGLIAMONTE, A., TAGLIAMONTE, P. & GESSA, G. L. (1972). Stimulation by dibutyryl cyclic AMP of serotonin synthesis and tryptophan transport in brain slices. *Nature, Lond., New Biol.*, **237**, 245–247.
949. FRANK, G. B. (1961). Effect of drugs on partially and completely denervated skeletal muscle of the frog. *Br. J. Pharmac.*, **17**, 59–69.
950. FRANK, K. & FUORTES, M. G. F. (1955). Potentials recorded from the spinal cord with microelectrodes. *J. Physiol.*, **130**, 625–654.
951. FRANK, M. & PFAFFMAN, C. (1969). Taste nerve fibres: a random distribution of sensitivities to four tastes. *Science, N.Y.*, **164**, 1183–1185.
952. FRANKENHAEUSER, B. (1960). Quantitative description of sodium currents in myelinated nerve fibres of *Xenopus laevis. J. Physiol.*, **151**, 491–501.
953. FRANKENHAEUSER, B. (1968). The ionic currents and the nervous impulse in myelinated nerve. *Prog. Biophys. molec. Biol.*, **18**, 99–105.

954. FRANKENHAEUSER, B. & HODGKIN, A. L. (1956). The after effects of impulses in the giant fibres of *Loligo*. *J. Physiol.*, **131**, 341–376.
955. FRANKENHAEUSER, B. & HODGKIN, A. L. (1957). The action of calcium on the electrical properties of squid axons. *J. Physiol.*, **137**, 217–241.
956. FRANKHAUSER, G., VERNON, J. A., FRANK, W. H. & SLACK, W. W. (1955). Effect of size and number of brain cells on learning in larvae of the salamander *Triturus viridescens*. *Science, N.Y.*, **122**, 692.
957. FREYGANG, W. H. & LANDAU, W. M. (1955). Some relations between resistivity and electrical activity in the cerebral cortex of the cat. *J. cell. comp. Physiol.*, **45**, 377–392.
958. FREYSZ, L., BIETH, R., JUDES, M., SENSENBRUNER, M., JACOB, M. & MANDELL, P. (1968). Distribution quantitative des divers phospholipides dans les neurones et les cellules gliales isolés du cérébral de rat adulte. *J. Neurochem.*, **15**, 307–313.
959. FREY-WYSSLING, A. & STEINMAN, E. (1948). Die Schichtendoppelbrechung grosser Chloroplasten. *Biochim. biophys. Acta*, **2**, 254–259.
960. FRIEDE, R. (1954). Die Bedeutung der Glia für den zentralen Kohlenhydratstoffwechsel. *Zentbl. Allg. Path. Anat.*, **92**, 65–74.
961. FRIEDE, R. L. (1961). A histochemical study of DPN diaphorase in human white matter; with some notes on myelination. *J. Neurochem.*, **8**, 17–31.
962. FRIEDE, R. L. (1962). The cytochemistry of normal and reactive astrocytes. *J. Neuropath. exp. Neurol.*, **21**, 471–478.
963. FRIEDE, R. L. (1964). The enzymatic response of astrocytes to various ions *in vitro*. *J. Cell Biol.*, **20**, 5–15.
964. FRIEDE, R. L. (1972). Control of myelin formation by axon calibre (with a model of the control mechanism). *J. comp. Neurol.*, **144**, 233–252.
965. FRIEDE, R. L. & FLEMING, L. M. (1962). A mapping of oxidative enzymes in the human brain. *J. Neurochem.*, **9**, 179–198.
966. FRIEDE, R. L. & MARTINEZ, A. J. (1970). Analysis of the process of sheath expansion in swollen nerve fibres. *Brain Res.*, **19**, 165–182.
967. FRIEDE, R. L. & MIYAGISHI, T. (1972). Adjustment of the myelin sheath to changes in axon calibre. *Anat. Rec.*, **172**, 1–14.
968. FRIEDE, R. L., MIYAGISHI, T. & HU, K. H. (1971). Axon calibre neurofilaments, microtubules, sheath thickness and cholesterol content in cat optic nerve fibers. *J. Anat.*, **108**, 365–373.
969. FRIEDE, R. L. & SAMORAJSKI, T. (1967). Relation between the number of myelin lamellae and axon circumference in fibers of vagus and sciatic nerves of mice. *J. comp. Neurol.*, **130**, 223–232.
970. FRIEDE, R. L. & SAMORAJSKI, T. (1968). Myelin formation in the sciatic nerve of the rat. *J. Neuropath. exp. Neurol.*, **27**, 546–570.
971. FRIEDLANDER, H. D., PERLMAN, I. & CHAIKOFF, I. L. (1941). The effects of denervation on phospholipid activity of skeletal muscle as measured with radioactive phosphorus. *Am. J. Physiol.*, **132**, 24–31.

972. FRIEDMAN, H. P. & WENGER, B. S. (1970). Accumulation of an organ-specific protein during development of the embryonic chick brain. *J. embryol. exp. Morph.*, **23**, 289–297.
973. FRIEND, D. S. & GILULA, N. B. (1970). Cell junctions of the rat epididymis. *J. Cell Biol.*, **47**, 66A.
974. FRIEDMANN, T., SEEGMILLER, J. E. & SUBAK-SHARPE, J. H. (1968). Metabolic cooperation between genetically marked human fibroblasts in tissue culture. *Nature, Lond.*, **220**, 272–274.
975. FRIEND, D. S. & GILULA, N. B. (1972). A distinctive cell contact in the rat adrenal cortex. *J. Cell Biol.*, **53**, 148.
976. FRIEND, D. S. & GILULA, N. B. (1972). Variation in tight and gap junctions in mammalian tissues. *J. Cell Biol.*, **53**, 758–776.
977. de FRIES, J. C., WILSON, J. R. & McCLEARN, G. E. (1970). Open field behaviour in mice: selection response and situational generality. *Behav. Genet.*, **1**, 195–211.
978. FRIESEN, A. J. & KHATTER, J. C. (1971). Effect of stimulation on synaptic vesicles in the superior cervical ganglion of the cat. *Experentia*, **27**, 285–287.
979. FRIZELL, M. & SJÖSTRAND, J. (1974). The axonal transport of slowly migrating ^{3}H-leucine labelled proteins and the regeneration rate in regenerating hypoglossal and vagus nerves of the rabbit. *Brain Res.*, **81**, 267–283.
980. FROMHERZ, P., PETERS, J., MÜLDNER, H. G. & OTTING, W. (1972). An infrared spectroscopic study of the lipid–protein interaction in an artificial lamellar system. *Biochim. biophys. Acta*, **274**, 644–648.
981. FRY, F. J. & COWAN, W. M. (1972). A study of retrograde cell degeneration in the lateral mammillary nucleus of the cat, with special reference to the role of axonal branching in the preservation of the cell. *J. comp. Neurol.*, **144**, 1–24.
982. FRYE, L. D. & EDIDIN, M. (1970). Mouse–human heterokaryons. *J. Cell Sci.*, **7**, 313–318.
983. FRYE, L. D. & EDIDIN, M. (1970). The rapid intermixing of cell surface antigens after formation of mouse–human heterokaryons. *J. Cell Sci.*, **7**, 319–335.
984. FUCHS, F. (1969). Inhibition of sarcotubular calcium transport by caffeine: species and temperature dependence. *Biochim. biophys. Acta*, **172**, 566–570.
985. FUJIMOTO, S. & MURRAY, R. G. (1970). Fine structure of degeneration and regeneration in denervated rabbit vallate buds. *Anat. Rec.*, **168**, 393–413.
986. FUJITA, S. (1963). Matrix cell and cytogenesis in the central nervous system. *J. comp. Neurol.*, **120**, 37–42.
987. FUJITA, S. & TAKEOKA, O. (1961). Morphologic and quantitative studies on cellular proliferation and differentiation in the central nervous system of the human embryo. *Acta path jap.*, **11**, 256.
988. FULKER, D. W., WILCOCK, J. & BROADHURST, P. L. (1972). Studies

in genotype-environment interactions. I. Methodology and preliminary multivariate analysis of a diallel cross of eight strains of rat. *Behav. Genet.*, **2**, 261–287.

989. FULLER, J. L. (1950). Genetic control of audiogenic seizures in hybrids between DBA subline 2 and C57 Black subline 6. *Genetics*, **35**, 106–107.
990. FULLER, J. L. (1967). Experiential deprivation and later behaviour. *Science, N.Y.*, **158**, 1645–1652.
991. FULLER, J. L., EASLER, C. & SMITH, M. E. (1950). Inheritance of audiogenic seizure susceptibility in the mouse. *Genetics*, **35**, 622–632.
992. FULLER, J. L. & WILLIAMS, E. (1951). Gene controlled time constants in convulsive behaviour. *Proc. natn. Acad. Sci. U.S.A.*, **37**, 349–356.
993. FURMANSKI, F., SIVERMAN, D. J. & LUBIN, M. (1971). Expression of dedifferentiated functions in mouse neuroblastoma mediated by dibutyryl-cyclic adenosine monophosphate. *Nature, Lond.*, **233**, 413–415.
994. FURSHPAN, E. J. (1964). 'Electrical transmission' at an excitatory synapse in a vertebrate brain. *Science, N.Y.*, **144**, 878–880.
995. FURSHPAN, E. J. & POTTER, D. D. (1959). Transmission at the giant motor synapses of the crayfish. *J. Physiol.*, **145**, 289–325.
996. FURSHPAN, E. J. & POTTER, D. D. (1968). Low resistance junctions between cells in embryo- and tissue culture. *In* Moscona, A. A. & Monroy, A., *Current topics in developmental biology*. Acad. Press; N.Y. pp. 95–136.
997. FURUKAWA, T. & PETER, J. B. (1971). Superprecipitation and adenosine triphosphatase activity of myosin B in Duchenne muscular dystrophy. *Neurology*, **21**, 920–924.
998. FURUKAWA, T. & PETER, J. B. (1971). Troponin activity of different types of muscle fibres. *Expl Neurol.*, **31**, 214–222.
999. FURUKAWA, T. & PETER, J. B. (1972). Comparative studies on superprecipitation in different types of skeletal muscle fibres. *Expl Neurol.*, **36**, 35–40.
1000. FURUKAWA, T. & PETER, J. B. (1972). Studies of structural and regulatory proteins in different types of skeletal muscle. *Proc. 2nd Int. Congr. Muscle Diseases*, Perth, Australia.
1001. FURUKAWA, T., SUGITA, H. & TOYOKORA, Y. (1972). Comparative studies on myofibrillar proteins in different types of skeletal muscle fibers. *Expl Neurol.*, **37**, 515–521.
1002. FUTTERMAN, & ROLLINS, M. H. (1971). Role of dihydroflavin in regeneration of rhodopsin *in vitro*. *Fedn Proc. Fedn Am. Socs exp. Biol.*, **30**, 1197.
1003. FUXE, K. (1966). Evidence for the existence of monoamine neurons in the central nervous system. IV. Distribution of monoamine nerve terminals in the central nervous system. *Acta physiol. scand.*, **64**, Suppl. 247, 36–85.
1004. GABALLAH, S. & POPOFF, C. (1971). Cyclic 3′5′nucleotide phosphodiesterase in nerve endings of developing rat brain. *Brain Res.*, **25**, 220–222.

1005. GABALLAH, S. & POPOFF, C. (1971). Localization of adenosine 3′5′monophosphate-dependent protein kinase in brain. *J. Neurochem.*, **18**, 1795–1797.
1006. GABALLAH, S., POPOFF, C. & SOOKNANDAN, G. (1971). Changes in cyclic 3′5′-adenosine monophosphate-dependent protein kinase levels in brain development. *Brain Res.*, **31**, 229–232.
1007. GABELLA, G. (1969). Taste buds and adrenergic fibres. *J. neurol. Sci.*, **9**, 237–242.
1008. GAGE, P. W. & ARMSTRONG, C. M. (1968). Miniature endplate potentials in voltage-clamped muscle fibre. *Nature, Lond.*, **218**, 363–365.
1009. GAITONDE, M. K. & RICHTER, D. (1966). Changes with age in the utilization of glucose carbon in liver and brain. *J. Neurochem.*, **13**, 1309–1318.
1010. GALLI, C. & GALLI, D. R. C. (1968). Cerebroside and sulphatide in the brain of 'Jimpy mice', a strain of mice exhibiting neurological symptoms. *Nature, Lond.*, **220**, 165–177.
1011. GALLUP, B. & DUBOWITZ, V. (1973). Failure of 'dystrophic' neurons to support functional regeneration of normal or dystrophic muscle in culture. *Nature, Lond.*, **243**, 287–289.
1012. GAMBETTI, P., GONOTAS, N. K. & FLEXNER, L. B. (1968). Puromycin: action on neural mitochondria. *Science, N.Y.*, **161**, 900–902.
1013. GAMBETTI, P., GONOTAS, N. K. & FLEXNER, L. B. (1968). The fine structure of puromycin-induced changes in mouse entorhinal cortex. *J. Cell Biol.*, **36**, 379–390.
1014. GAMBETTI, P., HIRST, L., STIEBER, A. & SHAFER, B. (1972). Distribution of puromycin peptides in mouse entorhinal cortex. *Expl Neurol.*, **34**, 223–228.
1015. GAMBLE, H. J. (1968). Axon ensheathing by ependymal cells in the human embryonic and foetal spinal cord. *Nature, Lond.*, **218**, 182–183.
1016. GAMBLE, H. J. & JHA, B. D. (1958). Some effects of temperature upon the rate and progress of Wallerian degeneration in mammalian nerve fibres. *J. Anat.*, **92**, 171–177.
1017. GANZ, L. (1968). An analysis of generalization behaviour in the stimulation-deprived organism. *In* Newton, G. & Levine, S. *Early experience and behaviour. The psychobiology of development.* C. C. Thomas; Springfield, Illinois. pp. 365–411.
1018. GANZ, L. & FITCH, M. (1968). The effect of visual deprivation on perceptual behaviour. *Expl Neurol.*, **22**, 638–660.
1019. GANZ, L., FITCH, M. & SATTERBERG, J. A. (1968). The selective effect of visual deprivation on receptive field shape determined neurophysiologically. *Expl Neurol.*, **22**, 614–637.
1020. GANZ, L., HIRSCH, H. B. & TIEMAN, S. B. (1972). The nature of perceptual deficits in visually deprived cats. *Brain Res.*, **44**, 547–568.
1021. GANZ, L. & RIESEN, A. H. (1962). Stimulus generalization to hue in the dark-reared macaque. *J. comp. Physiol. Psychol.*, **55**, 92–99.

1022. GARBER, B. (1963). Inhibition by glucosamine of aggregation of dissociated embryonic cells. *Devl Biol.*, **7**, 630–641.

1023. GARBER, B. B. & MOSCONA, A. A. (1972). Reconstitution of brain tissue from cell suspensions. I. Aggregation patterns of cells dissociated from different regions of the developing brain. *Devl Biol.*, **27**, 217–234.

1024. GARBER, B. B. & MOSCONA, A. A. (1972). Reconstruction of brain tissue from cell suspensions. II. Specific enhancement of aggregation of embryonic cerebral cells by supernatant from homologous cell cultures. *Devl Biol.*, **27**, 235–243.

1025. GARCIA-BUNUEL, L. & GARCIA-BUNUEL, V. M. (1967). Connective tissue and the pentose phosphate pathway in normal and denervated muscle. *Nature, Lond.*, **213**, 913–914.

1026. GAREY, L. J. (1971). A light and electron microscopic study of the visual cortex of the cat and the monkey. *Proc. R. Soc. B*, **179**, 21–42.

1027. GAREY, L. J., FISKIN, R. A. & POWELL, T. P. S. (1973). Effects of experimental deafferentiation on cells in the lateral geniculate nucleus of the cat. *Brain Res.*, **52**, 363–369.

1028. GAREY, L. J. & PETTIGREW, J. D. (1974). Ultrastructural changes in kitten visual cortex after environmental modification. *Brain Res.*, **66**, 165–172.

1029. GAREY, L. J. & POWELL, T. P. S. (1971). An experimental study of the termination of the lateral geniculo-cortical pathway in the cat and monkey. *Proc. R. Soc. B*, **179**, 41–63.

1030. GASSER, H. S. & GRUNDFEST, H. (1939). Axon diameters in relation to the spike dimensions and the conduction velocity in mammalian A fibres. *Am. J. Physiol.*, **127**, 393–414.

1031. GAUTHIER, G. F. (1969). On the relationship of ultrastructural and cytochemical features to color in mammalian skeletal muscle. *Z. Zellforsch. mikrosk. Anat.*, **95**, 462–482.

1032. GAUTHIER, G. F. (1973). Ultrastructural and cytochemical features of mammalian skeletal muscle fibres following denervation. *J. Cell Sci.*, **12**, 525–547.

1033. GAZE, R. M. (1960). Regeneration of the optic nerve in amphibia. *Int. Rev. Neurobiol.*, **2**, 1–40.

1034. GAZE, R. M. (1972). *The formation of nerve connections; a consideration of neural specificity, modulation and comparable phenomena.* Acad. Press; N.Y.

1035. GAZE, R. M., CHUNG, S.-H. & KEATING, M. J. (1972). Development of retinotectal projection in Xenopus. *Nature, Lond., New Biol.*, **236**, 133–135.

1036. GAZE, R. M. & JACOBSON, M. (1963). A study of the retinotectal projection during regeneration of the optic nerve in the frog. *Proc. R. Soc. B*, **157**, 420–448.

1037. GAZE, R. M. & KEATING, M. J. (1970). Further studies on the restoration of the contralateral retinotectal projection following regeneration of the optic nerve in the frog. *Brain Res.*, **21**, 183–195.

1038. GAZE, R. M. & KEATING, M. J. (1970). The restoration of the ipsilateral visual projection following regeneration of the optic nerve in the frog. *Brain Res.*, **21**, 207–216.
1039. GAZE, R. M. & KEATING, M. J. (1972). The visual system and 'neuronal specificity'. *Nature, Lond.*, **237**, 375.
1040. GAZE, R. M., KEATING, M. J., SZEKELY, G. & BEAZLEY, L. (1970). Binocular interaction in the formation of specific intertectal neuronal connections. *Proc. R. Soc. B,* **175**, 107–147.
1041. GAZE, R. M. & SHARMA, S. C. (1968). Axial differences in the reinnervation of the optic tectum by regenerating optic nerve fibres. *J. Physiol.*, **198**, 117P.
1042. GAZE, R. M. & SHARMA, S. C. (1970). Axial differences in the reinnervation of the goldfish optic tectum by regenerating optic nerve fibres. *Expl Brain Res.*, **10**, 171–181.
1043. GEEL, S. & TIMIRAS, P. S. (1967). The influence of neonatal hypothyroidism and of thyroxine on the ribonucleic acid and deoxyribonucleic acid concentration of rat cerebral cortex. *Brain Res.*, **4**, 135–142.
1044. GEEL, S. & TIMIRAS, P. S. (1967). The influence of neonatal hypothyroidism and of thyroxine on 1-^{14}C leucine incorporated in protein *in vivo* and the relationship to ionic levels in the developing brain of the rat. *Brain Res.*, **4**, 143–150.
1045. GEFFEN, L. B., DESCARRIES, L. & DROZ, B. (1971). Intra-axonal migration of ^{3}H-norepinephrine injected into the coeliac ganglion of cats: radioautographic study of the proximal segment of constricted splenic nerves. *Brain Res.*, **35**, 315–318.
1046. GEFFEN, L. B. & LIVETT, B. G. (1971). Synaptic vesicles in sympathetic neurones. *Physiol. Rev.*, **51**, 98–157.
1047. GEFFEN, L. B. & OSTBERG, A. (1969). Distribution of granular vesicles in normal and constricted sympathetic neurones. *J. Physiol.*, **204**, 583–592.
1048. GEFFEN, L. B. & RUSH, R. A. (1968). Transport of noradrenaline in sympathetic nerves and the effect of nerve impulses on its contribution to transmitter stores. *J. Neurochem.*, **15**, 925–930.
1049. GEISON, R. L. & WAISMAN, H. A. (1970). Effects of nutritional status on rat brain maturation as measured by lipid composition. *J. Nutr.*, **100**, 315–324.
1050. GENT, W. L. G., GRERSON, N. A., LOVELIDGE, C. A. & WINDER, A. F. W. (1971). Separation of lipid-protein complexes of rat brain myelin by isopycnic centrifugation. *Biochem. J.*, **122**, 63P.
1051. GEORGE, W. J., POLSON, J. B., O'TOOLE, A. G. & GOLDBERG, N. D. (1970). Elevation of guanosine 3′5′-cyclic phosphate in rat heart after perfusion with acetyl choline. *Proc. natn. Acad. Sci. U.S.A.*, **66**, 398–403.
1052. GEREN, B. B. (1954). The formation from the Schwann cell surface of myelin in the peripheral nerves of chick embryos. *Expl Cell Res.*, **7**, 558–562.

1053. GERSCH, I. (1939). The structure and function of the parenchymatous glandular cells in the neurohypophysis of the rat. *Am. J. Anat.*, **64**, 407–443.

1054. GERSCH, I. & BROOKS, C. M. (1941). Correlation of physiological and cytological changes in the neurohypophysis of rats with experimental diabetes insipidus. *Endocrinology*, **28**, 6–19.

1055. GERSCH, I. & CATCHPOLE, H. R. (1960). The nature of ground substance of connective tissue. *Perspect. biol. Med.*, **3**, 282–319.

1056. GESTELAND, R. C., LETTVIN, J. Y. & PITTS, W. H. (1965). Chemical transmission in the nose of the frog. *J. Physiol.*, **181**, 525.

1057. GIACOMETTI, L. & PARAKKAL, P. F. (1969). Skin transplantation: orientation of epithelial cells by the basement membrane. *Nature, Lond.*, **223**, 514–515.

1058. GIBBINS, J. R., TILNEY, L. G. & PORTER, K. R. (1969). Microtubules in the formation and development of the primary mesenchyme in *arbacia punctulata*. I. The development of microtubules. *J. Cell Biol.*, **41**, 201–226.

1059. GIBBONS, I. R. (1965). Chemical dissection of cilia. *Archs. Biol.*, **76**, 317–352.

1060. GIDGE, N. M. & ROSE, S. M. (1944). The role of larval skin in promoting limb regeneration in adult anura. *J. exp. Zool.*, **97**, 71–93.

1061. GIEBISCH, G. (1958). Electrical potential measurements on single nephrons of necturus. *J. cell. comp. Physiol.*, **51**, 221–239.

1062. de GIER, J., MANDERSLOOT, J. G. & van DEENEN, L. L. M. (1968). Lipid composition and permeability of liposomes. *Biochim. biophys. Acta*, **150**, 666–675.

1063. GILARDI, R. KARLE, I. L., KARLE, J. & SPERLING, W. (1971). Crystal structures of the visual chromophores, 11-*cis* and all-*trans* retinal. *Nature, Lond.*, **232**, 187–188.

1064. GILMAN, A. G. & NIRENBERG, M. (1971). Effect of catecholamines on the adenosine 3′5′-cyclic monophosphate concentrations of clonal satellite cells of neurones. *Proc. natn. Acad. Sci. U.S.A.*, **68**, 2165–2168.

1065. GILMORE, S. A. & DUNCAN, D. (1968). On the presence of peripheral-like nervous and connective tissue within irradiated spinal cord. *Anat. Rec.*, **160**, 675–690.

1066. GINSBORG, B. L. & HIRST, G. D. S. (1971). Theophylline and adenosine at the neuromuscular junction. *Br. J. Pharmac.*, **43**, 432–433P.

1067. GINSBURG, B. E. (1954). Genetics and physiology of the nervous system. Genetics and the inheritance of integrated neurological and psychiatric patterns. *Res. Publs Ass. Res. nerv. ment. Dis.*, **33**, 39–56.

1068. GINSBURG, B. E. (1958). Genetics as a tool in the study of behaviour. *Perspect. biol. Med.*, **1**, 397–424.

1069. GINSBURG, B. E. (1963). Causal mechanisms in audiogenic seizures. *Colloq. int. Centre nat. Rech. sci. Paris*, **112**, 227–240.

1070. GINSBURG, B. E. & FULLER, J. L. (1954). A comparison of chemical and mechanical alteration of seizure patterns in mice. *J. comp. Physiol. Psychol.*, **47**, 344–348.

1071. GINSBURG, B. E. & MILLER, D. S. (1963). Genetic factors in audiogenic seizures. *Colloq. Int. Centre nat. Rech. sci. Paris,* **112**, 217–225.

1072. GINSBURG, B. E., MILLER, D. S. & ZAMIS, M. J. (1949). On the mode of inheritance of susceptibility to sound induced seizures in the house mouse. *Rec. Genet. Soc. Am.,* **18**, 89.

1073. GINSBURG, B. E. & ROBERTS, E. (1951). Glutamic acid and central nervous system activity. *Anat. Rec.,* **111**, 492–493.

1074. GINSBURG, B. E., ROSS, S., ZAMIS, M. J. & PERKINS, A. (1951). Some effects of glutamic acid on sound induced seizures in mice. *J. comp. Physiol. Psychol.,* **44**, 134–141.

1075. GIROUX, E. L. & HENKIN, R. I. (1971). Oral effects of hydrolytic enzymes on taste acuity in man. *Life Sci.,* **10**, 361–370.

1076. GISIGER, V. (1971). Triggering of RNA synthesis by acetylcholine stimulation of the postsynaptic membrane in a mammalian sympathetic ganglion. *Brain Res.,* **33**, 139–146.

1077. GISPEN, W. H. & SCHOTMAN, P. (1970). Effect of hypophysectomy and conditioned avoidance behaviour on macromolecule metabolism in the brain stem of the rat. *In* de Wied & Weijnen, J. A. W. M. Pituitary, adrenal and the brain. *Prog. Brain Res.,* **32**, 236–244.

1078. GISPEN, W. H., de WIED, D., SCHOTMAN, P. & JANSZ, H. S. (1970). Effects of hypophysectomy on RNA metabolism in rat brain stem. *J. Neurochem.,* **17**, 751–761.

1079. GISPEN, W. H., de WIED, D., SCHOTMAN, P. & JANSZ, H. S. (1971). Brain stem polysomes and avoidance performance of hypophysectomized rats subjected to peptide treatment. *Brain Res.,* **31**, 341–351.

1080. GLADE, R. (1957). The effects of tail tissue on limb regeneration in *Triturus viridescens. J. Morph.,* **101**, 477–522.

1081. GLAESER, R. M., HAYES, T., MEL, H. & TOBIAS, C. (1966). Membrane structure of OsO_4-fixed erythrocytes viewed 'face on' by electron microscope techniques. *Expl Cell Res.,* **42**, 467–480.

1082. GLASER, M., SIMKINS, H., SINGER, S. J., SHEETZ, M. & CHAN, S. I. (1970). On the interactions of lipids and proteins in the red blood cell membrane. *Proc. natn. Acad. Sci. U.S.A.,* **65**, 721–728.

1083. GLASSMAN, E. (1969). The biochemistry of learning: an evaluation of the role of RNA and protein. *A. Rev. Biochem.,* **38**, 605–646.

1084. GLASSMAN, E. & WILSON, J. E. (1970). The incorporation of uridine into brain RNA during short experiences. *Brain Res.,* **21**, 157–168.

1085. GLEDHILL, R. F., HARRISON, B. M. & McDONALD, W. I. (1973). Morphological prerequisites for saltatory conduction present in remyelinated fibres. *Nature, Lond.,* **244**, 443–444.

1086. GLICK, J. L. & GITHENS, S. (1965). Role of sialic acid in potassium transport of L1210 leukaemia cells. *Nature, Lond.,* **208**, 88.

1087. GLOBUS, A. (1971). Neuronal ontogeny: its use in tracing connectivity. *In* Sterman, M. B., McGinty, D. J. & Adinolfi, A. M., *Brain development and behaviour.* Acad. Press; N.Y.

1088. GLOBUS, A., ROSENZWEIG, M. R., BENNETT, E. L. & DIAMOND, M. C. (1973). Effects of differential experience on dendritic spine counts in rat cerebral cortex. *J. comp. Physiol. Psychol.*, **82**, 175–181.
1089. GLOBUS, A. & SCHEIBEL, A. B. (1967). Synaptic loci on parietal cortical neurons: termination of corpus callosum fibres. *Science, N.Y.*, **156**, 1127–1129.
1090. GLOBUS, A. & SCHEIBEL, A. B. (1967). The effect of visual deprivation on cortical neurones: a Golgi study. *Expl Neurol.*, **19**, 331–345.
1091. GLOBUS, A. & SCHEIBEL, A. B. (1967). Synaptic loci on visual cortical neurones of the rabbit: the specific afferent radiation. *Expl Neurol.*, **18**, 116–131.
1092. GLÖTZNER, F. L. (1973). Membrane properties of neuroglia in epileptogenic gliosis. *Brain Res.*, **55**, 159–171.
1093. GODDARD, G. V. (1967). Development of epileptic seizures through brain stem stimulation at low intensity. *Nature, Lond.*, **214**, 1020–1021.
1094. GODDARD, G. V., McINTYRE, D. C. & LEECH, C. K. (1969). A permanent change in brain function resulting from daily electrical stimulation. *Expl Neurol.*, **25**, 295–330.
1095. GOGAN, P., GUERRTAUD, J. P., HORCHOLLE-BOSSAVIT, G. & TYC-DUMONT, S. (1974). Electronic coupling between motoneurones in the abducens nucleus of the cat. *Expl Brain Res.*, **21**, 139.
1096. GOLDBERG, A. L. & SINGER, J. J. (1969). Evidence for a role of cyclic AMP in neuromuscular transmission. *Proc. natn. Acad. Sci. U.S.A.*, **64**, 134–141.
1097. GOLDENSOHN, E. S. (1969). The epileptic neuron: experimental seizure mechanisms. *In* Jasper, H. H. *Basic mechanisms of the epilepsies.* Little, Brown; Boston. pp. 289–298.
1098. GOLDMAN, R. D. & FOLLET, E. A. C. (1969). The structure of the major cell processes of isolated BHK 21 fibroblasts. *Expl Cell Res.*, **57**, 263–276.
1099. GOLDSTEIN, A., LOWNEY, L. I. & PAL, B. K. (1971). Stereospecific and nonspecific interactions of the morphine congener levorphanol in subcellular fractions of mouse brain. *Proc. natn. Acad. Sci. U.S.A.*, **68**, 1744–1747.
1100. GOLDSTEIN, D. A. & SOLOMON, A. K. (1960). Determination of equivalent pore radius for human red cells by osmotic pressure measurement. *J. gen. Physiol.*, **44**, 1–17.
1101. GOMBOS, G., FILIPOWICZ, W. & VINCENDON, G. (1971). Fast and slow components of S-100 protein fractions: regional distribution in bovine central nervous system. *Brain Res.*, **26**, 475–479.
1102. GONOTAS, N. K., LEVINE, S. & SHOULSON, R. (1964). Phagocytosis and regeneration of myelin in an experimental leukoencephalopathy. An electron microscope study. *Am. J. Path.*, **44**, 565.
1103. GONZALEZ-SASTRE, F. (1970). The protein composition of isolated myelin. *J. Neurochem.*, **17**, 1049–1056.

1104. GONZALES-SERRATOS, H. (1971). Inward spread of activation in vertebrate muscle fibres. *J. Physiol.,* **212**, 777–799.
1105. GOODALL, M. C. (1970). Structural effects in the action of antibiotics on the ion permeability of lipid bilayers. III. Gramicidins 'A' and 'S' and lipid specificity. *Biochim. biophys. Acta,* **219**, 471–478.
1106. GOODENOUGH, D. A. & REVEL, J. P. (1970). A fine structural analysis of intercellular junctions in the mouse liver. *J. Cell Biol.,* **45**, 272–281.
1107. GOODENOUGH, D. A. & STOECKENIUS, W. (1972). The isolation of mouse hepatocyte gap junctions. Preliminary chemical characterization and X-ray diffraction. *J. Cell Biol.,* **54**, 646–656.
1108. GOODFRIEND, T. L. & KAPLAN, N. O. (1963). The influence of pO_2 on LDH isoenzymes. *J. Cell Biol.,* **19**, 28A.
1109. GOODWIN, B. C. & COHEN, M. H. (1969). A phase-shift model for the spatial and temporal organization of developing systems. *J. theoret. Biol.,* **25**, 49–107.
1110. GORIDIS, C. & MORGAN, I. G. (1973). Guanyl cyclase in rat brain subcellular fractions. *FEBS Lett.,* **34**, 71–73.
1111. GORIDIS, C. & VIRMAUX, N. (1974). Rapid, light-induced changes of retinal cyclic GMP levels. *FEBS Lett.,* **49**, 167–169.
1112. GORIDIS, C. & VIRMAUX, N. (1974). Light regulated guanosine 3′5′-monophosphate phosphodiesterase activity of bovine retina. *Nature, Lond.,* **248**, 57–58.
1113. GORIDIS, C., VIRMAUX, N., URBAN, P. F. & MANDEL, P. (1973). Guanyl cyclase in a mammalian photoreceptor. *FEBS Lett.,* **30**, 163–166.
1114. GOSOLOW, N. & GROBSTEIN, C. (1962). Epitheliomesenchymal interaction in pancreatic morphogenesis. *Devl Biol.,* **4**, 242–255.
1115. GOTTLIEB, D. I. & COWAN, W. M. (1972). Evidence for a temporal factor in the occupation of available synaptic sites during the development of the dentate gyrus. *Brain Res.,* **41**, 452–456.
1116. GOTTLIEB, D. I. & COWAN, W. M. (1973). Autoradiographic studies of the commisural and ipsilateral association connections of the hippocampus and dentate gyrus of the rat. I. The commisural connections. *J. comp. Neurol.,* **149**, 393–422.
1117. GOY, R. W. (1970). Experimental control of psychosexuality. *Phil. Trans. R. Soc. Ser. B.,* **259**, 149–162.
1118. GRAFF, G. L. A., HUDSON, A. Y. & STRICKLAND, K. P. (1965). Biochemical changes in denervated skeletal muscle. I. The effect of denervation atrophy on the main phosphate fractions of the rat-gastrocnemius muscle. *Biochim. biophys. Acta,* **104**, 524–531.
1119. GRAFF, G. L. A., HUDSON, A. Y. & STRICKLAND, K. P. (1965). Biochemical changes in denervated skeletal muscle. II. Labelling patterns of the main phosphate fractions in normal and denervated rat-gastrocnemius muscle using $^{32}P_i$ as indicator. *Biochim. biophys. Acta,* **104**, 532–542.
1120. GRAFF, G. L. A., HUDSON, A. Y. & STRICKLAND, K. P. (1965).

Biochemical changes in denervated muscle III. The effect of denervation atrophy on the concentration of ^{32}P labelling of the individual phospholipids of the rat-gastrocnemius muscle. *Biochim. biophys. Acta,* **104**, 543–553.

1121. GRAFSTEIN, B. (1969). Axonal transport; communication between soma and synapse. *Adv. Biochem. Psychopharmac.,* **1**, 11–25.

1122. GRAFSTEIN, B., McEWEN, B. S. & SHELANSKI, M. L. (1970). Axonal transport of neurotubule protein. *Nature, Lond.,* **227**, 289.

1123. GRAFSTEIN, B., MILLER, J. A., LEDEEN, R. W., HALEY, J. & SPECHT, S. C. (1975). Axonal transport of phospholipid in goldfish optic system. *Expl Neurol.,* **46**, 261–281.

1124. GRAHAM, D. E. & LEA, E. J. A. (1972). The effect of surface charge on the water permeability of phospholipid bilayers. *Biochim. biophys. Acta,* **274**, 286–293.

1125. GRAHAME-SMITH, D. G. (1972). The prevention by inhibitors of brain protein synthesis of the hyperactivity and hyperpyrexia produced in rats by monoamine oxidase inhibition and the administration of L-tryptophan or 5-methoxy-N,N-dimethyltryptamine. *J. Neurochem.,* **19**, 2409–2422.

1126. GRAMPP, W., HARRIS, J. B. & THESLEFF, S. (1972). Inhibition of denervation changes in skeletal muscle by blockers of protein synthesis. *J. Physiol.,* **221**, 743–754.

1127. GRANT, C. W. M. & McCONNELL, N. M. (1973). Fusion of phospholipid vesicles with viable *Acheloplasma laidlawii. Proc. natn. Acad. Sci. U.S.A.,* **70**, 1238–1240.

1128. GRAS, W. J. & WORTHINGTON, C. R. (1969). X-ray analysis of retinal photoreceptors. *Proc. natn. Acad. Sci. U.S.A.,* **63**, 233–240.

1129. GRAVES, H. B. & SIEGEL, P. B. (1969). Bidirectional selection for responses of *Gallus domesticus* chicks to an imprinting situation. *Anim. Behav.,* **17**, 683–691.

1130. GRAY, E. G. & HAMLYN, L. H. (1962). Electron microscopy of experimental degeneration in the avian optic tectum. *J. Anat.,* **96**, 309–316.

1131. GRAY, E. G. & WILLIS, R. A. (1970). On synaptic vesicles, complex vesicles and dense projections. *Brain Res.,* **25**, 241–253.

1132. GRAZIADEI, P. P. C. & DEHAN, R. S. (1973). Regeneration patterns in the olfactory bulb of frog and pigeon. *Anat. Rec.,* **175**, 332.

1133. GRAZIADEI, P. P. C. & DEHAN, R. S. (1973). Neuronal regeneration in frog olfactory system. *J. Cell. Biol.,* **59**, 529.

1134. GRAZIADEI, P. C. C. & METCALF, J. F. (1971). Neuronal dynamics in the olfactory mucosa of the adult vertebrates. *Anat. Rec.,* **169**, 328.

1135. GRAZIADEI, P. P. C. & METCALF, J. F. (1971). Autoradiographic and ultrastructural observations on the frog's olfactory mucosa. *Z. Zellforsch. mikrosk. Anat.,* **116**, 305–318.

1136. GRAZIANI, Y. & IVNE, A. (1971). Vasopressin and water permeability of artificial lipid membranes. *Biochem. biophys. Res. Commun.,* **45**, 321.

1137. GREASER, M. L. & GERGELY, J. (1970). Calcium binding component of troponin. *Fedn Proc. Fedn Am. Socs exp. Biol.*, **29**, 463.
1138. GREEN, D. E. (1972). Membrane proteins: a perspective. *Ann. N.Y. Acad. Sci.*, **195**, 150–172.
1138a GREEN, D. E. & FLEISCHER, S. (1963). The role of lipids in mitochondrial electron transfer and oxidative phosphorylation. *Biochim. biophys. Acta*, **70**, 554–582.
1139. GREEN, D. & PERDUE, J. F. (1966). Membranes as expressions of repeating units. *Proc. natn. Acad. Sci. U.S.A.*, **55**, 1295–1302.
1140. GREEN, D. E. & TZAGOLOFF, A. (1966). Role of lipids in the structure and function of biological membranes. *J. Lipid Res.*, **7**, 587–602.
1141. GREEN, R. D. (1969). The effect of denervation on the sensitivity of the superior cervical ganglion of the pithed cat. *J. Pharmac. exp. Ther.*, **167**, 143–150.
1142. GREENBERG, H. & PENMAN, S. (1966). Methylation and processing of ribosomal RNA in HeLa cells. *J. molec. Biol.*, **21**, 527–535.
1143. GREENE, H. S. N. (1951). Transplantation of tumors to brains of heterologous species. *Cancer Res.*, **11**, 529–534.
1144. GREENE, H. S. N. (1953). Transplantation of human brain tumors to brains of laboratory animals. *Cancer Res.*, **13**, 422–426.
1145. GREENE, H. S. N. (1957). Heterotransplantation of tumors. *Ann. N.Y. Acad. Sci.*, **69**, 818–829.
1146. GREENFIELD, S., NORTON, W. T. & MORELL, P. (1971). Quaking mouse: isolation and characterization of myelin protein. *J. Neurochem.*, **18**, 2119–2128.
1147. GREENGARD, P., KUO, J. F. & MIYAMOTO, E. (1971). Studies of the mechanism of action of cyclic AMP in nervous and other tissues. *Adv. Enzyme Regul.*, **9**, 113.
1148. GREENGARD, P. & McAFEE, D. A. (1972). Adenosine 3′5′-cyclic monophosphate as a mediator in the action of neurohumoral agents. *Biochem. Soc. Symp.*, **36**, 87–102.
1149. GREGORY, K. M. & DIAMOND, M. C. (1968). Acetyl cholinesterase and cholinesterase activities, protein content and wet weight measures in the rat brain after early hypophysectomy. *Expl Neurol.*, **21**, 502–511.
1150. GREGORY, K. M. & DIAMOND, M. C. (1968). The effect of early hypophysectomy on brain morphogenesis in the rat. *Expl Neurol.*, **20**, 394–414.
1151. GREGSON, N. A. & OXBERRY, J. M. (1972). The composition of myelin from the mutant mouse 'quaking'. *J. Neurochem.*, **19**, 1065–1071.
1152. van GREIDANUS, W. T. B. & de WIED, D. (1971). Effects of systemic and intracerebral administration of two opposite acting ACTH-related peptides on extinction of conditioned avoidance behaviour. *Neuroendocrinology*, **7**, 291–301.
1153. GRENELL, R. G., HAZAMA, H., NAKAZAWA, M. & EINBERG, E. (1968). Effect of gravitational changes on RNA of cerebral neurones and glia. I. RNA changes of Deiter's cells and glia. *Brain Res.*, **9**, 115–125.

1154. GRIFFIN, G. A. & HARLOW, H. F. (1966). Effects of three months of total social deprivation on social adjustment and learning in the rhesus monkey. *Child Dev.*, **37**, 533–548.
1155. GRIFFITH, O. H. & WAGGONER, A. S. (1969). The analysis of membrane structure with spin labeling. *Accounts Chem. Res.*, **2**, 17.
1156. GRIFFITHS, W. J. (1961). Effect of isolation on treadmill running in the albino rat. *Psychol. Rep.*, **8**, 243–250.
1157. GRIMM, L. M. (1971). An evaluation of myotypic respecification in axolotls. *J. exp. Zool.*, **178**, 479–496.
1158. GRINNELL, A. D. (1966). A study of the interaction between motoneurones in the frog spinal cord. *J. Physiol.*, **182**, 612–648.
1159. de GRIP, W. J., BONTING, S. L. & DAEMEN, F. J. M. (1973). The binding site of retinaldehyde in cattle rhodopsin. *Biochim. biophys. Acta*, **303**, 189–193.
1160. GROBSTEIN, C. (1953). Epithelio-mesenchymal specificity in the morphogenesis of mouse submandibular rudiments *in vitro*. *J. exp. Zool.*, **124**, 383–414.
1161. GROBSTEIN, C. (1955). Inductive interaction in the development of the mouse metanephros. *J. exp. Zool.*, **130**, 319–340.
1162. GROBSTEIN, C. (1956). Inductive tissue interaction in development. *Adv. Cancer Res.*, **4**, 187–236.
1163. GROBSTEIN, C. (1961). Cell contact in relation to embryonic induction. *Expl Cell Res.*, **Suppl. 8**, 234–245.
1164. GROBSTEIN, C. (1967). Mechanisms of organogenetic tissue interaction. *Natl Cancer Inst. Monographs*, **26**, 279–294.
1165. GROBSTEIN, C. & COHEN, J. (1965). Collagenase: effect on the morphogenesis of embryonic salivary epithelium *in vitro*. *Science, N.Y.*, **150**, 626–628.
1166. GROSSMAN, R. G. & HAMPTON, T. (1968). Depolarization of cortical glial cells during electrocortical activity. *Brain Res.*, **11**, 316–324.
1167. GROSSMAN, R. G., LYNCH, L. & SHIRES, G. T. (1968). Ionic content and membrane potentials of cortical neurons and glia. *Neurology*, **18**, 292.
1168. GROSSMAN, R. G. & ROSMAN, L. J. (1971). Intracellular potentials of inexcitable cells in epileptogenic cortex undergoing fibrillary gliosis after a local injury. *Brain Res.*, **28**, 181–201.
1169. GROSSMAN, R. G., WHITESIDE, L. & HAMPTON, T. L. (1969). The time course of evoked depolarization of cortical glial cells. *Brain Res.*, **14**, 401–415.
1170. GROVER, J. W. (1961). The enzymatic dissociation and reproducible reaggregation *in vitro* of 11-day embryonic chick lung. *Devl Biol.*, **3**, 555–568.
1171. GRUENDEL, A. D. & ARNOLD, W. J. (1969). Effects of early social deprivation on the reproductive behaviour of male rats. *J. comp. Physiol. Psychol.*, **67**, 123–128.
1172. GRUENER, N. & AVI-DOR, Y. (1966). Temperature-dependence of

activation and inhibition of rat-brain adenosine triphosphatase activated by sodium and potassium ions. *Biochem J.*, **100**, 762–767.

1173. GUERRERO-FIGUEROA, R., BARROS, A., HEATH, R. G. & GONZALEZ, G. (1964). Experimental subcortical epileptiform focus. *Epilepsia*, **5**, 112.
1174. GUILLERY, R. W. (1969). An abnormal retinogeniculate projection in Siamese cats. *Brain Res.*, **14**, 739–741.
1175. GUILLERY, R. W. (1973). The effect of lid suture upon growth of cells in the dorsal lateral geniculate nucleus of kittens. *J. comp. Neurol.*, **148**, 417–422.
1176. GUILLERY, R. W., SOBKOWICZ, H. M. & SCOTT, G. L. (1968). Light and electron microscopical observations of the ventral horn and ventral root in long term cultures of the spinal cord of the fetal mouse. *J. comp. Neurol.*, **134**, 433–476.
1177. GUILLOT, M. (1948). Anosmies partielles et odeurs fondamentales. *C.r. hebd. Séanc. Acad. Sci. Paris*, **226**, 1307–1309.
1178. GUILLOT, M. (1948). Sur quelques caractères des phénomènes d'anosmie partielle. *C.r. hebd. Séanc. Soc. Biol. Paris*, **142**, 161–162.
1179. GÜLDNER, F.-H. & WOLFF, J. R. (1973). Neurono-glial synaptoid contacts in the median eminence of the rat: ultrastructure, staining properties and distribution on tanycytes. *Brain Res.*, **61**, 217–234.
1180. GULIK-KRZYWICKI, T., SHECHTER, E., IWATSUBO, M., RANCK, J. L. & LUZZATI, V. (1970). Correlations between structure and spectroscopic properties in membrane model systems. Tryptophan and 1-anilino-8-naphthalene sulfonate fluorescence in protein-lipid-water phases. *Biochim. biophys. Acta*, **219**, 1–10.
1181. GURD, R. S., MAHLER, H. R. & MOORE, W. J. (1972). Differences in protein patterns on polyacrylamide gel electrophoresis of neuronal membranes from mice of different strains. *J. Neurochem.*, **19**, 553–556.
1182. GUTH, L. (1956). Regeneration in the mammalian peripheral nervous system. *Physiol. Rev.* **36**, 441–478.
1183. GUTH, L. (1956). Functional recovery following vagosympathetic anastomosis in the cat. *Am. J. Physiol.*, **185**, 205–208.
1184. GUTH, L. (1957). The effects of glossopharyngeal nerve transection on the circumvallate papillae of the rat. *Anat. Rec.*, **128**, 715–731.
1185. GUTH, L. (1958). Taste buds on the cat's circumvallate papilla after reinnervation by glossopharyngeal, vagus and hypoglossal nerves. *Anat. Rec.*, **130**, 25–37.
1186. GUTH, L. (1962). Neuromuscular function after regeneration of interrupted nerve fibres into partially denervated muscle. *Expl Neurol.*, **6**, 129–141.
1187. GUTH, L. (1963). The problem of selectivity between nerve and end-organ following nerve regeneration. *In* Gutmann, E. & Hnik, P. *Proceedings of the symposium on the effect of use and disuse on neuromuscular function.* Czechoslovak Acad. Sci; Prague. pp. 135–142.

1188. GUTH, L. (1968). 'Trophic' influences of nerve on muscle. *Physiol. Rev.,* **48**, 645–687.
1189. GUTH, L. (1971). Degeneration and regeneration of taste buds. *In* Beidler, L. M., *Taste. Handbook of sensory physiology IV,* (2), pp. 63–74.
1190. GUTH, L. (1974). Axonal regeneration and functional plasticity in the central nervous system. *Expl Neurol.,* **45**, 606.
1191. GUTH, L. & BERNSTEIN, J. J. (1961). Selectivity in the re-establishment of synapses in the superior cervical sympathetic ganglion of the cat. *Expl Neurol.,* **4**, 59–69.
1192. GUTH, L. & DENPSEY, P. J. (1970). Mechanism controlling glycogen levels in injured brain, normal brain, liver and heart (in mice). *Expl Neurol.,* **29**, 152–161.
1193. GUTH, L. & SAMAHA, F. J. (1969). Qualitative differences between actomyosin ATPase of slow and fast mammalian muscle. *Expl Neurol.,* **25**, 138–152.
1194. GUTH, L., SAMAHA, F. J. & ALBERS, R. W. (1970). The neural regulation of some phenotypic differences between the fiber types of mammalian skeletal muscle. *Expl Neurol.,* **26**, 126–135.
1195. GUTH, L. & WATSON, P. K. (1967). The influence of innervation on the soluble proteins of slow and fast muscles of the rat. *Expl Neurol.,* **17**, 107–117.
1196. GUTH, L. & WATSON, P. K. (1968). A correlated histochemical and quantitative study on cerebral glycogen after brain injury in the rat. *Expl Neurol.,* **22**, 590–602.
1197. GUTH, L. & WINDLE, W. F. (1970). The enigma of central nervous regeneration. *Expl Neurol.* Suppl. 5, 1–43.
1198. GUTH, L. & WINDLE. W. F. (1973). Physiological, molecular and genetic aspects of central nervous system regeneration. *Expl Neurol.,* **39**, iii–xvi.
1199. GUTH, L. & YELLIN, H. (1971). The dynamic nature of the so called 'fiber types' of mammalian skeletal muscle. *Expl Neurol.,* **31**, 277–300.
1200. GUTHRIE, D. M. (1962). Regenerative growth in insect nerve axons. *J. Insect. Physiol.,* **8**, 79–92.
1201. GUTHRIE, D. M. (1966). Physiological competition between host and inplanted ganglia in an insect. *Nature, Lond.,* **210**, 312–313.
1202. GUTHRIE, D. M. (1967). The regeneration of motor axons in an insect. *J. Insect Physiol.,* **13**, 1593–1611.
1203. GUTHRIE, D. M. & BANKS, J. R. (1969). The development of patterned activity by implanted ganglia and their peripheral connections in *Periplaneta Americana. J. exp. Biol.,* **50**, 255–273.
1204. GUTMANN, E. (1963). Evidence for the trophic function of the nerve cell in neuromuscular relations. *In* Gutmann, E. & Hnik, P., *The effect of use and disuse on neuromuscular function.* Elsevier; Amsterdam. pp. 29–34.
1205. GUTMANN, E. (1964). Neurotrophic relations in regeneration process. *Prog. Brain Res.,* **13**, 72–114.

1206. GUTMANN, E. (1972). Contractile properties and ATPase activity in fast and slow muscle of the rat during denervation. *Expl Neurol.,* **36**, 488–497.

1207. GUTMANN, E., HAJEK, I. & HORSKY, P. (1969). Effect of excessive use on contraction and metabolic properties of cross-striated muscle. *J. Physiol.,* **203**, 46–47P.

1208. GUTMANN, E., HAJEK, I., HORSKY, P., SYROVY, I. & ROHLICEK, V. (1968). The influence of stimulation *in vitro* on the mechanical properties of the muscle of rats. *Physiologia bohemoslov.,* **17**, 461–473.

1209. GUTMANN, E., HAJEK, I. & VITEK, V. (1970). Compensatory hypertrophy of the latissimus dorsi posterior muscle induced by elimination of the latissimus dorsi anterior of the chicken. *Physiologia bohemoslov.,* **19**, 483–489.

1210. GUTMANN, E. & SANDOW, A. (1965). Caffeine-induced contraction and potentiation of contraction in normal and denervated rat muscle. *Life Sci.,* **4**, 1149–1156.

1211. GUTMANN, E., SCHIAFFINO, S. & HANZLIKOVA, V. (1971). Mechanisms of compensatory hypertrophy in skeletal muscle of the rat. *Expl Neurol.,* **31**, 451–464.

1212. GYLLENSTEN, L., MALMFORS, T. & NORRLIN, M.-L. (1965). Effect of visual deprivation on the optic centres of growing and adult mice. *J. comp. Neurol.,* **124**, 149–160.

1213. GYLLENSTEN, L. MALMFORS, T. & NORRLIN, M. L. (1966). Growth activation in auditory cortex of visually deprived mice. *J. comp. Neurol.,* **126**, 463–470.

1214. HAFEMANN, D. R., COSTIN, A. & TARBY, T. J. (1970). Electrophysiological effects of enzymes introduced into the lateral geniculate body of the cat. *Expl Neurol.,* **27**, 238–247.

1215. HAGEDOORN, J. (1965). Seasonal changes in the ependyma of the skunk *Mephitis mephitis nigra. Anat. Rec.,* **151**, 453–454.

1216. HAGINS, W. A. & JENNINGS, W. H. (1960). Orientation of the photoreceptor. *Trans. Faraday Soc.,* **27**, 180.

1217. HAGINS, W. A. & McGAUCHY, R. E. (1968). Membrane origin of the fast photovoltage of squid retina. *Science, N.Y.,* **159**, 213–215.

1218. HAGIWARA, S. & MORITA, H. (1962). Electrotonic transmission between two nerve cells in leech ganglion. *J. Neurophysiol.,* **25**, 721–731.

1219. HAGIWARA, S. & TASAKI, I. (1958). A study of the mechanism of impulse transmission across the giant synapse of the squid. *J. Physiol.,* **143**, 114–137.

1220. HAGIWARA, S., TIYAMA, K. & HAYASHI, H. (1971). Mechanism of anion and cation permeations in the resting membrane of the barnacle muscle fiber. *J. gen. Physiol.,* **57**, 408–433.

1221. HAGLID, K. G. (1972). Protein synthesis *in vitro* in nuclei of human nervous tissue. *J. Neurochem.,* **19**, 19–25.

1222. HAIN, R. F. (1963). Discussion of paper *by* Königsmark, B. W. &

Sidman, R. L. Origin of gitter cells in the mouse brain. *J. Neuropath. exp. Neurol.*, **22**, 327–328.

1223. HAJEK, I., HANIKOVA, A. & GUTMANN, E. (1967). Changes of ATPase activity in the rat diaphragm undergoing denervation hypertrophy. *Physiologia bohemoslov.* **16**, 515–521.
1224. HALJAMÄE, H. & HAMBERGER, A. (1971). Potassium accumulation by bulk prepared neuronal and glial cells. *J. Neurochem.*, **18**, 1903–1912.
1225. HALL, M. O., BOK, D. & BACHARACH, A. D. E. (1969). Biosynthesis and assembly of rod outer segment membrane system. Formation and fate of visual pigment in the frog retina. *J. molec. Biol.*, **45**, 397–406.
1226. HALL, Z. W. (1972). Release of neurotransmitters and their interaction with receptors. *A. Rev. Biochem.*, **41**, 925–952.
1227. HALL-CRAGGS, E. G. B. (1968). The contraction times and enzyme activity of two rabbit laryngeal muscles. *J. Anat.*, **102**, 241.
1228. HALL-CRAGGS, E. G. B., GUTH, L. & YELLIN, H. (1972). The 'tonicophasic' muscle: an experimental model. *Expl Neurol.*, **35**, 398–401.
1229. HAMA, K. (1969). A study on the fine structure of the saccular macula of the goldfish. *Z. Zellforsch. mikrosk. Anat.*, **94**, 155–171.
1230. HAMBERGER, A., BLOMSTRAND, C. & LEHNINGER, A. L. (1970). Comparative studies on mitochondria isolated from neuron-enriched and glia-enriched fractions of rabbit and bovine brain. *J. Cell Biol.*, **45**, 221–234.
1231. HAMBURGER, A., BLOMSTRAND, C. & ROSENGREN, B. (1970). Effect of X-irradiation on respiration and protein synthesis in neuronal and neuroglia cell fractions. *Expl Neurol.*, **26**, 509–517.
1232. HAMBURGER, A., BLOMSTRAND, C. & YANAGIHARA, Y. (1971). Subcellular distribution of radioactivity in neuronal and glial-enriched fractions after incorporation of ^{3}H-leucine *in vivo* and *in vitro*. *J. Neurochem.*, **18**, 1469–1478.
1233. HAMBURGER, A. & HYDEN, H. (1963). Inverse enzymatic changes in neurones and glia during increased function and hypoxia. *J. Cell Biol.*, **16**, 521–525.
1234. HAMBURGH, M. (1960). Observations on the neuropathology of 'reeler', a neurological mutation in mice. *Experentia*, **16**, 460–461.
1235. HAMBURGH, M. (1963). Analysis of the postnatal developmental effects of 'reeler', a neurological mutation in mice. A study in developmental genetics. *Devl Biol.*, **8**, 165–185.
1236. HAMBURGH, M. (1968). An analysis of the action of thyroid hormone on development based on *in vivo* and *in vitro* studies. *J. comp. Endocrinol.*, **10**, 198–213.
1237. HAMBURGH, M., MENDOZA, L. A., KRUPA, P., GELFAND, D. & LEHRER, R. (1970). The effect of parabiosis on audiogenic convulsions in seizure-susceptible mice. *Expl Neurol.*, **26**, 283–290.

1238. HAMBURGH, M. & BORNSTEIN, M. K. (1970). Myelin synthesis in two demyelinating mutations in mice. *Expl Neurol.*, **28**, 471–476.
1239. HAMDI, F. A. & WHITTERIDGE, D. (1954). The representation of the retina on the optic tectum of the pigeon. *Q. Jl exp. Physiol.*, **39**, 111–119.
1240. HAMORI, J. (1973). The inductive role of presynaptic axons in the development of postsynaptic spines. *Brain Res.*, **62**, 337–344.
1241. HANAI, T., HAYDON, D. A. & TAYLOR, J. (1964). An investigation by electrical methods of lecithin-in-hydrocarbon films in aqueous solution. *Proc. R. Soc. A*, **281**, 377–390.
1242. HANSSON, H.-A. & SJÖSTRAND, J. (1971). Ultrastructural effects of colchicine on the hypoglossal and dorsal vagal neurones of the rabbit. *Brain Res.*, **35**, 379–396.
1243. HARDMAN, J. G., BEAVO, J. A., GRAY, J. P., CHRISMAN, T. D., PATTERSON, W. D. & SUTHERLAND, E. W. (1971). The formation and metabolism of cyclic GMP. *Ann. N.Y. Acad. Sci.*, **185**, 27–35.
1244. HÄRKÖNEN, M. (1964). Carboxylic esterases, oxidative enzymes and catecholamines in the superior cervical ganglion of the rat and the effect of pre- and post-ganglionic nerve division. *Acta physiol. scand.*, **63**, 1–94.
1245. HARLOW, H. (1958). The nature of love. *Am. Psychol.*, **13**, 673–685.
1246. HARLOW, H. F. (1959). The development of learning in the rhesus monkey. *Am. Sci.*, **47**, 459–479.
1247. HARLOW, H. F., ROWLAND, G. L. & GRIFFIN, G. A. (1964). The effect of total social deprivation on the development of monkey behaviour. *Psychiat. Res. Rep.*, **19**, 116–135.
1248. HARLOW, H. F. & ZIMMERMAN, R. R. (1959). Affectional responses in the infant monkey. *Science, N.Y.*, **130**, 421–432.
1249. van HARREVELD, A. (1957). Changes in volume of cortical neuronal elements during asphyxiation. *Am. J. Physiol.*, **191**, 233–242.
1250. van HARREVELD, A. (1961). Asphyxial changes in the cerebellar cortex. *J. cell. comp. Physiol.*, **57**, 101–110.
1251. van HARREVELD, A. (1966). *Brain tissue electrolytes.* Butterworth; London.
1252. van HARREVELD, A. & BIBER, M. D. (1962). Conductivity changes in some organs after circulatory arrest. *Am. J. Physiol.*, **203**, 609–614.
1253. van HARREVELD, A. & CROWELL, J. (1964). Extracellular space in central nervous tissue. *Fedn Proc. Fedn Am. Socs exp. Biol.*, **23**, 304.
1254. van HARREVELD, A., CROWELL, J. & MALHOTRA, S. K. (1965). A study of extracellular space in central nervous tissue by freeze-substitution. *J. Cell Biol.*, **25**, 117–139.
1255. van HARREVELD, A., HOOPER, N. K. & CUSICK, J. J. (1961). Brain electrolytes and cortical impedance. *Am. J. Physiol.*, **201**, 139–143.
1256. van HARREVELD, A. & MALHOTRA, S. K. (1967). Extracellular space in the cerebral cortex of the mouse. *J. Anat.*, **101**, 197–207.
1257. van HARREVELD, A., MURPHY, T. & NOBEL, K. W. (1963). Specific impedance of rabbit's cortical tissue. *Am. J. Physiol.*, **205**, 203–207.

1258. van HARREVELD, A. & OCHS, S. (1965). Cerebral impedance changes after circulatory arrest. *Am. J. Physiol.*, **187**, 180–192.

1259. van HARREVELD, A. & SCHADE, J. P. (1962). Changes in the electrical conductivity of cerebral cortex during seizure activity. *Expl Neurol.*, **5**, 383–400.

1260. van HARREVELD, A. & TACHIBANA, S. (1962). Recovery of cerebral cortex from asphyxiation. *Am. J. Physiol.*, **202**, 59–65.

1261. HARRIS, A. J. (1974). Inductive functions of the nervous system. *A. Rev. Physiol.*, **36**, 251–305.

1262. HARRIS, A. J., KUFFLER, S. W. & DENNIS, M. J. (1971). Differential chemosensitivity of synaptic and extrasynaptic areas on the neuronal surface membrane in parasympathetic neurones of the frog tested by microapplication of acetyl choline. *Proc. R. Soc. B*, **177**, 541–553.

1263. HARRIS, E. J. & NICHOLLS, J. G. (1956). The effect of denervation on the rate of entry of potassium into frog muscle. *J. Physiol.*, **131**, 473–476.

1264. HARRIS, J. B. (1971). The resting membrane potential of fibres of fast and slow twitch muscles in normal and dystrophic mice. *J. neurol. Sci.*, **12**, 45–52.

1265. HARRIS, J. B. & THESLEFF, S. (1972). Unpublished observations cited *by* Grampp, W., Harris, J. B. & Thesleff, S. (1972). Inhibition of denervation changes in skeletal muscle by blockers of protein synthesis. *J. Physiol.*, **221**, 743–754.

1266. HARRIS, J. B., WALLACE, C. & WING, J. (1972). Myelinated nerve counts in the nerves of normal and dystrophic mouse muscle. *J. neurol. Sci.*, **15**, 245–249.

1267. HARRIS, J. E., MORGENROTH, V. H., ROTH, R. H. & BALDESSARINI, R. J. (1974). Regulation of catecholamine synthesis in the rat brain *in vitro* by cyclic AMP. *Nature, Lond.*, **252**, 156–158.

1268. HARRISON, R. G. (1903). Experimentelle Untersuchungen über die Entwicklung der Sinnesorgane der Seitenlinie bei den Amphibien. *Arch. mikrosk. Anat. Entw Mech.*, **63**, 35–149.

1269. HARTSHORNE, D. J. & MUELLER, H. (1968). Fractionation of troponin into two distinct proteins. *Biochem. biophys. Res. Commun.*, **31**, 647–653.

1270. HARTSHORNE, D. J., THEINER, M. & MULLER, H. (1969). Studies on troponin. *Biochim. biophys. Acta*, **175**, 320–330.

1271. HARVEY, J. E. & SREBNIK, H. H. (1967). Locomotor activity and axon regeneration following spinal cord compression in rats treated with L-thyronine. *J. Neuropath. exp. Neurol.*, **26**, 661–668.

1272. HAY, E. D. (1959). Electron microscopic observations of muscle dedifferentiation in regenerating Amblystoma limbs. *Devl Biol.*, **1**, 555–585.

1273. HAY, E. D. (1964). Secretion of a connective tissue protein by developing epidermis. *In* Montagna, W. & Lobitz, W. C., *The epidermis.* Acad. Press; N.Y. pp. 97–116.

1274. HAY, E. D. & FISHMAN, D. A. (1960). Origin of the regeneration blastema of amputated *Triturus viridescens* limbs studied by autoradiography following injection of tritiated thymidine. *Anat. Rec.*, **136**, 208.
1275. HAY, E. D. & REVEL, J. P. (1963). Autoradiographic studies of the origin of the basement lamella in *amblyostoma. Devl Biol.*, **7**, 152–168.
1276. HAYDON, D. A. (1970). The organization and permeability of artificial lipid membranes. *In* Bittar, E. E., *Membranes and ion transport.* Wiley Interscience; N.Y. pp. 64–82.
1277. HAYNES, D. H., KOWALSKY, A. & PRESSMAN, B. C. (1969). Application of nuclear magnetic resonance to the conformational changes in valinomycin during complexation. *J. biol. Chem.*, **244**, 502–505.
1278. HAYASHI, H. & CAPALDI, R. A. (1972). The proteins of the outer membrane of beef heart mitochondria. *Biochim. biophys. Acta*, **282**, 166–173.
1279. HEBB, D. O. (1949). *The organization of behaviour.* Wiley; N.Y.
1280. HEDLEY-WHYTE, E. T., DARRAN, H. K., STENDLER, F. & UZMAN, B. G. (1968). The value of cholesterol-1,2,-^{3}H as a long term tracer for autoradiographic study of the nervous system of mice. *Lab. Invest.*, **19**, 526.
1281. HEDLEY-WHYTE, E. T., RAWLINS, F. A., SALPETER, M. M. & UZMAN, B. G. (1969). Distribution of cholesterol-1,2,-^{3}H during maturation of mouse peripheral nerve. *Lab. Invest.*, **21**, 536–547.
1282. HEILBRONN, E. & MATTSON, C. (1974). The nicotinic cholinergic receptor protein: improved purification method preliminary aminoacid composition and observed auto-immune response. *J. Neurochem.*, **22**, 315.
1283. HEINBECKER, P. (1944). The pathogenesis of Cushing's syndrome. *Medicine*, **23**, 225–247.
1284. HEINBECKER, P. & PFEIFFENBERGER, M. (1950). Further clinical and experimental studies on the pathogenesis of Cushing's syndrome. *Am. J. Med.*, **9**, 3–23.
1285. HEINER, L. (1971). Effect of denervation on the phospholipid classes in tonic and tetanic muscles. *Acta physiol. hung.*, **40**, 243–249.
1286. HEUSEY, S. R., HELD, D. & PAPPENHEIMER, J. R. (1962). Bulk flow and diffusion in the cerebrospinal fluid system of the goat. *Am. J. Physiol.*, **203**, 775–781.
1287. HEITZMAN, D. H. (1972). Rhodopsin is the predominant protein of rod outer segment membranes. *Nature, Lond., New Biol.*, **235**, 114.
1288. HELD, D., FENCL, V. & PAPPENHEIMER, J. R. (1964). Electrical potential of cerebrospinal fluid. *J. Neurophysiol.*, **27**, 942–959.
1289. HELD, I. & YOUNG, I. J. (1969). A comparative study of the somato-axonal flow of protein in the feline hypoglossal and vagal nerves. *Expl Brain Res.*, **8**, 150–162.
1290. HELLER, J., OSTWALD, T. J. & BOK, D. (1971). The osmotic behaviour of rod photoreceptor outer segment discs. *J. Cell Biol.*, **48**, 633.

1291. HELLUNG-LARSON, P. & ANDERSON, V. (1970). Lactate dehydrogenase isoenzymes of human lymphocytes cultured with phytohaemagglutinin at different oxygen tensions. *FEBS Lett.*, **18**, 163–167.
1292. HELMY, F. M. & HACK, M. H. (1963). Histochemical and lipid studies on human choroid plexus. *Proc. Soc. exp. Biol. Med.*, **114**, 361–362.
1293. HENDERSON, N. D. (1970). Brain weight increases resulting from environmental enrichment: a directional dominance in mice. *Science, N.Y.*, **169**, 776–778.
1294. HENDERSON, N. D. (1970). A genetic analysis of spontaneous alternation in mice. *Behav. Genet.*, **1**, 125–132.
1295. HENDERSON, N. D. (1970). Genetic influences on the behaviour of mice can be obscured by laboratory rearing. *J. comp. Physiol. Psychol.*, **72**, 505–511.
1296. HENDERSON, N. D. (1972). Relative effects of early rearing environment and genotype on discrimination learning in house mice. *J. comp. Physiol. Psychol.*, **79**, 243–253.
1297. HENDERSON, N. D. (1973). Brain weight changes resulting from enriched rearing conditions: a diallel analysis. *Devl Psychobiol.*, **6**, 367–376.
1298. HENDLER, R. W. (1971). Biological membrane studies. *Physiol. Rev.*, **51**, 66–97.
1299. HENDRICKSON, A. (1969). Electron microscopic radioautography: identification of origin of synaptic terminals in neuronal nervous tissue. *Science, N.Y.*, **165**, 194–196.
1300. HENDRICKSON, A. E. & COWAN, W. M. (1971). Changes in the rate of axoplasmic transport during postnatal development of the rabbit's optic nerve and tract. *Expl Neurol.*, **30**, 403–422.
1301. HENKIN, R. I. & BRADLEY, D. F. (1969). Regulation of taste acuity by thiols and metal ions. *Proc. natn. Acad. Sci. U.S.A.*, **62**, 30.
1302. HENKIN, R. I., GILL, J. R. & BARTER, F. C. (1963). Studies of taste thresholds in normal man and in patients with adrenal cortical insufficiency: the role of adrenal cortical steroids and of serum sodium concentration. *J. clin. Invest.*, **42**, 727–735.
1303. HENKIN, R. I., GRAZIADEI, P. P. G. & BRADLEY, D. F. (1964). The molecular basis of taste and its disorders. *Ann. intern. Med.*, **71**, 791–821.
1304. HENKIN, R. I., KEISER, H. R. & KARE, M. R. (1968). The effects of D-penicillamine and of copper repletion on salt preference and sugar intake. *Fedn Proc. Fedn Am. Socs exp. Biol.*, **27**, 583.
1305. HENKIN, R. I. & KOPIN, L. J. (1964). Abnormalities of taste and smell thresholds in familial dysautonomia: improvement with methacholine. *Life Sci.*, **3**, 1319–1325.
1306. HENN, F. A. & HAMBURGER, A. (1971). Glial cell function: uptake of transmitter substances. *Proc. natn. Acad. Sci. U.S.A.*, **68**, 2686–2690.
1307. HENN, F. A. & THOMPSON. T. E. (1969). Synthetic lipid bilayer membranes. *A. Rev. Biochem.*, **38**, 241–284.

1308. HERBERG, L. J., TRESS, K. H. & BLUNDELL, J. E. (1969). Raising the threshold in experimental epilepsy by hypothalamic and septal stimulation and by audiogenic seizures. *Brain,* **92**, 313–325.

1309. HERBERG, L. J. & WATKINS, P. J. (1966). Epileptiform seizures induced by hypothalamic stimulation in the rat: resistance to fits following fits. *Nature, Lond.,* **209**, 515–516.

1310. HERMAN, C. J. & LAPHAM, L. W. (1969). Neuronal polyploidy and nuclear volume in the cat central nervous system. *Brain Res.,* **15**, 35–48.

1311. HERRON, W. L., RIEGEL, B. W., MYERS, O. E. & RUBIN, M. L. (1969). Retinal dystrophy in the rat-pigment epithelium disease. *Invest. Ophthal.,* **8**, 595.

1312. HERSCHKOWITZ, N., VASSELA, F. & BISCHOFF, A. (1971). Myelin differences in the central and peripheral nervous system in the 'Jimpy' mouse. *J. Neurochem.,* **18**, 1361–1363.

1313. HERZ, R. & WEBER, A. (1965). Caffeine inhibition of Ca uptake by muscle reticulum. *Fedn Proc. Fedn Am. Socs exp. Biol.,* **24**, 208. (444).

1314. HESLOP, J. (1974). Axonal flow and fast transport in nerves. *Adv. comp. Physiol. Biochem.,* **5**, 432–437.

1315. HESS, A. (1960). The fine structure of degenerating nerve fibres, their sheaths and terminations in the CNS of the cockroach (*Periplaneta americana*). *J. biophys. biochem. Cytol.,* **7**, 339–344.

1316. HESS, A. (1961). Structural differences of fast and slow extrafusal fibres and their nerve endings in chickens. *J. Physiol.,* **157**, 221–231.

1317. HESS, A. (1970). Vertebrate slow muscle fibres. *Physiol. Rev.,* **50**, 40–62.

1318. HESS, A. & MACKIE, G. O. (1961). Is there a nervous system in Hydra? *In* Lenhoff, H. M. & Loomis, W. F., *The biology of Hydra.* University of Miami Press.

1319. HESS, A. & ROSSNER, S. (1970). The satellite cell bud and myoblast in denervated mammalian muscle fibres. *Am. J. Anat.,* **129**, 21–40.

1320. von HEUKELOM, J. S., van der GON, J. J. D. & PROP, F. J. A. (1972). Model approaches for evaluation of cell coupling in monolayers. *J. mem. Biol.,* **7**, 88–110.

1321. HEUSER, J. & REESE, T. S. (1972). Stimulation induced uptake and release of peroxidase from synaptic vesicles in frog neuromuscular junction. *Anat. Rec.,* **172**, 329.

1322. HIBBARD, E. (1963). Regeneration in the severed spinal cord of chordate larvae of *Petromyzon marinus. Expl Neurol.,* **7**, 175–185.

1323. HIGASHIDA, H. MIYAKE, A., TARAO, M. & WATANABE, S. (1971). Membrane potential changes of neuroglial cells during spreading depression in the rabbit. *Brain Res.,* **32**, 207–211.

1324. HIGMAN, H. B., PODLESKI, T. R. & BARTELS, E. (1963). Apparent dissociation constants between carbamyl choline, *d*-tubocurarine and the receptor. *Biochim. biophys. Acta,* **75**, 187–193.

1325. HIKIDA, R. S. (1972). Morphological transformation of slow to fast muscle fibres after tenotomy. *Expl Neurol.,* **35**, 265–273.

1326. HILD, W. (1966). Cell types and neuronal connections in cultures of mammalian central nervous tissue. *Z. Zellforsch. mikrosk. Anat.*, **69**, 155–188.

1327. HILD, W., CHANG, J. J. & TASAKI, I. (1958). Electrical responses of astrocytic glia from the mammalian central nervous system cultivated *in vitro. Experentia*, **14**, 220–221.

1328. HILD, W. & KLEE, M. R. (1968). Membrane properties of neurons and neuroglia *in vitro. Anat. Rec.*, **160**, 366.

1329. HILD, W., TAKENAKA, T. & WALKER, F. (1965). Electrophysiological properties of ependymal cells from the mammalian brain in tissue culture. *Anat. Rec.*, **151**, 361.

1330. HILD, W., TAKENAKA, T. & WALKER, F. (1965). Electrophysiological properties of ependymal cells from the mammalian brain in tissue culture. *Expl Neurol.*, **11**, 493–501.

1331. HILD, W. & TASAKI, I. (1962). Morphological and physiological properties of neurones and glial cells in tissue culture. *J. Neurophysiol.*, **25**, 277–304.

1332. HILDRETH, E. H. (1963). Prevention of rabies, or the decline of Sirius. *Ann. intern. Med.*, **58**, 883–896.

1333. HILL, M. (1967). The uptake of deoxyribonucleic acid released from damaged cells in tissue culture. *Expl Cell Res.*, **45**, 533–549.

1334. HILLARP, N. A. (1946). Structure of the synapse and the peripheral innervation apparatus of the autonomic nervous system. *Acta Anat.*, **2**, Suppl. 4, 1–153.

1335. HILLE, B. (1968). Pharmacological modification of the sodium channels of frog nerve. *J. gen. Physiol.*, **51**, 199.

1336. HILLE, B. (1968). Charges and potentials at the nerve surface. Divalent ions and pH. *J. gen. Physiol.*, **51**, 221.

1337. HILLE, B. (1970). Ionic channels in nerve membranes. *Prog. Biophys. molec. Biol.*, **21**, 3–32.

1338. HILLE, B. (1971). The permeability of the sodium channel to organic cations in myelinated nerve. *J. gen. Physiol.*, **58**, 599–619.

1339. HILLE, B. (1972). The permeability of the sodium channel to metal cations in myelinated nerve. *J. gen. Physiol.*, **59**, 637–658.

1340. HILLIER, J. & HOFFMAN, J. F. (1953). On the ultrastructure of the plasma membrane as determined by the electron microscope. *J. cell. comp. Physiol.*, **42**, 203–248.

1341. HINDE, R. A. & SPENCER-BOOTH, Y. (1968). The study of mother–infant interaction in captive group-living rhesus monkeys. *Proc. R. Soc. B*, **169**, 177–201.

1342. HINDE, R. A. & SPENCER-BOOTH, Y. (1971). Effects of brief separation from mother on Rhesus monkeys. *Science, N.Y.*, **173**, 111–118.

1343. HINDS, J. W. (1968). Autoradiographic study of histogenesis in the mouse olfactory bulb. I. Time of origin of neurons and neuroglia. *J. comp. Neurol.*, **134**, 287–304.

1344. HINDS, J. W. (1968). Autoradiographic study of histogenesis in the mouse olfactory bulb. II. Cell proliferation and migration. *J. comp. Neurol.*, **134**, 305–322.
1345. HINDS, J. W. (1971). Early neuroblast differentiation in the mouse olfactory bulb. *Anat. Rec.*, **169**, 340–341.
1346. HINES, H. M. & KNOWLTON, G. C. (1933). Changes in the skeletal muscle of the rat following denervation. *Am. J. Physiol.*, **104**, 379–391.
1347. HINKLEY, R. E. & SAMSON, F. E. (1972). Anaesthetic-induced transformation of axonal microtubules. *J. Cell Biol.*, **53**, 258–263.
1348. HINRICHSON, C. F. L. & LARRAMENDI, L. M. H. (1968). Synapses and cluster formation of the mouse mesencephalic fifth nucleus. *Brain Res.*, **7**, 296.
1349. HIRANO, A., LEVINE, S. & ZIMMERMAN, H. M. (1968). Remyelination in the central nervous system after cyanide intoxication. *J. Neuropath. exp. Neurol.*, **27**, 234–245.
1350. HIRANO, A. & ZIMMERMAN, H. M. (1967). Some new cytological observations of the normal rat ependymal cell. *Anat. Rec.*, **158**, 293–302.
1351. HIRONAKA, T. & MIYATA, Y. (1975). Transplantation of skeletal muscle in normal and dystrophic mice. *Expl Neurol.*, **47**, 1–15.
1352. HIRSCH, H. V. B. (1972). Visual perception in cats after environmental surgery. *Expl Brain Res.*, **15**, 405–423.
1353. HIRSCH, H. V. B. & SPINELLI, D. N. (1970). Visual experience modifies distribution of horizontally and vertically oriented receptive fields in cats. *Science, N. Y.*, **168**, 869–871.
1354. HIRSCH, J. (1963). Behaviour genetics and individuality. *Science, N. Y.*, **142**, 1436–1442.
1355. HJORTH-SIMONSEN, A. (1971). Hippocampal efferents to the ipsilateral entorhinal area: an experimental study in the rat. *J. comp. Neurol.*, **142**, 417–438.
1356. HJORTH-SIMONSEN, A. (1972). Projection of the part of the entorhinal area to the hippocampus and fascia dentata. *J. comp. Neurol.*, **146**, 219–232.
1357. HJORTH-SIMONSEN, A. (1973). Some intrinsic connections of the hippocampus in the rat: an experimental analysis. *J. comp. Neurol.*, **147**, 145–162.
1358. HJORTH-SIMONSEN, A. & JEUNE, B. (1972). Origin and termination of the hippocampal perforant path in the rat studied by silver impregnation. *J. comp. Neurol.*, **144**, 215–232.
1359. HLADKY, S. B. & HAYDON, D. A. (1970). Discreteness of conductance change in bimolecular lipid membranes in the presence of certain antibiotics. *Nature, Lond.*, **225**, 451–453.
1360. HLADKY, S. B. & HAYDON, D. A. (1972). Ion transfer across lipid membranes in the presence of gramicidin A. I. Studies of the unit conductance channel. *Biochim. biophys. Acta*, **274**, 294–312.
1361. HODGKIN, A. L. (1951). The ionic basis of electrical activity in nerve

and muscle. *Biol. Rev.*, **26**, 339–368.

1362. HODGKIN, A. L. & HUXLEY, A. F. (1952). Currents carried by sodium and potassium ions through the membrane of the giant axon of *Loligo. J. Physiol.*, **116**, 449–472.
1363. HODGKIN, A. L. & HUXLEY, A. F. (1952). The components of membrane conductance in the giant axon of *Loligo. J. Physiol.*, **116**, 473–496.
1364. HODGKIN, A. L. & HUXLEY, A. F. (1952). The dual effect of membrane potential on sodium conductance in the giant axon of *Loligo. J. Physiol.*, **116**, 497–506.
1365. HODGKIN, A. L. & HUXLEY, A. F. (1952). A quantitative description of membrane current and its application to conduction and excitation in nerve. *J. Physiol.*, **117**, 500–544.
1366. HODGKIN, A. L., HUXLEY, A. F. & KATZ, B. (1952). Measurement of current-votage relations in the membrane of the giant axon of *Loligo. J. Physiol.*, **117**, 424–448.
1367. HODGKIN, A. L. & KATZ, B. (1949). The effect of sodium ions on the electical activity of the giant axon of the squid. *J. Physiol.*, **108**, 37–77.
1368. HOFFER, B., SEIGER, A. LJUNGBERG, T. & OLSON, L. (1974). Electrophysiological and cytological studies of brain homografts in the anterior chamber of the eye: maturation of cerebellar cortex in oculo. *Brain Res.*, **79**, 165.
1369. HOFFMAN, H. (1951). Fate of interrupted nerve fibres regenerating into partially denervated muscles. *Aust. J. exp. Biol. med. Sci.*, **29**, 211–219.
1370. HOFFMAN, H. (1951). A study of the factors influencing innervation of muscles by implanted nerves. *Aust. J. exp. Biol. med. Sci.*, **29**, 289–308.
1371. HOFFMAN, H. (1952). Acceleration and retardation of the process of axon-sprouting in partially denervated muscles. *Aust. J. exp. Biol. med. Sci.*, **30**, 541–566.
1372. HOFFMAN, H. J. & OLSZEWSKI, J. (1961). Spread of sodium fluorescein in normal brain tissue. *Neurology*, **11**, 1081–1085.
1373. HOGAN, P. & ALBUQUERQUE, E. X. (1971). The pharmacology of batrachotoxin. III. Effects on the heart Purkinje fibres. *J. Pharmac. exp. Ther.*, **176**, 529–537.
1374. HOGAN, E. L., DAWSON, D. M. & ROMANUL, F. C. A. (1965). Enzymatic changes in denervated muscle II. Biochemical studies. *Archs. Neurol.*, **13**, 274–282.
1375. HOGAN, E. L. & JOSEPH, K. C. (1970). Composition of cerebral lipids in murine leucodystrophy: the quaking mutant. *J. Neurochem.*, **17**, 1209–1214.
1376. HOKFELT, T. & LJUNGDAHL, Å. (1971). Light and electron microscopic autoradiography on spinal cord slices after incubation with labeled glycine. *Brain Res.*, **32**, 189–194.
1377. HOKIN, L. E. & SHERWIN, A. L. (1957). Protein secretion and phosphate turnover in the phospholipids in salivary glands in vitro. *J. Physiol.*, **135**, 18–29.

1378. HOLLOWAY. R. L. (1966). Dendritic branching: some preliminary results of training and complexity in rat visual cortex. *Brain Res.*, **2**, 393–396.
1379. HOLMES, K. V. & CHOPPIN, P. W. (1968). On the role of microtubules in movement and alignment of nuclei in virus-induced syncytia. *J. Cell Biol.*, **39**, 526–543.
1380. HOLTON, F. A. & HOLTON, P. (1954). The capillary dilator substances in dry powders of spinal roots: a possible role of adenosine triphosphate in chemical transmission from nerve endings. *J. Physiol.*, **126**, 124–140.
1381. HOLTON, P. (1959). The liberation of adenosine triphosphate on antidromic stimulation of sensory nerves. *J. Physiol.*, **145**, 494–504.
1382. HOLTZER, S. W. (1956). The inductive activity of the spinal cord in urodele tail regeneration. *J. Morph.*, **99**, 1–39.
1383. HOLTZMAN, E., FREEMAN, A. R. & KASHNER, L. A. (1971). Stimulation-dependent alterations in peroxidase uptake at lobster neuromuscular junction. *Science, N.Y.*, **173**, 733–736.
1384. HOLTZMAN, E. & NOVIKOFF, A. B. (1965). Lysosomes in the rat sciatic nerve following crush. *J. Cell Biol.*, **27**, 651–669.
1385. HOLTZMAN, E., NOVIKOFF, A. B. & VILLAVERDE, H. (1967). Lysosomes and GERL in normal and chromatolytic neurones of the rat ganglion nodosum. *J. Cell Biol.*, **33**, 419–435.
1386. HONG, K., CHEN, Y.-S. & HUBBELL, W. L. (1973). The non-covalent coupling of rhodopsin and phosphatidylcholine. *Ann. N.Y. Acad. Sci.*, **222**, 527–529.
1387. HONG, K. & HUBBELL, W. L. (1972). Preparation and properties of phospholipid bilayers containing rhodopsin. *Proc. natn. Acad. Sci. U.S.A.*, **69**, 2617–2621.
1388. HONING, B. & KARPLUS, M. (1971). Implications of torsional potential of retinal isomers for visual excitation. *Nature, Lond.*, **229**, 558–560.
1389. HONJIN, R., NAKAMURA, T. & IMURA, M. (1959). Electron microscopy of peripheral nerve fibres. III. On the axoplasmic changes during Wallerian degeneration. *Okajima Folia Anat. Jap.*, **33**, 131–156.
1390. HOPFER, U. A., LEHNINGER, A. L. & LENNARTZ, W. J. (1970). Action of ionophores on membranes. *J. mem. Biol.*, **2**, 41–53.
1391. HOPFER, U. A., LEHNINGER, A. L. & LENNARTZ, W. J. (1970). Action of ionophores on reconstituted phospholipid bilayers. *J. mem. Biol.*, **3**, 142–155.
1392. HOPKINS, A. P. & LAMBERT, E. H. (1971). Conduction velocity of unmyelinated fibres in the growing rat and changes after crush injury. *4th Int. Congr. EMG. Proc.*, Brussels.
1393. HORN, A. S., CUELLO, A. C. & MILLER, R. J. (1974). Dopamine in the mesolimbic system of the rat brain: endogenous levels and the effects of drugs on the uptake mechanism and stimulation of adenylate cyclase activity. *J. Neurochem.*, **22**, 265–270.
1394. HORN, G. (1955). Thyroid deficiency and inanition: the effects of replacement therapy on the development of the cerebral cortex of young

albino rats. *Anat. Rec.*, **121**, 63–79.

1395. HORN, G., ROSE, S. P. R. & BATESON, P. P. G. (1973). Experience and plasticity in the central nervous system. *Science, N.Y.*, **181**, 506–514.

1396. HOROUPIAN, D. S., GHETTI, B. & WISNEWSKI, H. M. (1973). Retrograde transneuronal degeneration of optic fibres and their terminals in lateral geniculate nucleus of rhesus monkey. *Brain Res.*, **49**, 257.

1397. HORWITZ, A. F., HORSLEY, W. J. & KLEIN, M. P. (1972). Magnetic resonance studies on membranes and model membrane systems: proton magnetic relaxation rates in sonicated lecithin dispersions. *Proc. natn. Acad. Sci. U.S.A.*, **69**, 590–593.

1398. HORRIDGE, G. A. (1968). *Interneurones; their origin, action, specificity, growth and plasticity.* Freeman; San Francisco.

1399. HORSTMANN, E. & MEVES, H. (1959). Die Feinstruktur des molekularen Rindengraues und ihre physiologische Bedeutung. *Z. Zellforsch. mikrosk. Anat.*, **49**, 569–604.

1400. HÖSLI, E. & HÖSLI, L. (1970). The presence of acetylcholinesterase in cultures of cerebellum and brain stem. *Brain Res.*, **19**, 494–496.

1401. HOTTA, Y. & BENZER, S. (1969). Abnormal electroretinograms in visual mutants of Drosophila. *Nature, Lond.*, **222**, 354–356.

1402. HOTTA, Y. & BENZER, S. (1970). Genetic dissection of the *Drosophila* nervous system by means of mosaics. *Proc. natn. Acad. Sci. U.S.A.*, **67**, 1156–1158.

1403. HOTTA, Y. & BENZER, S. (1972). Mapping of behavior in *Drosophila* mosaics. *Nature, Lond.*, **240**, 527–535.

1404. HOWARD, E. (1965). Effects of corticosterone and food restriction on growth and on DNA, RNA and cholesterol contents of the brain and liver in infant mice. *J. Neurochem.*, **12**, 181–191.

1405. HOWARD, E. (1968). Reduction in size and total DNA in cerebrum and cerebellum in adult mice after corticosterone treatment in infancy. *Expl Neurol.*, **22**, 191–208.

1406. HOWARD, E. & GRANOFF, D. M. (1968). Increased voluntary running and decreased motor coordination in mice after neonatal corticosterone implantation. *Expl Neurol.*, **22**, 666–690.

1407. HOWELL, J. N., FAIRHURST, A. S. & JENDEN, D. J. (1966). Alterations of the calcium accumulation ability of striated muscle following denervation. *Life Sci.*, **5**, 439–446.

1408. HOWLAND, J. L. & CHALLBERG, M. D. (1973). Altered respiration and proton permeability in liver mitochondria from genetically dystrophic mice. *Biochem. biophys. Res. Commun.*, **50**, 547–580.

1409. HOY, R. R. (1969). Degeneration and regeneration in abdominal flexor motor neurones in the crayfish. *J. exp. Zool.*, **172**, 219–232.

1410. HOY, R. R. (1973). The curious nature of degeneration and regeneration in motor neurons and central connectives of the crayfish. *In* Young, D. (1973). *Developmental neurobiology of the arthropods.* Cambridge U.P.; pp. 203–232.

1411. HOY, R., BITTNER, C. D. & KENNEDY, D. (1967). Regeneration in crustacean motoneurones: evidence for axonal fusion. *Science, N.Y.*, **156**, 251–252.

1412. HSIE, A. W. & PUCK, T. T. (1971). Morphological transformation of Chinese hamster cells by dibutyryl adenosine cyclic 3′:5′-monophosphate and testosterone. *Proc, natn. Acad. Sci. U.S.A.*, **68**, 358–361.
1413. HSIA, J., SCHNEIDER, H. & SMITH, I. (1970). Spin label studies of oriented phospholipids: egg lecithin. *Biochim. biophys. Acta*, **202**, 399–402.
1414. HSIA, J., SCHNEIDER, H. & SMITH, I. (1970). A spin label study of the effects of cholesterol in liposomes. *Chem. Phys. Lipids*, **4**, 238–242.
1415. HSIA, J., SCHNEIDER, H. & SMITH, I. (1970). Spin label studies of oriented phospholipids: egg lecithin. *Chem. Phys. Lipids*, **4**, 120–125.
1416. HUANG, M., HO, A. K. S. & DALY, J. W. (1973). Accumulation of adenosine cyclic 3′5′monophosphate in rat cerebral cortical slices. *Molec. Pharmac.*, **9**, 711.
1417. HUANG, C. & THOMPSON, T. E. (1965). Properties of lipid bilayer membranes separating two aqueous phases: determination of membrane thickness. *J. molec. Biol.*, **13**, 183–193.
1418. HUBBARD, J. I. & KWANBUNBUMPEN, S. (1968). Evidence for the vesicle hypothesis. *J. Physiol.*, **194**, 407–422.
1419. HUBBARD, J. J. & WILLIS, W. D. (1962). Mobilization of transmitter by hyperpolarization. *Nature*, **193**, 174–175.
1420. HUBBARD, R. J. (1954). The molecular weight of rhodopsin-digitonin complex. *J. gen. Physiol.*, **37**, 381–398.
1421. HUBBARD, R. (1956). Retinine isomerase. *J. gen Physiol.*, **39**, 935–962.
1422. HUBBARD, R. (1958). The thermal stability of rhodopsin and opsin. *J. gen. Physiol.*, **42**, 259–270.
1423. HUBBARD, R. BOWNDS, D. & YOSHIZAWA, T. (1965). The chemistry of visual photoreception. *Cold Spring Harb. Symp. quant. Biol.*, **30**, 301–315.
1424. HUBBARD, R. & KROPF, A. (1958). The bleaching of rhodopsin. *Proc. natn. Acad. Sci. U.S.A.*, **44**, 130–135.
1425. HUBBARD, R. & WALD, G. (1952). *Cis-trans* isomers of Vitamin A and retinine in the rhodopsin system. *J. gen. Physiol.*, **36**, 269–274.
1426. HUBBARD, S. J. (1963). The electrical constants and component conductances of frog skeletal muscle after denervation. *J. Physiol.*, **165**, 443–456.
1427. HUBBELL, W. & McCONNELL, H. (1968). Spin-label studies of the excitable membranes of nerve and muscle. *Proc. natn. Acad. Sci. U.S.A.*, **61**, 12–16.
1428. HUBBELL, W. & McCONNELL, H. (1969). Motion of steroid spin labels in membranes. *Proc. natn Acad. Sci. U.S.A.*, **63**, 16–20.
1429. HUBBELL, W. & McCONNELL, H. (1969). Orientation and motion of amphiphilic spin labels in membranes. *Proc. natn. Acad. Sci. U.S.A.*, **64**, 20–27.
1430. HUBEL, D. H. & WIESEL, T. N. (1962). Receptive fields, binocular

interaction and functional architecture in the cat's visual cortex. *J. Physiol.*, **160**, 106–154.

1431. HUBEL, D. H. & WIESEL, T. N. (1963). Receptive fields of cells in striate cortex of very young, visually inexperienced kittens. *J. Neurophysiol.*, **26**, 994–1002.
1432. HUBEL, D. H. & WIESEL, T. N. (1965). Binocular interaction in striate cortex of kittens reared with artificial squint. *J. Neurophysiol.*, **28**, 1041–1059.
1433. HUBEL, D. H. & WIESEL, T. N. (1971). Abberant visual projections in the Siamese cats. *J. Physiol.*, **218**, 33–62.
1434. HUCHO, F. & CHANGEUX, J.-P. (1973). Molecular weight and quaternary structure of the cholinergic receptor protein extracted by detergents from *Electrophorus electricus* tissue. *FEBS Lett.*, **38**, 11–14.
1435. HUMPHREY, J. & DOURMASHKIN, R. (1965). In CIBA Fdn Symp. Complement. pp. 175–190.
1436. HUNEEUS, F. C. & DAVISON, P. F. (1970). Fibrillar proteins from squid axon. I. Neurofilament protein. *J. molec. Biol.*, **52**, 415–428.
1437. HUNEEUS, F. C. & DAVISON. P. F. (1970). Fibrillar proteins from squid axons. *J. molec. Biol.*, **52**, 429–439.
1438. von HUNGEN, K. & ROBERTS, S. (1973). Adenylate cyclase receptors for adrenergic neurotransmitters in rat cerebral cortex. *Europ. J. Biochem.*, **36**, 391–401.
1439. von HUNGEN, K., ROBERTS, S. & HILL, D. F. (1974). Development and regional variations in neurotransmitter-sensitive adenylcyclase systems in cell-free preparations from rat brain. *J. Neurochem.*, **22**, 811.
1440. HUNT, C. C. & NELSON, P. G. (1965). Structural and functional changes in the frog sympathetic ganglion following cutting of the presynaptic nerve fibres. *J. Physiol.*, **177**, 1–20.
1441. HUNT, C. C. & RIKER, W. K. (1966). Properties of frog sympathetic neurones in normal ganglia and after axon section. *J. Neurophysiol.*, **29**, 1096–1114.
1442. HUNT, R. K. & JACOBSON, M. (1970). Brain enhancement in tadpoles: increased DNA concentration after somatotropin or prolactin. *Science, N.Y.*, **170**, 342–344.
1443. HUNTINGTON, H. W. & TERRY, R. D. (1966). The origin of the reactive cells in cerebral stab wounds. *J. Neuropath. exp. Neurol.*, **25**, 646–653.
1444. HURNIK, J. F., BAILEY, E. D. & JEROME, F. N. (1973). Selection for divergent lines of mice based on their performance in a T-maze. *Behav. Genet.*, **3**, 45–55.
1445. HUXLEY, H. E. (1969). The mechanism of muscular contraction. *Science, N.Y.*, **164**, 1356–1366.
1446. HUXLEY, H. E. (1971). The structural basis of muscular contraction. *Proc. R. Soc. B*, **178**, 131–149.
1447. HUXLEY, H. E. (1973). Muscular contraction and cell motility. *Nature, Lond.*, **243**, 445–448.

1448. HYDEN, H. (1962). The neuron and its glia – a biochemical and functional unit. *Endeavor,* **21**, 144–155.
1449. HYDEN, H. (1967). Dynamic aspects of neuron-glia relationship. A study with microchemical methods. *In* Hyden, H., *The neuron.* Elsevier; Amsterdam. pp. 179–219.
1450. HYDEN, H. & LANGE, P. (1961). Differences in the metabolism of oligodendroglia and nerve cells in the vestibular area. *In* Kety, S. S. & Elkes, J., *Regional Neurochemistry.* Pergamon; N.Y. pp. 190–199.
1451. HYDEN, H. & LANGE, P. W. (1964). The steady state and endogenous respiration in neurons and glia. *Acta physiol. scand.,* **64**, 6–14.
1452. HYDEN, H. & LANGE, P. W. (1968). Protein synthesis in the hippocampal pyramidal cells of rats during a behavioural test. *Science, N.Y.,* **159**, 1370–1373.
1453. HYDEN, H. & LANGE, P. W. (1972). Protein synthesis in hippocampal nerve cells during re-reversal of handedness in rats. *Brain Res.,* **45**, 314–317.
1454. HYDEN, H., LØVTRUP, S. & PIGON, A. (1958). Cytochrome oxidase and succinoxidase activities in spinal ganglion cells and in glial capsule cells. *J. Neurochem.,* **2**, 304–311.
1455. HYDEN, H. & PIGON, A. (1960). A cytophysiological study of the functional relationship between oligodendroglial cells and nerve cells of Deiter's nucleus. *J. Neurochem.,* **6**, 52–72.
1456. IGGO, A. (1963). New specific sensory structures in hairy skin. *Acta neuroveg., Wien.,* **24**, 175–180.
1457. IGGO, A. (1963). An electrophysiological analysis of afferent fibres in primate skin. *Acta neuroveg., Wien.,* **24**, 225–240.
1458. IGGO, A. & MUIR, A. R. (1962). A cutaneous sense organ in hairy skin. *Excerpta Med. int. Congr. Ser.,* **48** (2), 1024.
1459. IIZUKA, E. & YANG, J. T. (1966). Optical rotatory dispersion and circular dichroism of the β-form of silk fibroin in solution. *Proc. natn. Acad. Sci. U.S.A.,* **55**, 1175–1178.
1460. IIZUKA, E. & YANG, J. T. (1968). The disordered and β-conformation of silk fibroin in solution. *Biochemistry,* **7**, 2218–2227.
1461. IKEDA, K. & CAMPBELL, J. B. (1971). Dorsal spinal nerve root regeneration verified by injection of leucine-^{3}H into the dorsal root ganglion. *Expl Neurol.,* **30**, 379–388.
1462. ILANI, A. & BERNS, D. S. (1972). Photoresponse of chlorophyll-containing bileaflet membranes and the effect of phycocyamine as extrinsic membrane protein. *J. mem. Biol.,* **8**, 333–356.
1463. ILLIANO, G., TELL, G. P. E., SIEGEL, M. I. & CUATRECASAS, P. (1973). Guanosine 3′5′-cyclic monophosphate and the action of insulin and acetylocholine. *Proc. natn. Acad. Sci. U.S.A.,* **70**, 2443–2447.
1464. INGRAM, V. M. (1969). A side view of moving fibroblasts. *Nature, Lond.,* **222**, 641–644.
1465. INOUE, H. & TIMASHEFF, S. N. (1968). The interaction of

beta-lactoglobulin with solvent components in mixed water – organic solvent systems. *J. Am. chem. Soc.*, **90**, 1890–1898.

1466. ISAACSON, A. & SANDOW, A. (1967). Caffeine effects on radiocalcium movement in normal and denervated rat skeletal muscle. *J. Pharmac. exp. Ther.*, **155**, 376–388.

1467. ISHIKAWA, H., BISCHOFF, R. & HOLTZER, H. (1969). Formation of arrowhead complexes with heavy meromyosin in a variety of cell types. *J. Cell Biol.*, **43**, 312–328.

1468. ITO, H. (1965). The neurosecretory apparatus in the ventricular wall of the reptilian brain. *J. Hirnforsch.*, **7**, 493–498.

1469. IVANOV, V. T., LAINE, I. A., ABDULAEV, N. D., SENYAVINA, L. B., POPV, E. M., OVCHINNOV, Y. A. & SHEMYAKIN, M. M. (1969). The physicochemical basis of the functioning of biological membranes. The conformation of valinomycin and its K^+ complex in solution. *Biochem. biophys. Res. Commun.*, **34**, 803–811.

1470. IWAYAMA, T. (1970). Ultrastructural changes in the nerves innervating the cerebral artery after sympathectomy. *Z. Zellforsch. mikrosk. Anat.*, **109**, 465–480.

1471. JACKLET, J. W. & COHEN, M. J. (1966). Nerve regeneration: correlation of electrical, histological and behavioural events. *Science, N.Y.*, **150**, 1640–1643.

1472. JACKLETT, J. W. & COHEN, M. J. (1967). Synaptic connections between a transplanted insect ganglion and muscle of the host. *Science, N.Y.*, **156**, 1638–1640.

1473. JACKLETT, J. W. & COHEN, M. J. (1967). Nerve regeneration: correlation of electrical, histological and behavioural events. *Science, N.Y.*, **156**, 1640–1643.

1474. JACOBSON, M. (1970). *Developmental neurobiology.* Holt, Rinehart and Winston; N.Y.

1475. JACOBSON, M. & GAZE, R. M. (1965). Selection of appropriate tectal connections by regenerating optic nerve fibres in adult goldfish. *Expl Neurol.*, **13**, 418–430.

1476. JACOBSON, M. (1969). Sex pheromone of the pink bollworm moth: biological masking by its geometrical isomer. *Science, N.Y.*, **163**, 190–191.

1477. JACQUE, C., BOURRE, J. M., MORENO, P. & BAUMANN, N. (1970). Etude des lipides cérébraux de la souris "quaking" au cours de la myélinisation. *C.r. hebd. Séanc. Acad. Sci., Paris, Ser. D,* **271**, 798–801.

1478. JACQUE, C. M., HARPIN, M. L. & BAUMANN, N. A. (1969). Brain lipid analysis of a myelin deficient mutant, the "quaking" mouse. *Europ. J. Biochem.*, **11**, 218–224.

1479. JAFFE, J. & SHARPLESS, S. (1967). Pharmacological denervation supersensitivity in the CNS: a theory of physical dependence. *In* Wickler, A. The addictive states. *Res. Publs Ass. Res. nerv. ment. Dis.*, **47**, 226.

1480. JAIN, M. K., WHITE, F. P., STRICKHOLM, A., WILLIAMS, E. & CORDES, E. H. (1972). Studies concerning the possible reconstitution of an active cation pump across an artificial membrane. *J. mem. Biol.*, **2**, 363–388.

1481. JAKINOVICH, W. & AGRANOFF, B. W. (1971). The stereospecificity of the inositol receptor of the silkworm *Bombyx Mori. Brain. Res.*, **33**, 173–180.
1482. JAKOUBEK, B. & EDSTRÖM, J. E. (1965). RNA changes in the Mauthner axon and myelin sheath after increased functional activity. *J. Neurochem.*, **12**, 845–849.
1483. JAKOUBEK, B. & SEMIGINOVSKY, B. (1970). The effect of increased functional activity on the protein metabolism of the nervous system. *Int. Rev. Neurobiol.*, **13**, 255–288.
1484. JAKOUBEK, B., SEMIGINOVSKY, B., KRAUS, M. & ERDOSSOVA, R. (1970). The alterations of protein metabolism of the brain cortex induced by anticipation stress and ACTH. *Life Sci.*, **9**, 1169–1179.
1485. JAMES, D. W. & TRESMAN, R. L. (1969). An electron microscopic study of the *de novo* formation of neuromuscular junctions in tissue culture. *Z. Zellforsch. mikrosk. Anat.*, **100**, 126–140.
1486. JANKOWSKA, E. & LINDSTRÖM, S. (1972). Morphology of interneurones mediating Ia recoprocal inhibition of motoneurones in the spinal cord of the cat. *J. Physiol.*, **226**, 805–823.
1487. JANSEN, J. K. S. & NICHOLLS, J. G. (1972). Regeneration and changes in synaptic connections between individual nerve cells in the central nervous system of the leech. *Proc. natn. Acad. Sci. U.S.A.*, **69**, 636–639.
1488. JARD, S., PREMONT, J. & BENDA, P. (1972). Adenylate cyclase, phosphodiesterase and protein kinase of rat glial cells in culture. *FEBS Lett.*, **26**, 344–348.
1489. JARLSTADT, J. & HAMBURGER, A. (1971). Patterns and labeling characteristics in neuronal and glial RNA. *J. Neurochem.*, **18**, 921–930.
1490. JARLSTEDT, J. & KARLSSON, J.-O. (1973). Evidence for axonal transport of RNA in mammalian neurones *Expl Brain Res.*, **16**, 501–606.
1491. JAYNES, J. (1958). Imprinting: the interaction of learned and innate behaviour. IV. Generalization and emergent discrimination. *J. comp. Physiol. Psychol.*, **51**, 238–242.
1492. JASINSKY, A., GORBMAN, A. & HARA, T. J. (1967). Rates of movement and redistribution of stainable neurosecretory granules in hypothalamic neurones. *Science, N.Y.*, **154**, 776–778.
1493. JEAN, D. H., GUTH, L. & ALBERS, R. W. (1973). Neural regulation of the structure of myosin. *Expl Neurol.*, **38**, 458–471.
1494. JENKINSON, D. H. (1960). The antagonism between tubocurarine and substances which depolarize the motor end plate. *J. Physiol.*, **152**, 309–324.
1495. JENKINSON, T. J., KAMAT, V. B. & CHAPMAN, D. (1969). Physical studies of myelin. II Proton magnetic resonance and infrared spectroscopy. *Biochim. biophys. Acta*, **183**, 427–433.
1496. JEWELL, P. A. & ZAIMIS, E. J. (1956). A differentiation between red

and white muscle in the cat based on responses to neuromuscular blocking agents. *J. Physiol.*, **124**, 417–428.

1497. JEWELL, P. A. & ZAIMIS, E. J. (1954). Changes at the neuromuscular junction of red and white muscle fibres in the cat induced by disuse atrophy and by hypertrophy. *J. Physiol.*, **124**, 429–442.

1498. JOHNS, T. R. & THESLEFF, S. (1961). Effects of motor inactivation on the chemical sensitivity of skeletal muscle. *Acta physiol. scand.*, **51**, 136–141.

1499. JOHNSON, E. M., MAENO, H. & GREENGARD, P. (1971). Phosphorylation of endogenous protein of rat brain by cyclic adenosine 3′5′monophosphate-dependent protein kinase. *J. biol. Chem.*, **246**, 7731–7739.

1500. JOHNSON, G. A., BOURKMA, S. J., LAHTI, R. A. & MATHEWS, J. (1972). Cyclic AMP and cylic nucleotide phosphodiesterase activity in synaptic vesicles. *Fedn Proc. Fedn Am. Socs exp. Biol.*, **31**, 513.

1501. JOHNSON, G. E. (1926). Studies on the function of the giant nerve fibres of crustaceans, with special reference to Cambarus and Palaemontes. *J. comp. Neurol.*, **42**, 19–33.

1502. JOHNSON, G. S., FRIEDMAN, R. M. & PASTAN, I. (1971). Restoration of several morphological characteristics of normal fibroblasts in sarcoma lines treated with adenosine 3′5′-cyclic monophosphate and its derivatives. *Proc. natn. Acad. Sci. U.S.A.*, **68**, 425–429.

1503. JOHNSON, P. N. & INESI, G. (1969). The effect of methylxanthines and local anaesthetics on fragments of sarcoplasmic reticulum. *J. Pharmac. exp. Ther.*, **169**, 308–314.

1504. JOHNSON, R. G. & SHERIDAN, J. (1971). Junctions between cancer cells in culture: ultrastructure and permeability. *Science, N.Y.*, **174**, 717–719.

1505. JOHNSTON, P. V. & ROOTS, B. I. (1972). *Nerve membranes. A study of the biological and chemical aspects of neuron-glia relationships.* Pergamon; Oxford. pp. 180–197.

1506. JONES, C. T. & BANKS, P. (1969). Inhibition of respiration by puromycin in slices of cerebral cortex. *J. Neurochem.*, **16**, 825–827.

1507. JONES, D. P. & SINGER, M. (1969). Neurotrophic dependence of the lateral-line sensory organs of the newt, *Triturus viridescens*. *J. exp. Zool.*, **171**, 433–442.

1508. JONES, R. & VRBOVA, G. (1970). Effect of muscle activity on denervation hypersensitivity. *J. Physiol.*, **210**, 144–145P.

1509. JONES, R. B. & NOWELL, N. W. (1973). The coagulating glands as a source of aversive and aggression-inhibiting pheromone(s) in the male albino mouse. *Physiol. Behav.*, **11**, 455–462.

1510. JONES, S. F. & KWANBUNBUMPEN, S. (1970). The effects of nerve stimulation and hemicholinium on synaptic vesicles at the mammalian neuromuscular junction. *J. Physiol.*, **207**, 31–50.

1511. JONES, S. F. & KWANBUNBUMPEN, S. (1970). Some effects of nerve stimulation and hemicholinium on quantal transmitter release at the

mammalian neuromuscular junction. *J. Physiol.*, **207**, 51–61.
1512. JONES, W. H. & THOMAS, D. B. (1962). Changes in dendritic organization of neurones in the cerebral cortex following deafferentation. *J. Anat.*, **96**, 375–381.
1513. JORDAN, H. (1925). The structure and staining reaction of the Reissner's fibre apparatus, particularly the subcommisural organ. *Am. J. Anat.*, **34**, 427–444.
1514. JOSEPH, B. S. (1973). Somatofugal events in Wallerian degeneration: a conceptual overview. *Brain Res.*, **59**, 1–18.
1515. JOSEPH, B. S. & WHITLOCK, G. G. (1972). The spatiotemporal course of Wallerian degeneration within the CNS of toads (Bufo marinus) as defined by the Nauta silver method. *Brain Behav. Evol.*, **5**, 1–17.
1516. JOSEPH, K. C. & HOGAN, E. L. (1971). Fatty acid composition of cerebrosides, sulfatides and ceramides in murine sudanophilic leucodystrophy: the 'Jimpy' mutant. *J. Neurochem.*, **18**, 1639–1645.
1517. JOSEPHSON, R. K. (1966). Mechanism of pacemaker and effector integration in coelenterates. *Symp. Soc. exp. Biol.*, **20**, 33–47.
1518. JOST, P., WAGGONER, A. S. & GRIFFITH, O. H. (1971). Spin labeling and membrane structure. *In* Rothfield, L. I., *Structure and function of biological membranes.* Acad. Press; N.Y. pp. 83–144.
1519. JUNGALWALA, F. B. & DAWSON, R. M. C. (1971). The turnover of myelin phospholipids in the adult and developing rat brain. *Biochem. J.*, **123**, 683–693.
1520. KADANOFF, D. (1971). Die Nervenfasergeflechte in den papillae vallatae et fungiformes der zunge bei Primaten. *Z. Zellforsch. mikrosk. Anat.*, **83**, 321–330.
1521. KAHAN, B. E., KRIGINAN, M. R., WILSON, J. E. & GLASSMAN, E. (1970). Brain function and macromolecules. VI. Autoradiographic analysis of the effect of a brief training experience on the incorporation of uridine into mouse brain. *Proc. natn. Acad. Sci. U.S.A.*, **65**, 300–304.
1522. KAHANE, I. & RAZIN, S. (1969). Synthesis and turnover of membrane protein and lipid in mycoplasma laidlawii. *Biochim. biophys. Acta,* **183**, 79–89.
1523. KAJIKAWA, K. & KAKIHARA, S. (1969). An electron microscope study of the basement membrane of proliferated bile ductules. *Expl molec. Path.*, **11**, 17–27.
1524. KAKIUCHI, S. & RALL, T. W. (1968). The influence of chemical agents on the accumulation of adenosine 3′5′-phosphate in slices of rabbit cerebellum. *Molec. Pharmac.*, 4, 369–378.
1525. KAKIUCHI, S. & RALL, T. W. (1968). Studies on adenosine 3′5′-monophosphate in rabbit cerebral cortex. *Molec. Pharmac.*, **4**, 379–388.
1526. KALISKER, A., RUTLEDGE, C. O. & PERKINS, J. P. (1973). Effect of nerve degeneration by 6-hydroxydopamine on catecholamine-stimulated 3′5′-monophosphate formation in rat cerebral cortex. *Molec. Pharmac.*, **9**, 619–633.

1527. KALMUS, H. & RIBBANDS, C. R. (1952). The origins of the odours by which honeybees distinguish their companions. *Proc. R. Soc. B,* **140**, 50–59.
1528. KAMRIN, R. P. & SINGER, M. (1953). Influence of sensory neurons isolated from central nervous system on maintainance of taste buds and regeneration of barbels in the catfish *Ameiurus nebulosus. Am. J. Physiol.,* **174**, 146–148.
1529. KANESEKI, T. & KADOTA, K. (1969). The 'vesicle in a basket'; a morphological study of the coated vesicle isolated from the nerve endings of the guinea pig brain, with special reference to the mechanism of membrane movements. *J. Cell Biol.,* **42**, 202–220.
1530. KANAZAWA, T., SAITO, M. & TONOMURA, Y. (1967). Properties of a phosphorylated protein as a reaction intermediate of Na^+-K^+ sensitive ATPase. *J. Biochem.,* **61**, 555–566.
1531. KANDEL, E. R. & TAUC, L. (1965). Heterosynaptic facilitation in neurones of the abdominal ganglion of *Aplysia depilans. J. Physiol.,* **181**, 1–27.
1532. KANDEL, E. R. & TAUC, L. (1965). Mechanisms of heterosynaptic facilitation in the giant cell of the abdominal ganglion of *Aplysia depilans. J. Physiol.,* **181**, 28–47.
1533. KANDUTSCH, A. A. & SAUCIER, S. E. (1969). Regulation of sterol synthesis in developing brains of normal and Jimpy mice. *Archs. Biochem. Biophys.,* **135**, 201–208.
1534. KANDUTSCH, A. A. & SAUCIER, S. E. (1972). Sterol and fatty acid synthesis in developing brains of three myelin-deficient mouse mutants. *Biochim. biophys. Acta,* **260**, 26–34.
1535. KAPELLER, K. & MAYOR, D. (1967). The accumulation of noradrenaline in constricted sympathetic nerves as studied by fluorescence and electron microscopy. *Proc. R. Soc. B,* **167**, 282–292.
1536. KAPELLER, K. & MAYOR, D. (1968). Accumulation of organelles distal to the site of constriction of postganglionic sympathetic nerves. *J. Physiol.,* **194**, 95–96P.
1537. KAPELLER, K. & MAYOR, D. (1969). An electron microscopic study of the early changes proximal to a constriction in sympathetic nerves. *Proc. R. Soc. B,* **172**, 39–51.
1538. KARAHASHI, Y. & GOLDRING, S. (1966). Intracellular potentials from 'idle' cells in cerebral cortex of cat. *Electroenceph. clin. Neurophysiol.,* **20**, 600–607.
1539. KARLIN, A. (1969). Chemical modification of the active site of the acetylcholine receptor. *J. gen. Physiol.,* **54**, 245–264S.
1540. KARLIN, A. & BARTELS, E. (1966). Effects of blocking sulfhydryl groups and of reducing disulfide bonds on the acetyl-choline-activiated system of the electroplax. *Biochim. biophys. Acta,* **126**, 525–535.
1541. KARLIN, A., PRIVES, J., DEAL, W. & WINNICK, M. (1971). Affinity labeling of the acetylcholine receptor in the electroplax. *J. molec. Biol.,* **61**, 175–188.

1542. KARLIN, A. & WINNICK, M. (1968). Reduction and specific alkylation of the receptor for acetylcholine. *Proc. natn. Acad. Sci. U.S.A.*, **60**, 668–674.
1543. KARLSON, U. (1967). Observations on the postnatal development of neuronal structures in the lateral geniculate nucleus of the rat by electron microscopy. *J. Ultrastruct. Res.*, **17**, 158–175.
1544. KARLSSON, J. O. & SJÖSTRAND, J. (1968). Transport of labeled proteins in the optic nerve and tract of the rabbit. *Brain Res.*, **11**, 431.
1545. KARLSSON, J.-O. & SJÖSTRAND. J. (1969). The effect of colchicine on the axonal transport of protein in the optic nerve and tract of the rabbit. *Brain Res.*, **13**, 617–619.
1546. KARLSSON, J.-O. & SJÖSTRAND, J. (1971). Transport of microtubular protein in axons of retinal ganglion cells. *J. Neurochem.*, **18**, 975–982.
1547. KARLSSON, J.-O. & SJÖSTRAND, J. (1972). Axonal transport of proteins in retinal ganglion cells. Amino acid incorporation into rapidly transported protein and distribution of radioactivity to the lateral geniculate body and the superior collicules. *Brain Res.*, **37**, 279–285.
1548. KARPATI, G. & ENGEL, W. K. (1967). Transformation of the histochemical profile of skeletal muscle by "foreign" innervation. *Nature, Lond.*, **215**, 1509–1510.
1549. KARPATI, G. & ENGEL, W. K. (1968). Correlative histochemical study of skeletal muscle after suprasegmental denervation, peripheral nerve section and skeletal fixation. *Neurology*, **18**, 681–692.
1550. KARPIAK, S. E., BOWEN, F. P. & RAPPORT, M. M. (1973). Epileptiform activity induced by antiserum to synaptic membrane. *Brain Res.*, **59**, 303–310.
1551. KARREMAN, G. (1973). Comparative specific adsorption. *Ann. N.Y. Acad. Sci.*, **204**, 393–409.
1552. KASAI, M. & CHANGEUX, J. P. (1971). *In vitro* excitation of purified membrane fragments by cholinergic agonists. I. Pharmacological properties of the excitable membrane fragements. *J. mem. Biol.*, **6**, 1–23
1553. KASAI, M. & GHANGEUX, J. P. (1971). *In vitro* excitation of purified membrane fragments by cholinergic agonists. II. The permeability change caused by cholinergic agonists. *J. mem. Biol.*, **6**, 24–57.
1554. KASAI, M. & CHANGEUX, J. P. (1971). *In vitro* excitation of purified membrane fragments by cholinergic agonists. III. Comparison of the dose-response curves to decamethonium with the corresponding binding curves of decamethonium to the cholinergic receptor. *J. mem. Biol.*, **6**, 58–80.
1555. KATCHALSKY, A. (1964). Polyelectrolytes and their biological interaction. *In* Katchalsky, A., *Connective tissue; intercellular macromolecules.* Little, Brown; Boston.
1556. KATO, T. & KUROKOWA, M. (1970). Studies on ribonucleic and homopoly ribonucleotide formation in neuronal glial and liver nuclei. *Biochem J.*, **116**, 599–609.
1557. KATO, G. YUNG, J. & IHNAT, M. (1970). Nuclear magnetic resonance

studies on acetylcholinesterase. *Molec. Pharmac.*, **6**, 588–596.

1558. KATZ, A. M. (1966). Purification and properties of a tropomyosin-containing fraction that sensitizes reconstituted actomyosin to calcium-binding agents. *J. biol Chem.*, **241**, 1522–1529.
1559. KATZMAN, R. (1961). Electrolyte distribution in mammalian central nervous system. Are glia high sodium cells? *Neurology, Minneap.*, **11**, 27–36.
1560. KATZMAN, R., BJORKLUND, A., OWMAN, C. H., STENEVI, U. & WEST, K. A. (1971). Evidence for regenerative axon sprouting of central catecholamine neurones in the rat mesencephalon. *Brain. Res.*, **25**, 579–596.
1561. KAUFFMAN, F. C. & ALBUQUERQUE, E. X. (1969). Substrate changes in degenerating extensor and soleus muscles in the rat. *Fedn Proc. Fedn Am. Socs exp. Biol.*, **28**, 541.
1562. KAUFMAN, F. C. & ALBUQUERQUE, E. X. (1970). Effect of ischaemia and denervation on metabolism of fast and slow mammalian skeletal muscle. *Expl Neurol.*, **28**, 46–63.
1563. KAUZMANN, W. (1959). Some factors in the interpretation of protein denaturation. *Adv. Protein Chem.*, **14**, 1–63.
1564. KAWIKITA, H. (1972). Immunochemical studies on the brain specific protein. *J. Neurochem.*, **19**, 87–93.
1565. KEATING, M. J. (1968). Functional interaction in the development of specific nerve connections. *J. Physiol.*, **198**, 75P.
1566. KEATING, M. J. & GAZE, R. M. (1970). The ipsilateral retinotectal pathway in the frog. *Q. Jl. exp. Physiol.*, **55**, 284.
1567. KEATING, M. J. & GAZE, R. M. (1970). The depth distribution of visual units in the contralateral optic tectum following regeneration of the optic nerve in the frog. *Brain Res.*, **21**, 197–206.
1568. KEBABIAN, J. W. & GREENGARD, P. (1971). Dopamine-sensitive adenyl cyclase: possible role in synaptic transmission. *Science, N.Y.*, **174**, 1346–1349.
1569. KEBABIAN, J. W., PETZOLD, G. L. & GREENGARD, P. (1972). Dopamine-sensitive adenylate cyclase in caudate nucleus of rat brain, and its similarity to the 'dopamine receptor'. *Proc. natn Acad. Sci. U.S.A.*, **69**, 2145–2149.
1570. KEFALIDES, N. A. (1968). Isolation and characterization of the collagen from glomerular basement membrane. *Biochemistry, N.Y.*, **7**, 3103–3112.
1571. KEFALIDES, N. A. (1973). Structure and biosynthesis of basement membranes. *Int. Rev. Conn. Tissue Res.*, **6**, 63–104.
1572. KEFALIDES, N. A. & WINZLER, R. J. (1966). The chemistry of glomerular basement membrane and its relation to collagen. *Biochemistry, N.Y.*, **5**, 702–713.
1573. KELLY, D. E. (1966). Fine structure of desmosomes, hemidesmosomes and an adepidermal globular layer in developing newt epidermis. *J. Cell. Biol.*, **28**, 51–72.

1574. KELTON, D. & RAUCH, H. (1962). Myelination and myelin degeneration in the central nervous system of dilute lethal mice. *Expl Neurol.*, **6**, 252–262.
1575. KEMP, G. & WENNER, C. E. (1972). Interaction of valinomycin with cations at the air-water interface. *Biochim. biophys. Acta*, **282**, 1–7.
1576. KEMP, N. E. (1959). Development of the basement lamellae of larval anuran skin. *Devl Biol.*, **1**, 459–476.
1577. KENNEDY, D. & BITTNER, G. D. (1974). Ultrastructural correlates of motor nerve regeneration in crayfish. *Cell Tissue Res.*, **148**, 97–110.
1578. KERKUT, G. A., SHAPIRA, A. & WALKER, R. J. (1967). The transport of labeled material from CNS ⇌ muscle along a nerve trunk. *Comp. Biochem. Physiol.*, **23**, 729–748.
1579. KEYNES, R. D. (1970). Evidence for structural changes during nerve activity and their relation to the conduction mechanism. *In* Schmitt, F. O., *The Neurosciences. Second study programme.* Rockefeller; N.Y. pp. 707–714.
1580. KEYNES, R. D. (1972). Excitable membranes. *Nature, Lond.*, **239**, 29–32.
1581. KEYNES, R. D., RITCHIE, J. M. & ROJAS, E. (1971). The binding of tetrodotoxin to nerve membranes. *J. Physiol.*, **213**, 235–254.
1582. KIERNAN, J. A. (1970). Two types of axonal regeneration in the neurohypophysis of the rat. *J. Anat.*, **107**, 187.
1583. KIERNAN, J. A. (1971). Pituicytes and the regenerative properties of neurosecretory and other axons in the rat. *J. Anat.*, **109**, 97–114.
1584. KIERNAN, J. A. & PETIT, D. R. (1971). Organ culture of the central nervous system of the adult rat. *Expl Neurol.*, **32**, 111–120.
1585. KIESOW, F. (1894). Beiträge zur physiologischen Psychologie des Geschmacksinnes. *Wunchs. Phil. Stud.*, **10**, 329–368.
1586. KILBOURN, B. T., DUNITZ, J. D., PIODA, L. A. R. & SIMON, W. (1967). Structure of K^+ complex of nonactin, a macrotetrolide antibiotic posessing highly specific K^+ transport properties. *J. molec. Biol.*, **30**, 559–563.
1587. KILHAM, L. & MARGOLIS, G. (1966). Viral etiology of spontaneous ataxis of cats. *Am. J. Path.*, **48**, 991–1011.
1588. KILKSON, R. (1969). Membrane structure and transitions. A molecular basis of regulation. *In* Enström, A. & Strandberg, B., *Symmetry and function of biological systems at the macromolecular level.* Wiley Interscience; N.Y. pp. 257–267.
1589. KIMURA, L. & BEIDLER, L. M. (1961). Microelectrode study of taste receptors of rat and hamster. *J. cell. comp. Physiol.*, **58**, 131–139.
1590. KIMURA, M. & KIMURA, I. (1973). Increase of nascent protein synthesis in neuromuscular junction of rat diaphragm induced by denervation. *Nature, Lond., New Biol.*, **241**, 114–115.
1591. KING, J. A. (1958). Parameters relevant to determining the effect of early experience upon the adult behaviour of animals. *Psychol. Bull.*, **55**, 46–58.

1592. KIRKLAND, W. L. & BURTON, P. R. (1972). Cyclic adenosine monophosphate-mediated stabilization of mouse neuroblastoma cell neurite microtubules exposed to low temperature. *Nature, Lond., New Biol.*, **240**, 205–207.

1593. KIRKPATRICK, J. B., BRAY, J. J. & PALMER, S. M. (1972). Visualization of axoplasmic flow *in vitro* by Nomarski microscopy. Comparison to rapid flow of radioactive proteins. *Brain Res.*, **43**, 1–10.

1594. KIRKPATRICK, J., HYAMS, L., THOMAS, J. & HOWLEY, P. (1970). Purification of intact microtubules from brain. *J. Cell Biol.*, **47**, 384–394.

1595. KIRSCHNER, D. A. & CASPAR, D. L. D. (1972). Comparative diffraction studies on myelin membranes. *Ann. N.Y. Acad. Sci.*, **195**, 309–320.

1596. KISHIMOTO, Y. (1971). Abnormality in sphingolipid fatty acids from sciatic nerve and brain of Quaking mice. *J. Neurochem.*, **18**, 1365–1368.

1597. KITCHEN, S. E. & WATTS, D. C. (1973). Comparison of turnover pattern of total and individual muscle proteins in normal mice and those with hereditary muscular dystrophy. *Biochem. J.*, **136**, 1017–1028.

1598. KITO, Y., AZUMA, M. & MAGDA, Y. (1968). Circular dichroism of squid rhodopsin. *Biochim. biophys. Acta*, **154**, 352–359.

1599. KIVALO, E., TALANTI, S. & RINNE, U. K. (1961). On the secretory phenomena in the subcommisural organ of the rat. Experimental studies with special reference to the possible relationship of the subcommisural organ to the hypothalamo-hypophyseal system. *Anat. Rec.*, **139**, 357–361.

1600. KJERULF, T. D. & LOESER, J. D. (1973). Neuronal hyperactivity following deafferentation of lateral cuneate nucleus. *Expl Neurol.*, **39**, 70–85.

1601. KJERULF, T. D., O'NEAL, J. T., CALVIN, W. H., LOESER, J. D. & WESTRUM, L. E. (1973). Deafferentation effects in lateral cuneate nucleus of the cat: correlation of structural alterations with firing pattern changes. *Expl Neurol.*, **39**, 86–102.

1602. KLAUS, W., LULLMANN, H. & MUSCHOLL, E. (1960). Potassium flux of normal and of denervated rat diaphragm. *Pflügers Arch. ges. Physiol.*, **271**, 761–775.

1603. KLINKERFUSS, G. H. (1964). An electron microscopic study of the ependyma and subependymal glia of the lateral ventricle of the cat. *Am. J. Anat.*, **115**, 71–100.

1604. KLIVINGTON, K. A. & GALAMBOS, R. G. (1967). Resistance shifts accompanying the evoked cortical response in the cat. *Science, N.Y.*, **157**, 211–213.

1605. KLIVINGTON, K. A. & GALAMBOS, R. G. (1968). Rapid resistance shifts in cat cortex during click-evoked responses. *J. Neurophysiol.*, **31**, 565–573.

1606. KNOTT, S., LEWIS, D. M. & LUCK, J. C. (1971). Motor unit areas in a cat limb. *Expl Neurol.*, **30**, 475–483.

1607. KOEHLER, J. K. (1968). Freeze etching on nucleated erythrocytes with special reference to the nuclear and plasma membranes. *Z. Zellforsch. mikrosk. Anat.*, **85**, 1–17.
1608. KOENIG, E. (1965). Synthetic mechanisms in the axon – II. RNA in myelin-free axons of the cat. *J. Neurochem.*, **12**, 357–361.
1609. KOENIG, E. (1967). Synthetic mechanisms in the axon – IV. *In vitro* incorporation of ^{3}H precursors into axonal protein and RNA. *J. Neurochem.*, **14**, 437.
1610. KOENIG, H. & BARRON, K. (1962). Morphologic and enzymic alterations of reacting glia in an experimental demyelinating lesion. *Anat. Rec.*, **142**, 249.
1611. KOENIG, H., BUNGE, R. P. & BUNGE, M. B. (1961). An autoradiographic study of nucleoprotein metabolism during demyelination and remyelination in the spinal cord of adult cats. *J. Neuropath. exp. Neurol.*, **20**, 279.
1612. KOENIG, H., BUNGE, M. B. & BUNGE, R. P. (1962). Nucleic acid and protein metabolism in white matter. Observation during experimental demyelination and remyelination. A histochemical and autoradiographic study of spinal cord of adult cat. *Archs. Neurol.*, **4**, 177.
1613. KOENIG, H. L., GIAMBERARDINO, L. D. & BENNETT, G. (1973). Renewal of protein and glycoprotein of synaptic constituents by means of axonal transport. *Brain Res.*, **62**, 413–417.
1614. KOEPPEN, A. H., BARRON, D. K. & MITZEN, E. J. (1971). Phospholipids, fatty acids and fatty aldehydes in rat brain subcellular fractions. *Brain Res.*, **35**, 199–214.
1615. KOHNO, K. (1969). Electron microscopic studies on Reissner's fibre and the ependymal cells in the spinal cord of the rat. *Z. Zellforsch. mikrosk. Anat.*, **94**, 565–573.
1616. KOIKE, H., EISENSTADT, M. & SCHWARTZ, J. H. (1972). Axonal transport of newly synthesized acetyl choline in an identified neuron of *Aplysia. Brain Res.*, **37**, 152–159.
1617. KOLAR, W., ZEMAN, W., DREW, A. & SPURGEON, C. (1968). Immunoglobulin synthesis in subacute sclerosing panencephalitis during X-irradiation of the brain. *Lancet*, **1**, 759.
1618. KOMIYA, Y. & AUSTIN, L. (1974). Axoplasmic flow of protein in the sciatic nerve of normal and dystrophic mice. *Expl Neurol.*, **43**, 1–12.
1619. KOMNICK, H., STOCKEM, W. & WOHLFARTH-BOTTERMAN, K. E. (1973). Cell motility: mechanisms in protoplasmic streaming and amoeboid movement. *Int. Rev. Cytol.*, **34**, 169–250.
1620. KÖNIGSMARK, B. W. & SIDMAN, R. L. (1963). Origin of brain macrophages in the mouse. *J. Neuropath. exp. Neurol.*, **22**, 643–676.
1621. KÖNIGSMARK, B. W. & SIDMAN, R. L. (1965). Response of astrocytes to brain injury. *J. Neuropath. exp. Neurol.*, **24**, 142.
1622. KONOPKA, R. J. & BENZER, S. (1971). Clock mutants of *Drosophila melanogaster. Proc. natn. Acad. Sci. U.S.A.*, **68**, 2112–2116.
1623. KOPEC, S. (1923). The influence of the nervous system on the

development and regeneration of muscles and integument in insects. *J. exp. Zool.*, **37**, 15–25.

1624. KORENBROT, J. I. & CONE, R. A. (1972). Dark ionic flux and the effects of light in isolated rod outer segments. *J. gen. Physiol.*, **60**, 20–45.
1625. KOREY, S. R. & ORCHEN, M. (1959). Relative respiration of neuronal and glial cells. *J. Neurochem.*, **3**, 277–285.
1626. KORN, E. D. & WEISSMAN, R. A. (1966). Loss of lipids during preparation of amoebae for electron microscopy. *Biochim. biophys. Acta*, **116**, 309–316.
1627. KORN, H. & BENNETT, M. V. L. (1972). Electrotonic coupling between teleost oculomotor neurons: restriction to somatic regions and functions of somatic and dendritic sites of impulse initiation. *Brain Res.*, **38**, 433–439.
1628. KORN, H. SOTELO, C. & CREPEL, F. (1973). Electrotonic coupling between neurones in the rat lateral vestibular nucleus. *Expl Brain Res.*, **16**, 255–275.
1629. KORNBERG, R. D. & McCONNELL, H. M. (1971). Inside – outside transitions of phospholipids in vesicle membranes. *Biochemistry*, **10**, 1111–1120.
1630. KORNBERG, R. D. & McCONNELL, H. M. (1971). Lateral diffusion of phospholipids in a vesicle membrane. *Proc. natn. Acad. Sci. U.S.A.*, **68** 2564–2568.
1631. KORNGOLD, L. (1956). The distribution of human tissue antigens in 5 human tumours grown in rats or hamsters. *Cancer Res.*, **16**, 956.
1632. KORNGUTH, S. E. & ANDERSON, J. W. (1965). Localization of a basic protein in the myelin of various species with the aid of fluorescence and electron microscopy. *J. Cell Biol.*, **26**, 157–166.
1633. KORNGUTH, S. E., ANDERSON, J. W. & SCOTT, G. (1966). Temporal relationship between myelinogenesis and the appearance of basic protein in the spinal cord of the white rat. *J. comp. Neurol.*, **127**, 1–18.
1634. KORNGUTH, S. E., KOZEL, L. R. & SMITHIES, O. (1972). Probable identity of tissue specific histone with encephalitogenic protein. *Nature, Lond., New Biol.*, **237**, 49–50.
1635. KORR, I. M. & APPELTAUER, G. S. L. (1971). Axonal transport of nerve cell proteins to muscle. *Fedn Proc. Fedn Am. Socs exp. Biol.*, **30**, 665.
1636. KORR, I. M. & APPELTAUER, G. (1974). The time course of axonal transport of neuronal proteins to muscle. *Expl Neurol.*, **43**, 452–463.
1637. KORR, H., SCHULTZE, B. & MAURER, W. (1975). Autoradiographic investigations of glial proliferation in the brain of adult mice. II. Cycle time and mode of proliferation of neuroglia and endothelial cells. *J. comp. Neurol.*, **160**, 477–490.
1638. KORR, I. M., WILKINSON, P. N. & CHORNOCK, F. W. (1967). Axonal delivery of neuroplasmic components to muscle cells. *Science, N.Y.*, **155**, 342–344.

1639. KOSUNEN, T. U., WAKSMAN, B. H., FLAX, M. H. & TIHEN, W. S. (1963). Radioautographic study of cellular mechanisms in delayed hypersensitivity. *Immunology,* **6**, 276–290.

1640. KOYAMA, N. & KURIHARA, K. (1971). Do unique proteins exist in taste buds? *J. gen. Physiol.,* **57**, 297–302.

1641. KOZAK, W. & WESTERMAN, R. A. (1961). Plastic changes of spinal monosynaptic responses from tenotomized muscles in cats. *Nature, Lond.,* **189**, 735–755.

1642. KRAUS-RUPPERT, R., LAISSUE, J., BÜRKI, H. & ODARTCHENKO, N. (1973). Proliferation and turnover of glial cells in the forebrain of young adult mice as studied by repeated injection of ^{3}H-thymidine over a prolonged period of time. *J. comp. Neurol.,* **148**, 211–216.

1643. KRAWIEC, L., ARGIZ, C. A. G., GOMEZ, C. J. & PASQUINI, J. M. (1969). Hormonal regulation of brain development. III. Effects of triiodothyronine and growth hormone on the biochemical changes in the cerebral cortex and cerebellum of neonatally thyroidectomized rats. *Brain Res.,* **15**, 209–218.

1644. KRECH, D., ROSENZWEIG, M. R. & BENNETT, E. L. (1962). Relations between brain chemistry and problem-solving among rats raised in enriched and impoverished environments. *J. comp. Physiol. Psychol.,* **55**, 801–807.

1645. KRECH, D., ROSENZWEIG, M. R. & BENNETT, E. L. (1963). Effects of complex environment and blindness on rat brain. *Archs. Neurol.,* **8**, 403–412.

1646. KRECHEVSKY, I. (1933). Hereditary nature of 'hypotheses', *J. comp. Psychol.,* **16**, 99–116.

1647. KREUTZIGER, G. O. (1968). Freeze-etching of intercellular junctions of mouse liver. *Proc. 26th Electron Microsc. Soc. Am. pp. 234.*

1648. KRIEBEL, M. E., BENNETT, M. V. L., WAXMAN, S. G. & PAPPAS, G. D. (1969). Oculomotor neurons in fish: electrotonic coupling and multiple sites of impulse initiation. *Science, N.Y.,* **166**, 520–524.

1649. KRISHNAN, A. & HSU, D. (1971). Binding of colchicine-^{3}H to vinblastine and vinblastine-induced crystals in mammalian tissue culture cells. *J. Cell Biol.,* **48**, 407–410.

1650. KRUGER, L. & MAXWELL, D. S. (1967). Comparative fine structure of vertebrate neuroglia: teleosts and reptiles. *J. comp. Neurol.,* **129**, 115–142.

1651. KRAMER, R., SCHLATTER, C. & ZAHLER, P. (1972). Preferential binding of sphingomyelin by membrane proteins of the sheep red cell. *Biochim. biophys. Acta,* **282**, 146–156.

1652. KRASNE, S., EISENMAN, G. & SZABO, G. (1971). Freezing and melting of lipid bilayers and the mode of action of nonactin, valinomycin and gramicidin. *Science, N.Y.* **174**, 412–415.

1653. KRECH, D., ROSENZWEIG, M. R. & BENNETT, E. L. (1962). Relation between brain chemistry and problem solving among rats raised in enriched and impoverished environments. *J. comp. Physiol. Psychol.,* **55**,

801–807.

1654. KREUTZBERG, G. W. (1967). Autoradiographic study of leucine-^{3}H in peripheral nerves during regeneration. *Experentia,* **23**, 33–34.

1655. KREUTZBERG. G. W. (1969). Neuronal dynamics and axonal flow, IV. Blockage of intraaxonal enzyme transport by colchicine. *Proc. natn. Acad. Sci. U.S.A.,* **62**, 722–728.

1656. KREUTZBERG, G. W. & SCHUBERT, P. (1971). Changes in axonal flow during regeneration of mammalian motor nerves. *Acta neuropath.,* Suppl. **5**, 70–75.

1657. KREUTZBERG, G. W., SCHUBERT, P., TOTH, L. & RIESKE, E. (1973). Intradendritic transport to postsynaptic sites. *Brain Res.,* **62**, 399–404.

1658. KRISTENSSON, K. & SJÖSTRAND, J. (1972). Retrograde transport of protein tracer in the rabbit hypoglossal nerve during regeneration. *Brain Res.,* **45**, 175–181.

1659. KRIVANEK, J. (1969). Changes in protein, electrolyte and water metabolism caused by prolonged spreading cortical depression in rats. *J. Neurobiol.,* **2**, 147–154.

1660. KRIVANEK, J. (1974). Changes in the neuronal protein labelling induced by potassium ions *in vivo* (spreading depression). *J. Neurochem.,* **23**, 1255.

1661. KRNJEVIC, K. (1974). Chemical nature of synaptic transmission in vertebrates. *Physiol. Rev.,* **54**, 418–540.

1662. KRNJEVIC, K. & SCHWARTZ, S. (1967). Some properties of unresponsive cells in the cerebral cortex of cat. *Expl Brain Res.,* **3**, 306–319.

1663. KROES, J., OSTWALD, R. & KEITH, A. (1972). Erythrocyte membranes – compression of lipid phases by increased cholesterol content. *Biochim. biophys. Acta,* **274**, 71–74.

1664. KRUGER, L. (1965). The spectrum of oligodendrocytes in normal rat cerebrum. *Anat. Rec.,* **151**, 375.

1665. KRUGER, L. & MAXWELL, D. S. (1966). Electron microscopy of oligodendrocytes in normal rat cerebrum. *Am. J. Anat.,* **118**, 411–436.

1666. KRUGER, L. & MAXWELL, D. S. (1966). The fine structure of ependymal processes in the teleost optic tectum. *Am. J. Anat.,* **119**, 479–497.

1667. KUBA, K. & KOKETSU, K. (1974). Ionic mechanism of sympathetic ganglion cells. *Brain Res.,* **81**, 338–342.

1668. KUBOTA, T. & KUBO, I. (1969). Bitterness and chemical structure. *Nature, Lond.,* **223**, 97–99.

1669. KUFFLER, S. W. (1967). Neuroglial cells: physiological properties and a potassium-mediated effect of neuronal activity on the glial membrane potential. *Proc. R. Soc. B,* **168**, 1–21.

1670. KUFFLER, S. W., DENNIS, M. J. & HARRIS, A. J. (1971). The development of chemosensitivity in extrasynaptic areas of the neuronal surface after denervation of parasympathetic ganglion cells in the heart of

the frog. *Proc. R. Soc. B,* **177**, 555–563.
1671. KUFFLER, S. W. & NICHOLLS, J. G. (1965). How do materials exchange between blood and nerve cells in the brain? *Perspect. biol. Med.,* **9**, 69–76.
1672. KUFFLER, S. W. & NICHOLLS, J. G. (1966). The physiology of neuroglial cells. *Ergebn. Physiol.,* **57**, 1–58.
1673. KUFFLER, S. W., NICHOLLS, J. G. & ORKAND, R. K. (1966). Physiological properties of glial cells in the central nervous system of amphibia. *J. Neurophysiol.,* **29**, 768–787.
1674. KUFFLER, S. W. & POTTER, D. D. (1964). Glia in the leech central nervous system. Physiological properties and neuron–glia relationship. *J. Neurophysiol.,* **27**, 290–320.
1675. KULENKAMPFF, H. (1952). Das Verhalten der Neuroglia in den Vorderhörnern des Rückenmarks der weissen Maus unter dem Reiz physiologischer Tatigkeit *Z. Anat. EntwGesch.,* **116**, 304–312.
1676. KULENKAMPFF, H. (1958). Untersuchungen zur Frage der Funktion des Ependyms im Zentralkanal des Rückenmarks der erwaschsenen weissen Maus. *Z. Anat. EntwGesch.,* **120**, 235–246.
1677. KUMAR, A. T. C. (1968). Sexual differences in the ependyma lining the third ventricle in the area of the anterior hypothalamus of the adult rhesus monkey. *Z. Zellforsch. mikrosk. Anat.,* **90**, 28–36.
1678. KUMAR, A. T. C. & KNOWLES, F. (1967). A system linking the third ventricle with the pars tuberalis of the rhesus monkey. *Nature, Lond.,* **215**, 54–55.
1679. KUMAR, A. T. C. & KNOWLES, F. (1968). Experimental modification of an area of specialized ependyma in the hypothalamus of the rhesus monkey. *Gen. comp. Endocr.,* **9**, 513.
1680. KUMAR, A. T. C. & THOMAS, G. H. (1968). Metabolites of ^{3}H-oestradiol-17β in the cerebrospinal fluid of the rhesus monkey. *Nature, Lond.,* **219**, 628–629.
1681. KUNO, M. & LLINAS, R. (1970). Alteration of synaptic action in chromatolyzed motoneurones of the cat. *J. Physiol.,* **210**, 823–838.
1682. KUNO, M. & LLINAS, R. (1970). Enhancement of synaptic transmission by dendritic potentials in chromatolyzed motoneurones of the cat. *J. Physiol.,* **210**, 807–821.
1683. KUNTZ, A. & SULKIN, N. M. (1947). The neuroglia in the autonomic ganglia. Cytologic structure and reaction to stimulation. *J. comp. Neurol.,* **86**, 467–477.
1684. KUNTZ, A. & SULKIN, N. M. (1947). Hyperplasia of peripheral neuroglia. A factor in pathological changes in autonomic ganglion cells. *J. Neuropath. exp. Neurol.,* **6**, 323–333.
1685. KUO, J. F. & KUO, W. N. (1973). Regulation by β-adrenergic receptor and muscarinic cholinergic receptor activation of intracellular cyclic AMP and cyclic GMP levels in rat lung slices. *Biochem. biophys. Res. Commun.,* **55**, 660–665.
1686. KUPFER, C. & PALMER, P. (1964). Lateral geniculate nucleus

histological and cytochemical changes following afferent denervation and visual deprivation. *Expl Neurol.*, **9**, 400–409.

1687. KURIHARA, K. (1969). Antisweet activity of gymnemic acid A_1 and its derivatives. *Life Sci.*, **8**, 537–543.
1688. KURIHARA, K. (1971). Taste modifiers. *In* Beidler, L. M., *Taste. Handbook of sensory physiology* IV, (2), 363–378.
1689. KURIHARA, K. (1972). Inhibition of cyclic 3′5′nucleotide phosphodiesterase in bovine taste papillae by bitter taste stimuli. *FEBS Lett.*, **27**, 279–281.
1690. KURIHARA, K. & BEIDLER, L. M. (1968). Taste-modifying protein from miracle fruit. *Science, N.Y.*, **161**, 1241–1243.
1691. KURIHARA, K. & BEIDLER, L. M. (1969). Mechanism of the action of taste modifying protein. *Nature, Lond.*, **222**, 1176–1179.
1693. KURIHARA, K. & KOYAMA, N. (1972). High activity of adenyl cyclase in olfactory and gustatory organs. *Biochem. biophys. Res. Commun.*, **48**, 30–34.
1694. KURIHARA, T., NUSSBAUM, J. L. & MANDEL, P. (1969). 2′3′-cyclic nucleotide 3′-phosphohydrolase in the brain of the 'Jimpy' mouse, a mutant with deficient myelination. *Brain Res.*, **13**, 401–403.
1695. KURIHARA, T., NUSSBAUM, J. L. & MANDEL, P. (1970). 2′3′-cyclic nucleotide 3′phosphohydrolase in brains of mutant mice with deficient myelination. *J. Neurochem.* **17**, 993–997.
1696. KURIHARA, T. & TSUKADA, Y. (1967). The regional and subcellular distribution of 2′3′-cyclic nucleotide 3′phosphohydrolase in the central nervous system. *J. Neurochem.*, **14**, 1167–1174.
1697. KURIHARA, T. & TSUKADA, Y. (1968). 2′3′-cyclic nucleotide 3′-phosphohydrolase in the developing chick brain and spinal cord. *J. Neurochem.*, **15**, 827–832.
1698. LAATSCH, R. H. (1969). Aggregates of irregular axoplasmic tubules in early Wallerian degeneration of guinea pig optic nerves. *Brain Res.*, **14**, 745–748.
1699. LADBROOKE, B. D. & CHAPMAN, D. (1969). Thermal analysis of lipids proteins and biological membranes. A review and summary of some recent studies. *Chem. Phys. Lipids*, **3**, 304–356.
1700. LADBROOKE, B. D., JENKINSON, T. J., KAMAT, V. B. & CHAPMAN, D. (1968). Physical studies of myelin. I. Thermal analysis. *Biochim. Biophys. Acta*, **164**, 101–109.
1701. LADBROOKE, B. D., WILLIAMS, R. M. & CHAPMAN, D. (1968). Studies on lecithin-cholesterol-water interaction by differential scanning calorimetry and X-ray diffraction. *Biochim. Biophys. Acta*, **150**, 333–340.
1702. LAGNADO, J. R., LYONS, C. A., WELLER, M. & PHILLIPSON, O. (1972). The possible significance of adenosine 3′5′cyclic monophosphate-stimulated protein kinase activity associated with purified microtubular protein preparations from mammalian brain. *Biochem. J.*, **128**, 95P.

1703. LAGNADO, J. R., LYONS, C. & WICKREMASINGHE, G. (1971). The subcellular distribution of colchicine-binding protein(s) ('microtubule protein') in rat brain. *Biochem. J.*, **122**, 56–57P.

1704. LAJTHA, A. (1964). Protein metabolism of nervous system. *Int. Rev. Neurobiol.*, **6**, 1–98.

1705. LAMB, A. (1974). The timing of the earliest motor innervation to the hind limb bud in the Xenopus tadpole. *Brain Res.*, **67**, 527–530.

1706. LAMPERT, P. W. & CRESSMAN, M. R. (1966). Fine structural changes in myelin sheaths after axonal degeneration in the spinal cord of rats. *Am. J. Path.*, **49**, 1139–1155.

1707. LANDE, S., FLEXNER, J. B. & FLEXNER, L. B. (1972). Effect of corticotropin and desglycinamide^{9}-lysine vasopressin on suppression of memory by puromycin. *Proc. natn. Acad. Sci. U.S.A.*, **69**, 558–560.

1708. LANDE, S., WITTER, A. & deWIED, D. (1971). Pituitary peptides. An octapeptide that stimulates conditioned avoidance acquisition in hypophysectomized rats. *J. biol. Chem.*, **246**, 2058–2062.

1709. LANGLEY, O. K. (1970). The interaction between peripheral nerve polyanion and alcian blue. *J. Neurochem.*, **17**, 1535–1541.

1710. LANGMAN, J. (1965). Cell proliferation in the developing spinal cord of the chick embryo examined by radioautography. *Anat. Rec.*, **151**, 375.

1711. LANGMAN, J. (1968). Histogenesis of the central nervous system. *In* Bourne, G. H., *The structure and function of nervous tissue.* Acad. Press; N.Y. pp. 33–66.

1712. LAPHAM, L. W. (1963). The occurrance of polyploidy in normal astrocytes of the Purkinje cell layer of the cerebellum. *J. Neuropath. exp. Neurol.*, **22**, 335.

1713. LAPHAM, L. W. & JOHNSTONE, M. A. (1963). Cytologic and cytochemical studies of neuroglia. II. The occurrance of two DNA classes among glial nuclei in the Purkinje cell layer of normal adult human cerebellar cortex. *Archs. Neurol.*, **9**, 194–202.

1714. LARSELL, 0. (1920). The cerebellum of amblystoma. *J. comp. Neurol.*, **31**, 259–282.

1715. LARSELL, O. (1925). The development of the cerebellum in the frog (Hyla Regilla) in relation to the vestibular and lateral line systems. *J. comp. Neurol.*, **39**, 249–289.

1716. LASEK, R. J. (1968). Axoplasmic transport in cat dorsal root ganglion cells: as studied with ^{3}H-L-leucine. *Brain Res.*, **8**, 360–377.

1717. LASEK, R. J. (1968). Axoplasmic transport of labelled proteins in rat ventral motoneurones. *Expl Neurol.*, **21**, 41–51.

1718. LASEK, R. J. (1970). Axonal transport of protein in dorsal root ganglion cells of the growing cat: a comparison of growing and mature neurons. *Brain Res.*, **20**, 121–126.

1719. LASEK, R. J. (1970). Protein transport in neurons. *Int. Rev. Neurobiol.*, **13**, 289–324.

1720. LASHER, R. S. & ZAGON, I. S. (1972). The effect of potassium on neuronal differentiation in cultures of dissociated newborn rat cerebellum. *Brain Res.*, **41**, 482–488.

1721. LASIEWSKI, R. C., GALEY, F. R. & VASQUEZ, C. (1965). Morphology and physiology of the pectoral muscles of humming birds. *Nature, Lond.*, **206**, 404.

1722. LAT, J., WIDDOWSON, E. M. & McCANCE, R. A. (1960). Some effects of accelerating growth. III. Behaviour and nervous activity. *Proc. R. Soc. B*, **153**, 347–356.

1723. LATORRE, R. EHRENSTEIN, G. & LECAR, H. (1972). Ion transport through excitability-inducing material (EIM) channels in lipid bilayer membranes. *J. gen. Physiol.*, **60**, 72–85.

1724. LAURENT, T. C. (1957). *Physical-chemical studies on hyaluronic acid.* Almqvist & Wiksell; Uppsala.

1725. LAURENT, T. C. (1966). *In vitro* studies on the transport of macromolecules through the connective tissue. *Fedn Proc. Fedn Am. Socs exp. Biol.*, **25**, 1128–1134.

1726. LAURENT, T. C. (1970). The structure and function of the intercellular polysaccharides in connective tissue. *In* Crone, C. & Thompson, A. M., *Capillary permeability.* Munksgaard; pp. 261–277.

1727. LAVAIL, J. H. & LAVAIL, M. M. (1972). Retrograde axonal transport in the central nervous system. *Science, N.Y.*, **176**, 1416–1417.

1728. LAVAIL, M., SIDMAN, R. L. & O'NEIL, D. (1972). Photoreceptor pigment epithelial cell relationship in rats with inherited retinal degeneration. *J. Cell Biol.*, **53**, 185–209.

1729. LAW, P. K. & ATWOOD, H. L. (1972). Cross-reinnervation of dystrophic mouse muscle. *Nature, Lond.*, **238**, 287–288.

1730. LAW, P. K. & ATTWOOD, H. L. (1972). Nonequivalence of surgical and natural denervation in dystrophic mouse muscle. *Expl Neurol.*, **34**, 200–209.

1731. LAW, P. K. & ATTWOOD, H. L. (1974). Does axonal sprouting occur in dystrophic mouse muscles? *Experentia*, **30**, 155–156.

1732. LAW, J. H. & REQUIER, F. E. (1971). Pheromones. *A. Rev. Biochem.*, **40**, 533–548.

1733. LEA, E. J. A. & GULIK-KRZYWICKI, T. (1972). Fluorescence in bimolecular phospholipid membranes. *Nature, Lond., New Biol.*, **237**, 95–96.

1734. LEATHERLAND, J. F. & DODD, J. M. (1968). Studies on the structure, ultrastructure and function of the subscommisural organ-Reissner's fibre complex of the European eel *Anguilla anguilla* L. *Z. Zellforsch. mikrosk. Anat.*, **89**, 533–549.

1735. LEBOWITZ, P. & SINGER, M. (1970). Neurotrophic control of protein synthesis in the regenerating limb of the newt *Triturus. Nature, Lond.*, **225**, 824–827.

1736. LECAR, H., EHRENSTEIN, G. & STILLMAN, I. (1971). Detection of molecular motion in lyophilized myelin by nuclear magnetic resonance. *Biophys. J.*, **10**, 140–145.

1737. LEE, C. J. (1974). Catecholamine-binding protein in mouse brain:

isolation and characterization. *Brain Res.*, **81**, 497.

1738. LEE, C. Y. & TSENG, L. F. (1966). Distribution of Bungarus multicinctus venom following evenomation. *Toxicology*, **3**, 281–290.
1739. LEE, C. Y., TSENG, L. F. & CHIU, T. H. (1967). Influence of denervation on localization of neurotoxins from clapid venoms in rat diaphragm. *Nature, Lond.*, **215**, 1177–1178.
1740. LEE, J. C. Y. (1963). Electron microscopy of Wallerian degeneration. *J. comp. Neurol.*, **120**, 65–71.
1741. LEE, P. A., BLASEY, K., GOLDSTEIN, I. J. & PIERCE, G. B. (1969). Basement membrane: carbohydrates and X-ray diffraction. *Expl molec. Path.*, **10**, 323–330.
1742. LEES, M. B. & SHEIN, H. M. (1970). Sodium and potassium content of normal and neoplastic rodent astrocytes in cell culture. *Brain Res.*, **23**, 280–283.
1743. LEE, T. P., KUO, J. F. & GREENGARD, P. (1972). Role of muscarinic cholinergic receptors in regulation of guanosine 3′5′-cyclic monophosphate content in mammalian brain, heart muscle and intestinal smooth muscle. *Proc. natn. Acad. Sci. U.S.A.*, **69**, 3287–3291.
1744. LEFKOWITZ, R. J. & HABER, E. (1971). A fraction of the ventricular myocardium that has the specificity of the cardiac beta adrenergic receptors. *Proc. natn Acad. Sci. U.S.A.*, **68**, 1773–1777.
1745. LEFKOWITZ, R. J., HABER, E. & O'HARA, D. (1972). Identification of the cardiac β-adrenergic receptor protein: solubilization and purification by affinity chromatography. *Proc. natn Acad. Sci. U.S.A.*, **69**, 2828–2832.
1746. LEHMANN, A. (1965). Contribution a l'étude psychophysiologique et neuropharmacologique de l'épilepsie acoustique de la souris et du rat. *J. Physiol. Paris*, **57**, 646–660.
1747. LEHMANN, A. (1967). Audiogenic seizures data in mice supporting new theories of biogenic amines mechanisms in the central nervous system. *Life Sci.*, **6**, 1423–1431.
1748. LEHRER, G. M., MAKER, H. S., SILIDES, D. J., WEISS, C. & SCHEINBERG, L. C. (1966). The quantitiative histochemistry of a chemically induced ependymoblastoma – I. Enzymes. *J. Neurochem.*, **13**, 1197–1206.
1749. LEITNER, J. M. & PERL, E. R. (1964). Receptors supplied by nerves which respond to cardiovascular changes and adrenaline. *J. Physiol.*, **175**, 254–274.
1750. LEMKEY-JOHNSTON, N. & LARRAMENDI, L. M. H. (1968). Morphological characteristics of mouse stellate and basket cells and their neuroglial envelope: an electron microscopic study. *J. comp. Neurol.*, **134**, 39–72.
1751. LENARD, J. & SINGER, S. J. (1966). Protein conformation in cell membrane preparations as studied by optical rotatory dispersion and circular dichroism. *Proc. natn. Acad. Sci. U.S.A.*, **56**, 1828–1835.
1752. LENNON, A. M., VERA, C. L., REX, A. L. & LUCR, J. V. (1967).

Cholinesterase activity of the nictitating membrane reinnervated by cholinergic fibres. *J. Neurophysiol.*, **30**, 1523–1530.

1753. LENTZ, T. L. (1963). Fine structure of the nervous system of hydra. *Anat. Rec.*, **145**, 334.

1754. LENTZ, T. L. (1965). Induction of supernumerary heads by isolated neurosecretory granules. *Science, N.Y.*, **150**, 633–635.

1755. LENTZ, T. L. (1966). Histochemical localization of neurohumors in a sponge. *J. exp. Zool.*, **162**, 171–180.

1756. LENTZ, T. L. (1971). Fine structure of nerves in the regenerating limb of the newt *Triturus. Am. J. Anat.*, **121**, 647.

1757. LENTZ, T. L. (1968). *Primitive nervous systems.* Yale Univ. Press; New Haven.

1758. LENTZ, T. L. (1969). Development of the neuromuscular junction. I. Cytological and cytochemical studies on the neuromuscular junction of differentiating muscle in regenerating limb of the newt *Triturus. J. Cell Biol.*, **42**, 431–443.

1759. LENTZ, T. L. (1970). Development of the neuromuscular junction. Cytochemical and cytological studies on the neuromuscular junction of dedifferentiating muscle in the regenerating limb of the newt *Triturus. J. Cell Biol.*, **47**, 423–436.

1760. LENTZ, T. L. (1971). Nerve trophic function: *in vitro* assay of effects of nerve tissue on muscle cholinesterase activity. *Science, N.Y.*, **171**, 187–189.

1761. LENTZ, T. L. (1972). Development of the neuromuscular junction. *J. Cell Biol.*, **55**, 93–103.

1762. LENTZ, T. L. (1972). Distribution of leucine-^{3}H during axoplasmic transport within regenerating neurons as determined by electron-microscopic autoradiography. *J. Cell Biol.*, **52**, 719–732.

1763. LENTZ, T. L. & Wood, J. G. (1963). Amines in the nervous system of coelenterates. *J. Histochem. Cytochem.*, **12**, 37P.

1764. LEONHARDT, H. (1966). Über ependymale Tanycyten des III Ventrikels beim Kaninchen in electroenmikroskopischen Betrachtung. *Z. Zellforsch. mikrosk. Anat.*, **74**, 1–11.

1765. LEONHARDT, H. (1968). Bukettförmige Strukturen im Ependym der Regio hypothalamica des III Ventrikels beim Kaninchen. Zur Neurosekretions und Rezeptorenfrage. *Z. Zellforsch. mikrosk. Anat.*, **88**, 297–317.

1766. LEONHARDT. H. (1968). Intraventrikuläre markhaltige Nervenfasern nahe der Apertura lateralis ventriculi quarti des Kaninchengehirns. *Z. Zellforsch. mikrosk. Anat.*, **84**, 1–8.

1767. LEONHARDT, H. & LINDNER, E. (1967). Marklose Nervenfasern im III und IV Ventrikel des Kaninchen- und Katzengehirns. *Z. Zellforsch. mikrosk. Anat.*, **78**, 1–18.

1768. LEONHARDT, H. & PRIEN, H. (1968). Eine weitere Art intraventrikulärer kolbenförmiger Axonendigung aus dem IV Ventrikel des Kaninchengehirns. *Z. Zellforsch. mikrosk. Anat.*, **92**, 394–399.

1769. LEONINI, J. (1969). Autoradiographic studies on incorporation of S^{35}-cysteine in the secretion of the subcommisural organ of white mouse. *Acta physiol. pol.*, **20**, 329–334.
1770. LESCH, M., PARMLEY, W. W., HAMOSH, M., KAUFMAN, S. & SONNENBLICK, E. J. (1969). Effects of acute hypertrophy on the contractile properties of skeletal muscle. *Am. J. Anat.*, **214**, 685–690.
1771. LESH, G. E. & BURNETT, A. L. (1966). An analysis of the chemical control of the polarized form in Hydra. *J. exp. Zool.*, **163**, 55–78.
1772. LESSLAUER, W. & BLASIE, J. K. (1972). Direct determination of the structure of barium stearate multilayers by X-ray diffraction. *Biophys. J.*, **12**, 175–190.
1773. LEA, A. A. & BUZINSKY, E. P. (1967). Cation specificity of model bimolecular phospholipid membranes with incorporated valinomycin. *Cytology*, **9**, 1–8.
1774. LEVEQUE, T. F. & SMALL. M. (1959). The relationship of the pituicyte to the posterior lobe hormones. *Endocrinology*, **65**, 909–915.
1775. LEVEY, G. S. (1971). Restoration of glucagon-responsiveness of solubilized myocardial adenyl cyclase by phosphatidyl serine. *Biochem. biophys. Res Commun.*, **43**, 108–113.
1776. LEVEY. G. S. (1971). Restoration of norepinephrine responsiveness of solubilized myocardial adenylate cyclase by phosphatidyl inositol. *J. biol. Chem.*, **246**, 7405–7407.
1777. LEVEY, G. S. (1973). The role of phospholipids in hormone activation of adenylate cyclase. *In* Greep, R. O., *Recent progress in hormone research.* Acad. Press; N.Y. pp. 361–382.
1778. LEVI, G. & MEYER, H. (1945). Reactive, regressive and regenerative processes of neurons, cultivated *in vitro* and injured with a micromanipulator. *J. exp. Zool.*, **99**, 141–181.
1779. LEVIN, E. & SCICLI, G. (1969). Brain barrier phenomena *Brain Res.*, **13**, 1–12.
1780. LEVIN, & FENSTERMACHER, J. D. (1969). Cortical ECS and diffusion profiles in cat, dog and monkey perfused through the subarachnoid space. *Fedn Proc. Fedn Am. Socs exp. Biol.*, **28**, 578.
1781. LEVINE, R., HECHTER, O. & SOSKIN, S. (1941). Biochemical characteristics of denervated skeletal muscle at rest and after direct stimulation. *Am. J. Physiol.*, **132**, 336–345.
1782. LEVINE, S. (1959). The effects of differential infantile stimulation on emotionality at weaning. *Can. J. Psychol.*, **13**, 243–247.
1783. LEVINE, S. & BRUSH, R. F. (1967). Adrenal cortical activity and avoidance learning as a function of time after avoidance training. *Physiol. Behav.*, **2**, 385–388.
1784. LEVINE, Y. K., BAILEY, A. I. & WILKINS, M. H. F. (1968). Multilayers of phospholipid bimolecular leaflets. *Nature, Lond.*, **220**, 577.
1785. LEVINE, Y. K. & WILKINS, M. H. F. (1971). Structure of oriented lipid bilayers. *Nature, Lond., New Biol.*, **230**, 69–72.
1786. LEVISON, S. R. & ELLORY, J. C. (1973). Molecular size of the

tetrodotoxin binding site estimated by irradiation inactivation. *Nature, Lond., New Biol.*, **245**, 122–123.

1787. LEVITAN, H., TAUC, L. & SEGUNDO, J. P. (1970). Electrical transmission among neurons in the buccal ganglion of a mollusc, *Nevanux inermis. J. gen. Physiol.*, **55**, 484–496.

1788. LEVITAN, I. B., MUSHYNSKI, W. E. & RAMIREZ, G. (1972). Effects of an enriched environment on amino acid incorporation into rat brain subcellular fractions *in vivo. Brain Res.*, **41**, 498–502.

1789. LEVITAN, I. B., MUSHNSKI, W. E. & RAMIREZ, G. (1972). Effects of environmental complexity on amino acid incorporation into rat cortex and hippocampus *in vivo. J. Neurochem.*, **19**, 2261–2630.

1790. LEVITAN, I. B., RAMIREZ, G. & MUSHYNSKI, W. E. (1972). Amino acid incorporation in the brain of rats trained to use the non-preferred paw in retrieving food. *Brain Res.*, **47**, 147–156.

1791. LEVY, M. G., TUNIS, M. & ISSEROFF, H. (1973). Population control in snails by natural inhibitors. *Nature, Lond.*, **241**, 65–66.

1792. LEWIS, D. M. (1972). The effect of denervation on the mechanical and electrical responses of fast and slow mammalian twitch muscle. *J. Physiol.*, **222**, 51–75.

1793. LEWIS, D. M., LUCK, J. C. & KNOTT, S. (1972). A comparison of isometric contractions of the whole muscle with those of motor units in a fast-twitch muscle of the cat. *Expl Neurol.*, **37**, 68–85.

1794. LEWIS, D. V. & SCHUETTE, W. H. (1975). NADH fluorescence and $(K)_0$ changes during hippocampal electrical stimulation. *J. Neurophysiol.*, **38**, 405–417.

1795. LEWIS, P. R. & SHUTE, C. C. D. (1967). The cholinergic limbic system. Projection to hippocampal formation, medial cortex, nuclei of ascending cholinergic reticular system and the subfornical organ and supraoptic crest. *Brain*, **90**, 521–537.

1796. LI, C. L. (1959). Cortical intracellular potentials and their responses to strychnine. *J. Neurophysiol.*, **22**, 436–450.

1797. LIBERTINI, L., WAGGONER, A., JOST, P. & GRIFFITH, O. (1969). Orientation of lipid spin labels in lecithin multilayers. *Proc. natn. Acad. Sci. U.S.A.*, **64**, 13–19.

1798. LIBET, B. (1964). Slow synaptic responses and excitatory changes in sympathetic ganglia. *J. Physiol.*, **174**, 1–25.

1799. LIBET, B. (1970). Generation of slow inhibitory and excitatory postsynaptic potentials. *Fedn Proc. Fedn Am. Socs exp. Biol.*, **29**, 1945–1956.

1800. LIEBERMAN, A. R. (1973). Some factors affecting retrograde neuronal responses to axonal lesions. *In* Bellairs, R. & Gray, E. G., *Essays on the nervous system.* Clarendon Press; Oxford. pp. 71–105.

1802. LIEBERMAN, A. R. (1971). The axon reaction: a review of the principal features of perikaryal responses to injury. *Int. Rev. Neurobiol.*, **14**, 49–124.

1804. LIEBERMAN, F. A. & TOPALY, V. P. (1968). Selective transport of ions through bimolecular phospholipid membranes. *Biochim. biophys. Acta,* **163**, 125–126.

1805. LIEBMAN, P. A. (1962). In situ microspectrophotometric studies on the pigments of single retinal rods. *Biophys. J.,* **2**, 161–178.
1806. LIEBMAN, P. A., JAGGER, W. S., KAPLAN, M. W. & BARGOOT, F. G. (1974). Membrane structure changes in rod outer segments associated with rhodopsin bleaching. *Nature, Lond.,* **251**, 31–36.
1807. LIEBMAN, P. A., RICE, R., CARROLL, S., ENTINE, G. & LATIES, A. (1967). Sensitive low light level microspectrophotometer; detection of photosensitive pigments of retinal cones. *Invest. Ophthal.,* **6**, 214–218.
1808. LIGHTBODY, J., PFEIFFER, S. E., KORNBLITH, P. L. & HERSCHMAN, H. R. (1970). Biochemically differentiated clonal human glial cells in tissue culture. *J. Neurobiol.,* **1**, 411–417.
1809. LILEY, A. W. (1956). The effects of presynaptic polarization on the spontaneous activity at the mammalian neuromuscular junction. *J. Physiol.,* **134**, 427–443.
1810. LIN, H. & INGRAM, W. R. (1973). Axonal degeneration in the peripheral optic pathway of the cat. *Expl Neurol.,* **39**, 234–248.
1811. LINDEMAN, B. & SOLOMAN, A. K. (1961). Permeability of luminal surface of intestinal mucosal cells. *J. gen. Physiol.,* **45**, 801–801.
1812. LINDSTROM, J. & PATRICK, J. (1973). Autoimmune response to acetylcholine receptor. *Science, N.Y.,* **180**, 871–872.
1813. LINDZEY, G., LOEHLIN, J., MANOSEVITZ, M. & THIESSEN, D. (1971). Behavioural genetics. *A. Rev. Psychol.,* **22**, 39–94.
1814. LIPPERT, J. L. & PETICOLAS, W. L. (1971). Laser Raman investigation of the effect of cholesterol on conformational changes in dipalmitoyl lecithin multilayers. *Proc. natn. Acad. Sci. U.S.A.,* **68**, 1572–1576.
1815. LIPPERT, J. L. & PETICOLAS, W. L. (1972). Raman active vibrations in long-chain fatty acids and phospholipid sonicates. *Biochim. biophys. Acta,* **282**, 8–17.
1816. LITCHY, W. J. (1973). Uptake and retrograde transport of horeseradish peroxidase in frog sartorius nerve *in vitro. Brain Res.,* **56**, 377–381.
1817. LIU, C. N. (1955). Time pattern in retrograde degeneration after trauma of central nervous system of mammals. *In* Windle, W. F., *Regeneration in the central nervous system.* C. C. Thomas; Springfield, Illinois. pp. 84–93.
1818. LIU, C. N. & LIU, C. Y. (1971). Role of afferents in maintainance of dendritic morphology. *Anat. Rec.,* **169**, 369–375.
1819. LIU, H. M. (1974). Schwann cell properties. II. The identity of phagocytes in the degenerating nerve. *Am. J. Path.,* **75**, 395–416.
1820. LIU, H. M. (1973). Schwann cell properties, I. Origin of Schwann cell during peripheral nerve regeneration. *J. Neuropath. exp. Neurol.,* **32**, 458.
1821. LIVETT, B. G., GEFFEN, L. B. & AUSTIN, L. (1968). Proximodistal transport of ^{14}C-noradrenaline and protein in sympathetic nerves. *J. Neurochem.,* **15**, 931–939.
1822. LOCKSHIN, R. A. & WILLIAMS, C. M. (1965). Programmed cell death.

III. Neural control of the breakdown of the intersegmental muscles of silkmoths. *J. Insect Physiol.*, **11**, 601–610.

1823. LOESER, J. & WARD, A. (1967). Some effects of deafferentation on neurons of the cat spinal cord. *Archs. Neurol. Psychiat., Chicago,* **17**, 629–636.

1824. LOEWENSTEIN, W. R. (1966). Permeability of membrane junctions. *Ann. N.Y. Acad. Sci.*, **137**, 441–449.

1825. LOEWENSTEIN, W. R. & ALTAMIRANO-ORREGO, R. (1956). Enhancement of activity in a Pacinian corpuscle by sympathomimetic agents. *Nature, Lond.*, **178**, 1292–1293.

1826. LOEWENSTEIN, W. R., GOTO, K. & NABACK. C. (1962). C fibre innervation of a mechanoreceptor. *Experentia,* **18**, 460.

1827. LOEWENSTEIN, W. R. & KANNO, Y. (1964). Studies on an epithelial (gland) cell junction. *J. Cell Biol.*, **22**, 565–586.

1828. LOEWENSTEIN, W. R., SOCOLAR, S. J., HIGASHINO, S., KANNO, Y. & DAVIDSON, N. (1965). Intercellular communication: renal, urinary bladder, sensory and salivary gland cells. *Science, N.Y.*, **149**, 295–298.

1829. LOEWY, A. D. (1973). Transneuronal changes in the gracile nucleus. *J. comp. Neurol.*, **174**, 497–510.

1830. LÖFGREN, B. (1951). The electrical impedance of a complex tissue and its relation to changes in volume and fluid distribution. A study on rat kidneys. *Acta physiol. scand.*, **23**, Suppl. 81.

1831. LOGAN, W. J. & SNYDER, S. H. (1971). Unique high affinity uptake systems for glycine, glutamic and aspartic acids in central nervous tissue of the rat. *Nature, Lond.*, **234**, 297–299.

1832. LONG, R. A., HRUSKA, F., GESSER, H. D., HSIA, J. C. & WILLIAMS, R. (1970). Membrane condensing effect of cholesterol and the role of its hydroxyl group. *Biochem. biophys. Res. Commun.*, **41**, 321–327.

1833. LONGNECKER, H. E., HURLBUT, W. P., MAURO, A. & CLARK, A. W. (1970). Effects of black widow spider venom on the frog neuromuscular junction. Effects on end-plate potential, miniature end-plate potential and nerve terminal spike. *Nature, Lond.*, **225**, 701.

1834. LONGO, A. M. & PENHOET, E. E. (1974). Nerve growth factor in rat glioma cells. *Proc. natn. Acad. Sci. U.S.A.*, **71**, 2347–2349.

1835. van der LOOS, H. (1965). The 'improperly' oriented pyramidal cell in the cerebral cortex and its possible bearing on problems of neuronal growth and cell orientation. *Bull. Johns Hopkins Hosp.*, **117**, 228–250.

1836. van der LOOS, H. & WOOLSEY, T. A. (1973). Somatosensory cortex: structural alterations following early injury to sense organs. *Science, N.Y.*, **179**, 395–397.

1837. LORENZ, K. (1935). Der Kumpan in der Unweit des Vogels. *J. Orn., Lpz.*, **83**, 137–213, 289–413.

1838. de LORENZO, A. J. D., BRZIN, M. & DETTBARN, W. D. (1968). Fine structure and organization of nerve fibers and giant axons in *Homanus americanus. J. Ultrastruct. Res.*, **24**, 367–384.

1839. LOWEY, S. & RISBY, D. (1971). Light chains from fast and slow muscle myosins. *Nature, Lond.*, **234**, 81–85.

1840. LUBINSKA, L., NIEMIERKO, S., ODERFELD, B. & SZWARC, L. (1964). Behaviour of acetylcholinesterase in isolated nerve segments. *J. Neurochem.*, **11**, 493–503.
1841. LUCO, J. V. & EYZAGUIRRE, C. (1955). Relation of fibrillation to acetylcholine and potassium sensitivity in denervated skeletal muscle. *J. Neurophysiol.*, **5**, 357–362.
1842. LUCY, J. A. (1964). Globular lipid micelles and cell membranes. *J. theor. Biol.*, **7**, 360–373.
1843. LUCY, J. A. (1968). Theoretical and experimental models for biological membranes. *In* Chapman, D., *Biological membranes, physical fact and function.* Acad. Press; London. pp. 233–285.
1844. LUCY, J. A. & GLAUERT, A. M. (1964). Structure of macromolecular lipid complexes composed of globular micelles. *J. molec. Biol.*, **8**, 727–740.
1845. LUDUENA, R. & WOODWARD, D. O. (1972). Microheterogeneity in the aminoacid sequence of microtubule protein. *Fedn Proc. Fedn Am. Socs exp. Biol.*, **31**, 882.
1846. LUDUENA, R. F. & WOODWARD, D. O. (1973). Isolation and partial characterization of α- and β-tubulin from outer doublets of sea-urchin sperm and microtubules of chick-embryo brain. *Proc. natn. Acad. Sci. U.S.A.*, **70**, 3594–3598.
1847. LUFT, J. H. (1966). Fine structure of nerve and muscle cell membrane permeability to ruthenium red. *Anat. Rec.*, **154**, 379.
1848. LUMSDEN, C. E., ROBERTSON, D. M. & BLIGHT, R. (1966). Chemical studies on experimental allergic encephalomyelitis. Peptide as the common denominator in all encephalitogenic 'antigens'. *J. Neurochem.*, **13**, 127–162.
1849. LUND, R. D. & LUND, J. S. (1971). Synaptic adjustment after deafferentation of the superior colliculus of the rat. *Science, N.Y.*, **171**, 804–807.
1850. LUND, R. D. & LUND, J. S. (1971). Modification of synaptic patterns in the superior colliculus of the rat during development and following deafferentation. *Vision Res.*, **11**, Suppl. 3, 281–298.
1851. LUNT, G. G., de ROBERTIS, E. & STEFANI, E. (1971). Proteolipid metabolism in denervated muscle. *Biochem. J.*, **121**, 23–24P.
1852. LUNT, G. G., STEFANI, E. & de ROBERTIS, E. (1971). Increased incorporation of G-^{3}H-leucine into a possible 'receptor' proteolipid in denervated muscle *in vivo*. *J. Neurochem.*, **18**, 1545–1553.
1853. LUSE, S. A. & McCAMAN, R. E. (1957). Electron microscopy and biochemistry of Wallerian degeneration in the optic and tibial nerves. *Am. J. Path.*, **33**, 586.
1853. LUTTGAU, H. D. (1958). Sprunghafte Schwankungen unter schwelliger potentiale an markhaltigen Nervenfasern. *Z. Naturf.*, **13b**, 692.
1854. LUX, H. D., SCHUBERT, P., KREUTZBERG, G. W. & GLOBUS, A. (1970). Excitation and axonal flow: autoradiographic study on

motoneurones intracellularly injected with a ^{3}H-amino acid. *Expl Brain Res.*, **10**, 197–204.

1855. LUZZATI, V. (1968). X-ray diffraction studies of lipid-water systems. *In* Chapman, D. *Biological membranes, physical fact and function.* Acad. Press. N.Y. pp. 71–123.

1856. LUZZATI, V., GULIK-KRZYWICKI, T., RIVAS, E., REISS-HUSSON, F. & RAND, R. P. (1968). X-ray study of model systems: structure of the lipid-water phases in correlation with the chemical composition of the lipids. *J. gen. Physiol.*, **51**, 37–43s.

1857. LYNCH, G. S., MATTHEWS, D. A., MOSKO, S., PARKS, T. & COTMAN, C. (1972). Induced acetylcholinesterase-rich layer in rat dentate gyrus following entorhinal lesions. *Brain Res.*, **42**, 311–318.

1858. LYNCH, G. S., MOSKO, S., PARKS, T. & COTMAN, C. W. (1973). Relocation and hyperdevelopment of the gyrus commisural system after entorhinal lesions in immature rats. *Brain Res.*, **50**, 174–178.

1859. LYSER, K. M. (1968). Early differentiation of motor neuroblasts in the chick embryo as studied by electron microscopy. II. Microtubules and neurofilaments. *Devl Biol.*, **17**, 117–124.

1860. LYSER, K. M. (1971). Microtubules and filaments in developing axons and optic stalk cells. *Tiss. Cell,* **3**, 395–404.

1861. LØMO, T. (1971). Potentiation of monosynaptic EPSPs in the perforant path – dentate granule cell synapse. *Expl Brain Res.*, **12**, 46–63.

1862. LØMO, T. (1971). Pattern of activation in a monosynaptic cortical pathway: the perforant path input to the dentate area of the hippocampal formation. *Expl Brain Res.*, **12**, 18–45.

1863. LØMO, T. & ROSENTHAL, J. (1972). Control of ACh sensitivity by muscle activity in the rat. *J. Physiol.*, **221**, 493–513.

1864. LØVTRUP-REIN, H. & McEWEN, B. S. (1966). Isolation and fractionation of rat brain nuclei. *J. Cell Biol.*, **30**, 405–415.

1865. LØVTRUP-REIN, H. & GRAHN, B. (1970). Newly synthesized RNA in nuclei isolated from nerve and glial cells. *J. Neurochem.*, **17**, 845–852.

1866. van MAANEN, E. F. (1950). The antagonism between acetylcholine and the curare alkaloids, d-tubocurarine, c-tubocurarine-I, c-toxiferine-II, and β-erythroidine in the rectus abdominis of the frog. *J. Pharmac. exp. Ther.*, **99**, 255–264.

1867. McAFEE, D. A., SCHORDERET, M. & GREENGARD, P. (1971). Adenosine 3′5′-monophosphate in nervous tissue: increase associated with synaptic transmission. *Science, N.Y.*, **171**, 1156–1158.

1868. McCAMAN, R. E. & ROBERTS, E. (1959). Quantitative biochemical studies of Wallerian degeneration in the peripheral and central nervous systems – II. *J. Neurochem.*, **5**, 32–42.

1869. McCOMAS, A. J. & MROZEK, K. (1967). Denervated muscle fibres in hereditary mouse dystrophy. *J. Neurol. Neurosurg. Psychiat.*, **30**, 526–530.

1870. McCONNELL, H. M. (1970). Molecular motion in biological membranes. *In* Schmitt, F. O., *The neurosciences. Second study programme.*

Rockefeller; N.Y. pp. 685–706.

1871. McCONNELL, H. M. & McFARLAND, B. G. (1970). Physics and chemistry of spin labels. *Q. Rev. Biophys.*, **3**, 91–136.

1872. McCONNELL, H. M. & McFARLAND, B. G. (1972). The flexibility gradient in biological membranes. *Ann. N.Y. Acad. Sci.*, **195**, 207–217.

1873. McDONALD, R. C. & BANGHAM, A. D. (1972). Comparison of double layer potentials in lipid monolayers and lipid bilayer membranes. *J. mem. Biol.*, **7**, 29–53.

1874. McDONALD, W. I. (1972). The time course of conduction failure during degeneration of a central tract. *Expl Brain Res.*, **14**, 550–556.

1875. McEWEN, B. S. & GRAFSTEIN, B. (1968). Fast and slow components in axonal transport of protein. *J. Cell. Biol.*, **38**, 494–508.

1876. McGEER, E. G., IKEDA, H., ASAKURA, T. & WADA, J. A. (1969). Lack of abnormality in brain aromatic amines in rats and mice susceptible to audiogenic seizure. *J. Neurochem.*, **16**, 945–950.

1877. McILWAIN, H. (1973). Regulatory significance of the release and action of adenine derivatives in cerebral systems. *Biochem. Soc. Symp.*, **36**, 69–85.

1878. McINNES, J. W. & LUTTGAU, M. W. (1972). Interaction of puromycin and cycloheximide with electroconvulsive shock in producing alterations of brain polysomes. *J. Neurochem.*, **19**, 1889–1892.

1879. McINTOSH, J. R., HEPLER, P. K. & van WIE, D. G. (1969). Model for mitosis. *Nature, Lond.*, **224**, 659–663.

1880. McINTOSH, J. R. & SANDIS, S. C. (1970). The distribution of spindle microtubules during mitosis in cultured human cells. *J. Cell Biol.*, **47**, 31–40.

1881. McINTYRE, A. K., BRADLEY, K. & BROCK, L. G. (1959). Responses of motoneurones undergoing chromatolysis. *J. gen Physiol.*, **42**, 931–958.

1882. MACKIE, G. O. (1965). Conduction in the nerve-free epithelia of siphonophores. *Am. Zool.*, **5**, 439–453.

1883. MACKIE, G. O. & PASSANO, L. M. (1968). Epithelial conduction in hydramedusae. *J. gen Physiol.*, **52**, 600.

1884. McLAUGHLIN, S. G. A., SZABO, G., CIANI, S. & EISENMAN, G. (1972). The effects of a cyclic polyether on the electrical properties of phospholipid bilayer membranes. *J. mem. Biol.*, **9**, 3–36.

1885. McLAUGHLIN, S. G. A., SZABO, G. & EISENMAN, G. (1971). Divalent ions and the surface potential of charged phospholipid membranes. *J. gen. Physiol.*, **58**, 667–687.

1886. McLAUGHLIN, S. G. A., SZABO, G. EISENMAN, G. & CIANI, S. M. (1970). Surface charge and the conductance of phospholipid membranes. *Proc. natn. Acad. Sci. U.S.A.*, **67**, 1268.

1887. McLENNAN, H. (1954). Acetylcholine metabolism of normal and axotomized ganglia. *J. Physiol.*, **124**, 113–116.

1888. McLENNAN, D. H., SEEMAN, P., ILES, G. H. & YIP, C. C. (1971). Membrane formation by the adenosine triphosphatase of sarcoplasmic reticulum. *J. biol. Chem.*, **246**, 2702.

1889. McMAHAN, V. J. & KUFFLER, S. W. (1971). Visual identification of synaptic boutons on living ganglion cells, and of varicosities in postganglionic axons in the heart of the frog. *Proc. R. Soc. B,* **177**, 485–508.

1890. McMASTERS, R. E. (1962). Regeneration of the spinal cord in the rat: effects of Piromen and ACTH upon the regenerative capacity. *J. comp. Neurol.*, **119**, 113–125.

1891. McMURRAY, W. C. & DAWSON, R. M. C. (1969). Phospholipid exchange reactions within the liver cell. *Biochem. J.*, **112**, 91–108.

1892. McNUTT, N. S. & WEINSTEIN, R. S. (1970). The ultrastructure of the nexus. A correlated thin section and freeze-cleave study. *J. Cell Biol.*, **47**, 666.

1893. McNUTT, N. S. & WEINSTEIN, R. S. (1973). Membrane ultrastructure at mammalian intercellular junctions. *Prog. Biophys. molec. Biol.*, **26**, 45–101.

1894. McPHERSON, A. & TOKUNAGA, J. (1967). The effects of cross innervation on the myoglobin concentration of tonic and phasic muscles. *J. Physiol.*, **188**, 121–129.

1895. McPHERSON, C. F. C. & CHINERMAN, J. (1971). Effect of intraventricular injection of brain isoantibodies on learning. *Expl Neurol.*, **31**, 45–52.

1896. McPHERSON, C. F. C. & SHEK, R. P. N. (1970). Effect of brain isoantibodies on learning and memory in the rat. *Expl Neurol.*, **29**, 1–15.

1897. MADDY, A. H. (1964). A fluorescent label for the outer components of the plasma membrane. *Biochim. biophys. Acta,* **88**, 390–399.

1898. MAENO, H., JOHNSON, E. M. & GREENGARD, P. (1971). Subcellular distribution of adenosine 3′5′-monophosphate dependent protein kinase in rat brain. *J. biol. Chem.*, **246**, 134–142.

1899. MAKMAN, M. H. (1971). Conditions leading to enhanced response to glucagon, epinephrine, or prostaglandins by adenylate cyclase of normal and malignant cultured cells. *Proc. natn. Acad. Sci. U.S.A.*, **68**, 2127–2130.

1900. MALHOTRA, S. K. (1970). Organization of cellular membranes. *Prog. Biophys. molec. Biol.*, **20**, 69–131.

1901. MALIS, L. I., BAKER, C. P., KRUGER, L. & ROSE, J. E. (1960). Effects of heavy ionizing monoenergetic particles on the cerebral cortex. *J. comp. Neurol.*, **115**, 219–242.

1902. MALVEY, J. E., SCHOTTELIUS, D. D. & SCHOTTELIUS, B. A. (1971). Energy reserves and chemical changes in denervated anterior and posterior latissimus dorsi muscles of the chicken. *Expl Neurol.*, **33**, 171–180.

1903. MANDELSTAMM, M. & KRYLOW, L. (1928). Vergleichende Untersuchungen über der Farbenspeicherung in Zentralnervensystem bei Injektionen der Farbe im Blut und in der Liquor Cerebrospinalis. *Z. ges. exp. Med.*, **60**, 63–85.

1904. MANERY, J. F. & HASTINGS, A. B. (1939). The distribution of electrolytes in mammalian tissues. *J. biol. Chem.*, **127**, 657–676.
1905. MANN, W. S. & SALAFSKY, B. (1970). Development of the differential response to succinylcholine in the fast and slow twitch skeletal muscle of the kitten. *J. Physiol.*, **210**, 581–592.
1906. MANNING, A. (1964). Drosophila and the evolution of behaviour. *Viewp. Biol.*, **4**, 125–168.
1907. MANNING, A. & HIRSCH, J. (1971). The effects of artificial selection for slow mating in *Drosophila simulans*. II. Genetic analysis of the slow mating line. *Anim. Behav.*, **19**, 448–453.
1908. MANOSEVITZ, M. & MONTEMAYOR, R. J. (1972). Interaction of environmental enrichment and genotype. *J. comp. Physiol. Psychol.*, **79**, 67–76.
1909. MAO, C. C., GUIDOTTI, A. & COSTA, E. (1974). Interaction between γ-aminobutyric acid and cyclic guanosine $3'5'$monophosphate in rat cerebellum. *Molec. Pharmac.*, **10**, 736–745.
1910. MAO, C. C., GUIDOTTI, A. & COSTA, E. (1974). The regulation of cyclic guanosine monophosphate in rat cerebellum: possible involvement of putative aminoacid transmitters. *Brain Res.*, **79**, 510.
1911. MARCHBANKS, R. M. (1966). Serotonin binding to nerve ending particles and other preparations from rat brain. *J. Neurochem.*, **13**, 1481–1493.
1912. MARES, P. (1973). Bioelectrical activity of an epileptogenic focus in rat neocortex. *Brain Res.*, **56**, 203–213.
1913. MARGOLIASH, E., SCHENTOK, E. R., MARGIC, M. P., BUROKAS, S., RICHTER, W. R., BARLOW, G. H. & MOSCONA, A. A. (1965). Characterization of specific cell aggregating materials from sponge cells. *Biochem. biophys. Res. Commun.*, **20**, 383–388.
1914. MARGOLIS, R. U. (1967). Acid mucopolysaccharides and proteins of bovine whole brain, white matter and myelin. *Biochim. biophys. Acta*, **141**, 91–102.
1915. MARGRETH, A., MUSCATELLO, U. & ANDERSON-CEDERGREN, E. (1963). A morphological and biochemical study on the regulation of carbohydrate metabolism in the muscle cell. *Expl Cell Res.*, **32**, 484.
1916. MARGRETH, A., SALVIATI, G., DIMAURO, S. & TURATI, G. (1972). Early biochemical consequences of denervation in fast and slow skeletal muscles, and their relationship to neural control over muscle differentiation. *Biochem. J.*, **126**, 1099–1110.
1917. MARGULIS, L. (1973). Colchicine-sensitive microtubules. *Int. Rev. Cytol.*, **34**, 333–362.
1918. MARIN-PADILLA, M. (1971). Ontogenesis of the cat neocortex, a Golgi study. *Anat. Rec.*, **169**, 375.
1919. MARK, R. F. (1965). Fin movements after regeneration of neuromuscular connections: an investigation of myotypic specificity. *Expl Neurol.*, **12**, 292–302.
1920. MARK, R. F. (1969). Matching muscles and motoneurones. A review of

some experiments on motor nerve regeneration. *Brain Res.*, **14**, 245–254.
1921. MARK, R. F., von CAMPENHAUSEN, G. & LISCHINSKY, D. J. (1966). Nerve muscle relations in the salamander; possible relevance to nerve regeneration and muscle specificity. *Expl Neurol.*, **16**, 438–499.
1922. MARK, R. F. & MAROTTE, L. R. (1972). The mechanism of selective reinnervation of fish eye muscles III. Functional, electrophysiological and anatomical analysis of recovery from section of the IIIrd and IVth nerves. *Brain Res.*, **46**, 131–148.
1923. MARK, R. F., MAROTTE, L. R. & MART, R. E. (1972). The mechanism of selective reinnervation of fish eye muscles. IV. Identification of repressed synapses. *Brain Res.*, **46**, 149–157.
1924. MARK, R. F. & WATTS, M. E. (1971). Drug inhibition of memory formation in chickens. I. Long-term memory. *Proc. R. Soc. B*, **178**, 439–454.
1925. MAROTTE, L. R. & MARK, R. F. (1970). The mechanism of selective reinnervation of fish eye muscles. Evidence from muscle function during recovery.*Brain Res.*, **19**, 41–52.
1926. MARR, J. N. & GARDNER, L. E. (1965). Early olfactory experience and later social behaviour in the rat: preference, sexual responsiveness and care of young. *J. genet. Psychol.*, **107**, 167–174.
1927. MARTENSON, R. E. & DEIBLER, G. E. (1975). Partial characterization of basic proteins of chicken, turtle and frog central nervous system myelin. *J. Neurochem.*, **24**, 79–88.
1928. MARTENSON, R. E., DEIBLER, G. E. & KIES, M. W. (1971). Electrophoretic characterization of basic proteins in acid extracts of central nervous system tissue. *J. Neurochem.*, **18**, 2417–2426.
1929. MARTENSON, R. E., DEIBLER, G. E. & KIES, M. W. (1971). The occurence of two myelin basic proteins in the central nervous system of rodents in the suborders *myomorphia* and *sciuromorphia*. *J. Neurochem.*, **18**, 2427–2433.
1930. MARTENSON, R. E., DEIBLER, G. E., KIES, M. W., McKNEALLY, S. S., SHAPIRA, R. & KIBLER, R. F. (1972). Differences between the two myelin basic proteins of the rat central nervous system. A deletion in the smaller protein. *Biochim. biophys. Acta*, **263**, 193–203.
1931. MARTENSON, R. E., DEIBLER, G. E., KREMER, A. J. & LEVINE, S. (1975). Comparative studies of guinea-pig and bovine myelin basic proteins: partial characterization of chemically derived fragments and their encephalitogenic activities in Lewis rats. *J. Neurochem.*, **24**, 173–180.
1932. MARTIN, A. R. & BRANCH, C. L. (1958). Spontaneous activity of Betz cells in cats with midbrain lesions. *J. Neurophysiol.*, **21**, 368–379.
1933. MARTIN, A. R. & PILAR, G. (1963). Dual mode of synaptic transmission in the avian ciliary ganglion. *J. Physiol.*, **168**, 443–463.
1934. MARTINEZ-PALOMO, A. (1970). The surface coats of animal cells. *Int. Rev. Cytol.*, **29**, 29–74.
1935. MASON, W. A., DAVENPORT, R. K. & MENZEL, E. W. (1968). Early

experience and the social development of rhesus monkeys and chimpanzees. *In* Newton, G. & Levine, S. *Early experience and behaviour. The psychobiology of development.* C. C. Thomas; Springfield, Illinois. pp. 440–480.

1936. MASON, W. A. & SPONHOLZ, R. R. (1963). Behaviour of rhesus monkeys raised in isolation. *J. Psychiat.*, **1**, 299–306.

1937. MASUROVSKY, E. B., BUNGE, M. B. & BUNGE, R. P. (1967). Cytological studies of organotypic cultures of rat dorsal root ganglia following X-irradiation in vitro. I. Changes in neurons and satellite cells. *J. Cell Biol.*, **32**, 467–496.

1938. MASUROVSKY, E. B., BUNGE, M. B. & BUNGE, R. P. (1967). Cytological studies of organotypic cultures of rat dorsal root ganglia following X-irradiation in vitro. II. Changes in Schwann cells, myelin sheaths and nerve fibres. *J. Cell Biol.*, **32**, 497–518.

1939. MATHAN, M., HERMOS, J. A. & TRIER, J. S. (1972). Structural features of the epithelio-mesenchymal interface of rat duodenal mucosa during development. *J. Cell. Biol.*, **52**, 577–588.

1940. MATSUMOTO, H. & AJMONE-MARSAN, C. (1964). Cortical cellular phenomena in experimental epilepsy: "interictal" manifestations. *Expl Neurol.*, **9**, 286–304.

1941. MATTHEWS, E. K. (1967). Membrane potential measurement in cells of the adrenal gland. *J. Physiol.*, **189**, 139–148.

1942. MATTHEWS, M. A. (1968). An electron microscopic study of the relationship between axon diameter and the initiation of myelin production in the peripheral nervous system. *Anat. Rec.*, **161**, 337–352.

1943. MATTHEWS, M. R. (1964). Further observations on transneuronal degeneration in the lateral geniculate nucleus of the macaque monkey. *J. Anat.*, **98**, 255–263.

1944. MATTHEWS, M. R., COWAN, W. M. & POWELL, T. P. S (1960). Transneuronal cell degeneration in the lateral geniculate nucleus of the macaque monkey. *J. Anat.*, **94**, 145–169.

1945. MATTHEWS, M. R. & NELSON, V. (1973). Ultrastructural changes at the synapse associated with depression of synaptic transmission in rat superior cervical ganglion after injury of postganglionic axons. *J. Physiol.*, **234**, 36–37P.

1946. MATTHEWS, M. R. & NELSON, V. H. (1975). Detatchment of structurally intact nerve endings from chromatolytic neurones of rat superior cervical ganglion during the depression of synaptic transmission induced by postganglionic axotomy. *J. Physiol.*, **245**, 91–135.

1947. MATTHEWS, M. R. & POWELL, T. P. S. (1962). Some observations on transneuronal cell degeneration in the olfactory bulb of the rabbit. *J. Anat.*, **96**, 89–102.

1948. MATTHEWS, M. R. & RAISMAN, G. (1972). A light and electron microscopic study of the cellular response to axonal injury in the superior cervical ganglion of the rat. *Proc. R. Soc. B*, **181**, 43–79.

1949. MATTHEWS, R. G., HUBBARD, R., BROWN, P. K. & WALD, G. (1963).

Tautomeric forms of metarhodopsin. *J. gen. Physiol.*, **47**, 215–240.

1950. MAUTNER, W. (1965). Studien an der Epiphysis cerebri und am Subcommisuralorgan der Frösche. *Z. Zellforsch. mikrosk. Anat.*, **67**, 234–250.

1951. MAUZERALL, D. & FINKELSTEIN, A. (1969). Light-induced changes in the conductivity of thin lipid membranes in the presence of iodine and iodide ion. *Nature, Lond.*, **224**, 690–692.

1952. MAX, S. R. & BRADY, R. O. (1971). Alteration of the ganglioside composition of skeletal muscle in murine muscular dystrophy. *Nature, Lond., New Biol.*, **233**, 55–56.

1953. MAX, S. R., NELSON, P. G. & BRADY, R. O. (1970). The effect of denervation on the composition of muscle gangliosides. *J. Neurochem.*, **17**, 1517–1520.

1954. MAXWELL, D. S. & KRUGER, L. (1965). The fine structure of astrocytes in the cerebral cortex and their response to focal injury produced by heavy ionizing particles. *J. Cell Biol.*, **25**, 141–158.

1955. MAXWELL, D. S. (1965). Electron microscopic observations on the origin of macrophages in degenerating cerebral cortex of the rat. *Anat. Rec.*, **151**, 383.

1956. MAXWELL, D. S. & KRUGER, L. (1965). Electron microscopy of normal and reactive astrocytes in rat cerebral cortex. *Anat. Rec.*, **148**, 310.

1957. MAXWELL, D. S. & KRUGER, L. (1966). The reactive oligodendrocyte. An electron microscopic study of cerebral cortex following alpha particle irradiation. *Am. J. Anat.*, **118**, 437–460.

1958. MEAD, J. F. & DHOPESHWARKAR, G. A. (1972). Types of fatty acids in brain lipids, their deviation and function. *In Lipids, malnutrition and the developing brain.* CIBA Fdn Symp. North Holland; Amst. pp. 59–72.

1959. MEDAWAR, P. B. (1948). Immunity to homologous grafted skin: III. Fate of skin homografts transplanted to brain, to subcutaneous tissue and to anterior chamber of eye. *Br. J. exp. Path.*, **29**, 58–69.

1960. MEDZIHRADSKY, F., NANDHASRI, P. S., IDOYAGA-VARGAS, V. & SELLINGER, O. Z. (1971). A comparison of the ATPase activity of the glial cell fraction and the neuronal perikaryal fraction isolated in bulk from rat cerebral cortex. *J. Neurochem.*, **18**, 1599–1603.

1961. MEDZIHRADSKY, F., SELLINGER, O. Z., NANDHASRI, P. S. & SANTIAGO, J. C. (1972). ATPase activity in glial cells and in neuronal perikarya of rat cerebral cortex during early postnatal development. *J. Neurochem.*, **19**, 543–545.

1962. MEHL, E. & HALARIS, A. (1970). Stoichiometric relation of protein components in cerebral myelin from different species. *J. Neurochem.*, **17**, 659–668.

1963. MEIER, H. & HOAG, W. G. (1962). The neuropathology of "reeler", a neuromuscular mutation in mice. *J. Neuropath. exp. Neurol.*, **21**, 649–654.

1964. MELLANBY, J. H. (1971). Presynaptic effects of tetanus toxin at the

neuromuscular junction. *J. Physiol.*, **218**, 68–69P.
1965. MELLER, K. (1974). The reaggregation of neurons and their satellite cells in cultures of trypsin-dissociated spinal ganglia. *Cell Tiss. Res.*, **152**, 175–189.
1966. MENKES, J. H., PHILIPPART, M. & CONCONE, M. C. (1966). Concentration and fatty acid composition of cerebrosides and sulfatides in mature and immature human brain. *J. Lipid Res.*, **7**, 479–486.
1967. MERRELL, R. & GLASER, L. (1973). Specific recognition of plasma membranes by embryonic cells. *Proc. natn. Acad. Sci. U.S.A.*, **70**, 2794–2798.
1968. MESZLER, R. M., PAPPAS, G. D. & BENNETT, M. V. L. (1972). Morphological demonstration of electrotonic coupling of neurons by way of presynaptic fibres. *Brain Res.*, **36**, 412–415.
1969. METZER, A. B. (1965). Diffusive transport rates in structured media. *Nature, Lond.*, **208**, 267–268.
1970. MEULDERS, M. & GODFRAIND, J. M. (1969). Influence du réveil d'origine réticulaire sur l'étendue des champs visuels des neurones de la région genouillée chez le chat avec cerveau intact ou avec cerveau isolé. *Expl Brain Res.*, **9**, 201–220.
1971. MEUNIER, J. C., SEALOCK, R., OLSEN, R. & CHANGEUX, J. P. (1974). Purification and properties of the cholinergic receptor protein from *Electrophorus electricus* electric tissue. *Europ. J. Biochem.*, **45**, 371.
1972. MEYER, K., HOFFMAN, P., LINKER, A., GRUMBACH, M. M. & SAMPSON, P. (1959). Sulfated mucopolysaccharides of urine and organs in Gargoylism (Hurler's Syndrome) II. Additional studies. *Proc. Soc exp. Biol. Med.*, **102**, 587–590.
1973. MEZEI, C. (1970). Inhibition by actinomycin D of labelling of phospholipids in the nervous system of chicks. *J. Neurochem.*, **17**, 691–693.
1974. MIALE, I. & SIDMAN, R. L. (1961). An autoradiographic analysis of histogenesis in the mouse cerebellum. *Expl Neurol.*, **4**, 277–296.
1975. MIANI, N. (1962). Evidence of a proximo-distal movement along the axon of phospholipid synthesized in the nerve cell body. *Nature, Lond.*, **193**, 887–888.
1976. MIANI, N. (1963). Analysis of the somato-axonal movement of phospholipids in the vagus and hypoglossal nerves. *J. Neurochem.*, **10**, 859–874.
1977. MIANI, N. (1971). Transport of S-100 protein in mammalian nerve fibres and transneuronal signals. *Acta neuropath.*, Suppl. **5**, 104–108.
1978. MIANI, N., de RENZIS, G., MICHETTI, F., CORRER, S., OLIVIERI, C., SANGIACOMO, O. & CANIGLIA, A. (1972). Axonal transport of S-100 protein in mammalian nerve fibres. *J. Neurochem.*, **19**, 1387–1394.
1979. MICHAEL, P. (1973). Pheromones: isolation of male sex attractant from a female primate. *Science, N.Y.*, **172**, 964–966.
1980. MICHAEL, R. P. & KEVERNE, E. B. (1970). Primate sex pheromones of

vaginal origin. *Nature, Lond.*, **225**, 84–85.
1981. MICHAELI, Z. (1960). Ion binding and the formation of insoluble polymethacrylic salts. *J. Polymer. Sci.*, **48**, 291–299.
1982. MICHAELSON, D. M. & RAFTERY, M. A. (1974). Purified acetylcholine receptor: its reconstruction to a chemically excitable membrane. *Proc. natn. Acad. Sci. U.S.A.*, **71**, 4768.
1983. MICKEL, H. S. & GILLES, F. H. (1970). Changes in glial cells during human telencephalic myelinogenesis. *Brain*, **93**, 337–346.
1984. MIHAILOVIC, L. T. & CUPIC, D. (1971). Epileptiform activity evoked by intracerebral injection of anti-brain antobodies. *Brain Res.*, **32**, 97–124.
1985. MIHAILOVIC, L., DIVAC, I., MITROVIC, K., MILOSEVIC, D. & JANKOVIC, B. D. (1969). Effects of intraventricularly injected anti-brain antibodies on delayed alternation and visual discrimination test performance in rhesus monkeys. *Expl Neurol.*, **24**, 325–336.
1986. MIHAILOVIC, L. & JANKOVIC, B. D. (1961). Effects of intraventricularly injected anti-nucl. caudatus antibody on the electrical activity of the cat brain. *Nature, Lond.*, **192**, 665–666.
1987. MIHAILOVIC, L. & JANKOVIC, B. D. (1965). Effects of anto-cerebral antibodies on electrical activity and behaviour. *Neurosci. Res. Prog. Bull.*, **3**, 8–17.
1988. MIHAILOVIC, L., JANKOVIC, B. D., BELESIN, B., MITROVIC, K. & KRZALIC, L. (1964). Effects of intraventricularly injected anti-cerebral antibodies on the histamine-like substance and potassium content of various regions of the cat brain. *Nature, Lond.*, **203**, 763–765.
1989. MIHAILOVIC, L., JANKOVIC, B. D., BELESLIN, B., MLOSEVIC, D. & CUPIC, D. (1965). Effects of antilobster nerve antibody on membrane potentials of the giant axon of *Palinurus vulgaris*. *Nature, Lond.*, **206**, 904–905.
1990. MILEDI, R. (1960). Properties of regenerating neuromuscular synapses in the frog. *J. Physiol.*, **154**, 190–205.
1991. MILEDI, R. (1960). Junctional and extrajunctional acetylcholine receptors in skeletal muscle fibres. *J. Physiol.*, **151**, 24–30.
1992. MILEDI, R. (1960). The acetylcholine sensitivity of frog muscle fibres after complete or partial denervation. *J. Physiol.*, **151**, 1–23.
1993. MILEDI, R. (1963). An influence of nerve not mediated by impulses. *In* Gutmann, E. & Hnik, P., *The effect of use and disuse on neuromuscular function.* Elsevier; Amsterdam. pp. 35–40.
1994. MILEDI, R., MOLINOFF, P. B. & POTTER, L. T. (1971). Isolation of the cholinergic receptor protein of *Torpedo* electric tissue. *Nature, Lond.*, **229**, 554–557.
1995. MILEDI, R. & POTTER, L. T. (1971). Acetyl choline receptors in muscle fibres. *Nature, Lond.*, **233**, 599–603.
1996. MILEDI, R. & SLATER, C. R. (1968). Some mitochondrial changes in denervated muscle. *J. Cell Sci.*, **3**, 49–54.
1997. MILEDI, R. & SLATER, C. R. (1970). On the degeneration of rat

neuromuscular junctions after nerve section. *J. Physiol.*, **207**, 507–528.

1998. MILEDI, R. & ZELENA, J. (1966). Sensitivity to acetyl choline in rat slow muscle. *Nature, Lond.*, **210**, 855–856.

1999. MILLER, D. S., GINSBURG, B. E. & POTAS, M. Z. (1956). The influence of castration on susceptibility to audiogenic siezures in DBA mice. *Anat. Rec.*, **124**, 336.

2000. MILLER, I. J. (1968). *Peripheral input from rat fungiform papillae to single chorda tympani fibres.* Doctoral Diss. Florida State University.

2001. MILLER, I. J. (1971). Peripheral interactions among single papillae inputs to gustatory nerve fibres. *J. gen. Physiol.*, **57**, 1–25.

2002. MILLER, R., VARON, S., KRUGER, L., COATES, P. W. & ORKAND, P. M. (1970). Formation of synaptic contacts on dissociated chick embryo sensory ganglion cells *in vitro. Brain Res.*, **24**, 356–358.

2003. MILLER, R. F. & DOWLING, J. E. (1970). Intracellular responses of the Müller (glial) cells of the mudpuppy retina: Their relation to b-wave of the electroretinogram. *J. Neurophysiol.*, **33**, 323–341.

2004. MILLER, R. J., HORN, A. S. & IVERSON, L. L. (1974). The action of neuroleptic drugs on dopamine-stimulated adenosine $3'5'$monophosphate production in rat neostriatum and limbic forebrain. *Molec. Pharmac.*, **10**, 759–766.

2005. MILLER, R., HORN, A., IVERSON, L. & PINDER, R. (1974). Effects of dopamine-like drugs on rat striatal adenyl cyclase have implications for CNS dopamine receptor topography. *Nature. Lond.*, **250**, 238.

2006. MINCHIN, M. C. W. & IVERSON, L. L. (1974). Release of ^{3}H-gamma-aminobutyric acid from glial cells in rat dorsal root ganglia. *J. Neurochem.*, **23**, 533–540.

2007. MINDICH, L. (1970). Membrane synthesis in Bacillus subtilis II. Integration of membrane proteins in the absence of lipid synthesis. *J. molec. Biol.*, **49**, 433–440.

2008. MINTZ, B. (1962). Experimental study of the developing mammalian egg: removal of the zona pellucida. *Science, N.Y.*, **138**, 594–595.

2009. MINTZ, B. (1965). Genetic mosaicism in adult mice of quadriparental lineage. *Science, N.Y.*, **148**, 1232–1233.

2010. MINTZ, B. (1967). Gene control of mammalian pigmentary differentiation. I. Clonal origin of melanocytes. *Proc. natn. Acad. Sci. U.S.A.*, **58**, 344–351.

2011. MINTZ, B. (1971). Genetic mosaicism *in vivo*: development and disease in allophenic mice. *Fedn Proc. Fedn Am. Socs exp. Biol.*, **30**, 935–943.

2012. MINTZ, B. & SILVERS, W. K. (1967). 'Intrinsic' immunological tolerance in allophenic mice. *Science, N.Y.*, **158**, 1484.

2013. MINTZ, B. & BAKER, W. (1967). Normal mammalian muscle differentiation and gene control of isocitrate dehydrogenase synthesis. *Proc. natn. Acad. Sci. U.S.A.*, **58**, 592–598.

2014. MISHRA, R. K., GARDNER, E. L., KATZMAN, R. & MAKMAN, M. H. (1974). Enhancement of dopamine-stimulated adenylate cyclase activity in rat caudate nucleus after lesions in substantia nigra: evidence for denervation supersensitivity. *Proc. natn. Acad. Sci. U.S.A.*, **71**, 3883.

2015. MISTRETTA, C. M. (1968). *The permeability of rat tongue epithelium.* Master's Thesis. Florida State University.

2016. MITCHELL, G. D., RAYMOND, E. J., RUPPENTHAL, G. C. & HARLOW, H. F. (1966). Long-term effects of total social isolation upon behaviour of rhesus monkeys. *Psychol. Rep.*, **18**, 567–580.

2017. MIYAMOTO, E., KUO, J. F. & GREENGARD, P. (1969). Adenosine 3′5′-monophosphate-dependent protein kinase from brain. *Science, N.Y.*, **165**, 63–65.

2018. MIYAMOTO, E., KUO, J. F. & GREENGARD, P. (1969). Cyclic nucleotide-dependent protein kinases III. Purification and properties of adenosine 3′5′-monophosphate dependent protein kinase from bovine brain. *J. biol. Chem.*, **244**, 6395–6402.

2019. MIYAMOTO, V. & THOMPSON, T. E. (1967). Some electrical properties of lipid membranes. *J. Colloid interfac. Sci.*, **25**, 16.

2020. MIZELL, M. (1968). Limb regeneration: induction in the newborn oppossum. *Science, N.Y.*, **161**, 283–286.

2021. MODEL, P. G., BORNSTEIN, M. B., CRAIN, S. M. & PAPPAS, G. D. (1971). An electron microscopic study of the development of synapses in cultured fetal mouse cerebrum continuously exposed by Xylocaine. *J. Cell Biol.*, **49**, 362–371.

2022. MOKRASCH, L. C. (1967). A rapid purification of proteolipid protein adaptable to large quantities. *Life Sci.*, **6**, 1905–1909.

2023. MOKRASCH, L. C. (1971). Biological chemistry and dynamics of myelin. *Neurosc. Res. Prog. Bull.*, **9**, 453–465.

2024. MOKRASCH, L. C. & MANNER, P. (1963). Incorporation of ^{14}C-aminoacids and ^{14}C-palmitate into proteolipids of rat brains in vitro. *J. Neurochem.*, **10**, 541–543.

2025. MOLTZ, H. (1963). Imprinting: an epigenetic approach. *Psychol. Rev.*, **70**, 123–138.

2026. MOLTZ, H. (1968). An epigenetic interpretation of the imprinting phenomenon. *In* Newton, G. & Levine, S. *Early experience and behaviour. The psychobiology of development.* C. C. Thomas; Springfield, Illinois.

2027. MOMMAERTS, W. F. H. M. (1969). Energetics of muscular contraction. *Physiol. Rev.*, **49**, 427–451.

2028. MOMMAERTS, W. F. H. M., BULLER, A. J. & SERAYDARIAN, K. (1969). The modification of some biochemical properties of muscle by cross innervation. *Proc. natn. Acad. Sci. U.S.A.*, **64**, 128–133.

2029. MONARD, D., RENTSCH, M., SOLOMON, F. & GYSIN, R. (1973). Astrocyte induced morphological differentiation in neuroblastoma cells. *Experentia*, **29**, 776.

2030. MONARD, D., SOLOMON, F., RENTSCH, M. & GYSIN, R. (1973). Glia-induced morphological differentiation in neuroblastoma cells. *Proc. natn. Acad. Sci. U.S.A.*, **70**, 1894–1897.

2031. MONROE, B. G. (1967). A comparative study of the ultrastructure of the median eminence, infundibular stem and neural lobe of the hypophysis of the rat. *Z. Zellforsch. mikrosk. Anat.* **76**, 405–432.

2032. MONTAL, M. (1972). Lipid-polypeptide interactions in bilayer lipid membranes. *J. mem. Biol.,* **7**, 245–266.
2033. MONTAL, M. & MUELLER, P. (1972). Formation of bimolecular membranes from lipid monolayers and a study of their electrical properties. *Proc. natn. Acad. Sci. U.S.A.,* **69**, 3561–3566.
2034. MOODY, M. F. (1963). X-ray diffraction pattern of nerve myelin: a method for determining the phases. *Science, N.Y.,* **142**, 1173–1174.
2035. MOODY, M. F. (1964). Photoreceptor organelles in animals. *Biol. Rev.,* **39**, 43–86.
2036. MOOR, H. (1966). The performance of freeze-etching and the interpretation of results concerning the surface structure of membranes and the fine structure of microtubules and spindle fibres. *Balzer's Rep.,* **9**, 1–11.
2037. MOOR, H. & MÜHLETHALER, K. (1963). Fine structure in freeze-etched yeast cells. *J. Cell Biol.,* **17**, 609–628.
2038. MOOR, H., RUSKA, C. & RUSCA, H. (1964). Elektronenmikroskopische Darstellung tierischer Zellen mit der Gefreierätztechnik. *Z. Zellforsch. mikrosk. Anat.,* **62**, 581–601.
2039. MOORE, A. U. (1968). Effects of modified maternal care in the sheep and goat. *In* Newton, G. & Levine, S. *Early experience and behaviour. The psychobiology of development.* C. C. Thomas; Springfield, Illinois. pp. 481–529.
2040. MOORE, R. Y., BJÖRKLUND, A. & STENEVI, U. (1971). Plastic changes in the adrenergic innervation of the rat septal area in response to dennervation. *Brain Res.,* **33**, 13–35.
2041. MOORE, W. M., HOLLADAY, L. A., PUETT, D. & BRADY, R. N. (1974). On the conformation of the acetylcholine receptor protein from *Torpedo nobiliana. FEBS Lett.,* **45**, 145.
2042. MOORE, W. V. & WETLAUFER, D. B. (1971). The circular dichroism of synaptosomal membranes: studies of interactions with neuropharmacological agents. *J. Neurochem.,* **18**, 1167–1178.
2043. MOREST, D. K. (1970). A study of neurogenesis in the forebrain of opossum pouch young. *Z. Anat. EntwGesch.,* **130**, 265–305.
2044. MORETON, R. A. & PITT, G. A. J. (1955). Studies on rhodopsin. 9. pH and the hydrolysis of indicator yellow. *Biochem J.,* **59**, 128–134.
2045. MORETZ, R. C., AKERS, C. K. & PARSONS, D. F. (1969). Use of small angle X-ray diffraction to investigate disordering of membranes during preparation for electron microscopy. *Biochim. biophys. Acta,* **193**, 1–11.
2046. MORETZ, R., AKERS, C. K. & PARSONS, D. F. (1969). Use of small angle X-ray diffraction to investigate disordering of membranes during preparation for electron microscopy. *Biochim. biophys. Acta.,* **193**, 12–23.
2047. MORI, S. & LEBLOND, C. P. (1969). Identification of microglia in light and electron microscopy. *J. comp. Neurol.,* **135**, 57–80.

2048. MORI, S. & LEBLOND, C. P. (1969). Electron microscopic features and proliferation of astrocytes in the corpus callosum of the rat. *J. comp. Neurol.*, **137**, 197–226.
2049. MOROSOW, I. I. (1938). Die Hemmung und Wiederherstellung des Regenerationsprocess der Extremität beim Axolotl. *C. r. Acad. Sci. URSS*, **20**, 207–210.
2050. MORRELL, F. (1969). Physiology and histochemistry of the mirror focus. *In* Jasper, H. H., Ward, A. A. & Pope, A., *Basic mechanisms of the epilepsies.* Little, Brown; Boston. pp. 357–370.
2051. MORRELL, F., PROCTOR, F. & PRINCE, D. A. (1965). Epileptogenic properties of subcortical freezing. *Neurology*, **15**, 744.
2052. MORSE, P. F. & HOWLAND, J. L. (1971). Erythrocytes from animals with genetic muscular dystrophy. *Nature, Lond.*, **245**, 156–157.
2053. MOSCONA, A. (1961). Rotation-mediated histogenetic aggregation of dissociated cells. A quantifiable approach to cell interactions *in vitro. Expl Cell Res.*, **22**, 455–475.
2054. MOSCONA, A. A. & MOSCONA, M. H. (1967). Comparison of aggregation of embryonic cells dissociated with trypsin or versine. *Expl Cell Res.*, **45**, 239–243.
2055. MOSKO, S., LYNCH, G. & COTMAN, C. W. (1973). The distribution of septal projections to the hippocampus of the rat. *J. comp. Neurol.*, **152**, 633–674.
2056. MOSSAKOWSKI, M. J. (1962). The activity of succinic dehydrogenase in glial tumours. *J. Neuropath. exp. Neurol.*, **21**, 137–146.
2057. MOSSAKOWSKI, M. J. (1963). The activity of succinic dehydrogenase in the reactive glia. *Acta neuropath.*, **2**, 282–290.
2058. MOYER, E. K., KIMMEL, D. L. & WINBORNE, L. W. (1953). Regeneration of sensory spinal nerve roots in young and in senile rats. *J. comp. Neurol.*, **98**, 283–308.
2059. MUELLER, P. & RUDIN, D. O. (1963). Induced excitability in reconstituted cell membrane structure. *J. theor. Biol.*, **4**, 268–287.
2060. MUELLER, P. & RUDIN, D. O. (1967). Action potential phenomena in experimental bimolecular lipid membranes. *Nature, Lond.*, **213**, 603.
2061. MUELLER, P. & RUDIN, D. O. (1967). Development of K^+-Na^+ discrimination in experimental bimolecular lipid membranes by macrocyclic antibiotics. *Biochem. biophys. Res. Commun.* **26**, 398.
2062. MUELLER, P. & RUDIN, D. O. (1968). Action potentials induced in bimolecular lipid membranes. *Nature, Lond.*, **217**, 713.
2063. MUELLER, P. & RUDIN, D. O. (1969). Translocators in bimolecular lipid membranes: their role in dissipative and conservative bioenergy transductions. *In* Sanadi, D. R. *Curr. Topics Bioenerg.*, **3**, Acad. Press; N.Y. pp. 157–182.
2064. MUELLER, P., RUDIN, D. O., TIEN, H. T. & WESTCOTT, W. C. (1962). Reconstitution of cell membrane structure in vitro and its transformation into an excitable system. *Nature, Lond.*, **194**, 979–980.
2065. MUELLER, P., RUDIN, D. O., TIEN, H. T. & WESTCOTT, W. C. (1962).

Reconstitution of excitable cell membrane structure in vitro. *Circulation,* **26**, 1167–1172.

2066. MUGNAINI, E. & WALBERG, F. (1965). The fine structure of the capillaries and their surroundings in the cerebral hemispheres of *myxine glutinosa.* L. *Z. Zellforsch. mikrosk. Anat.,* **66**, 333–351.

2067. MÜHLETHALER, K. (1966). Freeze fracture studies of chloroplasts. *In* Goodwin, T. W. *Biochemistry of chloroplasts.* Acad. Press; London. pp. 49–73.

2068. MÜHLETHALER, K. (1972). Studies on freeze-etching of cell membranes. *Int. Rev. Cytol.,* **31**, 1–42.

2069. MÜHLETHALER, K., MOORE, H. & SZAKOWSKI, J. W. (1965). The ultrastructure of the chloroplast lamellae. *Planta,* **67**, 305–323.

2070. MUKERJEE, H., RAM, S. J. & PIERCE, G. B. (1965). Basement membranes V. Chemical composition of neoplastic basement membrane mucoprotein. *Am. J. Path.,* **46**, 49–57.

2071. MUNCK, A. (1957). The interfacial activity of steroid hormones and synthetic estrogens. *Biochim. biophys. Acta,* **24**, 507–514.

2072. MUNGER, B. L. (1966). The intraepidermal innervation of the snout skin of the opossum. *In* CIBA Fdn Symp. *Touch, heat and pain.* Churchill; pp. 129–130.

2073. MURDAUGH, H. V., ROBIN, E. D., SOTERES, P. & WEISS, E. (1965). Intracellular electrolyte patterns in the dogfish. II. Muscle and brain. *Bull. Mt. Desert Island Biol. Labs.,* **5**, 14–15.

2074. MURPHY, J. B. (1914). Factors of resistance to heteroplastic tissue grafting. Studies in tissue specificity III. *J. exp. Med.,* **19**, 515–522.

2075. MURPHY, J. B. & STRUM, E. (1923). Conditions determining the transplantability of tissues in the brain. *J. exp. Med.,* **38**, 183–197.

2076. MURRAY, A. W. & FROSCHIO, M. (1971). Cyclic adenosine 3′5′-monophosphate and microtubule function: specific interaction of the phosphorylated protein subunits with a soluble brain component. *Biochem. biophys. Res. Commun.,* **44**, 1089–1095.

2077. MURRAY, J. G. & THOMPSON, J. W. (1957). The occurrence and function of collateral sprouting in the sympathetic nervous system of the cat. *J. Physiol.,* **135**, 133–162.

2078. MURRAY, M. (1968). Effects of dehydration on the rate of proliferation of hypothalamic neuroglia cells. *Expl Neurol.,* **20**, 460–468.

2079. MURRAY, R. G. (1970). Ultrastructure of taste receptors. *In* Beidler, L. M. *Handbook of sensory physiology IV* (2). Springer Verlag; pp. 31–50.

2080. MURRAY, R. G. & MURRAY, A. (1969). Degeneration in taste buds of rabbit foliate papillae after section of the glossopharyngeal nerve. *Anat. Rec.,* **163**, 234.

2081. MUSCATELLO, V., MARGRETH, A. & ALOISI, M. (1965). On the differential response of sarcoplasm and myoplasm to denervation in frog muscle. *J. Cell Biol.,* **27**, 1–24.

2082. MYER, F. & BLOCH, K. (1963). Metabolism of stearolic acid in yeast. *J. biol. Chem.,* **238**, 2654–2659.

2083. MØLLGAARD, K., DIAMOND, M. C., BENNETT, E. L., ROSENZWEIG, M. R. & LINDNER, B. (1971). Quantitative synaptic changes with differential experience in rat brain. *Int. J. Neurosci.,* **2**, 113–128.

2084. NAFSTAD, P. H. & ANDERSEN, A. E. (1970). Ultrastructural investigation on the innervation of the Herbst corpuscle. *Z. Zellforsch. mikrosk. Anat.,* **103**, 109–114.

2085. NAGAI, R. & REBHUN, L. I. (1966). Cytoplasmic microfilaments in streaming *Nitella* cells. *J. Ultrastruct. Res.,* **14**, 571–589.

2086. NAGAKI, J. YAMASHITA, S. & SATO, M. (1964). Neural response of cat to taste stimuli of varying temperatures. *Jap. J. Physiol.,* **14**, 67–89.

2087. NAHORSKY, S. R., ROGERS, K. J. & SMITH, B. M. (1974). Regulation of the cyclic AMP concentration in chick brain by β-adrenoreceptors and histamine H_2 receptors. *Br. J. Pharmac.,* **52**, 121P.

2088. NAKAI, J. (1956). Dissociated dorsal root ganglia in tissue culture. *Am. J. Anat.,* **99**, 81–130.

2089. NAKAI, J. (1965). Tridimensional nerve formation in vitro. *Texas Rep. Biol. Med.,* **23**, 371–375.

2090. NAKAO, A., DAVIS, W. J. & EINSTEIN, E. R. (1966). Basic proteins from the acidic extract of bovine spinal cord. I. Isolation and characterization. *Biochim. biophys. Acta,* **130**, 163–170.

2091. NAKAO, A., DAVIS, W. J. & EINSTEIN, E. R. (1966). Basic proteins from the acidic extract of bovine spinal cord. II. Encephalitogenic, immunologic and structural interrelationships. *Biochim. biophys. Acta,* **130**, 171–179.

2092. NAKAO, A. & ROBOZ-EINSTEIN, E. (1961). Chemical and immunochemical studies with dialyzable encephalitogenic compound from bovine spinal cord. *Ann. N.Y. Acad. Sci.,* **122**, 171–179.

2093. NAMBA, T. & GROB, D. (1967). Cholinergic receptors in skeletal muscle. Isolation and properties of muscle ribonucleoprotein with affinity for d-tubocurarine and acetylcholine, and binding activity of the subneural apparatus of motor end plates with divalent metal ions. *Ann N.Y. Acad. Sci.,* **144**, 772–800.

2094. NANDY, K. & BOURNE, G. H. (1964). A histochemical study of the localization of the oxidative enzymes in the neurones of the spinal cord in rats. *J. Anat.,* **98**, 647–654.

2095. NAPOLITANO, L., LEBARON, F. & SCALETTI, J. (1967). Preservation of myelin lamellar structure in the absence of lipid. *J. Cell Biol.,* **34**, 817–826.

2096. NAHARASHI, T., ALBUQUERQUE, E. X. & DEGUCHI, T. (1971). Effects of batrachotoxin on ionic conductances of giant axon membranes. *J. gen. Physiol.,* **58**, 54–70.

2097. NARAHASHI, T. & MOORE, J. W. (1968). Neuroactive agents and nerve membrane conductances. *J. gen. Physiol.,* **51**, 93S.

2098. NATHANIEL, E. J. H. (1965). Fine structure of the posterior horn of the rat spinal cord. *Anat. Rec.,* **151**, 392–393.

2099. NATHANIEL, E. J. H. & NATHANIEL, D. R. (1966). Fine structure of the posterior horn in the rat spinal cord. *Anat. Rec.*, **155**, 629–642.

2100. NATHANIEL, E. J. H. & NATHANIEL, D. R. (1973). Degeneration of dorsal roots in the adult rat spinal cord. *Expl Neurol.*, **40**, 316–332.

2101. NATHANIEL, E. J. H. & NATHANIEL, D. R. (1973). Regeneration of dorsal root fibres into the adult rat spinal cord. *Expl Neurol.*, **40**, 333–350.

2102. NATHANIEL, E. J. H. & PEASE, D. C. (1963). Collagen and basement membrane formation by Schwann cells during nerve regeneration *J. Ultrastruct. Res.*, **9**, 550–560.

2103. NATHANIEL, E. J. H. & PEASE, D. C. (1963). Degenerative changes in rat dorsal roots during Wallerian degeneration. *J. Ultrastruct. Res.*, **9**, 511–532.

2104. NATHANIEL, E. J. H. & PEASE, D. C. (1963). Regenerative changes in rat dorsal roots following Wallerian degeneration. *J. Ultrastruct. Res.*, **9**, 533–549.

2105. NATHANS, D. (1964). Puromycin inhibition of protein synthesis: incorporation of puromycin into peptide chains. *Proc. natn. Acad. Sci U.S.A.*, **51**, 585–589.

2106. NATHANSON, J. A. & GREENGARD, P. (1974). Serotonin-sensitive adenylate cyclase in neural tissue and its similarity to the serotonin receptor. A possible site of action of lysergic acid diethyl amide. *Proc. natn. Acad. Sci. U.S.A.*, **71**, 797–783.

2107. NAUMANN, R. A. (1963). A unique intercellular material in the brain. *Anat. Rec.*, **145**, 266.

2108. NAUMANN, W. (1968). Histochemische Untersuchungen am Subcommisuralorgan und am Reissnerschen Faden von *Lampetra planeri* (Bloch). *Z. Zellforsch. mikrosk. Anat.*, **87**, 571–591.

2109. NAUTA, W. J. H. (1956). An experimental study of the fornix system in the rat. *J. comp. Neurol.*, **104**, 247–272.

2110. NAUTA, W. J. H. (1958). Hippocampal projections and related neural pathways to the midbrain in the cat. *Brain*, **81**, 319–340.

2111. NEET, K. E. & FREISS, S. L. (1962). Curare-binding macromolecules from medullated nerve tissue. *Archs. Biochem. Biophys.*, **99**, 484–493.

2112. NELSON, P. G. (1969). Functional consequences of tenotomy in hind limb muscles of the cat. *J. Physiol.*, **201**, 321–333.

2113. NELSON, P. G. & PEACOCK, J. H. (1973). Electrical activity in dissociated cell cultures from fetal mouse cerebellum. *Brain Res.*, **61**, 163–174.

2114. NESKOVIC, N., NUSSBAUM, J. L. & MANDEL, P. (1970). Enzymatic deficiency in neurological mutants. Brain uridine diphosphate galactose: ceramide galactosyl transferase in Jimpy mouse. *FEBS Lett.*, **8**, 213–216.

2115. NESKOVIC, N. M., NUSSBAUM, J. L. & MANDEL, P. (1970). A study of glycolipid metabolism in myelination disorder of Jimpy and Quaking mice. *Brain Res.*, **21**, 39–53.

2116. NESKOVIC, N. M., SARLIEVE, L. L. & MANDEL, P. (1972). Biosynthesis of glycolipids in myelin deficient mutants: brain glycosyl transferases in Jimpy and Quaking mice. *Brain Res.*, **42**, 147–157.

2117. NEUFELD, A. H., MILLER, W. H. & BITENSKY, M. W. (1972). Calcium binding to retinal rod disc membranes. *Biochim. biophys. Acta,* **266**, 67–71.

2118. NEUFELD, A. J. & LEVY, H. M. (1970). The steady state level of phosphorylated intermediate in relation to the two sodium-dependent adenosine triphosphatases of calf brain microsomes. *J. biol. Chem.*, **245**, 4962–4967.

2119. NEWTH, D. & USHERWOOD, P. N. R. (1974). *'Simple' nervous systems.* Edward Arnold; London.

2120. NEWTON, G. & LEVINE, S. (1968). *Early experience and behaviour. The psychobiology of development.* C. C. Thomas; Springfield, Illinois.

2121. NGUYEN-THUONG-VAN & TIEN, H. T. (1970). Black lipid membranes (BLM) in aqueous media. Photoelectric spectroscopy. *J. phys. Chem.*, **74**, 3559.

2122. NICHOLLS, J. G. (1956). The electrical properties of denervated skeletal muscle. *J. Physiol.*, **131**, 1–12.

2123. NICHOLLS, J. G. & BAYLOR, D. A. (1968). Long lasting hyperpolarization after activity of neurones in leech central nervous system. *Science, N.Y.*, **162**, 279–281.

2124. NICHOLLS, J. G. & KUFFLER, S. W. (1964). Extracellular space as a pathway for exchange between blood and neurons in the central nervous system of the leech: ionic composition of glial cells and neurons. *J. Neurophysiol.*, **27**, 645–671.

2125. NICHOLLS, J. G. & BAYLOR, D. A. (1968). Specific modalities and receptive fields of sensory neurones in the CNS of the leech. *J. Neurophysiol.*, **31**, 740–756.

2126. NICHOLLS, J. & BAYLOR, D. A. (1969). The specificity and functional role of individual cells in a simple central nervous system. *Endeavour,* **28**, 3–7.

2127. NICHOLLS, J. G. & KUFFLER, S. W. (1965). Na and K content of glial cells and neurons determined by flame photometry in the central nervous system of the leech. *J. Neurophysiol.*, **28**, 519–525.

2128. NICHOLLS, J. G. & PURVES, D. (1970). Monosynaptic chemical and electrical connections between sensory and motor cells in the central nervous system of the leech. *J. Physiol.*, **209**, 647–667.

2129. NISHI, S. & KOKETSU, K. (1966). Late after discharge of sympathetic postganglionic fibres. *Life Sci.*, **5**, 1991–1997.

2130. NISHI, S. & KOKETSU, K. (1967). Origin of ganglionic inhibitory postsynaptic potential. *Life Sci.*, **6**, 2049–2055.

2131. NISHI, S. & KOKETSU, K. (1968). Analysis of slow inhibitory postsynaptic potential of bullfrog sympathetic ganglion. *J. Neurophysiol.*, **31**, 717–728.

2132. NISHI, S., SOEDA, H. & KOKETSU, K. (1969). Unusual nature of ganglionic slow EPSP studied by a voltage-clamp method. *Life Sci.*, **8**, 33–42.

2133. NOBIN, A. BAUMGARTEN, H. G., BJÖRKLUND, A., LACHENMAYER, L. & STENEVI, U. (1973). Axonal degeneration and regeneration of the bulbospinal indolamine neurons after 5,6-dihydroxytryptamine treatment. *Brain Res.*, **56**, 1–24.
2134. NORDLANDER, R. H. & SINGER, M. (1972). Electron microscopy of severed motor fibres in the crayfish. *Z. Zellforsch. mikrosk. Anat.*, **126**, 157–181.
2135. NORSTRÖM, A., HANSSON, H.-A. & SJÖSTRAND, J. (1971). Effects of colchicine on axonal transport and ultrastructure of the hypothalamoneurohypophyseal system of the rat. *Z. Zellforsch. mikrosk. Anat.*, **113**, 271–293.
2136. NORTON, W. T. & PODUSLO, S. E. (1973). Myelination in rat brain: changes in myelin composition during brain maturation. *J. Neurochem.*, **21**, 759–773.
2137. NORTON, W. T. & PODUSLO, S. E. (1973). Myelination in rat brain: method of myelin isolation. *J. Neurochem.*, **21**, 749–757.
2138. NOVAKOVA, V., SANDRITTER, W. & SCHLUETER, G. (1970). DNA content of neurons in rat central nervous system. *Expl Cell Res.*, **60**, 454–456.
2139. NOZAWA, Y. & THOMPSON, G. A. (1971). Studies of membrane formation in tetrahymena pyriformis. *J. Cell Biol.*, **49**, 722–730.
2140. NOZAWA, Y. & THOMPSON, G. A. (1972). Studies of membrane formation in *tetrahymena pyriformis*. V. Lipid incorporation into various cellular membranes of stationary phase cells, starving cells and cells treated with metabolic inhibitors. *Biochim. biophys. Acta*, **282**, 93–104.
2141. NÜESCH, H. (1952). Über den Einfluss der Nerven auf die Muskelentwicklung bei *Telea polyphemus* (Lepid). *Revue suisse Zool.*, **59**, 294–301.
2142. NÜESCH, H. (1954). Segmentierung und Muskelinnervation bei *Telea polyphemus* (Lepid). *Revue. suisse Zool.*, **61**, 420–428.
2143. NÜESCH, H. (1968). The role of the nervous system in insect morphogenesis and regeneration. *A. Rev. Entomol.*, **13**, 27–44.
2144. NUSSBAUM, J. L., NESKOVIC, N. & MANDEL, P. (1969). A study of lipid components in the brain of the Jimpy mouse, a mutant with myelin deficiency. *J. Neurochem.*, **16**, 927–934.
2145. NUSSBAUM, J. L., NESKOVIC, N. & MANDEL, P. (1971). The fatty acid composition of phospholipids and glycolipids in Jimpy mouse brain. *J. Neurochem.*, **18**, 1529–1543.
2146. NYGREN, L.-G., FUXE, K., JONSSON, G. & OLSON, L. (1974). Functional regeneration of 5-hydroxytryptamine nerve terminals in the rat spinal cord following 5,6-dihydroxytryptamine induced degeneration. *Brain Res.*, **78**, 377–394.
2147. NYMAN, A. J. (1967). Problem solving in rats as a function of experience at different ages. *J. genet. Psychol.*, **110**, 31–39.
2148. OAKLEY, B. (1967). Altered temperature and taste responses from cross-regenerated sensory nerves in the rat's tongue. *J. Physiol.*, **188**,

353–371.

2149. OAKLEY, B. & BENJAMIN, R. M. (1966). Neural mechanisms of taste. *Physiol. Rev.*, **46**, 173–211.

2150. OBATA, K. (1974). Transmitter sensitivities of some nerve and muscle cells in culture. *Brain Res.*, **73**, 71–88.

2151. O'BRIEN, J. S. (1964). A molecular defect of myelination. *Biochem. biophys. Res. Commun.*, **15**, 484–490.

2152. O'BRIEN, J. S., SAMPSON, E. L. & STERN, M. B. (1967). Lipid composition of myelin from the peripheral nervous system. Intradural spinal roots. *J. Neurochem.*, **14**, 357–365.

2153. O'BRIEN, R. D., ELDEFRAWI, M. E. & ELDEFRAWI, A. T. (1972). Isolation of acetyl choline receptors. *A. Rev. Pharmacol.*, **12**, 19–34.

2154. O'BRIEN, R. D. & GOLMOUR, L. P. (1969). A muscarone-binding material in electroplax and its relation to the acetylcholine receptor. I. Centrifugal assay. *Proc. natn. Acad. Sci. U.S.A.*, **43**, 496–503.

2155. OCHS, S. (1971). Local supply of energy to the fast axoplasmic transport mechanism. *Proc. natn. Acad. Sci. U.S.A.*, **68**, 1279–1282.

2156. OCHS, S. (1971). The dependence of fast transport in mammalian nerve fibres on metabolism. *Acta neuropath.*, Suppl. 5, 86–96.

2157. OCHS, S. & RANISH, N. (1969). Characteristics of the fast transport system in mammalian nerve fibres. *J. Neurobiol.*, **1**, 247.

2158. OCHS, S., RANISH, N., HOLLINGWORTH, D. & HELMER, E. (1970). Dependence of fast axoplasmic transport in mammalian nerve on oxidative metabolism. *Fedn Proc. Fedn Am. Socs exp. Biol.*, **29**, 264.

2159. ODLAND, G. & ROSS, R. (1968). Human wound repair. I. Epidermal regeneration. *J. Cell Biol.*, **39**, 135–151.

2160. OEHRWALL, H. (1891). Untersuchungen über den Geschmacksinn. *Skand. Arch. Physiol.*, **2**, 1096.

2161. OGAWA, H., SATO, M. & YAMASHITA, S. (1968). Multiple sensitivity of chorda tympani fibres of the rat and hamster to gustatory and thermal stimuli. *J. Physiol.*, **199**, 223–240.

2162. OGAWA, K. & ZIMMERMANN, H. (1959). The activity of succinic dehydrogenase in the experimental ependymoma of C3H mice. *J. Histochem. Cytochem.*, **7**, 342–349.

2163. OGDEN, R. E. & MILLER, R. F. (1966). Studies of the optic nerve of the rhesus monkey: nerve fibre spectrum and physiological properties. *Vision Res.*, **6**, 485–506.

2164. OHMI, S. (1961). Electron microscopic study on Wallerian degeneration of the peripheral nerve. *Z. Zellforsch. mikrosk. Anat.*, **54**, 39–67.

2165. OHNISHI, M. & URRY, D. W. (1970). Solution conformation of valinomycin-potassium ion complex. *Science, N.Y.*, **168**, 1091–1092.

2166. OKSCHE, A. (1961). Der histochemisch rachweisbare Glycogenaufbau aud Abbau in den Astrocyten und Ependymzellen als Beispiel einer Funktionsabhängigen Stoffwechselaktivität der Neuroglia. *Z. Zellforsch. mikrosk. Anat.*, **54**, 307–361.

2167. OKUN, L. M., ONTKEAN, F. K. & THOMAS, C. A. (1972). Removal of non neuronal cells from suspensions of dissociated embryonic dorsal root ganglia. *Expl Cell. Res.*, **73**, 226–229.

2168. OLAFSON, R. W. & DRUMMOND, G. I. (1969). Subcellular distribution of 2′3′-cyclic nucleotide-3′-phosphohydrolase in brain. *Fedn Proc. Fedn Am. Socs exp. Biol.*, **28**, 733.
2169. OLAFSON, R. W., DRUMMOND, G. I. & LEE, J. F. (1969). Studies on 2′3′-cyclic nucleotide 3′-phosphohydrolase from brain. *Can. J. Biochem.*, **47**, 961–966.
2170. OLDENDORF, W. H. & DAVSON, H. (1967). Brain extracellular space and the sink action of cerebrospinal fluid. *Archs. Neurol.*, **17**, 196–205.
2171. OLIVEROS, N. L. (1965). Observations on the lining of the cavum septum pellucidi in the brain of newborn and adult man. *Anat. Rec.*, **151**, 468.
2172. OLMSTED, J. B. & BORISY, G. G. (1973). Microtubules. *A. Rev. Biochem.*, **42**, 507–540.
2173. OLMSTED, J. B., WITMAN, G. B., CARLSON, K. & ROSENBAUM. J. L. (1971). Comparison of the microtubule proteins of neuroblastoma cells, brain and *chlamydomonas* flagella. *Proc. Natn. Acad. Sci. U.S.A.*, **68**, 2273–2277.
2174. OLSON, C. B. & SWETT, C. P. (1966). A functional and histochemical characterization of motor units in a heterogeneous muscle (flexor digitorum longus) of the cat. *J. comp. Neurol.*, **128**, 475–498.
2175. OLSON, L. & MALMFORS, T. (1970). Growth characteristics of adrenergic nerves in the adult rat. Fluorescence, histochemical and ^{3}H-noradrenaline uptake studies using tissue transplantation to the anterior chamber of the eye. *Acta physiol. scand.* Suppl. **348**, 1–112.
2176. OLSON, M. I. & BUNGE, R. P. (1973). Anatomical observations on the specificity of synapse formation in tissue culture. *Brain Res.*, **59**, 19–33.
2177. OLSSON, R. (1956). The development of Reissner's fibre in the brain of the salmon. *Acta zool. Scand.*, **37**, 235.
2178. OLSSON, R. (1957). An experimental breakage of Reissner's fibre in the central canal of the pike (Esox lucius). *Z. Zellforsch. mikrosk. Anat.*, **46**, 12–17.
2179. OLSSON, R. (1958). Studies on the subcommisural organ. *Acta zool. Scand.*, **39**, 71–102.
2180. OLSSON, Y. & SJÖSTRAND, J. (1969). Origin of macrophages in Wallerian degeneration of peripheral nerves demonstrated autoradiographically. *Expl Neurol.*, **23**, 102–112.
2181. ONODERA, K., HIRANO, S., HORIUCHI, F. & KASHIMURA, N. (1966). A comparative study of some animal brains with regard to content of acidic mucopolysaccharide. *Carbohyd. Res.*, **3**, 234.
2182. OPPEL, A. (1900). *Lehrbuch der vergleichenden Mikroskopischen Anatomie der Wirbeltiere.* Fisher; Jena.
2183. ORDMANN, L. J. & GILLMAN, T. (1966). Studies on the healing of cutaneous wounds. I. The healing of incisions through the skin of pigs. *Archs. Surg.*, **93**, 857–882.

2184. ORKAND, P. M., BRACHIO, H. & ORKAND, R. K. (1973). Glial metabolism: alterations by potassium levels comparable to those during neural activity. *Brain Res.*, **55**, 467–471.

2185. ORKAND, R. K., NICHOLLS, J. G. & KUFFLER, S. W. (1966). Effect of nerve impulses on the membrane potential of glia cells in the central nervous system of amphobia. *J. Neurophysiol.*, **29** , 788–806.

2186. ORTIZ, J. R., YAMADA, T. & HSIE, A. W. (1973). Induction of the stellate configuration in cultured iris epithelial cells by adenosine and compounds related to adenosine 3′:5′-cyclic monophosphate. *Proc. natn. Acad. Sci. U.S.A.*, **70**, 2286–2290.

2187. ORTMANN, R. (1951). Morphologische und experimentelle Untersuchungen über das hypothalamisch-hypophysäre System. *Z. Zellforsch. mikrosk. Anat.*, **36**, 92–99.

2188. ORTMAN, L. L. & CRAIG, J. V. (1968). Social dominance in chickens modified by genetic selection – physiological mechanism. *Anim. Behav.*, **16**, 33–37.

2189. OSEROFF, A. R., ROBBINS, P. W. & BURGER, M. M. (1973). The cell surface membrane: biochemical aspects and biophysical probes. *A. Rev. Biochem.*, **42**, 647–682.

2190. OSHIRO, Y. & EYLAR, E. H. (1970). Allergic encephalomyelitis: a comparison of the encephalitogenic A1 protein from human and bovine brain. *Archs. Biochem. Biophys.*, **138**, 606–613.

2191. OSTERBERG, K. & WATTENBERG, L. (1962). Inductive factors in gliosis. *Proc. Soc. exp. Biol. Med.*, **111**, 452–455.

2192. OSTERBERG, K. A. & WATTENBERG, L. W. (1962). Oxidative histochemistry of reactive astrocytes. *Archs. Neurol.*, **7**, 211–218.

2193. OSTERBERG, K. & WATTENBERG, L. (1963). The age dependency of enzymes in reactive glia. *Proc. Soc. exp. Biol. Med.*, **113**, 145–147.

2194. OSTROY, S. E., RUDNEY, H. & ABRAHAMSON, E. W. (1966). The sulphydryl groups of rhodopsin. *Biochim. biophys. Acta,* **126**, 409–412.

2195. ÖTZAN, N. (1967). Neurosecretory processes projecting from the preoptic nucleus into the third ventricle of *Zoaries viviparus* (L). *Z. Zellforsch. mikrosk. Anat.*, **80**, 458–460.

2196. OWENS, K. & HUGHES, B. P. (1970). Lipids of dystrophic and normal mouse muscle: whole tissue and particulate fraction. *J. Lipid Res.*, **11**, 486–495.

2197. OZEKI, M. (1971). Conductance changes associated with receptor potentials of gustatory cells in rat. *J. gen. Physiol.*, **58**, 688–699.

2198. PADYKULA, H. A. & GAUTHIER, G. F. (1967). The ultrastructure of neuromuscular junctions of mammalian red and white skeletal muscle fibres. *J. Cell Biol.*, **35**, 155A.

2199. PAGANELLI, C. V. & SOLOMON, A. K. (1957). The rate of exchange of tritiated water across the human red cell membrane. *J. gen Physiol.*, **41**, 259–278.

2200. PAGE, S. (1968). Structure of the sarcoplasmic reticulum in vertebrate muscle. *Br. med. Bull.*, **24**, 170–173.

2201. PAGGI, P. & ROSSI, A. (1971). Effect of *Latrodectus mactans tredecimguttatus* venom on sympathetic ganglion isolated *in vitro. Toxicology,* **9**, 265.

2202. PALAY, S. L. (1966). The role of neuroglia in the organization of the central nervous system. *In* Rodahl, K. & Issekutz, B., *Nerve as a tissue.* Hoeber; N.Y. pp. 3–10.

2203. PALKOVITS, M. & FÖLDVARI, I. P. (1962). Effect of the subcommisural organ and the pineal body on the adrenal cortex. *Endocrinology,* **72**, 28–32.

2204. PANNBACKER, R. G. (1973). Control of guanylate cyclase activity in the rod outer segment. *Science, N.Y.,* **182**, 1138–1140.

2205. PANNBACKER, R. G., FLEISCHMAN, D. E. & REED, D. W. (1972). Cyclic nucleotide phosphodiesterase: high activity in mammalian photoreceptor. *Science, N.Y.,* **175**, 757–758.

2206. PANNESE, E. & FERANNINI, E. (1967). Nuclear pyknosis in neuroglia cells of normal mammals. *Acta neuropath.,* **8**, 309–319.

2207. PAPACHARALAMPOUS, N. X., SCHWINK, A. & WETZSTEIN, R. (1968). Elektronmikroskopische Untersuchungen am Subcommisuralorgan des Meerschweinchens. *Z. Zellforsch. mikrosk. Anat.,* **90**, 202–229.

2208. PAPE, L. G. & KATZMAN, R. (1972). K^{42} distribution in brain during simultaneous ventriculocisternal and arachnoid perfusion. *Brain Res.,* **38**, 49–69.

2209. PAPE, L. G. & KATZMAN, R. (1972). Response of glia in cat sensorimotor cortex to increased extracellular potassium. *Brain Res.,* **38**, 71–92.

2210. PAPPAS, G. D., ASADA, Y. & BENNETT, M. V. L. (1971). Morphological correlates of increased coupling resistance at an electrotonic synapse. *J. Cell Biol.,* **49**, 173–188.

2211. PAPPAS, G. D. & PURPURA, D. P. (1972). *Structure and function of synapses.* Appleton, Century, Crofts; N.Y.

2212. PAPPAS, G. D. & WAXMAN, S. G. (1972). Synaptic fine structure: morphological correlates of chemical and electrotonic transmission. *In* Pappas, G. D., & Purpura, D. P., *Structure and function of synapses.* Appleton, Century, Crofts; N.Y.

2213. PAPPENHEIMER, J. R. (1970). Transport of HCO_3^- between brain and blood. *In* Crone, C. & Thompson, A. M., *Capillary permeability.* Munksgaard; pp. 454–467.

2214. PARDUCZ, A. & FEHER, O. (1970). Fine structural alterations of presynaptic endings in the superior cervical ganglion of the cat after exhausting preganglionic stimulation. *Experentia,* **26**, 629–630.

2215. PARISH, H. J. & CANNON, D. A. (1962). *Antisera, toxoids, vaccines and tuberculins in prophylaxis and treatment.* Livingstone; Edinburgh.

2216. PARK, R. B. (1965). Substructure of chloroplast lamellae. *J. Cell Biol.,* **27**, 151–161.

2217. PARK, R. B. & BIGGINS, J. (1964). Quantasome: size and composition.

Science, N. Y., **144**, 1009–1011.

2218. PARK, R. B. & PON, N. G. (1963). Chemical composition and the substructure of lamellae isolated from Spinacea Oleracea chloroplasts. *J. molec. Biol.*, **6**, 105–114.

2219. PARSONS, P. J. & SPEAR, N. E. (1972). Long-term retention of avoidance learning by immature and adult rats as a function of environmental enrichment. *J. comp. Physiol. Psychol.*, **80**, 297–303.

2220. PASQUINI, J., KAPLUN, B., ARGIZ, G. C. A. & GOMEZ. C. J. (1967). Hormonal regulation of brain development. I. The effect of neonatal thyroidectomy upon nucleic acids, proteins and two enzymes in developing cerebral cortex and cerebellum of the rat. *Brain Res.*, **6**, 621–634.

2221. PASSANO, L. M. & McCULLOUGH, C. B. (1962). The light response and the rhythmic potential in *Hydra*. *Proc. natn. Acad. Sci. U.S.A.*, **48**, 1376–1382.

2222. PATERSON, P. Y. & DIDAKOW, N. C. (1961). Transfer of allergic encephalomyelitis using splenectomized albino rats. *Proc. Soc. exp. Biol. Med.*, **108**, 768.

2223. PATON, W. D. M. (1961). A theory of drug action based on the rate of drug-receptor combination. *Proc. R. Soc. B*, **154**, 21–69.

2224. PATON, W. D. M. (1967). Kinetic theories of drug action with special reference to the acetyl choline group of agonists and antagonists. *Ann. N.Y. Acad. Sci.*, **144**, 869–876.

2225. PATON, W. D. M. & RANG, H. P. (1965). The uptake of atropine and related drugs by intestinal smooth muscle of the guinea pig in relation to acetylcholine receptors. *Proc. R. Soc. B.*, **163**, 1–44.

2226. PATON, W. D. M. & ZAIMIS, E. J. (1951). The action of D-tubocurarine and decamethonium on respiratory and other muscles in the cat. *J. Physiol.*, **112**, 311–331.

2227. PATRICK, J., HEINEMANN, S. F., LINDSTROM, J., SCHUBERT, D. & STEINBACH, J. H. (1972). Appearance of acetylcholine receptors during differentiation of a myogenic muscle line. *Proc. natn. Acad. Sci. U.S.A.*, **69**, 2762–2766.

2228. PATRIZI, G. & MUNGER, B. L. (1965). The cytology of encapsulated nerve endings in the rat penis. *J. Ultrastruct. Res.*, **12**, 500–515.

2229. PATTERSON, P. H. & CHUN, L. L. Y. (1974). The influence of non-neural cells on catecholamine and acetylcholine synthesis and accumulation in cultures of dissociated sympathetic neurons. *Proc. natn. Acad. Sci. U.S.A.*, **71**, 3607.

2230. PAULING, L. (1949). Zur *cis-trans* Isomerisierung von Carotinoiden. *Helv. Chim. Acta*, **32**, 2241–2246.

2231. PAULSEN, R. & SCHWEMER, J. (1972). Studies on the insect visual pigment sensitive to ultraviolet light: retinal as the chromophore group. *Biochim. biophys. Acta*, **283**, 520–524.

2232. PAXTON, H. D. (1959). Quantitative histochemistry of brain tumours and analogous normal tissue. *Neurology, Minneap.*, **9**, 367–370.

2233. PAYTON, B. W., BENNETT, M. V. L. & PAPPAS, G. D. (1969). Temperature dependence of resistance at an electrotonic synapse. *Science, N.Y.*, **165**, 594–597.
2234. PEACHEY, L. D. (1968). Muscle. *A. Rev. Physiol.*, **30**, 401–433.
2235. PEARCE, T. L. & ZWAAN, J. (1970). A light and electron microscopic study of cell behaviour and microtubules in the embryonic chicken lens using colcemid. *J. Embryol. exp. Morph.*, **23**, 491–507.
2236. PEARSON, K. G. & BRADLEY, A. B. (1972). Specific regeneration of excitatory motorneurones to leg muscles in the cockroach. *Brain Res.*, **47**, 492–496.
2237. PEASE, D. C. (1966). Polysaccharides associated with the exterior surface of epithelial cells:— kidney, intestine, brain. *Anat. Rec.*, **154**, 400.
2238. PEASE, D. C. (1966). Polysaccharides associated with the exterior surface of epithelial cells: kidney, intestine, brain. *J. Ultrastruct. Res.*, **15**, 555–588.
2239. PEASE, D. C. & QUILLIAM. T. A. (1957). Electron microscopy of Pacinian corpuscle. *J. biophys. biochem. Cytol.*, **3**, 331–342.
2240. PELLEGRINO, C. & FRANZINI, C. (1963). An electron microscope study of denervation atrophy in red and white skeletal muscle fibres. *J. Cell Biol.*, **17**, 327–349.
2241. PELLEGRINO, C., VILLANI, G. & FRANZINI, C. (1957). Beta-glicuronidasi e fosfatasi acida nella atrofia da denervazione de muscolo tibiale anteriore di ratto. Archs. Sci. biol., **41**, 339–348.
2242. PERKINS, J. P., McINTYRE, E. H. & RILEY, W. E. (1971). Studies on adenyl cyclase, phosphodiesterase and protein kinase from cultured glial cells. *Fedn Proc. Fedn Am. Socs exp. Biol.*, **30**, 330.
2243. PERKINS, J. P., McINTYRE, E. H., RILEY, W. D. & CLARK, R. B. (1971). Adenyl cyclase, phosphodiesterase and cyclic AMP dependent protein kinase of malignant glial cells in culture. *Life Sci.*, **10** (i), 1069–1080.
2244. PERKINS, L. C., SOLOW, E. & FREEMAN, L. W. (1970). The effect of enzymatic debridement on scar formation and cavitation in experimental spinal cord transection. *Neurology, Minneap.*, **20**, 1185–1187.
2245. PERRI, V. SACCHI, O., RAVIOLA, E. & RAVIOLA, G. (1972). Evaluation of the number and distribution of synaptic vesicles at cholinergic nerve endings after sustained stimulation. *Brain Res.*, **39**, 526–529.
2246. PERRY, M. M., JOHN, H. A. & THOMAS, N. S. T. (1971). Actin-like filaments in the cleavage furrow of newt egg. *Expl Cell Res.*, **65**, 249–253.
2247. PERRY, W. L. M. & REINERT, H. (1954). The effects of preganglionic denervation on the reaction of ganglion cells. *J. Physiol.*, **126**, 101.
2248. PERT, C. B., SNOWMAN, A. M. & SNYDER, S. H. (1974). Localization of opiate receptor in synaptic membranes of rat brain. *Brain Res.*, **70**, 184–188.
2249. PERT, C. B. & SNYDER, S. H. (1973). Opiate receptor: demonstration

in nervous tissue. *Science, N.Y.*, **179**, 1011–1014.
2250. PERT, C. B. & SNYDER, S. H. (1973). Properties of opiate-receptor binding in rat brain. *Proc. natn. Acad. Sci. U.S.A.*, **70**, 2243–2247.
2251. PETERS, A. (1960). The structure of myelin sheaths in the central nervous system. *J. Anat.*, **94**, 288–289.
2252. PETERS, A. (1964). Further observations on the structure of myelin sheaths in the central nervous system. *J. Cell Biol.*, **20**, 281–296.
2253. PETERS, A. (1964). Observations on the connexions between myelin sheaths and glial cells in the optic nerve of young rats. *J. Anat.*, **98**, 125–134.
2254. PETERS, A. (1969). The ratio between myelin segments and oligodendrocytes in the optic nerve of the adult rat. *Anat. Rec.*, **169**, 243.
2255. PETERS, A. & PALAY, S. L. (1965). An electron microscopic study of the distribution and patterns of astroglial processes in the central nervous system. *J. Anat.*, **99**, 419.
2256. PETERS, A. & PALAY, S. L. (1966). The morphology of laminae A and A1 of the dorsal nucleus of the lateral geniculate body of the cat. *J. Anat.*, **100**, 451–486.
2257. PETERS, A., PALAY, S. L. & WEBSTER, H. deF. (1970). *The fine structure of the nervous system. The cells and their processes.* Hoeber, Harper and Row; N.Y.
2258. PETERS, A. & VAUGHN, J. E. (1967). Microtubules and filaments in the axons and astrocytes of early postnatal rat optic nerves. *J. Cell Biol.*, **32**, 113–119.
2259. PETERSON, E. R. & CRAIN, S. M. (1972). Regeneration and innervation in cultures of adult mammalian skeletal muscle coupled with fetal rodent spinal cord. *Expl Neurol.*, **36**, 136–159.
2260. PETERSON, E. R., CRAIN, S. M. & MURRAY, M. R. (1965). Differentiation and prolonged maintainance of bioelectrically active spinal cord cultures (rat, chick and human). *Z. Zellforsch. mikrosk. Anat.*, **66**, 130.
2261. PETERSON, N. (1962). Effect of monochromatic rearing on the control of responding by wavelength. *Science, N.Y.*, **136**, 774–775.
2262. PETERSON, R. P. & KERNELL, D. (1970). Effects of nerve stimulation on the metabolism of ribonucleic acid in a molluscan giant neurone. *J. Neurochem.*, **17**, 1075–1085.
2263. PETICOLAS, B. J. & LIPPERT, J. L. (1971). Laser Raman investigation of the effect of cholesterol on conformational changes in dipalmitoyl lecithin multilayers. *Proc. natn. Acad. Sci. U.S.A.*, **68**, 1572–1576.
2264. de PETRIS, S. & RAFF, M. C. (1972). Distribution of immunoglobulin on the surface of mouse lymphoid cells as determined by immunoferritin electron microscopy. *Europ. J. Immunol.*, **2**, 523–535.
2265. de PETRIS, S. & RAFF, M. C. (1973). Normal distribution, patching and capping of lymphocyte surface immunoglobulin studied by electron microscopy. *Nature, Lond., New Biol.*, **241**, 257–259.
2266. de PETRIS, S. & RAFF, M. C. (1973). Fluidity of the plasma membrane

and its implications for cell movement. *In Locomotion of tissue cells* CIBA Fdn Symp. 14. Elsevier, North Holland; Amsterdam. pp. 27–51.

2267. PETTIGREW, J. D. & DANIELS, J. D. (1973). GABA antagonism in visual cortex: differential effect on simple, complex and hypercomplex neurons. *Science, N.Y.*, **182**, 81–83.

2268. PETTIGREW, J. D. & GAREY, L. J. (1974). Selective modification of single neuron properties in the visual cortex of kittens. *Brain Res.*, **66**, 160–164.

2269. PETTIGREW, J. D., OLSON, C. & BARLOW, H. B. (1973). Kitten visual cortex: short-term, stimulus-induced changes in connectivity. *Science, N.Y.*, **180**, 1202–1203.

2270. PETTIGREW, J. D., OLSON, C. & HIRSCH, H. V. B. (1973). Cortical effect of selective visual experience: degeneration or reorganization? *Brain Res.*, **51**, 345–351.

2271. PFAFFMAN, C. (1941). Gustatory afferent impulses. *J. cell comp. Physiol.*, **17**, 243–258.

2272. PFAFFMAN, C. (1955). Gustatory nerve impulses in rat, cat and rabbit. *J. Neurophysiol.*, **18**, 429–440.

2273. PFAFFMAN, C. (1959). *The sense of taste.* Handbook Physiol. Neurophysiol. Sec. 1. Am. Physiol Soc., Washington D.C.

2274. PFAFFMAN, C. (1970). Physiological and behavioural processes of the sense of taste. *In Taste and smell in vertebrates.* CIBA Fdn Symp. Churchill; London. pp. 31–45.

2275. PFENNINGER, K. H. (1971). The cytochemistry of synaptic densities. I. An analysis of the bismuth iodide impregnation method. *J. Ultrastruct. Res.*, **34**, 103–122.

2276. PFENNINGER, K. (1972). Freeze-cleaving of outgrowing nerve fibres in tissue culture. *J. Cell Biol.*, **55**, 203A.

2277. PFENNINGER, K., AKERT, K., MOOR, H. & SANDRI, C. (1971). Freeze fracturing of presynaptic membranes in the central nervous system. *Phil. Trans. R. Soc. Ser. B*, **216**, 387.

2278. PFENNINGER, K., AKERT, K. MOOR, H. & SANDRI, C. (1972). The fine structure of freeze fractured presynaptic membranes. *J. Neurocytol.*, **1**, 129–149.

2279. PFENNINGER, K. H. & BUNGE, R. P. (1974). Freeze-fracturing of nerve growth cones and young fibres: a study of developing plasma membrane. *J. Cell Biol.*, **63**, 180.

2280. PFENNINGER, K. H. & RINDERER, E. R. (1975). Methods for the freeze-fracturing of nerve tissue cultures and cell monolayers. *J. Cell Biol.*, **65**, 15–28.

2281. PFENNINGER, K. H. & ROVAINEN, C. M. (1974). Stimulation and calcium-dependence of vesicle attachment sites in the presynaptic membrane; a freeze-cleave study on the lamprey spinal cord. *Brain Res.*, **72**, 1–23.

2282. PHELPS, C. H. (1969). Comparative ultrastructure of spinal cord regeneration in teleosts and mammals. *Anat. Rec.*, **169**, 244.

2283. PHILLIPS, C. G. (1956). Intracellular records from Betz cells in the cat. *Q. Jl exp. Physiol.*, **41**, 58–69.
2284. PIATT, J. & PIATT, M. (1958). Transection of the spinal cord in the adult frog. *Anat. Rec.*, **131**, 81–95.
2285. PICKEL, V. M., SEGAL, M. & BLOOM, F. E. (1974). Axonal proliferation following lesions of cerebellar peduncles. A combined fluorescence microscopic and radioautographic study. *J. comp. Neurol.*, **155**, 43.
2286. PIERCE, G. B. (1970). Epithelial basement membrane: origin, development and role in disease. *In* Balazs, E. A., *Chemistry and molecular biology of the intercellular matrix.* Acad Press. pp. 471–506.
2287. PIERCE, G., BARRY, J. & NAKANE, P. K. (1967). Antigens of epithelial basement membranes of mouse, rat and man. *Lab. Invest.*, **17**, 499.
2288. PIERCE, G. B., BEALS, T. F., RAM, J. S. & MIDGLEY, A. R. (1964). Basement membranes. IV. Epithelial origin and immunologic cross reaction. *Am. J. Path.*, **45**, 929–961.
2289. PINCHING, A. J. & DØVING, K. B. (1974). Selective degeneration in the rat olfactory bulb following exposure to different odours. *Brain Res.*, **82**, 195–204.
2290. PINCHING, A. J. & POWELL, T. P. S. (1971). Ultrastructural features of transneuronal cell degeneration in the olfactory system. *J. Cell Sci.*, **8**, 253–287.
2291. PINCHING, A. J. & POWELL, T. P. S. (1972). A study of terminal degeneration in the olfactory bulb of the rat. *J. Cell Sci.*, **10**, 585.
2292. PINKERTON, M. L. K., STEINRAUF, L. K. & DAWKINS, P. (1969). The molecular structure and some transport properties of valinomycin. *Biochem. biophys. Res. Commun.* **35**, 512–518.
2293. PIPA, R. L. (1963). Studies on the hexapod nervous system. VI. Ventral nerve cord shortening. *Biol. Bull.*, **124**, 293–302.
2294. PIPA, R. L. (1967). Insect neurometamorphosis. III. Nerve cord shortening in a moth, Galleria mellonella (L) may be accomplished by humoral potentiation of neuroglial motility. *J. exp. Zool.*, **164**, 47–60.
2295. PIPA, R. L. (1969). Insect neurometamorphosis. IV. Effects of the brain and synthetic α-ecdysone upon interganglionic connective shortening in *Galleria mellonella* (L) (Lepidoptera). *J. exp. Zool.*, **170**, 181.
2296. PIPA, R. L. & WOOLEVER, P. S. (1964). Insect metamorphosis. I. Histological changes during ventral nerve cord shortening in *Galleria mellonella* (L) (Lepidoptera). *Z. Zellforsch. mikrosk Anat.*, **63**, 405–417.
2297. PIPA, R. L. & WOOLEVER, P. S. (1965). Insect metamorphosis II. The fine structure of perineurial connective tissue, adipohemocytes and the shortening ventral nerve cord of a moth, *Galleria mellonella* (L). *Z. Zellforsch. mikrosk. Anat.*, **68**, 80.
2298. PISETSKY, D. & TERRY, T. M. (1974). Are mycoplasma membrane proteins affected by variations in membrane fatty acid composition? *Biochim. biophys. Acta*, **274**, 95–104.

2299. PITOTTI, A., CONTESSA, A. R., DABBENI-SALA, F. & BRUNI, A. (1972). Activation by phospholipids of particulate mitochondrial ATPase from rat liver. *Biochim. biophys. Acta,* **274**, 528–535.

2300. de PLAZAS, S. F. & de ROBERTIS, E. (1972). Binding of α-bungarotoxin to the cholinergic receptor proteolipid from *electrophorus* electroplax. *Biochim. biophys. Acta,* **274**, 258–265.

2301. de PLAZAS, S. F. & de ROBERTIS, E. (1974). Isolation of hydrophobic proteins binding neurotransmitter amino acids. Glutamate receptor of the shrimp muscle. *J. Neurochem.,* **23**, 1115–1120.

2302. PLEASURE, D., BORA, F. W., LANE, J. & PROCKOP, D. (1974). Regeneration after nerve transection: effect of inhibition of collagen synthesis. *Expl Neurol.,* **45**, 72–78.

2303. PLOTNIKOFF, N. (1963). A neuropharmacological study of escape from audiogenic seizures. *Colloq. int. Centre nat. Rech. sci., Paris,* **112**, 429–446.

2304. PODELSKI, T. R. (1967). Distinction between the active sites of acetyl choline receptor and acetylcholine esterase. *Proc. natn Acad. Sci. U.S.A.,* **58**, 268–273.

2305. PODLESKI, T. R. & CHANGEUX, J.-P. (1969). The acetyl choline receptor. *In* Danielli, J. F., Moran, J. F. & Triggle, D. J. *Fundamental concepts in drug-receptor interactions.* Acad. Press; N.Y. pp. 93–120.

2306. POINCELOT, R. P. & ABRAHAMSON, E. W. (1970). Fatty acid composition of bovine rod outer segments and rhodopsin. *Biochim. biophys. Acta,* **202**, 382–385.

2307. POINCELOT, R. P., MILLER, P. G., KIMBEL, R. L. & ABRAHAMSON, E. W. (1969). Lipid to protein chromophore transfer in the photolysis of visual pigments. *Nature, Lond.,* **221**, 256–257.

2308. POINCELOT, R. P., MILLER, P. G., KIMBEL, R. L. & ABRAHAMSON, E. W. (1970). Determination of the chromophore binding site in native bovine rhodopsin. *Biochemistry, N.Y.,* **9**, 1809–1816.

2309. POINCELOT, R. P. & ZULL, J. E. (1969). Phospholipid composition and extractability of light and dark adapted bovine retinal rod outer segments. *Vision Res.,* **9**, 647–651.

2310. POLACEK, P. & MAZANEK, K. (1966). Ultrastructure of mature pacinian corpuscles from mesentery of adult cat. *Z. mikrosk.–anat. Forsch.,* **75**, 343–354.

2311. POLEZAJEV, L. W. (1939). Über die Bedeutung des Nervensystems bei der Regeneration der Extremitäten bei den Anuran. *C.r. Acad. Sci. URSS,* **25**, 543–546.

2312. POLLARD, T. D. & KORN, E. D. (1971). Filaments of *amoeba proteus.* II. Binding of heavy meromyosin by thin filaments in motile cytoplasmic extracts. *J. Cell Biol.,* **48**, 216–219.

2313. POLLARD, T. D., SHELTON, E. & KORN, E. D. (1970). The validity of identifying filaments in amoebas as F-actin by their ability to form an arrowhead complex with muscle heavy meromyosin. *J. Cell Biol.,* **47**, 159–160A.

2314. POLLARD, T. D., SHELTON, E., WEIHING, R. R. & KORN, E. D. (1970). Ultrastructural characterization of F-actin isolated from acanthamoeba castellani and identification of cytoplasmic filaments as F-actin by reaction with rabbit heavy meromyosin. *J. molec. Biol.*, **50**, 91–97.
2315. POLLAY, M. & KAPLAN, R. J. (1970). Diffusion of non-electrolyte in brain tissue. *Brain Res.*, **17**, 407–416.
2316. POLLEN, D. A. & TRACHTENBERG, M. L. (1970). Neuroglia: gliosis and focal epilepsy. *Science, N.Y.*, **167**, 1252–1253.
2317. POLYZHAYEV, L. V. & ERMAKOVA, N. I. (1960). Restoration of regenerative capacity of the extremities in axolotls depressed by roentgen radiation. *Dokl. Akad. Nauk SSSR.*, **131**, 209–212.
2318. POLYZHAYEV, L. V. & FAWORINA, W. N. (1935). Über die Rolle des Epithels in den anfänglichen Entwicklungsstadien einer Regenerationsanlage der Extremität beim Axolotl. *Arch. EntwMech. Org.*, **133**, 701–727.
2319. POMERAT, C. M. (1951). Rhythmic contraction of Schwann cells. *Science, N.Y.*, **130**, 1759–1760.
2320. PORITSKY, R. L. & SINGER, M. (1963). The fate of taste buds in tongue transplants to the orbit in the urodele, *Triturus*. *J. exp. Zool.*, **153**, 211–218.
2321. PORTER, K. R. (1966). Cytoplasmic microtubules and their function. *In Principles of biomolecular organization.* CIBA Fdn Symp. Churchill; London. pp. 308–343.
2322. POST, R. L., SEN, A. K. & ROSENTHAL, A. S. (1965). A phosphorylated intermediate in adenosine triphosphate-dependent sodium and potassium transport across kidney membranes. *J. biol. Chem.*, **240**, 1437–1445.
2323. POSTE, G. (1970). Virus-induced polykaryocytosis and the mechanism of cell fusion. *Adv. Virus Res.*, **16**, 303–356.
2324. POSTE, G. (1971). Sublethal autolysis. Modification of cell periphery by lysosomal enzymes. *Expl Cell Res.*, **67**, 11–16.
2325. POSTE, G. (1971). Tissue dissociation with proteolytic enzymes. Adsorption and activity of enzymes at the cell surface. *Expl Cell Res.*, **65**, 359–367.
2326. POSTE, G. (1972). Mechanisms of virus-induced cell fusion. *Int. Rev. Cytol.*, **33**, 157–252.
2327. de POTTER, W. P., SMITH, A. D. & de SCHAEP-DRYVER, A. F. (1970). Subcellular fractionation of splenic nerve: ATP, chromogranin A and dopamine-β-hydroxylase in noradrenergic vesicles. *Tiss. Cell,* **2**, 529–546.
2328. POWELL, J. A. (1973). Development of normal and genetically dystrophic mouse muscle in tissue culture. *Expl Cell Res.*, **80**, 251–264.
2329. POWELL, J. A. & POWERS, C. (1973). Effect on lens regeneration of implantation of spinal ganglia into the eyes of the newt *Notophthalmus*. *J. exp. Zool.*, **183**, 95–114,

2330. POWELL, T. P. S. & ERULKAR, S. D. (1962). Transneuronal cell degeneration in the auditory relay nuclei of the cat. *J. Anat.*, **96**, 249–268.
2331. PRATT, R. T. C. (1967). *The genetics of neurological disorders.* Oxford Univ. Press.
2332. PRESTEGARD, J. H. & CHAN, S. I. (1969). Proton magnetic resonance studies of the cation-binding properties of nonactin. I. The K^+-nonactin complex. *Biochemistry, N.Y.*, **8**, 3921–3927.
2333. PRESTIGE, M. (1974). Axon and cell numbers in the developing nervous system. *Br. med. Bull.*, **30**, 107–111.
2334. PREWITT, M. A. & SALAFSKY, B. (1967). Effect of cross-innervation on biochemical characteristics of skeletal muscles. *Am. J. Physiol.*, **213**, 295–300.
2335. PRICE, J. L. & POWELL, T. P. S. (1971). Certain observations on the olfactory pathway. *J. Anat.*, **110**, 105–126.
2336. PRICER, W. E. & ASHWELL, G. (1971). The binding of desialated glycoproteins by plasma membranes of rat liver. *J. biol. Chem.*, **246**, 4825–4833.
2337. PRIETO, A., KORNBLITH, P. L. & POLLEN, D. A. (1967). Electrical recording from meningioma cells during cytolytic action of antibody and complement. *Science, N.Y.*, **157**, 1185–1187.
2338. PRINCE, D. A. (1968). The depolarization shift in "epileptic" neurons. *Expl Neurol.*, **21**, 467–485.
2339. PRINCE, D. A. (1968). Inhibition in "epileptic" neurons. *Expl Neurol.*, **21**, 307–321.
2340. PRINCE, D. A. & FUTAMACHI, K. J. (1970). Intracellular recording from chronic epileptogenic foci in the monkey. *Electroenceph. clin. Neurophysiol.*, **29**, 496–509.
2341. PRINCE, D. A., LUX, H. D. & NEHER, E. (1973). Measurement of extracellular potassium activity in cat cortex. *Brain Res.*, **50**, 489–495.
2342. PRIVAT, A., DRIAN, M. J. & MANDON, P. (1973). The outgrowth of rat cerebellum in organized culture. *Z. Zellforsch. mikrosk. Anat.*, **146**, 45–67.
2343. PROCTOR, F., PRINCE, D. & MORRELL, F. (1966). Primary and secondary spike foci following depth lesions. *Archs. Neurol.*, **15**, 151.
2344. PULL, I. & McILWAIN, H. (1972). Metabolism of ^{14}C-adenine and derivatives by cerebral tissues, superfused and electrically stimulated. *Biochem. J.*, **126**, 965–973.
2345. PULL, I. & McILWAIN, H. (1972). Output of ^{14}C-adenine derivatives on electrical excitation of tissues from the brain: calcium ion-sensitivity and an accompanying reuptake process. *Biochem. J.*, **127**, 91P.
2346. PURPURA, D. P. & HOUSEPIAN, E. M. (1961). Morphological and physiological properties of chronically isolated immature neocortex. *Expl Neurol.*, **4**, 377–401.
2347. PURPURA, D. P. & McMURTY, J. G. (1965). Intracellular activities and evoked potential changes during polarization of motor cortex. *J.*

Neurophysiol., **28**, 166–185.
2348. PUSZKIN, S. & BERL, S. (1970). Actin-like properties of colchicine-binding protein isolated from brain. *Nature, Lond.* **225**, 558–559.
2349. PUSZKIN, S. & BERL. S. (1972). Actomyosin-like protein from brain. *Biochim. biophys. Acta,* **256**, 695–709.
2350. PUSZKIN, S., BERL, S., PUSZKIN, E. & CLARKE, D. D. (1968). Actomyosin-like protein isolated from mammalian brain. *Science, N.Y.*, **161**, 170–171.
2351. PUSZKIN, S., NICKLAS, W. J. & BERL, S. (1972). Actomyosin – like protein in brain: subcellular distribution. *J. Neurochem.*, **19**, 1319–1333.
2352. PYSH, J. J. & KHAN, T. (1972). Variations in mitochondrial structure and content of neurons and neuroglia in rat brain: an electron microscopic study. *Brain Res.*, **36**, 1–18.
2353. PYSH, J. J. & WILEY, R. G. (1972). Morphologic alterations of synapses in electrically stimulated superior cervical ganglia of the cat. *Science, N.Y.*, **176**, 191–193.
2354. PYSH, J. J. & WILEY, R. G. (1974). Synaptic vesicle depletion and recovery in cat synpathetic ganglia electrically stimulated in vivo. *J. Cell Biol.*, **60**, 365–374.
2355. QUARTERMAIN, D., McEWEN, B. S. & AZMITIA, E. C. (1970). Amnesia produced by electroconvulsive shock or cycloheximide: conditions for recovery. *Science, N.Y.*, **169**, 683–686.
2356. QUARTERMAIN, D., McEWEN, B. S. & AZMITIA, E. C. (1972). Recovery of memory following amnesia in the rat and mouse. *J. comp. physiol. Psychol.*, **79**, 360–370.
2357. QUAY, W. B. (1960). Experimental and comparative studies of succinic dehydrogenase activity in mammalian choroid plexus, ependyma and pineal organ. *Physiol. Zool.*, **33**, 206–212.
2358. QUILLIAM, T. A. (1962). Growth, degrowth and regrowth in the Herbst corpuscle. *Anat. Rec.*, **142**, 322.
2359. QUILLIAM, T. A. (1966). Structure of receptor organs. Unit design and array patterns in receptor organs. *In Touch, heat and pain.* CIBA Fdn Symp. Churchill; London. pp. 86–112.
2360. RACHMANOW, A. (1913). Beiträge zur vitalen Farbung des Centralnervensystems. *Folia neurol. biol.*, **7**, 750–771.
2361. RACINE, R. J. (1972). Modification of seizure activity by electrical stimulation. I. After discharge threshold. *Electroenceph. clin. Neurophysiol.*, **32**, 269–279.
2362. RACINE, R. J. (1972). Modification of seizure activity by electrical activity. II. Motor seizure. *Electroenceph. clin. Neurophysiol.*, **32**, 281–294.
2363. RACINE, R. J., GARTNER, J. G. & BURNHAM, W. M. (1972). Epileptiform activity and neural plasticity in limbic structures. *Brain Res.*, **47**, 262–268.
2364. RACINE, R., OKUJAVA, V. & CHIPASHVILI, S. (1972). Modification of seizure activity by electrical stimulation. III. Mechanisms. *Electro-*

enceph. clin. Neurophysiol., **32**, 295–299.

2365. RADDING, C. M. & WALD, G. (1955). Acid-base properties of rhodopsin and opsin. *J. gen. Physiol.*, **39**, 909–922.
2366. RADDING, C. M. & WALD, G. (1956). The stability of rhodopsin and opsin. *J. gen. Physiol.*, **39**, 923–940.
2367. RAINE, C. S. & BORNSTEIN, M. B. (1970). Experimental allergic encephalomyelitis: an ultrastructural study of experimental demyelination in vitro. *J. Neuropath. exp. Neurol.*, **29**, 177–191.
2368. RAINE, C. S., GHETTI, B. & SHELANSKI, M. L. (1971). On the association between microtubules and mitochondria within axons. *Brain Res.*, **34**, 389–393.
2369. RAINE, C. S., PODUSLO, S. E. & NORTON, W. T. (1971). The ultrastructure of purified preparations of neurons and glial cells. *Brain Res.*, **27**, 11–24.
2370. RAISMAN, G. (1966). The connections of the septum. *Brain*, **89**, 317–348.
2371. RAISMAN, G. (1969). Neuronal plasticity in the septal nuclei of the adult rat. *Brain Res.*, **14**, 25–48.
2372. RAISMAN, G. (1969). A comparison of the mode of termination of the hippocampal and hypothalamic afferents to the septal nuclei as revealed by electron microscopy of degeneration. *Expl Brain Res.*, **7**, 317–343.
2373. RAISMAN, G. (1970). An evaluation of the basic pattern of connections between the limbic system and the hypothalamus. *Am. J. Anat.*, **129**, 197–202.
2374. RAISMAN, G. (1972). An experimental study of the projection of the amygdala to the accessory olfactory bulb and its relationship to the concept of a dual olfactory system. *Expl Brain Res.*, **14**, 395–408.
2375. RAISMAN, G. (1973). Electron microscopic studies of the development of new neurohaemal contacts in the median eminence of the rat after hypophysectomy. *Brain Res.*, **55**, 245–261.
2376. RAISMAN, G. (1973). An ultrastructural study of the effects of hypophysectomy on the supraoptic nucleus of the rat. *J. comp. Neurol.*, **147**, 181–208.
2377. RAISMAN, G., COWAN, W. M. & POWELL, T. P. S. (1965). The extrinsic afferent, commisural and association fibres of the hippocampus. *Brain*, **88**, 963–996.
2378. RAISMAN, G. & FIELD, P. M. (1971). Sexual dimorphism in the preoptic area of the rat. *Science, N.Y.*, **173**, 731–733.
2379. RAISMAN, G. & FIELD, P. M. (1973). A quantitative investigation of the development of collateral innervation after partial deafferentation of the septal nuclei. *Brain Res.*, **50**, 241–264.
2380. RAISMAN, G. & FIELD, P. M. (1973). Sexual dimorphism in the neuropil of the preoptic area of the rat and its dependence on neonatal androgen. *Brain Res.*, **54**, 1–29.
2381. RAJAM, P. C., GAUDREAU, C. J., GRADY, A. & RUNDLETT, S. T. (1969). Preparation, derivation and partial characterization of organ

specific antigen from human brain. *Immunology,* **17**, 367.

2382. RAJAM, P. C., GAUDREAU, C. J., GRADY, A. & RUNDLETT, S. T. (1969). Species distribution of ribonucleoprotein-derived brain antigen. *Immunology,* **17**, 813.

2383. RAKIC, P. (1965). Mesocolic recess in the human brain. *Neurology, Minneap.,* **15**, 708–715.

2384. RAKIC, P. (1971). Neuron-glia relationship during granule cell migration in developing cerebellar cortex. A Golgi and electronmicroscopic study in Macacus Rhesus. *J. comp. Neurol.,* **141**, 283–312.

2385. RAKIC, P. (1971). Guidance of neurons migrating to the fetal monkey neocortex. *Brain Res.,* **33**, 471–476.

2386. RAKIC, P. (1972). Mode of cell migration to superficial layers of fetal monkey neocortex. *J. comp. Neurol.,* **145**, 61–84.

2387. RAKIC, P. (1972). Extensive cytological determinants of basket and stellate cell dendritic pattern in the cerebellar molecular layer. *J. comp. Neurol.,* **146**, 335–354.

2388. RAKIC, P. & SIDMAN, R. L. (1968). Subcommisural organ and adjacent ependyma: autoradiographic study of their origin in the mouse brain. *Am. J. Anat.,* **122**, 317–336.

2389. RAKIC, P. & SIDMAN, R. L. (1973). Organization of cerebellar cortex secondary to deficit of granule cells in weaver mutant mice. *J. comp. Neurol.,* **152**, 133–162.

2390. RAKIC, P. & SIDMAN, R. L. (1973). Weaver mutant mouse cerebellum: defective neuronal migration secondary to abnormality of Bergman glia. *Proc. natn. Acad. Sci. U.S.A.,* **70**, 240–244.

2391. RALL, D. P. (1967). Transport through the ependymal linings. *In* Lajtha, A. & Ford, D. H. (1967). *Prog. Brain Res.* **29**, 159–167. Elsevier. Amsterdam.

2392. RALL, D. P., OPPELT, W. W. & PATLAK, C. S. (1962). Extracellular space of the brain as determined by diffusion of inulin from ventricular system. *Life Sci.,* **2**, 43–48.

2393. RALL, T. W. (1971). Studies on the formation and metabolism of cyclic AMP in the mammalian central nervous system. *Ann N.Y. Acad. Sci.,* **185**, 520–530.

2394. RALSTON, H. J. & CHOW, K. L. (1973). Synaptic reorganization in the degenerating lateral geniculate nucleus of the rabbit. *J. comp. Neurol.,* **147**, 321–350.

2395. RAMBOURG, A. & LEBLOND, C. P. (1967). Staining of basement membranes and associated structures by the periodic acid–Schiff and periodic acid – silver methenamine techniques. *J. Ultrastruct Res.,* **20**, 306–309.

2396. RAMBOURG, A. & LEBLOND, C. P. (1967). Electron microscope observations on the carbohydrate-rich cell coat present at the surface of cells in the rat. *J. Cell Biol.,* **32**, 27–53.

2397. RANCK, J. B. (1963). Specific impedance of rabbit cerebral cortex. *Expl Neurol.,* **7**, 144–152.

2398. RANCK, J. B. (1963). Analysis of specific impedance of rabbit cerebral cortex. *Expl Neurol.*, 7, 153–174.
2399. RANCK, J. B. (1966). Electrical impedance in the subicular area of rats during paradoxical sleep. *Expl Neurol.*, **16**, 416–437.
2400. RANDALL, W. C. (1969). Effects of denervation on insect muscle. *27th Ann. Proc. Electron Microscop. Soc. Am.*
2401. RANDALL, W. C. & PIPA, R. L. (1969). Ultrastructural and functional changes during metamorphosis of a proleg muscle and its innervation in *Galleria mellonella* (L) (Lepidoptera: Pyralididae). *J. Morph.*, **128**, 171–193.
2402. RANG, H. P. & RITTER, J. M. (1969). A new kind of drug antagonism: evidence that agonists cause a molecular change in acetylcholine receptors. *Molec. Pharmac.* **5**, 394–411.
2403. RANG, H. P. & RITTER, J. M. (1970). On the mechanism of desensitization at cholinergic receptors. *Molec. Pharmac.* **6**, 357–382.
2404. RANG, H. P. & RITTER, J. M. (1970). The relationship between desensitization and metaphilic effect at cholinergic receptors. *Molec. Pharmac.* **6**, 383–390.
2405. RANSOM, B. R. & GOLDRING, S. (1973). Ionic determinants of membrane potential of cells presumed to be glia in cerebral cortex of cat. *J. Neurophysiol.*, **36**, 855–868.
2406. RANSOM, B. R. & GOLDRING, S. (1973). Slow depolarization in cells presumed to be glia in cerebral cortex of the cat. *J. Neurophysiol.*, **36**, 869–878.
2407. RANSOM, B. R. & GOLDRING, S. (1973). Slow hyperpolarization in cells presumed to be glia in cerebral cortex of cat. *J. Neurophysiol.*, **36**, 879–892.
2408. RANSON, S. W. (1912). Degeneration and regeneration of nerve fibres. *J. comp. Neurol.*, **22**, 487–545.
2409. RANVIER, M. (1878). *Leçon sur l'histologie du système nerveux.* Paris.
2410. RAPPAPORT, L., LETERRIER, J. F. & NUNEZ, J. (1972). Nonphosphorylation *in vitro* of the 6S tubulin from brain and thyroid tissue. *FEBS Lett.*, **26**, 349–352.
2411. RAPPUZZI, G. & CASELLA, C. (1965). Innervation of the fungiform papillae in the frog tongue. *J. Neurophysiol.*, **28**, 154–165.
2412. RAWLINS, F. A. (1973). A time-sequence autoradiographic study of the in vivo incorporation of 1–2 (^{3}H) cholesterol into peripheral nerve myelin. *J. Cell Biol.*, **58**, 42–53.
2413. RAWLINS, F. A. (1973). A quantitative electron microscopic analysis of myelination in the optic nerve of suckling rats treated with an inhibitor of cholesterol biosynthesis. *Z. Zellforsch. mikrosk. Anat.*, **140**, 9–23.
2414. RAWLINS, F. A., HEDLEY-WHYTE, E. T., VILLEGAS, G. M. & UZMAN, B. G. (1970). Reutilization of cholesterol-1,2-^{3}H in the regeneration of peripheral nerve: an autoradiographic study. *Lab. Invest.*, **22**, 237.

2415. RAWLINS, F. A. & SMITH, M. E. (1971). Myelin synthesis *in vitro:* a comparative study of central and peripheral nervous tissue. *J. Neurochem.*, **18**, 1861–1870.
2416. RAWLINS, F. A. & UZMAN, B. G. (1970). Retardation of peripheral nerve myelination in mice treated with inhibitors of cholesterol biosynthesis. A quantitative electron microscopic study. *J. Cell Biol.*, **46**, 505–517.
2417. RAWLINS, F. A., VILLEGAS, G. M., HEDLEY-WHYTE, E. T. & UZMAN, B. G. (1972). Fine structural localization of cholesterol-1,2-^{3}H in degenerating and regenerating mouse sciatic nerve. *J. Cell Biol.*, **52**, 615–625.
2418. RAZIN, S., NE'EMAN, Z. & OHAD, I. (1969). Selective reaggregation of solubilized mycoplasma membrane proteins and the kinetics of membrane reformation. *Biochim. biophys. Acta*, **193**, 277–293.
2419. REASOR, M. J. & KANFER, J. N. (1969). Alterations of the sphingolipid composition in the brains of the "quaking" mouse and the "jimpy" mouse. *Life Sci.*, **8**, 1055–1060.
2420. REBHUN, L. I. (1972). Polarized intracellular particle transport: saltatory movement and cytoplasmic streaming. *Int. Rev. Cytol.*, **32**, 93–131.
2421. REDFERN, P., LUNDH, H. & THESLEFF, S. (1970). Tetrodotoxin resistant action potentials in denervated rat skeletal muscle. *Europ. J. Pharmac.*, **11**, 263–265.
2422. REDFERN, P. & THESLEFF, S. (1971). Action potential generation in denervated rat skeletal muscle. I. Quantitative aspects. *Acta physiol. scand.*, **81**, 557–564.
2423. REDFERN, P. & THESLEFF, S. (1971). Action potential generation in denervated rat skeletal muscle. II. The action of tetrodotoxin. *Acta physiol. scand.*, **82**, 70–78.
2424. REES, R. & BUNGE, R. P. (1974). Morphological and cytochemical studies of synapses formed in culture between isolated rat superior cervical ganglion neurones. *J. comp. Neurol.*, **157**, 1–12.
2425. REESE, T. S. & KARNOVSKY, M. J. (1967). Fine structural localization of a bloodbrain barrier for exogenous peroxidase. *J. Cell Biol.*, **34**, 207–217.
2426. REEVES, J. P., SCHECHTER, E., WEIL, R. & KABACK, H. R. (1973). Dansyl-galactoside, a fluorescent probe of active transport in bacterial membrane vesicles. *Proc. natn. Acad. Sci. U.S.A.*, **70**, 2722–2726.
2427. REID, G. (1941). A comparison of the effects of disuse and denervation upon skeletal muscle. *Med. J. Aust.*, **2**, 165–167.
2428. REITSMA, W. (1970). Some structural changes in skeletal muscles of the rat after intensive training. *Acta morph. neerl.–Scand.*, **7**, 229–249.
2429. RENNELS, E. G. & DRAGER, G. A. (1955). The relationship of pituicytes to neurosecretion. *Anat. Rec.*, **122**, 193–200.
2430. REVEL, J. P. (1962). The sarcoplasmic reticulum of the bat cricothyroid muscle. *J. Cell Biol.*, **12**, 571–588.
2431. REYNOLDS, J. A. (1972). Are inorganic cations essential for the

stability of biological membranes. *Ann. N.Y. Acad. Sci.*, **195**, 75–85.

2432. REZNIK, M. (1969). Origin of myoblasts during skeletal muscle regeneration. Electron microscopic observations. *Lab. Invest.*, **20**, 353–363.
2433. RICHARDS, W. & KALIL, R. (1974). Dissociation of retinal fibres by degeneration rate. *Brain Res.*, 72, 288.
2434. RICHARDSON, K. C. (1966). Electron microscopic identification of autonomic nerve endings. *Nature, N.Y.*, **210**, 756.
2435. RICHARDSON, S. H., HULTIN, H. O. & GREEN, D. E. (1963). Structural proteins of membrane systems. *Proc. natn. Acad. Sci. U.S.A.*, **50**, 821–827.
2436. RIDDIFORD, L. M. (1970). Antennal proteins of saturniid moths – their possible role in olfaction. *J. Insect Physiol.*, **16**, 653–660.
2437. RIDLEY, A. (1970). A biopsy study of the innervation of forearm skin grafted to the finger tip. *Brain*, **93**, 547–554.
2438. RIDLEY, A. & CAVANAGH, J. B. (1969). The cellular reactions to heterologous, homologous and autologous skin implanted into brain. *J. Path. Bact.*, **99**, 193–203.
2439. RIEGE, W. H. (1971). Environmental influences on brain and behaviour of old rats. *Devl Psychobiol.*, **4**, 157–167.
2440. RIEGE, W. H. & MORIMOTO, H. (1970). Effects of chronic stress and differential environments upon brain weights and biogenic amine levels. *J. comp. physiol. Psychol.*, **71**, 396–404.
2441. RIESSEN, A. H. (1961). Excessive arousal effects of stimulation after early sensory deprivation. *In* Solomon, P., Kubzansky, P. E., Liederman, P. H., Mendelson, J. H., Trumbull, R. & Wexler, D., *Sensory deprivation.* Cambridge Harvard Univ. Press. pp. 34–40.
2442. RIESSEN, A. H. & MELLINGER, J. C. (1956). Interocular transfer of habits in cats after alternating monocular visual experience. *J. comp. physiol. Psychol.*, **49**, 516–520.
2443. RILEY, W. J. (1966). Nerve degeneration and regeneration *in vitro*. An experimental study. *Diss. Abstr. Sec. B.*, **27**, 675.
2444. RINGO, D. L. (1967). The arrangement of subunits in flagellar fibres. *J. Ultrastruct. Res.*, **17**, 266.
2445. RITCH, R. & PHILPOTT, C. W. (1969). Repeating particles associated with an electrolyte-transport membrane. *Expl Cell Res.*, **55**, 17–24.
2446. RITCHIE, J. M. (1967). On the role of acetylcholine in conduction in mammalian non-myelinated nerve fibres. *Ann. N.Y. Acad. Sci.*, **144**, 505–516.
2447. RITCHIE, J. M. & ARMETT, C. J. (1963). On the role of acetylcholine in conduction in mammalian non-myelinated nerve fibres. *J. Pharmac. exp. Ther.*, **139**, 201–205.
2448. ROBBINS, N. & FISCHBACH, G. D. (1971). Effect of chronic disuse of rat soleus neuromuscular junctions on presynaptic function. *J. Neurophysiol.*, **34**, 570–578.
2449. ROBBINS, N. & YONEZAWA, T. (1971). Development of neuro-

muscular junctions: first signs of chemical transmission during formation in tissue culture. *Science,* **172**, 395–397.

2450. ROBBINS, N. & YONEZAWA, T. (1971). Physiological studies during formation and development of rat neuromuscular junction in tissue culture. *J. gen. Physiol.,* **58**, 467–481.

2451. ROBERT, E. D. & OESTER, Y. T. (1970). Absence of supersensitivity to acetylcholine in innervated muscle subjected to prolonged pharmacologic block. *J. Pharmac. exp. Ther.,* **174**, 133–140.

2452. de ROBERTIS, E., FISZER, S., PASQUINI, J. M. & SOTO, E. G. (1969). Isolation and chemical nature of the receptor for d-tubocurarine in nerve-ending membranes of the cerebral cortex. *J. Neurobiol.,* **1**, 41–52.

2453. de ROBERTIS, E., FISZER, S. & SOTO, E. F. (1967). Cholinergic binding capacity of proteolipids from isolated nerve-ending membranes. *Science, N.Y.,* **158**, 928–929.

2454. de ROBERTIS, E. & GERSCHENFELD, H. M. (1961). Submicroscopic morphology and function of glial cells. *Int. Rev. Biol.,* **3**, 1–65.

2455. ROBERTS, A. (1969). Conducted impulses in the skin of young tadpoles. *Nature, Lond.,* **222**, 1265–1266.

2456. ROBERTS, A. (1971). The role of propagated skin impulses in the sensory system of young tadpoles. *Z. vergl. Physiol.,* **75**, 388–401.

2457. ROBERTS, B. L. (1972). Activity of lateral-line sense organs in swimming dogfish. *J. exp. Biol.,* **56**, 105.

2458. ROBERTS, B. L. & RUSSELL, I. J. (1970). Efferent activity in the lateral line nerve of the dogfish. *J. Physiol.,* **208**, 37P.

2459. ROBERTS, R. B. & FLEXNER, L. B. (1969). The biochemical basis of long term memory. *Q. Rev. Biophys.,* **2**, 135–173.

2460. ROBERTS, R. B., FLEXNER, J. B. & FLEXNER, L. B. (1970). Some evidence for the involvement of adrenergic sites in the memory trace. *Proc. natn. Acad. Sci. U.S.A.,* **66**, 310–313.

2461. ROBERTSON, J. & PIPA, R. (1973). Metamorphic shortening of interganglionic connectives of *Galleria mellonella* (Lepidoptera) *in vitro*: stimulation by ecdysone analogues. *J. Insect Physiol.,* **19**, 673–679.

2462. ROBERTSON, J. D. (1957). New observations on the ultrastructure of the membranes of frog peripheral nerve fibres. *J. biophys. biochem. Cytol.,* **3**, 1043.

2463. ROBERTSON, J. D. (1964). Structure of cell membranes. *In* Locke, M., *Cellular membranes in development.* Acad. Press; N.Y.

2464. ROBERTSON, J. D. (1966). Design principles of the unit membrane. *In Principles of biomolecular organization.* CIBA Fdn Symp. Little, Brown; Boston.

2465. ROBERTSON, J. D. (1966). Granulo-fibrillar substructure in unit membranes. *Ann N.Y. Acad. Sci.,* **137**, 421–440.

2466. ROBERTSON, J. D., BODENHEIMER, T. S. & STAGE, D. E. (1963). The ultrastructure of Mauthner cell synapses and nodes in goldfish brains. *J. Cell Biol.,* **19**, 159–199.

2467. ROBINSON, W. E., GORDON-WALKER, A. & BOWNDS, D. (1972).

Molecular weight of frog rhodopsin. *Nature, Lond., New Biol.*, **235**, 112–114.

2468. ROBOZ-EINSTEIN, E., ROBERTSON, D. M., DiCAPRIO, J. M. & MOORE, W. (1962). The isolation from bovine spinal cord of a homogeneous protein with encephalitogenic activity. *J. Neurochem.*, **9**, 353–361.
2469. RODBELL, M. (1972). Cell surface receptor sites. *In* Anfison, C. B., Goldberger, R. F. & Schechter, A. N., *Current topics in biochemistry.* Acad. Press; pp. 187–218.
2470. RODBELL, M., BIRNBAUMER, L. & POHL, S. L. (1970). Adenyl cyclase in fat cells. III. Stimulation by secretin and the effects of trypsin on the receptors for lipolytic hormones. *J. biol. Chem.*, **245**, 718–722.
2471. RODRIGUEZ, J. G. la TORRE, J. L. & de ROBERTIS, E. (1970). The interaction between atropine sulphate and a proteolipid from cerebral cortex studied by polarization of fluorescence. *Molec. Pharmac.* **6**, 122–127.
2472. ROELOFS, W. L. & COMEAU, A. (1968). Sex pheromone perception. *Nature, Lond.*, **220**, 600.
2473. ROELOFS, W. L. & COMEAU, A. (1969). Sex pheromone specificity: taxonomic and evolutionary aspects in Lepidoptera. *Science, N.Y.*, **165**, 398.
2474. ROESSMAN, U. & FRIEDE, R. L. (1967). Entry of labelled donor cells from the blood stream into the CNS. *J. Neuropath. exp. Neurol.*, **26**, 144–145.
2475. ROESSMAN, U. & FRIEDE, R. L. (1968). Entry of labelled monocytes into the central nervous system. *Acta neuropath.*, **10**, 359–362.
2476. ROISEN, F. J. & MURPHY, R. A. (1973). Neurite developments in vitro: II. The role of microfilaments and microtubules in dibutyryl adenosine 3′5′-cyclic monophosphate and nerve growth factor stimulated maturation. *J. Neurobiol.*, **4**, 397.
2477. ROISEN, F. J., MURPHY, R. A. & BRADEN, W. G. (1972). Neurite development *in vitro.* I. The effect of adenosine 3′5′-cyclic monophosphate (cyclic AMP). *J. Neurobiol.*, **3**, 347–368.
2478. ROMANUL, F. C. A. (1965). Capillary supply and metabolism of muscle fibres. *Archs. Neurol.*, **12**, 497–509.
2479. ROMANUL, F. C. A. & HOGAN, E. L. (1965). Enzymatic changes in denervated muscle. I. Histochemical studies. *Archs. Neurol.*, **13**, 263–273.
2480. ROMANUL, F. C. A. & van der MEULEN, J. P. (1966). Reversal of the enzyme profiles of muscle fibres in fast and slow muscles by cross innervation. *Nature, Lond.*, **212**, 1369–1370.
2481. ROMINE, W. O., GOODALL, M. C., PETERSON, J. & BRADLEY, R. J. (1974). The acetyl choline receptor: isolation of a brain nicotine receptor and its preliminary characterization in lipid bilayer membranes. *Biochim. biophys. Acta*, **367**, 316.
2482. ROSE, B. & LOEWENSTEIN, W. R. (1971). Junctional membrane

permeability. Depression by substitution of Li for extracellular Na and by long-term lack of Ca^{++} and Mg^{++}: restoration by cell repolarization. *J. mem. Biol.*, **5**, 20–50.

2483. ROSE, J. E., MALIS, L. I., KRUGER, L. & BAKER, C. P. (1960). Effects of heavy ionizing monoenergetic particles on the cerebellar cortex. II. Histological appearance of laminar lesions and growth of nerve fibres after laminar destruction. *J. comp. Neurol.*, **115**, 243–297.
2484. ROSE, S. M. (1944). Methods of initiating limb regeneration in adult anura. *J. exp. Zool.*, **95**, 149–170.
2485. ROSE, S. M. (1960). Regulation of growth of organisms. *Ecology*, **41**, 188–200.
2486. ROSE, S. M. & ROSE. F. C. (1961). Growth-controlling exudates of tadpoles. *Symp. Soc. exp. Biol.*, **15**, 207–218.
2487. ROSE, S. P. R. (1967). Preparation of enriched fraction from cerebral cortex containing isolated, metabolically active neuronal and glial cells. *Biochem. J.*, **102**, 33–43.
2488. ROSE, S. P. R. (1968). Glucose and associated amino acid metabolism in isolated neuronal and glial cell fractions *in vitro*. *J. Neurochem.*, **15**, 1415–1429.
2489. ROSE, S. P. R. & SINHA, A. K. (1969). Separation of neuronal and neuropil fractions: a modified procedure. *Life Sci.*, **9**, 907–915.
2490. ROSENBAUM, J. L. & CHILD, F. (1967). Flagellar regeneration in protozoan flagellates. *J. Cell Biol.*, **34**, 345–364.
2491. ROSENBAUM, J. L., MOULDER, J. E. & RINGO, D. L. (1969). Flagellar elongation and shortening in *chlamydomonas*. *J. Cell Biol.*, **41**, 600.
2492. ROSENBERG, P. & ECHLIN, F. (1968). Time course of changes in cholinesterase activity of chronically isolated cortex. *J. nerv. ment. Dis.*, **147**, 56–60.
2493. ROSENZWEIG, M. R. (1966). Environmental complexity, cerebral change and behaviour. *Am. Psychol.*, **21**, 321–332.
2494. ROSENZWEIG, M. R. & BENNETT, E. L. (1969). Effects of differential environments on brain weights and enzyme activities in gerbils, rats and mice. *Devl Psychobiol.*, **2**, 87–95.
2495. ROSENZWEIG, M. R. & BENNETT, E. L. (1972). Cerebral changes in rats exposed individually to an enriched environment. *J. comp. physiol. Psychol.*, **80**, 304–313.
2496. ROSENZWEIG, M. R., BENNETT, E. L. & DIAMOND, M. C. (1970). Cerebral effects of differential environments. *Fedn Proc. Fedn Am. Socs exp. Biol.*, **29**, 264.
2497. ROSENZWEIG, M. R., BENNETT, E. L. & DIAMOND, M. C. (1972). Chemical and anatomical plasticity of brain. Replication and extension. *In* Gaito, J., *Macromolecules and behaviour.* Appleton, Century, Crofts; N.Y.
2498. ROSENZWEIG, M. R., BENNETT, E. L. & DIAMOND, M. C. (1972). Cerebral effects of differential experience in hypophysectomized rats. *J. comp. physiol. Psychol.*, **79**, 56–66.

2499. ROSENZWEIG, M. R., BENNET, E. L., DIAMOND, M. C., WU, S. Y. SLAGLE, R. W. & SAFFRAN, E. (1969). Influences of environmental complexity and visual stimulation on development of occipital cortex in rat. *Brain Res.*, **14**, 427–445.

2500. ROSENZWEIG, M. R., KRECH, D., BENNETT, E. L. & DIAMOND, M. C. (1962). Effects of environmental complexity and training on brain chemistry and anatomy. A replication and extension. *J. comp. physiol. Psychol.*, **55**, 429–437.

2501. ROSENZWEIG, M. R., LOVE, W. & BENNETT, E. L. (1968). Effects of a few hours a day of enriched experience on brain chemistry and brain weights. *Physiol. Behav.* **3**, 819–825.

2502. ROSENZWEIG, M. R., MØLLGAARD, K., BENNETT, E. L. & DIAMOND, M. C. (1972). Negative as well as positive synaptic changes may store memory. *Psychol. Rev.*, **79**, 93–96.

2503. ROSIN, S. (1946). Über Bau und Wachstum der Grezlamella der Epidermis bei Amphibienlauren; Analyse einer orthogonalen Fibrillenstruktur. *Revue suisse Zool.*, **53**, 133–201.

2504. ROSS, L. L. & BORNSTEIN, M. B. (1969). An electron microscopic study of synaptic alteration in cultured mammalian central nervous tissue exposed to serum from animals with experimental allergic encephalomyelitis. *Lab. Invest.*, **20**, 26–35.

2505. ROSS, R. (1968). The fibroblast and wound repair. *Biol. Rev.*, **43**, 51–96.

2506. ROSS, R., EVERETT, N. B. & TYLER, R. (1970). Wound healing and collagen formation. VI. The origin of the wound fibroblast studied in parabiosis. *J. Cell Biol.*, **44**, 645–654.

2507. ROSS, R. & ODLAND, G. (1968). Human wound repair. II. Inflammatory cells; epithelio-mesenchymal interrelations and fibrogenesis. *J. Cell Biol.*, **39**, 152–168.

2508. ROSS, S., SAWIN, P. B., DENENBERG, V. H. & VOLOW, M. (1963). Effects of previous experience and age on sound-induced seizures in rabbits. *Int. J. Neuropharmac.*, **2**, 255–258.

2509. ROSVOLD, H. E. & DELGADO, J. M. R. (1956). The effect on delayed alternation test performance of stimulating or destroying electrically structures within the frontal lobes of the monkey's brain. *J. comp. physiol. Psychol.*, **40**, 365–372.

2510. ROTH, S. (1968). Studies on intercellular adhesive selectivity. *Devl Biol.*, **18**, 602–631.

2511. ROTH, S. (1973). A molecular model for cell interactions. *Q. Rev. Biol.*, **48**, 541–563.

2512. ROTH, S., McGUIRE, E. J. & ROSEMAN, S. (1971). Evidence for cell surface glycosyltransferases. Their potential role in cellular recognition. *J. Cell Biol.*, **51**, 536–547.

2513. ROTH, S. & WHITE, W. (1972). Intercellular contact and cell surface galactosyl transferase activity. *Proc. natn. Acad. Sci. U.S.A.*, **69**, 485–489.

2514. ROTH, T. F. & PORTER, K. R. (1962). Specialized sites on the cell surface for protein uptake. *In* Breese, S. S., *Electron microscopy.* Fifth Int. Congress for Electron Microscopy. Philadelphia. Acad. Press; N.Y. pp. 224–225.

2515. ROTH, W. D. (1956). Some evolutionary aspects of neurosecretion in the sea lamprey *Petromyzon marinus. Anat. Rec.*, **124**, 437.

2516. ROTHFIELD, L. I. (1971). *Structure and function of biological membranes.* Acad. Press; N.Y.

2517. ROTHFIELD, L. & FINKELSTEIN, A. (1965). Membrane biochemistry. *A. Rev. Biochem.*, **37**, 463–480.

2518. ROTHFIELD, L. & PEARLMAN, M. (1966). The role of the cell envelope phospholipid in the enzymatic synthesis of bacterial lipopolysaccharide: structural requirement of the phospholipid molecule. *J. biol. Chem.*, **241**, 1386.

2519. ROTHFIELD, L., WEISER, M. & ENDO, A. (1969). Dissociation and reassociation of bacterial membrane components. *J. gen. Physiol.*, **51**, 27–37.

2520. ROTMANS, J. P., DAEMEN, F. J. M. & BONTING, S. L. (1971). Retinaldehyde isomerization in rhodopsin regeneration. *Fedn Proc. Fedn Am. Socs exp. Biol.*, **30**, 1197.

2521. ROTMANS, J. P., DAEMEN, F. J. M. & BONTING, S. L. (1972). Biochemical aspects of the visual process. XIX. Formation of isorhodopsin from photolyzed rhodopsin by bacterial action. *Biochim. biophys. Acta*, **267**, 583–587.

2522. ROTTEM, S., HUBBELL, W. L., HAYFLICK, L. & McCONNELL, H. M. (1970). Motion of fatty acid spin labels in the plasma membrane of mycoplasma. *Biochim. biophys. Acta*, **219**, 105–107.

2523. ROTTMAN, A. R. (1967). Variations in the morphology of the subcommisural organ and secretion of Reissner's fibre in response to isotonic, hypertonic and hyperosmotic injections in rats. *Abstr. 4th Conf. Europ. comp. Endocrin.*, pp. 154.

2524. ROTTMAN, W. L., HELLERQUIST, C., UMBREIT, J. & WALTHER, B. T. (1973). A quantitative assay for intercellular adhesion and its application to cell aggregation by lectins. *Fedn Proc. Fedn Am. Socs exp. Biol.*, **32**, 556.

2526. ROUSSER, G., NELSON, G. J., FLEISCHER, S. & SIMON, G. (1968). Lipid composition of animal cell membranes, organelles and organs. *In* Chapman, D. *Biological membranes, physical fact and function.* Acad. Press; N.Y.

2526. ROWELL, C. H. F. & DOREY, A. E. (1967). The number and size of axons in the thoracic connectives of the desert locust *Schistocerca gregaria* Forsk. *Z. Zellforsch. mikrosk. Anat.*, **83**, 288–294.

2527. ROWEN, J., BRUNISH, R. & BISHOP, F. W. (1956). Form and dimensions of isolated hyaluronic acid. *Biochim. biophys. Acta*, **19**, 480–489.

2528. RUBALCAVA, B., de MUNOZ, M. D. & GITLER, C. (1969). Interaction of fluorescent probes with membranes. I. Effect of ions on erythrocyte membranes. *Biochemistry, N.Y.*, **8**, 2742–2747.

2529. RUNION, H. I. & PIPA, R. L. (1970). Electrophysiological and endocrinological correlates during the metamorphic degeneration of a muscle fibre in *Galleria mellonella* (L) (Lepidoptera). *J. exp. Biol.*, **53**, 9–24.

2530. RUSHTON, W. A. H. (1951). A theory of the effects of fibre size in medullated nerve. *J. Physiol.*, **115**, 101–122.

2531. RUSSELL, I. J. (1968). Influence of efferent fibres on a receptor. *Nature, Lond.*, **219**, 177–178.

2532. RUSSELL, I. J. & ROBERTS, B. L. (1971). Efferent innervation of a mechanoreceptor. *Proc. 1st Phys. Congr.*, **9**, p. 485.

2533. RUSSELL, P. A. (1970). Effects of maternal deprivation treatments in the rat. *Anim. Behav.*, **18**, 700–702.

2534. RUSSELL, P. S. & BILLINGHAM, R. E. (1962). Some aspects of the repair process in mammals. *Prog. Surg.*, **2**, 1–72.

2535. RUTLEDGE, L. T., RANK, J. B. & DUNCAN, J. A. (1967). Prevention of supersensitivity in partially isolated cerebral cortex. *Electroenceph. clin. Neurophysiol.*, **23**, 256–262.

2536. RZEHAK, K. & SINGER, M. (1966). Limb regeneration and nerve fibre number in *Rana sylvatica* and *xenopus laevis*. *J. exp. Zool.*, **162**, 15–21.

2537. SABATINI, M. T., DIPOLO, R. & VILLEGAS, R. (1968). Adenosine triphosphatase activity in the membranes of the squid nerve fibre. *J. Cell Biol.*, **38**, 176–183.

2538. SACKETT, G. P. (1967). Responses to stimulus novelty and complexity as a function of rats' early rearing experiences. *J. comp. physiol. Psychol.*, **63**, 369–375.

2539. SACKETT, G. P. (1968). The persistence of abnormal behaviour in monkeys following isolation rearing. *In The role of learning in psychotherapy.* CIBA Fdn Symp. Churchill; London.

2540. SACKLER, A. M., WELTMAN, A. S., BRADSHAW, M. & JURTSHUK, P. (1959). Endocrine changes due to auditory stress. *Acta endocrinol.*, **31**, 405–418.

2541. SACKLER, A. M., WELTMAN, A. S. & JURTSHUK, P. (1960). Endocrine aspects of auditory stress. *Aerospace Med.*, **31**, 749–769.

2542. SAFALSKY, B. (1971). Functional studies of regenerated muscles from normal and dystrophic mice. *Nature, Lond.*, **229**, 270–272.

2543. SAFALSKY, B. & STIRLING, C. A. (1973). Altered neural protein in murine muscular dystrophy. *Nature, Lond., New Biol.*, **246**, 126–128.

2544. SAFFITZ, J. E., BURNETT, A. L. & LESH, G. E. (1972). Nervous system transplantation in hydra. *J. exp. Zool.*, **179**, 215–226.

2545. SARKAR, S., SRETER, F. A. & GERGELY, J. (1971). Light chains of myosins from white, red and cardiac muscles. *Proc. natn. Acad. Sci. U.S.A.*, **68**, 946–950.

2546. SALEM, L. (1962). The role of long range forces in the cohesion of lipoproteins. *Can. J. Physiol. Biochem.*, **40**, 1287–1298.

2547. SALVATERRA, P. M. & MOORE, W. J. (1973). Binding of ^{125}I-α-bungarotoxin to particulate fractions of rat and guinea pig brain. *Biochem. biophys. Res. Commun.*, **55**, 1311–1318.
2548. SALZEN, E. A. & SLUCKIN, W. (1959). The incidence of the following response and duration of responsiveness in domestic fowl. *Anim. Behav.*, **7**, 172–179.
2549. SAMAHA, F. J., GUTH, L. & ALBERS, R. W. (1970). The neural regulation of gene expression in the muscle cell. *Expl Neurol.*, **27**, 276–282.
2550. SAMAHA, F. J. & THIES, W. H. (1973). Superprecipitation and ATPase activity in normal, newborn and denervated muscle. *Expl Neurol.*, **38**, 398–405.
2551. SANCHEZ, Y. & SANCHEZ, D. (1919). Sobre el desarollo de los elementos nerviosos en la retina del *Pieris brassicae* L. (continuacion). *Trab. Lab. Invest. biol. Univ. Madr.*, **17**, 1–63.
2552. SANCHEZ, Y. & SANCHEZ, D. (1919). Sobre el desarollo de los elementos nerviosos en la retina del *Pieris brassicae* L. (continuacion). *Trab. Lab Invest. biol. Univ. Madr.*, **17**, 117–180.
2553. SANDERS, F. V. (1948). The thickness of the myelin sheath of normal and regenerating nerve fibres. *Proc. R. Soc. B*, **135**, 323–357.
2554. SANDRI, C., AKERT, K., LIVINGSTON, R. B. & MOOR, H. (1972). Particle aggregation at specialized sites in freeze-etched postsynaptic membrane. *Brain Res.*, **41**, 1–16.
2555. SANDRITTER, W., NOVAKOVA, V., PILNY, J. & KIEFER, G. (1967). Cytophotometric measurements of the amount of nucleic acids and proteins in neurons of the rat during postnatal development and in old age. *Z. Zellforsch. mikrosk. Anat.*, **80**, 145–152.
2556. SANGIACOMO, C. O. (1969). Denervated taste buds. I. General ultrastructural observations. *Experentia*, **25**, 1323–1324.
2557. SANTINI, M. (1969). New fibres of sympathetic nature in the inner core region of Pacinian corpuscles. *Brain Res.*, **16**, 535–538.
2558. SANTINI, M. & BERL, S. (1970). Levels of amino acids and proteins in Pacinian corpuscles of the cat. *J. Neurochem.*, **17**, 789–793.
2559. SANUI, H. (1970). pH dependence of the effect of adenosine triphosphate and ethylenediamine tetraacetate on sodium and magnesium binding by cellular membrane fragments. *J. Cell Physiol.*, **75**, 361–368.
2560. SANUI, H., CARVALHO, A. P. & PACE, N. (1962). Relationship of hydrogen ion binding to sodium and potassium binding by rat liver cell microsomes and human erythrocyte ghosts. *J. cell comp. Physiol.*, **59**, 241–250.
2561. SANUI, H. & PACE, N. (1959). Sodium and potassium binding by rat liver cell microsomes. *J. gen. Physiol.*, **42**, 1325–1345.
2562. SANUI, H. & PACE, N. (1965). Mass law effect of adenosine triphosphate on Na, K, Mg, Ca binding by rat liver microsomes. *J. cell comp. Physiol.*, **65**, 27–30.
2563. SARLIEVE, L. L., NESKOVIC, N. M. & MANDEL, P. (1971).

PAPS-cerebroside sulphotransferase activity in brain and kidney of neurological mutants. *FEBS Lett.*, **19**, 91–95.

2564. SASAKI, H., SAITO, Y., BEAR, D. M. & ERVIN, F. R. (1971). Quantitative variation in striate receptive fields of cats as a function of light and dark adaptation. *Expl Brain Res.*, **13**, 273–293.

2565. SATAKE, M. & ABE, S. (1966). Preparation and characterization of nerve cell perikarya from rat cerebral cortex. *J. Biochem.*, **59**, 72.

2566. SATAKE, M., HASEGAWA, S.-I., ABE, S. & TANAKA, R. (1968). Preparation and characterization of nerve cell perikaryon from pig brain stem. *Brain Res.*, **11**, 246–250.

2567. SATO, M. (1971). Neural coding in taste as seen from recordings from peripheral receptors and nerves. *In* Beidler, L. M., *Taste. Handbook of sensory physiology IV*, (2), 116–147.

2568. SATO, T., YAMAMOTO, M. & NAKAHAMA, H. (1973). Influence of synchronized sleep upon spontaneous and induced discharges of single units in visual system. *Expl Brain Res.*, **16**, 533–541.

2569. SATTIN, A. & RALL, T. W. (1970). The effect of adenosine and adenine nucleotides on the cyclic adenosine 3′5′-phosphate content of guinea pig cerebral cortex slices. *Molec. Pharmac.*, **6**, 13–23.

2570. SAWAKI, S. & PETER, J. B. (1972). Occurrence and distribution of isoenzymes in different types of skeletal muscle fibres. *Expl Neurol.*, **35**, 421–430.

2571. SAXOD, R. (1972). Role du nerf et du territoire cutané dans le developpment des corpuscles de Herbst et de Grandy. *J. Embryol. exp. Morph.*, **27**, 277–300.

2572. SCHAEFER, T. (1968). Some methodological implications of the research on 'early handling' in the rat. *In* Newton, G. & Levine, S., *Early experience and behaviour. The psychobiology of development.* C. C. Thomas; Springfield, Illinois. pp. 102–141.

2573. SCHALLER, H. C. (1973). Isolation and characterization of a low molecular weight substance activating head and bud formation in hydra. *J. Embryol. exp. Morph.*, **29**, 27–38.

2574. SCHALLER, H. & GIERER, A. (1973). Distribution of the head-activating substance in hydra and its localization in membranous particles in nerve cells. *J. Embryol. exp. Morph.*, **29**, 39–52.

2575. SCHAPIRO, S., SALAS, M. & VUKOVICH, K. (1970). Hormone effects on ontogeny of swimming ability in the rat: assessment of central nervous system development. *Science, N.Y.*, **168**, 147–151.

2576. SCHAPIRO, S. & VUKOVICH, K. R. (1970). Early experience effects upon cortical dendrites. A proposed model for development. *Science, N.Y.*, **167**, 292–294.

2577. SCHAPIRO, S., VUKOVICH, K., & GLOBUS, A. (1973). Effects of neonatal thyroxine and hydrocortisone administration on the development of dendritic spines in the visual cortex of rats. *Expl Neurol.*, **40**, 286–302.

2578. SCHAUBLE, M. K. (1972). Seasonal variation of newt forelimb regeneration under controlled environmental conditions. *J. exp. Zool.*, **181**, 281–286.

2579. SCHECHTER, J. & WIENER, R. (1972). Ultrastructural changes in the ependymal lining of the median eminence following the intraventricular administration of catecholamine. *Anat. Rec.*, **172**, 643–650.

2580. SCHEINBERG, L. C. (1967). Is the brain an immunologically privileged site? Studies based on homologous skin grafts to brain and subcutaneous tissues. *J. Neuropath. exp. Neurol.*, **26**, 144–152.

2581. SCHEINBERG, L. C., EDELMAN, F. L. & LEVY, W. A. (1964). Is the brain 'an immunologically privileged site'? *Archs. Neurol.*, **11**, 248–264.

2582. SCHEINBERG, L. C., KOTSILUMBAS, D. G., KARPF, R. & MAYOR, N. (1966). Is the brain an immunologically privileged site? *Archs. Neurol.*, **15**, 62–70.

2583. SCHIAFFINO, S. & HANZLIKOVA, V. (1970). On the mechanism of compensatory hypertrophy in skeletal muscles. *Experentia*, **26**, 152–153.

2584. SCHIAFFINO, S., HANZLIKOVA, V. & PIEROBON, S. (1970). Relations between structure and function in rat skeletal muscle fibres. *J. Cell Biol.*, **47**, 107–119.

2585. SCHIFFER, D. & VESCO, D. (1963). Histochemical observations about the pattern of tetrazolium reduction with different substrates, in glia cells of normal and pathological human nervous tissue. *J. Histochem. Cytochem.*, **11**, 335.

2586. SCHIILLING, J. A. (1968). Wound healing. *Physiol. Rev.*, **48**, 374–423.

2587. SCHIMMER, B. P. (1971). Effect of catecholamines and monovalent cations on adenylate cyclase activity in cultured glial tumour cells. *Biochim. biophys. Acta*, **252**, 567–573.

2588. SCHIMRIGK, K. (1966). Über die Wandstruktur der Seitenventrikel und des dritten Ventrikels beim Menschen. *Z. Zellforsch. mikrosk. Anat.*, **70**, 1–20.

2589. SCHLAEPFER, W. W. (1971). Experimental alteration of neurofilaments and neurotubules by calcium and other ions. *Expl Cell Res.*, **67**, 73–80.

2590. SCHLAEPFER, W. W. (1971). Vincristine-induced axonal alterations in rat peripheral nerve. *J. Neuropath. exp. Neurol.*, **30**, 488–505.

2591. SCHLAEPFER, W. W. (1974). Calcium-induced degeneration of axoplasm in isolated segments of rat peripheral nerve. *Brain Res.*, **69**, 203.

2592. SCHLAEPFER, W. W. (1974). Effects of energy deprivation on Wallerian degeneration in isolated segments of rat peripheral nerve. *Brain Res.*, **78**, 71.

2593. SCHLAEPFER, W. W. & BUNGE, R. P. (1973). Effects of calcium ion concentration on the degeneration of amputated axons in tissue culture. *J. Cell Biol.*, **59**, 456.

2594. SCHLAEPPI, U. (1963). Comparison of serotonin with directly and indirectly acting compounds in their effect on the nictitating membrane of spinal cats. *J. Pharmac. exp. Ther.*, **139**, 330–336.

2595. SCHLESINGER, K., BOGGAN, W. & FREEDMAN, D. (1965). Genetics of audiogenic seizures. Relation to brain serotonin and norepinephrine in mice. *Life Sci.*, **4**, 2345–2351.

2596. SCHLESINGER, K. & WIMER, R. (1967). Genotype and conditioned avoidance learning in the mouse. *J. comp. physiol. Psychol.*, **63**, 139–141.

2597. SCHMIDT, M. J. & ROBISON, G. A. (1970). Cyclic AMP, adenyl cyclase and the effect of norepinephrine in the developing rat brain. *Fedn Proc. Fedn Am. Socs exp. Biol.*, **29**, 479.

2598. SCHMIDT, M. J. & ROBISON, G. A. (1971). The effect of norepinephrine on cyclic AMP levels in discrete regions of the developing rabbit brain. *Life Sci.*, **10**, 459–464.

2599. SCHMIDT, M. J. & ROBISON, G. A. (1972). The effect of neonatal thyroidectomy on the development of the adenosine 3′5′monophosphate system in the rat brain. *J. Neurochem.*, **19**, 937–947.

2600. SCHMIDT, M. J. & SOKOLOFF, L. (1973). Activity of cyclic AMP-dependent microsomal protein kinase and phosphorylation of ribosomal protein in rat brain during postnatal development. *J. Neurochem.*, **21**, 1193.

2601. SCHMIDT, W. J. (1938). Polarisationoptische Analyse eines Eiweiss-Lipoid-Systems, Erläutert am Aussenglied der Schzellen. *Kolloid Z.*, **85**, 137–150.

2602. SCHMIDT, W. J. (1937). *Die Doppelbrechung von Karyoplasma, Zytoplasma und Metaplasma.* Bornträger; Berlin.

2603. SCHMITT, F. O. (1968). Fibrous proteins – neuronal organelles. *Proc. natn. Acad. Sci. U.S.A.*, **60**, 1091–1101.

2604. SCHMITT, F. O., BEAR, R. S. & PALMER, K. J. (1941). X-ray diffraction studies on the structure of the nerve myelin sheath. *J. cell comp. Physiol.*, **18**, 31–42.

2605. SCHMITT, F. O., BEAR, R. S. & CLARK, G. L. (1935). X-ray diffraction studies on nerve. *Radiology*, **25**, 131–151.

2606. SCHNEIDER, D. (1969). Insect olfaction: deciphering system for chemical messages. *Science, N.Y.*, **163**, 1031.

2607. SCHOENFELD, R. I. & SEIDEN, L. S. (1969). Effect of α-methyltyrosine on operant behaviour and brain catecholamine levels. *J. Pharmac. exp. Ther.*, **167**, 319–327.

2608. SCHON, F. & IVERSON, L. L. (1972). Selective accumulation of ^{3}H-GABA by stellate cells in rat cerebellar cortex in vivo. *Brain Res.*, **42**, 503–507.

2609. SCHONBACH, J., HU, K. H. & FRIEDE, R. L. (1968). Cellular and chemical changes during myelination: histologic, autoradiographic histochemical and biochemical data on myelination in the pyramidal tract and corpus callosum of rat. *J. comp. Neurol.*, **134**, 21–38.

2610. SCHOTMAN, P., GISPEN, W. H., JANSZ, H. S. & de WIED, D. (1972). Effects of ACTH analogues on macromolecule metabolism in the brain stem of hypophysectomized rats. *Brain Res.*, **46**, 349–369.

2611. SCHOTTE, O. E. & BUTLER, E. G. (1944). Phases in regeneration of the urodele limb and their dependence upon the nervous system. *J. exp. Zool.*, **97**, 95–121.

2612. SCHRÖDER, J. M. (1970). Feinstruktur und quantitative Auswertung regenerierter peripherer Nervenfasern. *6th Int. Congr. Neuropath.*, pp. 628–646.

2613. SCHRÖDER, J. M. (1972). Altered ratio between axon diameter and myelin sheath thickness in regenerated nerve fibres. *Brain Res.*, **45**, 49–65.

2614. SCHRODT, G. R. & WALKER, S. W. (1966). Ultrastructure of membranes in denervation atrophy. *Am. J. Path.*, **49**, 33.

2615. SCHROEDER, T. E. (1969). The role of 'contractile ring' filaments in dividing *Arbacia* egg. *Biol. Bull.*, **137**, 413.

2616. SCHROEDER, T. E. (1970). Functional and biochemical aspects of contractile ring filaments in HeLa cells. *J. cell Biol.*, **47**, 183A.

2617. SCHROEDER, T. E. (1970). The contractile ring. I. Fine structure of dividing mammalian (HeLa) cells and the effects of cytochalasin B. *Z. Zellforsch. mikrosk. Anat.*, **109**, 431–449.

2618. SCHROEDER, T. E. (1972). Actin in dividing cells: contractile ring filaments bind heavy meromysin. *Proc. natn. Acad. Sci. U.S.A.*, **70**, 1688–1692.

2619. SCHROEDER, T. E. (1972). The contractile ring. II. Determining the brief existence, volumetric changes, and the vital role of cleaving Arbacia eggs. *J. cell Biol.*, **53**, 419–434.

2620. SCHULTZE, B., NOWAK, B. & MAURER, W. (1974). Cycle times of the neural epithelial cells of various types of neuron in the rat. An autoradiographic study. *J. comp. Neurol.*, **158**, 207.

2621. SCHUBERT, P. & KREUTZBERG, G. W. (1974). Axonal transport of adenosine and uridine derivatives and transfer to postsynaptic neurones. *Brain Res.*, **76**, 525–530.

2622. SCHUBERT, P. & KREUTZBERG, G. W. (1975). ^{3}H-adenosine, a tracer for neuronal connectivity. *Brain Res.*, **85**, 317–319.

2623. SCHUBERT, P., KREUTZBERG, G. W. & LUX, H. D. (1972). Neuroplasmic transport in dendrites: effect of colchicine on morphology and physiology of motoneurones in the cat. *Brain Res.*, **47**, 331–343.

2624. SCHULTZ, J. & DALY, J. W. (1973). Accumulation of cyclic adenosine 3′5′monophosphate in cerebral cortical slices from rat and mouse: stimulatory effect of α and β adrenergic agents and adenosine. *J. Neurochem.*, **21**, 1319–1326.

2625. SCHULTZ, R. L. (1964). Macroglial identification in electron micrographs. *J. comp. Neurol.*, **122**, 281–296.

2626. SCHULTZ, R. L., MAYNARD, E. A. & PEASE, D. C. (1957). Electron microscopy of neurons and neuroglia of cerebral cortex and corpus callosum. *Am. J. Anat.*, **100**, 369–407.

2627. SCHWARK, W. S., SINGHAL, R. L. & LING, G. M. (1972). Metabolic control mechanisms in mammalian systems. Regulation of key glycolytic enzymes in developing brain during experimental cretinism. *J. Neuro-*

chem., **19**, 1171–1182.

2628. SCHWARK, W. S., SINGHAL, R. L. & LING, G. M. (1972). Metabolic control mechanisms in mammalian systems. XVII. Thyroid hormone control of brain hexose monophosphate shunt enzymes during experimental cretinism. *Brain Res.*, **42**, 103–116.
2629. SCHWARTZKROIN, P. A. (1972). The effect of body tilt on the directionality of units in cat visual cortex. *Expl Neurol.*, **36**, 498–506.
2630. SCHWARZACHER, H. G. (1954). Markscheidendicke und Achsenzylunderdurchmesser in peripheren menschlichen Nerven. *Acta anat.*, **21**, 26–46.
2631. SCHWYN, R. C. (1967). An autoradiographic study of satellite cells in autonomic ganglia. *Am. J. Anat.*, **121**, 727–740.
2632. SCHWYN, R. C. & HALL, J. L. (1965). Studies of neuroglial activity in autonomic ganglia during electrical stimulation and drug administration. *Anat. Rec.*, **151**, 414.
2633. SCHWYN, R. C. & HALL, J. L. (1966). An autoradiographic study of neuroglia in autonomic ganglia. *Anat. Rec.*, **154**, 419.
2634. SCOTT, D. E., PAULL, W. K. & DUDLEY, G. K. (1972). A comparative scanning electron microscopic analysis of the human cerebral ventricular system. I. The third ventricle. *Z. Zellforsch. mikrosk. Anat.*, **132**, 203–215.
2635. SCOTT, D. E., KOZLOWSKI, G. P., PAULL, W. K., RAMALINGAM, S. & KROBISCH-DUDLEY, G. (1973). Scanning electron microscopy of the human cerebral ventricular system. II. The fourth ventricle. *Z. Zellforsch. mikrosk. Anat.*, **139**, 61–68.
2636. SCOTT, D. E., KOZLOWSKI, G. P. & SHERIDAN, M. N. (1974). Scanning electron microscopy in the ultrastructural analysis of the mammalian cerebral ventricular system. *Int. Rev. Cytol.*, **37**, 349–388.
2637. SCOTT, J. E. (1968). Ion binding in solution containing acid mucopolysaccharides. *In* Quintarelli, G., *The chemical physiology of mucopolysaccharides.* Churchill; London. pp. 171–187.
2638. SCOTT, J. P. (1962). Critical periods in behavioural development. *Science, N.Y.*, **138**, 949–958.
2639. SCOTT, J. P. (1968). The process of primary socialization in the dog. *In* Newton, G. & Levine, S., *Early experience and behaviour. The psychobiology of development.* C. C. Thomas; Springfield, Illinois. pp. 412–439.
2640. SCOTT, J. P. & FREDERICSON, E. (1951). The causes of fighting in mice and rats. *Physiol. Zool.*, **24**, 273–309.
2641. SCOTT, R. E., CARTER, R. L. & KIDWELL, W. R. (1971). Structural changes in membranes of synchronized cells demonstrated by freeze-cleavage. *Nature, Lond., New Biol.*, **233**, 219–220.
2642. SEECOFF, R. L., DONADY, J. J. & FIORUM, P. T. (1973). Formation of axon to myocyte contacts in *Drosophila* cell cultures. *Am. Zool.*, **13**, 331–336.
2643. SEECOFF, R. L., TEPLITZ, R. L., GERSON, I., IKEDA, K. &

DONADY, J. J. (1972). Differentiation of neuromuscular junctions in cultures of embryonic *Drosophila* cells. *Proc. natn. Acad. Sci. U.S.A.*, **69**, 566–570.

2644. SEEDS, N. W. (1971). Biochemical differentiation in reaggregating brain cell culture. *Proc. natn. Acad. Sci. U.S.A.*, **68**, 1858–1861.

2645. SEEDS, N. W. & VATTER, A. E. (1971). Synaptogenesis in reaggregating brain cell cultures. *Proc. natn. Acad. Sci. U.S.A.*, **68**, 3219–3222.

2646. SEELIG, J., AXEL, F. & LIMACHER, H. (1973). Molecular architecture of bilayer membranes. *Ann. N.Y. Acad. Sci.*, **222**, 588–596.

2647. SEELIG, J. & HASSELBACH, W. (1971). A spin label study of sarcoplasmic vesicles. *Europ. J. Biochem.*, **21**, 17–21.

2648. SEGAL, D. S., SQUIRE, L. R. & BARONDES, S. H. (1971). Cycloheximide: its effects on activity are dissociable from its effect on memory. *Science, N.Y.*, **172**, 82–84.

2649. SEIDEL, J. G., SRETER, F. A., THOMPSON, M. M. & GERGELY, J. (1964). Comparative studies of myofibrils, myosin and actomyosin from red and white rabbit skeletal muscle. *Biochem. biophys. Res. Commun.*, **17**, 662–667.

2650. SEIDEN, L. S. & PETERSON, D. D. (1968). Reversal of the reserpine-induced suppression of the conditioned avoidance response by L-Dopa: correlation of behaviour and biochemical differences in two strains of mice. *J. Pharmac. exp. Ther.*, **159**, 422–428.

2651. SEIL, F. J. (1972). Neuronal groups and fiber patterns in cerebellar tissue cultures. *Brain Res.*, **42**, 33–51.

2652. SELYE, H. & HALL, C. E. (1943). Further studies concerning the action of sodium chloride on the pituitary. *Anat. Rec.*, **86**, 579–583.

2653. SERAVIN, L. N. (1971). Mechanisms and coordination of cellular locomotion. *Adv. comp. Physiol. Biochem.*, **4**, 37–111.

2654. SEROTA, R. G. (1971). Acetoxycycloheximide and transient amnesia in the rat. *Proc. natn. Acad. Sci. U.S.A.*, **68**, 1249–1250.

2655. SEROTA, R. G., ROBERTS, R. B. & FLEXNER, L. B. (1972). Acetoxycycloheximide-induced transient amnesia: protective effects of adrenergic stimulants. *Proc. natn. Acad. Sci. U.S.A.*, **69**, 340–342.

2656. SEUFERT, W. D. (1965). Induced permeability changes in reconstituted cell membrane structure. *Nature, Lond.*, **207**, 174.

2657. SHAH, D. O. (1969). Lipid-protein interaction in monolayers. Effect of conformation of poly-L-lysine on stearic acid monolayers. *Biochim. biophys. Acta*, **193**, 217.

2658. SHAH, D. O. (1970). Lipid-polymer interaction in monolayers: effect of conformation of poly-L-lysine on stearic acid monolayers. *In* Blank, M., *Surface chemistry of biological systems.* Plenum Press; N.Y. pp. 101–117.

2659. SHALLENBERGER, R. S. & ACREE, T. E. (1967). Molecular theory of sweet taste. *Nature, Lond.*, **216**, 480–482.

2660. SHALLENBERGER, R. S., ACREE, T. E. & LEE, C. Y. (1969). Sweet taste of D and L-aminosugars and aminoacids and the steric nature of their chemoreceptor site. *Nature, Lond.*, **221**, 555–556.

2661. SHARPLESS, S. K. (1964). Reorganization of functions in the nervous system – use and disuse. *A. Rev. Physiol.*, **26**, 357–388.

2662. SHARPLESS, S. K. (1969). Isolated and deafferented neurons: disuse sensitivity. *In* Jasper, H. H., Ward, A. A. & Pope, A., *Basic mechanisms of the epilepsies.* Little, Brown; Boston. pp. 329–348.

2663. SHARPLESS, S. K. & HALPERN, L. M. (1962). The electrical excitability of chronically isolated cortex studied by means of permanently implanted electrodes. *Electroenceph. clin. Neurophysiol.*, **14**, 244–255.

2664. SHARPLESS, S. K. & HALPERN, L. (1962). The electrical excitability of chronically isolated cortex during barbiturate withdrawal. *J. Pharmac. exp. Ther.*, **151**, 321.

2665. SHARPLESS, S. K. & JAFFE, J. (1966). The electrical excitability of isolated cortex during barbiturate withdrawal. *J. Pharmac. exp. Ther.*, **151**, 321.

2666. SHASHOUA, V. E. (1968). RNA changes in goldfish brain during learning. *Nature, Lond.*, **217**, 238–240.

2667. SHASHOUA, V. E. (1970). RNA metabolism in goldfish brain during acquisition of new behavioural patterns. *Proc. natn. Acad. Sci. U.S.A.*, **65**, 160–167.

2668. SHASHOUA, V. E. (1971). Dibutyryl adenosine cyclic 3′5′-monophosphate effects on goldfish behaviour and brain RNA metabolism. *Proc. natn. Acad. Sci. U.S.A.*, **68**, 2835–2838.

2669. SHEIN, H. M., BRITVA, A., HESS, H. & SELKOE, D. J. (1970). Isolation of hamster brain astroglia by *in vitro* cultivation and subcutaneous growth, and content of cerebroside, ganglioside, RNA and DNA. *Brain Res.*, **19**, 497–501.

2670. SHEMYAKIN, M. M., OVCHINNIKOV, Y. A., IVANOV, V. O., ANTONOV, V. K., VINOGRADOVA, E. I., SHKROB, A. M., MALENKOV, G. G., EVSTRATOV, A. V., LAINE, I. A., MELNIK, E. I. & RYABOVA, I. D. (1969). Mass spectrometric determination of amino-acids in antibiotic peptides. *J. mem. Biol.*, **1**, 402–430.

2671. SHEPPARD, H. & BURGHARDT, C. R. (1974). The dopamine-sensitive adenylate cyclase of the rat caudate nucleus. I. A comparison with the β type cyclase of the rat erythrocyte in the responses to dopamine derivatives. *Molec. Pharmac.*, **10**, 721–726.

2672. SHEREBRIN, M. H., McCLEMENT, B. A. E. & FRANKO, A. J. (1972). Electric-field-induced shifts in the infrared spectrum of conducting nerve axons. *Biophys. J.*, **12**, 977–989.

2673. SHERMAN, G. & FOLCH-PI, J. (1970). Rotary dispersion and circular dichroism of brain 'proteolipid' protein *J. Neurochem.*, **17**, 597–605.

2674. SHIMADA, Y., FISCHMAN, D. A. & MOSCONA, A. A. (1969). Formation of neuromuscular junctions in embryonic cultures. *Proc. natn. Acad. Sci. U.S.A.*, **62**, 715–721.

2675. SHIMIZU, N. & HAMURO, Y. (1958). Deposition of glycogen and changes in some enzymes in brain wounds. *Nature, Lond.*, **181**, 781–782.

2676. SHIMIZU, N. & ISHII, S. (1964). Fine structure of the area postrema of the rabbit brain. *Z. Zellforsch. mikrosk. Anat.*, **64**, 462–473.

2677. SHIPLEY, W. U. (1963). The demonstration in the domestic guinea pig of a process resembling classical imprinting. *Anim. Behav.*, **11**, 470–474.

2678. SHLAER, R. (1971). Shift in binocular disparity causes compensatory change in the cortical structure of kittens. *Science, N.Y.*, **173**, 638–641.

2679. SHOOTER, E. M. & EINSTEIN, E. R. (1971). Proteins of the nervous system. *A. Rev. Biochem.*, **40**, 635–652.

2680. SHULMAN, S., MILGROM, F. & SWANBORG, R. H. (1969). Immunochemical studies of an organ-specific thermostable antigen of bovine brain. *Immunology*, **16**, 25–33.

2681. SHURALEFF, N. & THORNTON, C. S. (1965). The effects of superinnervation on the regeneration of the hindlimb of the axolotl *Ambystoma mexicanum*. *Am. Zool.*, **5**, 723.

2682. SIBAOKA, T. (1966). Action potentials in plant organs. *Symp. Soc. exp. Biol.*, **20**, 49.

2683. SIDMAN, R. L. (1968). Development of interneuronal connections in brains of mutant mice. *In* Carlson, F. D., *Physiological and biochemical aspects of nervous integration.* Soc. Gen. Physiol. Monogr. Prentice Hall; N.J. pp. 163–188.

2684. SIDMAN, R. L. (1970). Cell proliferation, migration and interaction in the developing mammalian nervous system. *In* Schmitt, F. O., *The neurosciences. Second study programme.* Rockefeller U. P; N.Y. pp. 100–115.

2685. SIDMAN, R. L., DICKIE, M. M. & APPEL, S. H. (1964). Mutant mice (Quaking and Jimpy) with deficient myelination in the central nervous system *Science, N.Y.*, **144**, 309–311.

2686. SIDMAN, R. L., GREEN, M. C. & APPEL, S. H. (1965). *Catalog of the neurological mutants of the mouse.* Harvard Univ. Press.

2687. SIDMAN, R. L. & SINGER, M. (1951). Stimulation of forelimb regeneration in the newt, *Triturus viridescens* by a sensory nerve supply isolated from the central nervous system. *Am. J. Physiol.*, **165**, 257–260.

2688. SIDMAN, R. L. & SINGER, M. (1960). Limb regeneration without innervation of the apical epidermis in the adult newt *Triturus. J. exp. Zool.*, **144**, 105–109.

2689. SIEGEL, M. R. & SISLER, H. D. (1963). Inhibition of protein synthesis in vitro by cycloheximide. *Nature, Lond.*, **200**, 675–676.

2690. SIGGINS, G. R., HOFFER, B. J. & BLOOM, F. E. (1969). Cyclic adenosine monophosphate: possible mediator for norepinephrine effects on cerebellar Purkinje cells. *Science, N.Y.*, **165**, 1018–1020.

2691. SIGGINS, G., HOFFER, B. & BLOOM, F. (1971). Prostaglandin-norepinephrine interactions in brain : microelectrophoresis and histochemical correlates. *Ann. N.Y. Acad. Sci.*, **180**, 302–323.

2692. SIGGINS, G. R., HOFFER, B. J. & UNGERSTEDT, U. (1974). Electrophysiological evidence for involvement of cyclic adenosine monophosphate in dopamine responses of caudate neurons. *Life Sci.*, **15**,

779–792.
2693. SIGGINS, G. R., OLIVER, A. P., HOFFER, B. J. & BLOOM, F. E. (1971). Cyclic adenosine monophosphate and norepinephrine: effect on transmembrane properties of cerebellar Purkinje cells *Science, N.Y.*, **171**, 192–194.
2694. da SILVA, P. (1971). *The use of membrane markers in freeze etching.* Ph. D. Diss. Univ. Calif., Berkely.
2695. da SILVA, P. & BRANTON, D. (1970). Membrane splitting in freeze etching. Covalently bound ferritin as a membrane marker. *J. Cell Biol.*, **45**, 598.
2696. da SILVA, P. & BRANTON, D. (1972). Membrane intercalated particles. The plasma membrane as a planar fluid domain. *Chem. Phys. Lipids*, **8**, 265.
2697. da SILVA, P., DOUGLAS, S. D. & BRANTON, D. (1971). Localization of A antigen sites on human erythrocyte ghosts. *Nature, Lond.*, **232**, 194–195.
2698. da SILVA, P. & GILULA, N. B. (1972). Gap junctions in normal and transformed fibroblasts in culture. *Expl Cell Res.*, **73**, 519–523.
2699. SIMON, R. G., WADE, R. R. de LARCO, J. E. & BAKER, M.-L. (1969). Wallerian degeneration: a sequential process. *J. Neurochem.*, **16**, 1435–1438.
2700. SIMONSON, M., SHERWIN, R. W., ANILANE, J. K., YU, W. Y. & CHOW, B. F. (1969). Neuromotor development in progeny of underfed mother rats. *J. Nutr.*, **98**, 18–24.
2701. SIMANTOV, R. & SACHS, L. (1973). Regulation of acetylcholine receptors in relation to acetylcholinesterase in neuroblastoma cells. *Proc. natn. Acad. Sci. U.S.A.*, **70**, 2902–2905.
2702. SIMON, E. J., HILLER, J. M. & EDELMAN, I. (1973). Stereospecific binding of the potent narcotic analgesic ^{3}H etorphine to rat-brain homogenates. *Proc. natn. Acad. Sci. U.S.A.*, **71**, 1947–1949.
2703. SIMPSON, L. L. (1974). The use of neuropoisons in the study of cholinergic transmission. *A. Rev. Pharmac.*, **14**, 305–330.
2704. SIMPSON, S. (1961). Induction of limb regeneration in the lizard *Lygosoma laterale* by augmentation of the nerve supply. *Proc. Soc., exp. Biol. Med.*, **107**, 108–111.
2705. SIMPSON, S. B. (1964). Analyse of tail regeneration in the lizard *Lygosoma laterale*. I. Initiation of regeneration and cartilage differentiation: the role of ependyma. *J. Morph.*, **114**, 425–436.
2706. SIMPSON, S. B. (1965). Regeneration of the lizard tail *In*, Kiortis, K. & Trampusch, T., *Regeneration in animals and related problems.* North Holland; Amsterdam. pp. 431–443.
2707. SIMPSON, S. B. (1968). Morphology of the regenerated spinal cord in the lizard *Anolis carolensis. J. comp. Neurol.*, **134**, 193–210.
2708. SINCLAIR, D. (1967). *Cutaneous sensation.* Oxford Univ. Press.
2709. SINGER, M. (1952). The influence of the nerve in regeneration of the amphibian extremity. *Q. Rev. Biol.*, **27**, 169–200.

2710. SINGER, M. & BRYANT, S. V. (1969). Movement in the myelin sheath of the vertebrate axon. *Nature, Lond.*, **221**, 1148–1150.
2711. SINGER, M. & CASTON, J. D. (1972). Neurotrophic dependence of macromolecuar synthesis in the early limb regeneration of the newt *Triturus. J. Embryol. exp. Morph.*, **28**, 1–11.
2712. SINGER, M. & CRAVEN, L. (1948). The growth and morphogenesis of the regenerating forelimb of adult *Triturus* following denervation at various stages of development. *J. exp. Zool.*, **108**, 279–308.
2713. SINGER, M. & GREEN, M. R. (1968). Autoradiographic studies of uridine incorporation in peripheral nerve of the newt *Triturus. J. Morph.*, **124**, 321–343.
2714. SINGER, M. & INOUE, S. (1964). The nerve and the epidermal apical cap in regeneration of the forelimb of adult *Triturus. J. exp. Zool.*, **155**, 105–116.
2715. SINGER, M. & MUTTERPERL, E. (1963). Nerve fibre requirements for regeneration in forelimb transplants of the newt *Triturus. Devl Biol.*, **7**, 180–191.
2716. SINGER, M. & STEINBERG, M. C. (1972). Wallerian degeneration. A reevaluation based on transected and colchicine-poisoned nerves in the amphibian, *Triturus. Am. J. Anat.*, **133**, 51–84.
2717. SINGER, S. J. (1971). The molecular organization of biological membranes. *In* Rothfield, L. I., *Structure and function of biological membranes.* Acad. Press. pp. 145–222.
2719. SINGER, S. J. (1972). A fluid lipid-globular protein mosaic model of membrane structure. *Ann. N.Y. Acad. Sci.*, **195**, 16.
2720. SINGER, S. J. & NICOLSON, G. L. (1972). The fluid mosaic model of the structure of cell membranes. *Science, N.Y.*, **175**, 720–731.
2721. SINGH, M. & BACHHAWAT, B. K. (1965). The distribution and variation with age of different uronic acid-containing mucopolysaccharides in brain. *J. Neurochem.*, **12**, 519–525.
2722. SINGH, M. & BACHHAWAT, B. K. (1968). Isolation and characterization of glycosoaminoglycans in human brains of different age groups. *J. Neurochem.*, **15**, 249–258.
2723. SINGH, M., CHANDRASEKARAN, E. V., CHERIAN, R. & BACHHAWAT, B. K. (1969). Isolation and characterization of glycosoaminoglycans in brains of different species. *J. Neurochem.*, **16**, 1157–1162.
2724. SJÖSTRAND, F. S. (1959). Structure of myelin. *Ergebn. Biol.*, **21**, 128–140.
2725. SJÖSTRAND, J. (1970). Fast and slow components of axoplasmic transport in the hypoglossal and vagus nerves of the rabbit. *Brain Res.*, **18**, 461–467.
2726. SJÖSTRAND, J., FRIZELL, M. & HASSELGREN, P.-O. (1970). Effects of colchicine on axonal transport in peripheral nerves. *J. Neurochem.*, **17**, 1563–1570.
2727. SJÖSTRAND, J. & KARLSSON, J.-O. (1969). Axoplasmic transport in the optic nerve and tract of the rabbit: a biochemical and radioauto-

graphic study. *J. Neurochem.*, **16**, 833–844.

2728. SJÖSTRAND, J., KARLSSON, J.-O. & MARCHISIO, P. C. (1973). Axonal transport in growing and mature retinal ganglion cells. *Brain Res.*, **62**, 395–397.

2729. SKARF, B. (1973). Development of binocular single units in the optic tectum of frogs raised with disparate stimulation to the eyes. *Brain Res.*, **51**, 352–357.

2730. SKOFF, R. P. & VAUGHN, J. E. (1970). An autoradiographic study of cellular proliferation in degenerating rat optic nerve. *J. comp. Neurol.*, **141**, 133–156.

2731. SLAUGHTERBUCK, D. B. & FAWCETT, D. W. (1959). The development of the cnidoblasts of hydra. An electron microscope study of cell differentiation. *J. biophys. biochem. Cytol.*, **5**, 441–452.

2732. SLOPER, J. J. (1972). Gap junctions between dendrites in the primate neocortex. *Brain Res.*, **44**, 641–646.

2733. SLUCKIN, W. (1965). *Imprinting and early learning.* Aldine; Chicago.

2734. SMALL, J. V. & SQUIRE, F. M. (1972). Structural basis of contraction in vertebrate smooth muscle. *J. molec. Biol.*, **67**, 117–149.

2735. SMART, I. (1960). DNA synthesis in the brain of juvenile and adult white mice. *Anat. Rec.*, **136**, 279–280.

2736. SMART, I. (1961). The subependymal layer of the mouse brain and its cell production as shown by radioautography after thymidine-H^3 injection. *J. comp. Neurol.*, **116**, 325–347.

2737. SMART, I. & LEBLOND, C. P. (1961). Evidence for division and transformation of neuroglia cells in the mouse brain as derived from radioautography after injection of thymidine-H^3. *J. comp. Neurol.*, **116**, 349–367.

2738. SMART, I. (1964). The evolution of neurone production in the vertebrate central nervous system. *J. Anat.*, **98**, 466–467.

2739. SMART, I. (1965). The operation of the ependymal 'choke'. *J. Anat.*, **99**, 941–942.

2740. SMART, I. (1968). The location of mitosis-capable neuron precursor cells in the mouse neural epithelium. *J. Anat.*, **99**, 212.

2741. SMART, I. M. H. (1972). Proliferative characteristics of the ependymal layer during the early development of the mouse diencephalon, as revealed by recording the number location and plane of cleavage of mitotic figures. *J. Anat.*, **113**, 109–130.

2742. SMART, J. L. (1970). Trial and error behaviour of inbred and F_1 hybrid mice. *Anim. Behav.*, **18**, 445–453.

2743. SMART, J. L. (1971). Long lasting effects of early nutritional deprivation on the behaviour of rodents. *Psychiat. Neurol. Neurochir.*, **74**, 443–452.

2744. SMART, J. L. & DOBBING, J. (1971). Vulnerability of developing brain. II. Effects of early nutritional deprovation on reflex ontogeny and development of behaviour in the rat. *Brain Res.*, **28**, 85–95.

2745. SMART, J. L. & DOBBING, J. (1971). Vulnerability of developing brain. VI. Relative effects of foetal and early postnatal undernutrition on reflex

ontogeny and development of behaviour in the rat. *Brain Res.*, **33**, 303–314.
2746. SMITH, A. A., FARBMAN, A. & DANCIS, J. (1965). Tongue in familial dysautonomia. A diagnostic sign. *Am. J. Dis. Child.*, **110**, 152–153.
2747. SMITH, A. A., FARBMAN, A. & DANCIS, J. (1965). Absence of taste bud papillae in familial dysautonomia. *Science, N.Y.*, **147**, 1040–1041.
2748. SMITH, D. E. (1973). The location of neurofilaments and microtubules during the postnatal development of Clarke's nucleus in the kitten. *Brain Res.*, **55**, 41–53.
2749. SMITH, H. V. (1972). Effects of environmental enrichment on open field activity and Hebb-Williams problem solving in rats. *J. comp. physiol. Psychol.*, **80**, 163–168.
2750. SMITH, I. C. P. (1968). The spin label method. *In* Swartz, H. M., Bolton, J. R. & Borg, D. C., *Biological applications of electron spin resonance.* Wiley Interscience; N.Y.
2751. SMITH, M. E. (1967). Metabolism of myelin lipids. *Adv. Lipid Res.*, **5**, 241–278.
2752. SMITH, M. E. & ENG, L.-F. (1965). The turnover of the lipid components of myelin. *J. Am. Oil Chem. Soc.*, **42**, 1013–1018.
2753. SMITH, S. B. & WAKSMAN, B. H. (1969). Passive transfer and labelling studies on the cell infiltrate in experimental allergic encephalomyelitis. *J. Path. Bact.*, **99**, 237–244.
2754. SOBKOWICZ, H. M., GOLUEKE, D. & LOWRY, B. L. (1973). Interaction between the nerve fibre and the glial cell during growth in cultures of the central nervous tissue. *Soc. Neurosci. 3rd Ann. Meeting.*
2755. SOBKOWICZ, H. M., HARTMANN, H. A., MONZAIN, R. & DESNOYERS, P. (1973). Growth, differentiation and ribonucleic acid content of the fetal rat spinal ganglion cells in culture. *J. comp. Neurol.*, **148**, 249–284.
2756. SOIFER, D. (1975). Enzymatic activity in tubulin preparations: cyclic-AMP dependent protein kinase activity of brain microtubule protein. *J. Neurochem.*, **24**, 21–33.
2757. SOKOL, H. W. & VALTIN, H. (1965). Morphology of the neurosecretory system in rats homozygous and heterozygous for hypothalamic diabetes insipidus (Brattleboro rats). *Endocrinology*, **77**, 692–700.
2758. SOLANDT, D. Y., PARTRIDGE, R. C. & HUNTER, J. (1943). The effect of skeletal fixation on skeletal muscle. *J. Neurophysiol.*, **6**, 17–22.
2759. SOMJEN, G. G. (1969). Sustained evoked potential changes of the spinal cord. *Brain Res.*, **12**, 268–272.
2760. SOMJEN, G. G. (1970). Evoked sustained focal potentials and membrane potentials of neurons and of unresponsive cells of the spinal cord. *J. Neurophysiol.*, **33**, 562–582.
2761. SOMJEN, G. G. (1973). Electrogenesis of sustained potential shifts of the central nervous system. *Prog. Neurobiol.*, **1**, 199–235.
2762. SOOD, C. K., SWEET, C. & ZULL, J. E. (1972). Interaction of kidney Na^+-K^+ ATPase with phosphlipid model membrane systems. *Biochim.*

biophys. Acta, **282**, 429–434.

2763. SOTELO, C. (1968). Permanence of postsynaptic specializations in the frog sympathetic ganglion cells after denervation. *Expl Brain Res.,* **6**, 294–305.
2764. SOTELO, C. (1973). Permanence and fate of paramembranous synaptic specializations in mutants and experimental animals. *Brain Res.,* **62**, 345–351.
2765. SOTELO, C. & CHANGEUX, J. P. (1974). Transynaptic degeneration 'en cascade' in the cerebellar cortex of staggerer mutant mice. *Brain Res.,* **67**, 519.
2766. SOTELO, C. & LLINAS, R. (1972). Specialized membrane junctions between neurons in vertebrate cerebellar cortex. *J. Cell Biol.,* **53**, 271–289.
2767. SOTELO, C. & PALAY, S. L. (1968). The fine structure of the lateral vestibular nucleus in the rat. *J. Cell Biol.,* **36**, 151–179.
2768. SOTELO, C. & PALAY, S. L. (1970). The fine structure of the lateral vestibular nucleus in the rat. II. Synaptic organization. *Brain Res.,* **18**, 93–115.
2769. SOUTHWICK, C. H. & CLARK, L. H. (1968). Interstrain differences in aggressive behaviour and exploratory activity of inbred mice. *Communs. behav. Biol.,* **1**, 45–59.
2770. de SOUZA, S. W. & DOBBING, J. (1971). Cerebral edema in developing brain. I. Normal water and cation content in developing rat brain and postmortem changes. *Expl Neurol.,* **32**, 431–438.
2771. SPACEK, J. (1971). Three dimensional reconstruction of astroglia and oligodendroglia cells. *Z. Zellforsch. mikrosk. Anat.,* **112**, 430–442.
2772. SPATARO, J. (1966). Anoxic-ischaemic encephalopathy of the rat brain. *Expl Neurol.,* **16**, 16–27.
2773. SPECHT, S. & GRAFSTEIN, B. (1973). Accumulation of radioactive protein in mouse cerebral cortex after injection of ^{3}H-fucose into the eye. *Expl Neurol.,* **41**, 705–722.
2774. SPEHLMANN, R. (1971). Acetylcholine and epileptiform activity of chronically isolated cortex. *Archs. Neurol.,* **24**, 495–502.
2775. SPEIDEL, C. C. (1944). The regeneration of denervated lateral line organs. *Anat. Rec.,* **68**, 456.
2776. SPEIDEL, C. C. (1944). The trophic influence of specific nerve supply on special sensory organs, as revealed by prolonged survival of denervated lateral line organs. *Anat. Rec.,* **88**, 459.
2777. SPEIDEL, C. C. (1946). Cine-photomicrographs of tadpole cells *in vivo*, with special reference to experiments on nerves and special sense organs of the lateral line. *Anat. Rec.,* **94**, 551.
2778. SPEIDEL, C. C. (1946). Correlated histories of individual sense organs and their nerves, as seen in living frog tadpoles. *Biol. Bull.,* **91**, 224–225.
2779. SPEIDEL, C. C. (1947). Correlated studies of sense organs and nerves of the lateral line in living tadpoles. I. Regeneration of denervated organs. *J. comp. Neurol.,* **87**, 29–55.

2780. SPEIDEL, C. C. (1948). Correlated studies of sense organs and nerves of the lateral line in living frog tadpoles. II. The trophic influence of specific nerve supply as revealed by prolonged observations of denervated and reinnervated organs. *Am. J. Anat.*, **82**, 277–320.
2781. SPEIDEL, C. C. (1964). *In vivo* studies of myelinated nerve fibres. *Int. Rev. Cytol.*, **16**, 173–231.
2782. SPEIDEL, C. C. (1964). Correlated studies of sense organs and nerves of the lateral line in living frog tadpoles. IV. Patterns of vagus nerve regeneration after single and multiple operations. *Am. J. Anat.*, **114**, 133–160.
2783. SPERLING, W. & RAFFERTY, C. N. (1969). Relationship between absorption spectrum and molecular conformation of 11-cis-retinal. *Nature, Lond.*, **224**, 591.
2784. SPERRY, R. W. (1940). The functional results of muscle transposition in the hind limbs of the rat. *J. comp. Neurol.*, **73**, 379–404.
2785. SPERRY, R. W. (1941). The effect of crossing nerves to antagonistic muscles in the hind limb of the rat. *J. comp. Neurol.*, **75**, 1–19.
2786. SPERRY, R. W. (1943). Effect of 180 degree rotation of the retinal field on visuomotor coordination. *J. exp. Zool.*, **92**, 263–279.
2787. SPERRY, R. W. (1943). Visuomotor coordination in the newt (*Triturus viridescens*) after regeneration of the optic nerve. *J. comp. Neurol.*, **79**, 33–55.
2788. SPERRY, R. W. (1944). Optic nerve regeneration with return of vision in anurans. *J. Neurophysiol.*, **7**, 57–69.
2789. SPEERY, R. W. (1945). Restoration of vision after crossing of optic nerves and after contralateral transplantation of the eye. *J. Neurophysiol.*, **8**, 15–28.
2790. SPERRY, R. W. (1945). The problem of central nervous reorganization after nerve regeneration and muscle transposition. *Q. Rev. Biol.*, **20**, 311–369.
2791. SPERRY, R. W. (1948). Orderly patterning of synaptic associations in regeneration of intracentral nerve tracts mediating visuomotor coordination. *Anat. Rec.*, **102**, 63–75.
2792. SPERRY, R. W. (1963). Chemoaffinity in the orderly growth of nerve fibre patterns and connections. *Proc. natn. Acad. Sci. U.S.A.*, **50**, 703–710.
2793. SPERRY, R. W. (1965). Embryogenesis of behavioural nerve nets. *In* de Haan, R. L. & Ursprung, H. *Organogenesis.* Holt, Rinehart and Winston. N.Y; pp. 161–186.
2794. SPERRY, R. W. & ARORA, H. L. (1965). Selectivity in the regeneration of the oculomotor nerve in the cichlid fish *Astronotus ocellatus. J. Embryol. exp. Morph.*, **14**, 307–317.
2795. SPINELLI, D. N. & BARRETT, T. W. (1969). Visual receptive field organizations of single units in the cat's visual cortex. *Expl Neurol.*, **24**, 76–98.
2796. SPIRA, M. E. & BENNETT, M. V. L. (1972). Penicillin induced seizure

activity in the hatchet fish. *Brain Res.*, **43**, 235–241.

2797. SPIRA, M. E. & BENNETT, M. V. L. (1972). Synaptic control of electrotonic coupling between neurons. *Brain Res.*, **37**, 294–300.
2798. SPIRO, R. G. (1970). Biochemistry of basement membranes. *In* Balazs, E. A., *Chemistry and molecular biology of the extracellular matrix.* **1**, pp. 511–534. Acad. Press; London.
2799. SPOONER, B. S., ASH, J. F. WRENN, J. T., FRATER, R. B. & WESSELS, N. K. (1973). Heavy meromyosin binding to microfilaments involved in cell and morphogenetic movements. *Tiss. Cell,* **5**, 37–46.
2800. SPOONER, B. S., LUDUENA, M. A. & WESSELLS, N. K. (1974). Membrane fusion in the growth cone – microspike region of embryonic nerve cells undergoing elongation in tissue culture. *Tiss. Cell,* **6**, 399–409.
2801. SPOONER, B. S. & WESSELLS, N. K. (1970). Effects of cytochalasin B upon microfilaments involved in morphogenesis of salivary epithelium. *Proc. natn. Acad. Sci. U.S.A.*, **66**, 360–364.
2802. SPOONER, B. S., YAMADA, K. M. & WESSELLS, N. K. (1971). Microfilaments and cell locomotion. *J. Cell Biol.*, **49**, 595–613.
2803. SPUDICH, J. A. & LIN, S. (1972). Cytochalasin B, its interaction with actin and actomyosin from muscle. *Proc. natn. Acad. Sci. U.S.A.*, **69**, 442–446.
2804. SQUIRE, L. R. & BARONDES, S. H. (1972). Variable decay of memory and its recovery in cycloheximide-treated mice. *Proc. natn. Acad. Sci. U.S.A.*, **69**, 1416–1420.
2805. SRETER, F. A. (1970). Effect of denervation on fragmented sarcoplasmic reticulum of white and red muscle. *Expl Neurol.*, **29**, 52–64.
2806. SRETER, F. A. & GERGELY, J. (1964). Comparative studies of the Mg activated ATPase activity and Ca^{++} uptake of fractions of white and red muscle homogenates. *Biochem. biophys. Res. Commun.*, **16**, 438.
2807. SRETER, F. A., SEIDEL, J. C. & GERGELY, J. (1966). Studies on myosin from red and white skeletal muscle of the rabbit. I. Adenosine triphosphatase activity. *J. biol. Chem.*, **241**, 5772–5776.
2808. SRIVASTAVA, U. (1972). Biochemical changes in progressive muscular dystrophy. IX. Synthesis of native myosin, actin and tropomyosin in the skeletal muscle of mouse as a function of muscular dystrophy. *Can. J. Biochem.*, **50**, 409–415.
2809. STAPRANS, I., KENNEY, W. C. & DIRKSEN, E. R. (1975). Calcium affinity of chick brain tubulin. *Biochem. biophys. Res. Commun.*, **62**, 92.
2810. STAVRAKY, G. W. (1961). *Supersensitivity following lesions of the nervous system.* Univ. Toronto Press; Toronto.
2811. STEEN, T. P. & THORNTON, C. S. (1963). Tissue interaction in amputated aneurogenic limb of *ambystoma* larvae. *J. exp. Zool.*, **154**, 207–221.
2812. STEERE, R. L. (1969). Freeze-etching simplified. *Cryobiol.*, **5**, 306.
2813. STEERS, E. & MARCHESI, V. T. (1969). Studies on a protein component of guinea pig erythrocyte membranes. *J. gen. Physiol.*, **51**, 65–71S.

2814. STEINBACH, J. H. (1974). Role of muscular activity in nerve muscle interaction *in vitro*. *Nature, Lond.*, **248**, 70–71.
2815. STEINBACH, J. H., HARRIS, A. J., PATRICK, J., SCHUBERT, D. & HEINEMANN, S. (1973). Nerve-muscle interaction in vitro. Role of acetylcholine. *J. gen. Physiol.*, **62**, 255–270.
2816. STENEVI, U. BJERRE, B., BJÖRKLUND, A. & MOBLEY, W. (1974). Effects of localized intracerebral injections of nerve growth factor on the regenerative growth of lesioned central noradrenergic neurons. *Brain Res.*, **69**, 217–234.
2817. STENEVI, U., BJÖRKLUND, A. & MOORE, R. Y. (1973). Morphological plasticity of central adrenergic neurons. *Brain Behav. Evol.*, **8**, 110.
2818. STENSAAS, L. J. & GILSON, B. C. (1972). Ependymal and subependymal cells of the caudato-pallial junction in the lateral ventricle of the neonatal rabbit. *Z. Zellforsch. mikrosk. Anat.*, **132**, 297–322.
2819. STENSAAS, L. J. & STENSAAS, S. S. (1968). An electron microscopic study of cells in the matrix and intermediate laminae of the cerebral hemisphere of the 45 mm rabbit embryo. *Z. Zellforsch. mikrosk. Anat.*, **91**, 341.
2820. STERBA, G. (1967). Incorporation of sulphur 35 into the subcommisural organ and Reissner's fibre in Cyprinus carpio. *Nature, Lond.*, **216**, 504.
2821. STERBA, G., HESS, J. & ERMISCH, A. (1965). Reissner's fibre as receptor model for biogenic amines. II. Tyrosine binding to Reissner's fibre of cats after intraventricular ^{3}H-tyrosine injection. *Pflügers. Arch. ges. Physiol.*, **310**, 277–286.
2822. STEWART, D. M. (1962). Protein composition of denervated muscle. *Am. J. Physiol.*, **202**, 281–284.
2823. STEWART, D. M. (1968). Effect of age on response of four muscles of the rat to denervation. *Am. J. Physiol.*, **214**, 1139–1146.
2824. STEWART, W. W. (1972). Comments on the chemistry of scotophobin. *Nature, Lond.*, **238**, 202–209.
2825. STOCH, M. B. & SMYTHE, P. M. (1963). Does undernutrition during infancy inhibit brain growth and subsequent intellectual development? *Archs. Dis. Child.*, **38**, 546–552.
2826. STOCUM, D. (1968). The urodele limb regeneration blastema: a self organizing system. I. Differentiation *in vitro*. *Devl Biol.*, **18**, 441–456.
2827. STOCUM. D. L. (1968). The urodele limb regeneration blastema. II. Morphogenesis and differentiation of autografted whole and fractional blastemas. *Devl Biol.*, **18**, 457–480.
2828. STOCUM, D. L. & DEARLOVE, G. E. (1972). Epidermal-mesodermal interaction during morphogenesis of the limb regeneration blastema in larval salamanders. *J. exp. Zool.*, **181**, 49–62.
2829. STOKER, M. G. P., SHEARER, M. & O'NIELL, L. (1966). Growth inhibition of polyoma-transformed cells by contact with static normal fibroblasts. *J. Cell Sci.*, **1**, 297–310.
2830. van STONE, J. (1964). The relation of nerve number to regenerative

capacity in the developing hind limb of *Rana sylvatica*. *J. exp. Zool.*, **155**, 293–302.

2831. STONE, L. S. (1931). Studies on the migratory lateral line primordia in *amblystoma*. *Anat. Rec.*, **48**, 64.

2832. STONE, L. S. (1933). Independence of taste organs with respect to their nerve fibres demonstrated in living salamanders. *Proc. Soc exp. Biol.*, **30**, 1256–1257.

2833. STONE, L. S. (1933). Developmental changes in primary lateral line organs studied in living larvae of Anurans and Urodeles. *Proc. Soc. exp. Biol.*, **30**, 1258–1259.

2834. STONE, L. S. (1933). The development of the lateral line organs in amphibians observed in vital-stained preparations. *J. comp. Neurol.*, **57** 507–540.

2835. STONE, L. S. (1937). Further experimental studies of the development of lateral line organs in amphibians observed in living preparations. *J. comp. Neurol.*, **68**, 83–115.

2836. STONE, L. S. & ZAUER, I. (1940). Reimplantation and transplantation of larval eyes in the salamander (Amblystoma punctatum). *J. exp. Zool.*, **85**, 243–269.

2837. STONER, J., MANGANIELLO, V. C. & VAUGHAN, M. (1974). Guanosine cyclic 3′5′-monophosphate and guanylate cyclase activity in the guinea pig lung: effect of acetylcholine and cholinesterase inhibition. *Molec. Pharmac.*, **10**, 155–161.

2838. STORM-MATHISEN, J. (1973). Increase of choline acetylase and cholinesterase in stratum moleculare fasciae dentatae following degeneration of the perforant path. *Acta physiol. scand.*, **396**, 33.

2839. STRACHAR, A. (1969). Evidence for the involvement of light chains in the biological functioning of myosin. *Biochem. biophys. Res. Commun.*, **35**, 519.

2840. STRASSMAN, R. J., LETOURNEAU, P. C. & WESSELLS, N. K. (1973). Elongation of axons in an agar matrix that does not support cell locomotion. *Expl Cell Res.*, **81**, 482–487.

2841. STRAUSS, A. J. L., SEEGAL, B. C., HSU, K. G., BURKHOLDER, P. M. NASTUR, W. L. & OSSERMAN, K. E. (1972). Immunofluorescent demonstration of a muscle-binding, complement-fixing serum globulin fraction in myasthenia gravis. *Proc. Soc. exp. Biol. Med.*, **105**, 184.

2842. STREIT, P., AKERT, K., SANDRI, C., LIVINGSTON, R. B. & MOOR, H. (1972). Dynamic ultrastructure of presynaptic membranes at nerve terminals in the spinal cord of rats. Anaesthetized and unanaesthetized preparations compared. *Brain. Res.*, **48**, 11–26.

2843. STRETTON, A. O. W. & KRAVITZ, G. A. (1968). Neuronal geometry: determination with a technique of intracellular dye injection. *Science, N.Y.*, **162**, 132–134.

2844. STROM, T. B., DEISSEROTH, A., MORGANROTH, J., CARPENTER, C. B. & MERRILL, J. P. (1972). Alteration of the cytotoxic action of sensitized lymphocytes by cholinergic agents and activators of adenylate cyclase. *Proc. natn. Acad. Sci. U.S.A.*, **69**, 2995–2999.

2845. STUART, A. E. (1970). Physiological and morphological properties of motoneurones in the central nervous system of the leech. *J. Physiol.*, **209**, 627–646.

2846. STUDITSKY, A., ZHENEVSKAYA, R. & RUMYANTSEVA, O. (1963). The role of neurotrophic influences upon the restitution of structure and function of regenerating muscles. *In* Gutmann, E. & Hnik, P., *The effect of use and disuse on neuromuscular function.* Elsevier; Amsterdam. pp. 71–81.

2847. STUTINSKY, F. (1951). Sur l'origine de la substance Gomori-positive du complexe hypothalamo-hypophysaire. *C. r. Seanc. Soc. Biol.* **145**, 367–370.

2848. SUBAK-SHARPE, J. H., BÜRK, R. R. & PITTS, J. D. (1966). Metabolic cooperation by cell to cell transfer between genetically different mammalian cells in tissue culture. *Heredity, Lond.*, **21**, 342–343.

2849. SUBAK-SHARPE, J. H., BÜRK, R. R. & PITTS, J. D. (1969). Metabolic cooperation between biochemically marked mammalian cells *J. Cell Sci.*, **4**, 353–367.

2850. SUGAYA, E., GOLDRING, S. & O'LEARY, J. L. (1964). Intracellular potentials associated with direct cortical response and seizure discharge in cat. *Electroenceph. clin. Neurophysiol.*, **17**, 661–669.

2851. SULZMANN, R. (1959). Ein Beitrag zum Problem des Verhältnisses zwischen Markscheidendicke und Achsenzylinderdurchmesser bei den peripheren Nervenfasern des Schäferhundes. *Jb. Morph. mikrosk. Anat.*, **65**, 259–281.

2852. SUMMERS, K. E. & GIBBONS, I. R. (1971). Adenosine triphosphate induced sliding of tubulin in trypsin-treated flagella of sea urchin sperm. *Proc. natn. Acad. Sci. U.S.A.*, **68**, 3092–3096.

2853. SUMNER, B. E. H. (1974). The effect of injury on two hydrolases in the hypoglossal nucleus, with quantitative data on N-acetyl-β-glucosaminidase. *Brain Res.*, **68**, 157–166.

2854. SUMNER, B. E. H. (1975). A quantitative study of subsurface cisterns and their relationships in normal and axotomized hypoglossal neurones. *Brain Res.*, **22**, 175–186.

2855. SUMNER, B. E. H. (1975). A quantitative analysis of the response of presynaptic boutons to postsynaptic motor neurone axotomy. *Expl Neurol.*, **46**, 605–615.

2856. SUMNER, B. E. H. (1975). The uptake of exogenous peroxidase by vesicles of presynaptic boutons after axotomy. *J. Physiol.* (in press).

2857. SUMNER, B. E. H. & SUTHERLAND, F. (1973). Quantitative electron microscopy on the injured hypoglossal nucleus in the rat. *J. Neurocytol.*, **2**, 315–328.

2858. SUMNER, B. E. H. & WATSON, W. E. (1971). Retraction and expansion of the dendritic tree of motor neurones of adult rats produced in vivo. *Nature, Lond.*, **233**, 273–275.

2859. SUMNER, B. E. H & WATSON, W. E. (1972). A simple method for obtaining autoradiographs of single isolated neuroglial cells. *J. Physiol.*, **222**, 119–120P.

2860. SUZUKI, K. (1970). Formation and turnover of myelin ganglioside. *J. Neurochem.*, **17**, 209–213.

2861. SUZUKI, K. & SUZUKI, Y. (1970). Globoid cell leucodystrophy (Krabbe's disease): deficiency of galactocerebroside-β-galactosidase. *Proc. natn. Acad. Sci. U.S.A.*, **66**, 302.

2862. SVEJDA, J. & SKACH, M. (1971). Die Zunge der Ratte im Raster-Elektronenmikroskop (Stereoscan). *Z. mikrosk.-anat. forsch.*, **84**, 101–116.

2863. SWANBORG, R. H. & SHULMAN, S. (1970). Purification of an organ-specific thermostable antigen found in bovine brain. *Immunology*, **19**, 31–40.

2864. SWANSON, H. H. (1970). Effects of castration at birth in hamsters of both sexes on luteinization of ovarian implants, oestrus cycles and sexual behaviour. *J. Reprod. Fert.*, **21**, 183–186.

2865. SWANSON, P. D. (1966). Temperature dependence of sodium ion activation of the cerebral microsomal adenosine triphosphatase. *J. Neurochem.*, **13**, 229–236.

2866. SWEET, C. & ZULL, J. E. (1970). The binding of serum albumin to phospholipid liposomes. *Biochim. biophys. Acta*, **219**, 253–262.

2867. SYROVY, I., GUTMANN, E. & MELCHINA, J. (1971). Differential response of myosin-ATPase activity and contraction properties of fast and slow rabbit muscles following denervation *Experentia*, **27**, 1426–1427.

2868. SYTKOWSKI, A. J., VOGEL, Z. & NIRENBERG, M. W. (1973). Development of acetylcholine receptor clusters on cultured muscle cells. *Proc. natn. Acad. Sci. U.S.A.*, **70**, 270–274.

2869. SZABO, G., EISENMAN, G. & CIANI, S. (1969). The effects of the macrotetralide antibiotics on the electrical properties of phospholipid bilayer membranes. *J. mem. Biol.*, **1**, 346–382.

2870. SZABO, M. & ROBOZ-EINSTEIN, E. (1962). Acid polysaccharides in the central nervous system. *Archs. Biochem.*, **98**, 406–412.

2871. SZENTAGOTHAI, J. (1964). The parvicellular neurosecretory system. *Prog. Brain Res.*, **5**, 1–32.

2872. TAGLIAMONTE, A., TAGLIAMONTE, P., FORN, J., PEREZ-CRUET, J., KRISHNA, G. & GESSA, G. L. (1971). Stimulation of brain serotonin synthesis by dibutyryl-cyclic AMP in rats. *J. Neurochem.*, **18**, 1191–1196.

2873. TAGLIAMONTE, A., TAGLIAMONTE, P., PEREZ-CRUET, J. & GESSA, G. L. (1971). Increase of brain tryptophan caused by drugs which stimulate serotonin synthesis. *Nature, Lond., New Biol.*, **229**, 125–126.

2874. TAKAGI, S. F. (1973). Degeneration and regeneration of the olfactory epithelium. *In* Beidler, L. M., *Handbook of sensory physiology IV* (2), Springer, Verlag; Berlin.

2875. TAKAHASHI, Y., HSU, C. S. & HONMA, S. (1970). Potassium and glutamate effects on protein synthesis in isolated neuroglial cells. *Brain Res.*, **23**, 284–287.
2876. TAKENAKA, T. (1963). Effects of temperature and metabolic inhibitors on the active Na transport in frog skin. *Jap. J. Physiol.*, **13**, 208–218.
2877. TAKESHIGE, C. & VOLLE, R. L. (1963). Choliceptive sites in denervated sympathetic ganglia. *J. Pharmac. exp. Ther.*, **141**, 206–213.
2878. TAKEZAKI, M. & KITO, Y. (1967). Circular dichroism of rhodopsin and isorhodopsin. *Nature, Lond.*, **215**, 1197–1199.
2879. TAMIR, H. & KUHAR, M. J. (1975). Association of serotononin binding protein with projections of the midbrain raphe nuclei. *Brain Res.*, **83**, 169–172.
2880. TANFORD, C. (1972). Structure of membrane proteins. *Ann. N.Y. Acad. Sci.*, **195**, 126–127.
2881. TANG, B. Y., KOMIYA, Y. & AUSTIN, L. (1974). Axoplasmic flow of phospholipids and cholesterol in the sciatic nerve of normal and dystrophic mice. *Expl Neurol.*, **43**, 13–20.
2882. TANI, E. & AMETANI, T. (1971). Extracellular distribution of ruthenium-red positive substance in the cerebral cortex. *J. Ultrastruct. Res.*, **34**, 1–14.
2883. TANIGUCHI, K. & IIDA, S. (1971). The binding of ouabain to Na^+-K^+ dependent ATPase treated with phospholipase. *Biochim. biophys. Acta*, **233**, 831–833.
2884. TANIGUCHI, K. & IIDA, S. (1973). The role of phospholipids in the binding of ouabain to sodium- and potassium-dependent adenosine triphosphatase. *Molec. Pharmac.*, **9**, 350–359.
2885. TANIGUCHI, K. & TONOMURA, K. (1971). Inactivation of Na^+K^+-dependent ATPase by phospholipase treatment and its reactivation by phospholipids. *J. Biochem., Tokyo*, **69**, 543–557.
2886. TARDENT, P. (1963). Regeneration in the hydrozoa. *Biol. Rev.*, **38**, 293–333.
2887. TARDENT, P. (1965). Developmental aspects of regeneration in coelenterates. *In* Kiortis, V. & Trampusc, H. A. L., *Regeneration in animals and related problems.* North Holland; Amsterdam. pp. 71–88.
2888. TARIN, D. & CROFT, C. B. (1970). Ultrastructural studies of wound healing in mouse skin. II. Dermo-epidermal interrelationship. *J. Anat.*, **106**, 79–91.
2889. TASAKI, I. (1968). *Nerve excitation.* C. C. Thomas; Springfield. Illinois.
2890. TASAKI, I. (1965). Excitability of neurons and glial cells. *In* de Robertis, E. & Carrea, R., Biology of neuroglia. *Prog. Brain Res.*, **15**, 234–242.
2891. TASAKI, I., CARNAY, L., SANDLIN, R. & WATANABE, A. (1969). Fluorescence changes during conduction in nerves stained with acridine orange. *Science, N.Y.*, **163**, 683–685.
2892. TASAKI, I. & CHANG, J. J. (1958). Electric response of glia cells in cat brain. *Science, N.Y.*, **128**, 1209.
2893. TASAKI, I. & KAMIYA, N. (1950). Membrane potentials of the amoeba.

Protoplasma, **39**, 333–343.

2894. TASAKI, I., SISCO, K. & WARASHINA, A. (1974). Alignment of anilinonaphthalene-sulfonate and related fluorescent probes in squid axon membrane and in synthetic polymers. *Biophys. Chem.,* **2**, 316.

2895. TASAKI, I., WATANABE, A. & HALLETT, M. (1972). Fluorescence of squid axon membrane labelled with hydrophobic probes. *J. mem. Biol.,* **8**, 109–132.

2896. TASAKI, I., WATANABE, A., SANDLIN, R. & CARAY, L. (1968). Changes in fluorescence, turbidity and birefringence associated with nerve excitation. *Proc. natn. Acad. Sci. U.S.A.,* **61**, 883–888.

2897. TATEDA, H. & BEIDLER, L. M. (1964). The receptor potential of the taste cell of the rat. *J. gen. Physiol.,* **47**, 479–486.

2898. TAUC, L. (1965). Presynaptic inhibition in the abdominal ganglion of Aplysia. *J. Physiol.,* **181**, 282–307.

2899. TAXI, J. & SOTELO, C. (1973). Cytological aspects of the axonal migration of catecholamines and of their storage material. *Brain Res.,* **62**, 431–437.

2900. TAYLOR, A. C. (1966). Microtubules in the microspikes and cortical cytoplasm of isolated cells. *J. Cell Biol.,* **28**, 155–168.

2901. TAYLOR, A. C. & WEISS, P. (1965). Demonstration of axonal flow by measurement of tritium-labelled protein in mature optic nerve fibres. *Proc. natn. Acad. Sci. U.S.A.,* **54**, 1521–1527.

2902. TAYLOR, G. W. (1942). The correlation between sheath thickness and conduction velocity with special reference to cat nerve fibres. *J. cell. comp. Physiol.,* **20**, 359–372.

2903. TAYLOR, M. H. & HOWARD, E. (1971). Impaired glucose homeostasis in adult rats after corticosterone treatment in infancy. *Endocrinology,* **88**, 1190–1202.

2904. TENENBAUM, D. & FOLCH-PI, J. (1966). The preparation and characterization of water-soluble proteolipid protein from bovine brain white matter. *Biochim biophys. Acta,* **115**, 141–147.

2905. TENNYSON, V. M. (1970). The fine structure of the axon and growth cone of the dorsal root neuroblast of the rabbit embryo. *J. Cell Biol.,* **45**, 62–79.

2906. TERRY, T. M., ENGELMAN, D. M. & MOROWITZ, H. J. (1967). Characterization of the plasma membrane of mycoplasma laidlawii. II. Modes of aggregation of solubilized membrane components. *Biochim. biophys. Acta,* **135**, 391–405.

2907. THESLEFF, S. (1960). Sensitivity of skeletal muscle produced by botulinum toxin. *J. Physiol.,* **151**, 598–607.

2908. THESLEFF, S. (1963). Spontaneous electrical activity in denervated rat skeletal muscle. *In* Gutmann, E. & Hnik, P., *The effect of use and disuse on neuromuscular functions.* Elsevier; Amsterdam. pp. 41–51.

2909. THOA, N. B., WOOTEN, G. F., AXELROD, J. & KOPEN, I. J. (1972). Inhibition of release of dopamine-β-hydroxylase and norepinephrine from sympathetic nerves by colchicine, vinblastine or cytochalasin B. *Proc.*

natn. Acad. Sci. U.S.A., **69**, 520–522.

2910. THOMAS, E. & PEARSE, A. G. E. (1961). The fine structural localization of dehydrogenase in the nervous system. *Histochemie*, **2**, 266–282.
2911. THOMPSON, W. R. & HERON, W. (1954). The effects of restricting early experience on the problem solving capacity of dogs. *Can. J. Psychol.*, **8**, 17–31.
2912. THOMPSON, W. R. & MELZACK, R. (1956). Early environment. *Scient. Am.*, **194**, 38–42.
2913. THORNTON, C. S. (1962). Influence of head skin on limb regeneration in urodele amphibians. *J. exp. Zool.*, **150**, 5–16.
2914. THORNTON, C. S. (1965). Influence of wound skin on blastemal cell aggregation. *In* Kiortis, V. & Trampusch, H. A. L., *Regeneration in animals and related problems.* North Holland; Amsterdam. pp. 333–340.
2915. THORNTON, C. S. (1968). Amphibian limb regeneration. *Adv. Morphogen.*, **7**, 205–250.
2916. THORNTON, C. S. (1968). Recuperation of ability to regenerate denervated larval urodele limb. *Am. Zool.*, **8**, 785.
2917. THORNTON, C. S. (1970). Amphibian limb regeneration and its relation to nerves. *Am. Zool.*, **10**, 113–118.
2918. THORNTON, C. S. & THORNTON, M. T. (1970). Recuperation of regeneration in denervated limbs of *amblystoma* larvae. *J. exp. Zool.*, **173**, 293–302.
2919. TIEFER, L. (1970). Gonadal hormones and mating behaviour in the adult golden hamster. *Horm. Behav.*, **1**, 189–202.
2920. TIEN, H. T. (1968). Photoelectric effects in thin and bilayer lipid membranes in aqueous media. *J. phys. Chem.*, **72**, 4512.
2921. TIEN, H. T. (1970). The effect of modifiers on the intrinsic properties of bilayer lipid membranes. *In* Blank, M., *Surface chemistry of biological systems.* Plenum Press; N.Y. pp. 135–154.
2922. TIEN, H. T. & VERMA, S. P. (1970). Electronic processes in bilayer lipid membranes. *Nature, Lond.*, **227**, 1232.
2923. TILLACK, T. W., SCOTT, R. E. & MARCHESI, V. T. (1972). The structure of erythrocyte membranes studied by freeze-etching. *J. exp. Med.*, **135**, 1209–1227.
2924. TILNEY, L. G. (1968). The assembly of microtubules and their role in the development of cell form. *In* Locke, M., *The emergence of order in developing systems.* Acad. Press; N.Y. pp. 63–102.
2925. TILNEY, L. G., BRYAN, J., BUSH, D. J., FURIJAWA, K., MOOSERER, M. S., MURPHY, D. B. & SNYDER, D. H. (1973). Microtubules: evidence for 13 protofilaments. *J. Cell Biol.*, **59**, 267–275.
2926. TILNEY, L. G. & GIBBINS, J. R. (1969). Microtubules in the formation and development of the primary mesenchyme in *Arbacia punctata.* II. An experimental analysis of their role in development and maintainance of cell shape. *J. Cell Biol.*, **41**, 227–250.
2927. TILNEY, L. G. & MOOSEKER, M. (1971). Actin in the brush border of

epithelial cells of the chicken intestine. *Proc. natn. Acad. Sci. U.S.A.*, **68**, 2611–2615.

2928. TIMASHEFF, S. N. & INOUE, H. (1968). Preferential binding of solvent components to proteins in mixed water-organic solvent systems. *Biochemistry, N.Y.*, 7, 2501–2513.

2929. TIMIRAS, P. S., WOODBURY, D. M. & GOODMAN, L. S. (1954). Effect of adrenalectomy, hydrocortisone acetate and desoxycorticosterone acetate on brain excitability and electrolyte distribution in mice. *J. Pharmac.*, **112**, 80–93.

2930. TINOCO, J., GHOSH, D. & KEITH, A. D. (1972). Interactions of spin-labelled lipid molecules with natural lipids in monolayers at the air-water interface. *Biochim. biophys. Acta*, **274**, 279–285.

2931. TIPLADY, B. & ROSE, S. P. R. (1971). Amino acid incorporation into protein in neuronal cell body and neuropil fractions *in vitro*. *J. Neurochem.*, **18**, 549–558.

2932. TODARO, G. T., LAZAR, G. K. & GREEN, H. (1965). The initiation of cell division in a contact-inhibited mammalian cell line. *J. cell. comp. Physiol.*, **66**, 325–334.

2933. TORDA, C. (1972). Cyclic AMP-dependent diphosphoinositide kinase. *Biochim. biophys. Acta*, **286**, 389–395.

2934. TORDA, C. (1972). Hyperpolarization by cyclic AMP (Activation of diphosphoinositide kinase). *Experentia*, **28**, 1438–1439.

2935. TORDA, C. (1974). Model of molecular mechanism able to generate a depolarization-hyperpolarization cycle. *Int. Rev. Neurobiol.*, **16**, 1–66.

2936. TORDA, C. (1974). Restorative effect of cyclic AMP on the bioelectric processes of calcium deprived ganglia. *Experentia*, **30**, 1154.

2937. LA TORRE, J. L., LUNT, G. S. & de ROBERTIS, E. (1970). Isolation of a cholinergic receptor protein from electric tissue. *Proc. natn. Acad. Sci. U.S.A.*, **65**, 716.

2938. TORREY, T. W. (1934). Temperature coefficient of nerve degeneration. *Proc. natn. Acad. Sci. U.S.A.*, **20**, 303–305.

2939. TORVIK, A. (1956). Transneuronal changes in the inferior olive and pontine nuclei in kittens. *J. Neuropath. exp. Neurol.*, **15**, 119–145.

2940. TORVIK, A. & HEDING, A. (1967). Histological studies on the effect of Actinomycin D on retrograde nerve cell reaction in the facial nucleus of mice. *Acta neuropath.*, **9**, 146–157.

2941. TORVIK, A. & HEDING, A. (1969). Effect of actinomycin D on retrograde nerve cell reaction. Further observations. *Acta neuropath.*, **14**, 62–71.

2942. TORVIK, A. & SKJÖRTEN, F. (1971). Electron microscopic observations on nerve cell regeneration and degeneration after axon lesions. Changes in glial cells. *Acta neuropath.*, **17**, 265–282.

2943. TORVIK, A. & SÖREIDE, A. J. (1972). Nerve cell regeneration after axon lesions in newborn rabbits. Light and electron microscopic study. *J. Neuropath. exp. Neurol.*, **31**, 683–695.

2944. TOSTESON, D. C. (1968). Kinetic characteristics of the activation of sheep red cell membrane ATPase by magnesium ions. *Fedn Proc. Fedn Am. Socs exp. Biol.*, **27**, 234.

2945. TOUTELLOTTE, M. E., BRANDON, D. & KEITH, A. (1970). Membrane structure: spin labeling and freeze-etching of mycoplasma laidlawii. *Proc. natn. Acad. Sci. U.S.A.*, **66**, 909–916.
2946. TOWER, S. S., HOWE, H. & BODIAN, D. (1941). Fibrillation in skeletal muscle in relation to denervation and to inactivation without denervation. *J. Neurophysiol.*, **4**, 398–401.
2947. TRACHTENBERG, M. C. & POLLEN, D. A. (1970). Neuroglia: biophysical properties and physiological function. *Science, N.Y.*, **167**, 1248–1251.
2948. TRAMS, E. G. (1964). Properties of electroplax protein II. *Biochim. biophys. Acta*, **79**, 521–530.
2949. TRÄUBLE, H. & SACKMANN, E. (1973). Estimation of the rotational relaxation time of rhodopsin in the visual receptor membrane. *Nature, Lond.*, **245**, 210–211.
2950. TRENDELENBERG, U. (1962). The action of acetylcholine on the nictitating membrane of the cat. *J. Pharmac. exp. Ther.*, **135**, 39–44.
2951. TRENDELENBERG, U. (1963). Supersensitivity and subsensitivity to sympathetic amines. *Pharmac. Rev.*, **15**, 225–276.
2952. TRENDELENBERG, U. (1963). Time course of changes in sensitivity after denervation of the nictitating membrane of the spinal cat. *J. Pharmac. exp. Ther.*, **148**, 329–338.
2953. TRENDELENBERG, U. (1966). Mechanisms of supersensitivity and subsensitivity to sympathomimetic amines. *Pharmac. Rev.*, **18**, 629–640.
2954. TRENDELENBERG, U., MUSKUS, A., FLEMING, W. W. & de la SIERRA, G. A. (1962). Effect of cocaine, denervation and decentralization on the response of the nictitating membrane to various sympathomimetic amines. *J. Pharmac. exp. Ther.*, **138**, 181–193.
2955. TRENDELENBERG, U. & WIENER, N. (1962). Sensitivity of the nictitating membrane after various procedures and agents. *J. Pharmac. exp. Ther.*, **136**, 152–161.
2956. TRESS, K. H. & HERBERG, L. J. (1972). Permanent reduction in seizure threshold resulting from repeated electrical stimulation. *Expl Neurol.*, **37**, 347–359.
2957. TRISSL, H. W. & LÄUGER, P. (1970). Photoelectric effects in thin chlorophyll films. *Z. Naturf.*, **25B**, 1059.
2958. TRISSL, H.-W. & LÄUGER, P. (1972). Photoelectric effects at lipid bilayer membranes; theoretical models and experimental observations. *Biochim. biophys. Acta*, **282**, 40–54.
2959. TROSPER, T. (1972). Some properties of chlorophyll *a* at hydrocarbon-water interfaces and in black lipid membranes. *J. mem. Biol.*, **8**, 133–148.
2960. TRUOG, P. & WASER, P. G. (1970). Einflüse der Denervation von Muskelendplatten auf die Lokalisation der Acetylcholinesterase und die Bindung von Decamethonium. *Arch. Pharmak.*, **226**, 101–112.

2961. TSCHIRGI, R. D., INANAGA, K., TAYLOR, J. L., WALKER, R. M. & SONNENSCHEIN, R. R. (1957). Changes in the cortical pH and blood flow accompanying spreading cortical depression and convulsion. *Am. J. Physiol.*, **190**, 557–562.
2962. TU, A. T. (1973). Neurotoxins of animal venoms: snakes. *A. Rev. Biochem.*, **42**, 235–258.
2963. TUNG, A. S. C. & PIPA, R. L. (1971). Fine structure of transected interganglionic connectives and degenerating axons of wax moth larvae. *J. Ultrastruct. Res.*, **36**, 694–707.
2964. TUNG, A. S. C. & PIPA, R. L. (1972). Insect neurometamorphosis. V. Fine structure of axons and neuroglia in the transforming interganglionic connectives of *Galleria mellonella* (L) (Lepidoptera). *J. Ultrastruct. Res.*, **39**, 556–567.
2965. TUPIKOVA, N. & GERARD, R. W. (1937). Salt content of neural tissues. *Am. J. Physiol.*, **119**, 414–415.
2966. TURNER, L. V. & MANCHESTER, K. L. (1971). Enzyme changes indicative of fibre types in the denervated rat hemidiaphragm. *Biochem. J.*, **121**, 24P.
2967. TURNER, L. V. & MANCHESTER, K. L. (1972). Effects of denervation on the glycogen content and on the activities of enzymes of glucose and glycogen metabolism in rat diaphragm muscle. *Biochem J.*, **128**, 789–801.
2968. TURNER, L. V. & MANCHESTER, K. L. (1972). Effects of denervation on the activities of some tricarboxylic acid cycle associated dehydrogenases and adenine-metabolizing enzymes in rat diaphragm muscle. *Biochem J.*, **128**, 803–809.
2969. TURNER, P. T. & HARRIS, A. B. (1973). Ultrastructure of synaptic vesicle formation in cerebral cortex. *Nature, Lond.*, **242**, 57–58.
2970. TWEEDLE, C. (1971). Transneuronal effects on amphibian limb regeneration. *J. exp. Zool.*, **177**, 13–30.
2971. UEHARA, S. (1973). Disc electrophoresis of extracts from the taste buds located in circuvallate papillae in rat tongues. *J. gen. Physiol.*, **61**, 290–304.
2972. UNGAR, G. (1970). Molecular mechanisms in information processing. *Int. Rev. Neurobiol.*, **13**, 223–253.
2973. UNGAR, G. & BURZYNSKI, S. R. (1973). Isolation and purification of a habituation-producing peptide from trained rat brain. *Fedn Proc. Fedn Am. Socs exp. Biol.*, **32**, 367.
2974. UNGAR, G., DESIDERIO, D. M. & PARR, W. (1972). Isolation, identification and synthesis of a specific behaviour inducing brain peptide. *Nature, Lond.*, **238**, 198–202.
2975. UNGAR, G. GALVAN, L. & CLARK, R. H. (1968). Chemical transfer of learned fear. *Nature, Lond.*, **217**, 1259–1260.
2976. URRY, D. W. (1972). A molecular theory of ion-conducting channels: a fluid-dependent transition between conducting and nonconducting conformation. *Proc. natn. Acad. Sci. U.S.A.*, **69**, 1610–1614.

2977. URRY, D. W. (1972). Conformation of protein in biological membranes and a model transmembrane channel. *Ann. N. Y. Acad. Sci.*, **195**, 108–125.

2978. URRY, D. W., GOODALL, M. C., CLICKSON, J. D. & MAYERS, D. F. (1971). The gramicidin A transmembrane channel: characteristics of head to head dimerized $\pi_{(L,D)}$ helices. *Proc. natn. Acad. Sci. U.S.A.*, **68**, 1907–1911.

2979. USHERWOOD, P. N. R. (1963). Response of insect muscle to denervation, I. Resting potential changes. *J. Insect Physiol.*, **9**, 247–255.

2980. USHERWOOD, P. N. R. (1963). Response of insect muscle to denervation. II. Changes in neuromuscular transmission. *J. Insect Physiol.*, **9**, 811–825.

2981. USHERWOOD, P. N. R. (1973). Release of transmitter from degenerating locust motoneurones. *J. exp. Biol.*, **59**, 1–16.

2982. USHERWOOD, P. N. R., COCHRANE, D. G. & REES, D. (1968). Changes in insect excitatory nerve-muscle synapses after motor nerve section. *Nature, Lond.*, **218**, 589–591.

2983. USHERWOOD, P. N. R. & WOOD, M. R. (1973). Structure and physiology of denervated and reinnervated cockroach muscle. *J. Physiol.*, **230**, 1–2P.

2984. UYEDA, C. T., ENG, L. F. & BIGNAMI, A. (1972). Immunological study of the glial fibrillary acidic protein. *Brain Res.*, **37**, 81–89.

2985. UYEMURA, K., VINCENDON, G., GOMBOS, G. & MANDEL, P. (1971). Purification and some properties of S-100 protein fractions from sheep and pig brain. *J. Neurochem.*, **18**, 429–438.

2986. UZMAN, L. L. & RUMLEY, M. K. (1958). Changes in the composition of the developing mouse brain during early myelination. *J. Neurochem.*, **3**, 171.

2987. VALE, J. R., VALE, C. A. & HARLEY, J. P. (1971). Interaction of genotype and population number with regard to agressive behaviour, social grooming and adrenal and gonadal weight in male mice. *Communs. behav. Biol.*, **6**, 209–221.

2988. VALENTINE, R. C., SHAPIRO, B. M. & STADTMAN, E. R. (1968). Regulation of glutamine synthetase. XII. Electron microscopy of the enzyme from Escherichia coli. *Biochemistry, N.Y.*, **7**, 2143–2152.

2989. VALTIN, H., SAWYER, W. H. & SOKOL, H. W. (1965). Neurohypophyseal principles in rats homozygous and heterozygous for hypothalamic diabetes insipidus (Brattleboro strain). *Endocrinology*, **77**, 701–706.

2990. VALTIN, H. & SCHROEDER, H. A. (1964). Familial hypothalamic diabetes insipidus in rats (Brattleboro strain). *Am. J. Physiol.*, **206**, 425–430.

2991. VALVERDE, F. (1968). Structural changes in the area striata of the mouse after enucleation. *Expl Brain Res.*, **5**, 274–292.

2992. VALVERDE, F. & MARCOS, A. R. (1967). Light deprivation and the spines of apical dendrites in the visual cortex of the mouse. *Anat. Rec.*,

157, 392.

2993. VANDENHEUVEL, F. A. (1965). Structural studies of biological membranes: the structure of myelin. *Ann. N.Y. Acad. Sci.*, **122**, 57–76.

2994. VANDERKOOI, G. (1972). Molecular architecture of biological membranes. *Ann. N.Y. Acad. Sci.*, **195**, 6–15.

2995. VANDERKOOI, G. & CAPALDI, R. A. (1972). A comparative study of the amino acid composition of membrane proteins and other proteins. *Ann. N.Y. Acad. Sci.*, **195**, 135–138.

2996. VANDERKOOI, G., SENIOR, A. E., CAPALDI, R. A. & HAYASHI, H. (1972). Biological membrane structure. III. The lattice structure of membranous cytochrome oxidase. *Biochim. biophys., Acta,* **274**, 38–48.

2997. VARGA, L. (1955). Studies on hyaluronic acid prepared from the vitreous body. *J. biol. Chem.*, **217**, 651–658.

2998. VARON, S. & RAIBORN, C. W. (1969). Dissociation, fractionation and culture of embryonic brain cells. *Brain Res.*, **12**, 180–199.

2999. VASQUEZ, C., BARRANTES, F. J. La TORRE, J. C. & de ROBERTIS, E. (1970). Electron microscopy of proteolipid molecules from cerebral cortex. *J. molec. Biol.*, **52**, 221–226.

3000. VAUGHN, J. E., HINDS, P. L. & SKOFF, R. P. (1970). Electron microscopic studies of Wallerian degeneration in rat optic nerves. I. The multipotential glia. *J. comp. Neurol.*, **140**, 175–206.

3001. VAUGHN, J. E. & PEASE, D. C. (1967). Electron microscopy of classically stained astrocytes. *J. comp. Neurol.*, **131**, 143–154.

3002. VAUGHN, J. E. & PETERS, A. (1967). Electron microscopy of the early postnatal development of fibrous astrocytes. *Am. J. Anat.*, **121**, 131–152.

3003. VAUGHN, J. E. & PETERS, A. (1968). A third neuroglial cell type. An electron microscopic study. *J. comp. Neurol.*, **133**, 269–288.

3004. VENTILLA, M., CANTOR, C. R. & SHELANSKI, M. L. (1972). A circular dichroism study of microtubule protein. *Biochemistry, N.Y.*, **11**, 1554–1561.

3005. VERA, C. L. & LUCO, J. V. (1967). Reinnervation of smooth and striated muscle by sensory nerve fibres. *J. Neurophysiol.*, **30**, 620–627.

3006. VERA, C. L., VIALL, J. D. & LUCO, J. V. (1957). Reinnervation of nictitating membrane of cat by cholinergic fibres. *J. Neurophysiol.*, **20**, 365–373.

3007. VERNON, J. A. & BUTSCH, J. (1957). Effect of tetraploidy on learning and retention in the salamander. *Science, N.Y.*, **125**, 1033–1034.

3008. VIALL, J. D. (1955). The influence of axon length on the course of Wallerian degeneration of the motor endplate. *Acta physiol. lat. am.*, **5**, 95–103.

3009. VIALL, J. D. (1958). The early changes in the axoplasm during Wallerian degeneration. *J. biophys. biochem. Cytol.*, **4**, 551–555.

3010. VIGH, B., RÖHLICH, P., TEICHMANN, I. & AROS, B. (1967). Ependymosecretion (ependymal neurosecretion). VI. Light and electron microscopic examination of the subcommisural organ of the guinea pig.

Acta biol. Hung., **18**, 53–66.

3011. VIGH-TEICHMANN, I., VIGH, B. & AROS, B. (1973). CSF contacting axons and synapses in the lumen of the pineal organ. *Z. Zellforsch. mikrosk. Anat.*, **144**, 139–152.
3012. VILLEGAS, G. M. & FERNANDEZ, J. (1966). Permeability to thorium dioxide of the intracellular spaces of the frog cerebral hemisphere. *Expl Neurol.*, **15**, 18–36.
3013. VILLEGAS, R. & BARNOLA, F. V. (1960). Characterization of the resting axolemma in the giant axon of the squid. *J. gen. Physiol.*, **44**, 963–977.
3014. VILLEGAS, R. & BARNOLA, F. V. (1972). Ionic channels and nerve membrane constituents. Tetrodotoxin-like interaction of saxitoxin with cholesterol monolayers. *J. gen. Physiol.*, **59**, 33–46.
3015. VILLEGAS, R., BARNOLA, F. V. & CAMEJO, G. (1970). Ionic channels and nerve membrane lipids. *J. gen. Physiol.*, **55**, 548–561.
3016. VILLEGAS, R., BARNOLA, F. V. & CAMEJO, G. (1973). Action of proteases and phospholipids on tetrodotoxin binding to axolemma preparations isolated from lobster nerve fibres. *Biochim. biophys. Acta*, **318**, 61–68.
3017. VILLEGAS, R., BRUZUAL, I. B. & VILLEGAS, G. M. (1968). Equivalent pore radius of the axolemma of resting and stimulated squid axon. *J. gen. Physiol*, **51**, 81–92S.
3018. VILLEGAS, R. & CAMEJO, G. (1968). Tetrodotoxin interaction with squid nerve membrane lipids. *Biochim. biophys. Acta*, **163**, 421–423.
3019. VILLEGAS, R., VILLEGAS, L., GIMENEZ, M. & VILLEGAS, G. M. (1963). Schwann cell and axon electrical potential differences: squid nerve structure and excitable membrane location. *J. gen. Physiol.*, **46**, 1047–1064.
3020. VOGEL, F. S. & KEMPER, L. (1962). A modification of Hortega's silver impregnation method to assist in the identification of astrocytes with electron microscopy. *J. Neuropath. exp. Neurol.*, **21**, 147–154.
3021. VOGEL, Z., SYTKOWSKI, A. J. & NIRENBERG, M. W. (1972). Acetyl choline receptors of muscle grown *in vitro*. *Proc. natn. Acad. Sci. U.S.A.*, **69**, 3180–3184.
3023. VOLKMAN, A. & GOWANS, J. L. (1965). The production of macrophages in the rat. *Br. J. exp. Path.*, **46**, 50–70.
3024. VOLKMAR, F. R. & GREENOUGH, W. T. (1972). Rearing complexity affects branching of dendrites in the visual cortex of the rat. *Science, N.Y.*, **176**, 1445–1447.
3025. VOLLE, R. L. & KOELLE, G. B. (1961). The physiological role of acetylcholinesterase in sympathetic ganglia. *J. Pharmac. exp. Ther.*, **133**, 223–240.
3026. VRBOVA, G. (1963). The effect of motoneurone activity on the speed of contraction of striated muscle. *J. Physiol.*, **169**, 513–526.
3027. WADA, J. A. & ASAKURA, T. (1970). Circadian alteration of audiogenic seizure susceptibility in rats. *Expl Neurol.*, **29**, 211–214.

3028. WADA, J. A. & CORNELIUS, L. A. (1960). Functional alteration of deep structures in cats with chronic focal cortical irritative lesion. *Archs. Neurol.*, **3**, 425.

3029. WAKSMAN, B. & ADAMS, R. (1965). A comparative study of experimental allergic neuritis in the rabbit, guinea pig and mouse. *J. Neuropath. exp. Neurol.*, **15**, 293–334.

3030. WALD, G. (1934). Carotenoids and vitamin A cycle in vision. *Nature Lond.*, **134**, 65.

3031. WALD, G. (1968). The molecular basis of visual excitation. *Nature, Lond.*, **219**, 800–807.

3032. WALD, G. (1968). Molecular basis of visual excitation. *Science, N.Y.*, **162**, 230.

3033. WALD, G. & BROWN, P. K. (1951). The role of sulfhydryl groups in the bleaching and synthesis of rhodopsin. *J. gen. Physiol.*, **35**, 797–821.

3034. WALD, G., BROWN, P. K. & GIBBONS, I. R. (1963). The problem of visual excitation. *J. opt. Soc. Am.*, **53**, 20–26.

3035. WALKER, B. E. (1963). Fate of labelled cells in injured spinal cord. *Anat. Rec.*, **145**, 296.

3036. WALLACE, B. J., VOLK, B. W. & LAZARUS, S. S. (1963). Glial cell enzyme alterations in infantile amaurotic idiocy (Tay Sachs Disease). *J. Neurochem.*, **10**, 439–446.

3037. WALLACH, D. F. H. (1969). Membrane lipids and the conformation of membrane proteins. *J. gen. Physiol.*, **54**, 3–26S.

3038. WALLACH, D. F. H. & GORDON, A. S. (1968). Lipid protein interactions in cellular membranes. *Fedn Proc. Fedn Am. Socs exp. Biol.*, **27**, 1263.

3039. WALLACH, D. F. H. & ZAHLER, P. H. (1966). Protein conformations in cellular membranes. *Proc. natn. Acad. Sci. U.S.A.*, **56**, 1552–1559.

3040. WALSH, R. N., BUDTZ-OLSEN, O. E., PENNY, J. E. & CUMMINS, R. A. (1969). The effects of environmental complexity on the histology of the rat hippocampus. *J. comp. Neurol.*, **137**, 361–366.

3041. WALTHER, B. T., OHNAN, R. & ROSEMAN, S. (1973). A quantitative assay for intercellular adhesion. *Proc. natn. Acad. Sci U.S.A.*, **70**, 1569–1573.

3042. WANG, H. H. & ADEY, W. R. (1969). Effects of cations and hyaluronidase on cerebral electrical impedance. *Expl Neurol.*, **25**, 70–84.

3043. WANG, H. H., KADO, R. T. & ADEY, W. R. (1968). Calcium mucopolysaccharides and cerebral impedance. *Fedn Proc. Fedn Am. Socs exp. Biol.*, **27**, 749.

3044. WANG, H. H., TARBY, T. J., KADO, R. T. & ADEY, W. R. (1966). Periventricular impedance after intraventricular injection of calcium. *Science, N.Y.*, **154**, 1183–1185.

3045. WANG, M. B. & BERNARD, R. A. (1969). Characterization and interaction of taste responses in chorda tympani fibres of the cat. *Brain Res.*, **15**, 657–670.

3046. WARD, R. (1972). Peripheral auditory sensitivity in mice prone to

audiogenic seizure. *Expl Neurol.*, **37**, 236–239.
3047. WARDELL, W. M. (1966). Electrical and pharmacological properties of mammalian neuroglial cells in tissue culture. *Proc. R. Soc. B*, **165**, 326–361.
3048. WARDEN, C. J., ROSS, S. & ZAMENHOF, S. (1942). Effect of artificial changes in brain of maze learning in white rat. *Science, N.Y.*, **95**, 414–415.
3049. WARECKA, K. (1970). Isolation of brain-specific glycoprotein. *J. Neurochem.*, **17**, 829–830.
3050. WARECKA, K. & BAUER, H. (1968). Untersuchungen über ein hirnspzifischer alpha2glykoprotein in der fetalen und postnatalen Lebensperiode beim Menschen. *Dt. Z. Nervenheilk.*, **194**, 66–75.
3051. WARECKA, K. & MÜLLER, D. (1969). The appearance of human "brain-specific" glycoprotein in ontogenesis. *J. neurol. Sci.*, **8**, 329–345.
3052. WARNER, F. & MEZA, I. (1972). Macromolecular and biochemical aspects of microtubule protofilaments. *J. Cell Biol.*, **55**, 273A.
3053. WARNICK, J. E. & ALBUQUERQUE, E. X. (1971). Batrachotoxin and tetrodotoxin action on chronically denervated mammalian skeletal muscle. *Pharmacologist*, **13**, 217.
3054. WARNICK, J. E., ALBUQUERQUE, E. X. & SANSONE, F. M. (1971). The pharmacology of batrachotoxin. I. Effects on the contractile mechanism and on neuromuscular transmission of mammalian skeletal muscle. *J. Pharmac. exp. Ther.*, **176**, 497–510.
3055. WASER, P. G. (1967). Receptor localization by autoradiographic techniques. *Ann. N.Y. Acad. Sci.*, **144**, 737–753.
3056. WASER, P. G. (1970). On receptors in the postsynaptic membrane of the motor end plate. *In* Porter, R. & O'Connor, M., *Molecular properties of drug receptors.* Churchill; London.
3057. WASHIZU, Y. (1960). Single spinal motoneurones excitable from two antidromic pathways. *Jap. J. Physiol.*, **10**, 121–131.
3058. WASHIZU, Y., BONEWELL, G. W. & TERZUOLO, C. A. (1961). Effect of strychnine upon the electrical activity of an isolated nerve cell. *Science, N.Y.*, **133**, 333–334.
3059. WATANABE, A. & BULLOCK, T. H. (1961). Modulations of activity of one neurone by subthreshold slow potentials in another in lobster cardiac ganglion. *J. gen. Physiol.*, **43**, 1031–1046.
3060. WATANABE, S. (1970). Relation between superprecipitation and ATPase of myosin B. *J. Biochem.*, **68**, 913–921.
3061. WATANABE, S., MITARI, G. & TAKENAKA, S. (1968). The glial cell in the cerebral cortex of the cat. *Proc. int. Union physiol. Sci.* XXIV Int Congr. pp. 59.
3062. WATSON, W. E. (1965). An autoradiographic study of the incorporation of nucleic acid precursors by neurones and glia during nerve regeneration. *J. Physiol.*, **180**, 741–753.
3062a WATSON, W. E. (1965). An autoradiographic study of the incorporation of nucleic acid precursors by neurones and glia during nerve stimulation

J. Physiol., **180**, 754–765.

3063. Watson, W. E. (1966). Some quantitative observations upon the oxidation of substrates of the tricarboxylic acid cycle in injured neurones. *J. Neurochem.*, **13**, 849–856.
3064. WATSON, W. E. (1966). Quantitative observations upon acetylcholine hydrolase activity of nerve cells after axotomy. *J. Neurochem.*, **13**, 1549–1550.
3065. WATSON, W. E. (1966). Alteration of the adherence of glia to neurones following nerve injury. *J. Neurochem.*, **13**, 536–537.
3066. WATSON, W. E. (1971). Some metabolic responses of rat neuroglial cells to perfusion of the cerebral venricles with artificial cerebrospinal fluid of abnormal composition. *J. Physiol.*, **218**, 88–89P.
3067. WATSON, W. E. (1972). Some responses of neuroglial cells to stimulation of adjacent neurones. *J. Physiol.* **225**, 54–56.
3068. WATSON, W. E. (1972). Some quantitative observations upon the reponses of neuroglial cells which follow axotomy of adjacent neurons. *J. Physiol.*, **225**, 415–435.
3069. WATSON, W. E. (1972). A quantitative study of the response of neuroglial cells of the albino rat to portocaval anastomosis and to associated procedures. *J. Physiol.*, **225**, 10–11P.
3070. WATSON, W. E. (1969). The response of motor neurones to intramuscular injection of botulinum toxin. *J. Physiol.*, **202**, 611–630.
3071. WATSON, W. E. (1974). The binding of Actinomycin D to the nuclei of axotomized neurones. *Brain Res.*, **65**, 317–322.
3072. WATSON, W. E. (1974). Physiology of neuroglia. *Physiol. Rev.*, **54**, 245–271.
3073. WATSON, W. E. (1974). Cellular responses to axotomy and to related procedures. *Br. med. Bull.*, **30**, 112–115.
3074. WATTS, M. E. & MARK. R. F. (1971). Separate actions of ouabain and cycloheximide on memory. *Brain Res.*, **40**, 420–423.
3075. WATTS, M. E. & MARK, R. F. (1971). Drug inhibition of memory in chickens. II. Short-term memory. *Proc. R. Soc. B*, **178**, 455–464.
3076. WATTS, M. E., MARK. R. F., JEFFREY, P. L. & AUSTIN, L. (1971). Ionic movements, protein synthesis and memory. *Proc. Aust. Physiol. Pharmac. Soc.* p. 2.
3077. WAZIRI, R., KANDEL, E. R. & FRAZIER, W. T. (1969). Organization of inhibition in abdominal ganglion of Aplysia. II Posttetanic potentiation, heterosynaptic depression, and increase in frequency of inhibitory postsynaptic potentials. *J. Neurophysiol.*, **32**, 509–519.
3078. WEAKLY, J. N. (1969). Effect of barbiturates on 'quantal' synaptic transmission in spinal motoneurones. *J. Physiol.*, **204**, 63–77.
3079. WEBB, N. G. (1972). X-ray diffraction from outer segments of visual cells in intact eyes of frog. *Nature, N. Y.*, **235**, 44–46.
3080. WEBER, G., BORRIS, D. P., de ROBERTIS, E., BARRANTES, F. J. La TORRE, J. L. & deCARLIN, M. C. L. (1971). The use of a cholinergic fluorescent probe for the study of the receptor proteolipid. *Molec.*

Pharmac., 7, 530–537.

3081. WEBER, G. & LAURENCE, D. J. R. (1951). Fluorescent indicators of adsorption in aqueous solution and on the solid phase. *Biochem. J.*, **56**, xxxi.

3082. WEBER, M. & CHANGEUX, J. P. (1974). Binding of *Naja nigricollis* ^{3}H α-toxin to membrane fragments from *Electrophorus* and *Torpedo* electric organs. *Molec. Pharmac.*, **10**, 1–14.

3083. WEBER, M. & CHANGEUX, J. P. (1974). Binding of *Naja nigricollis* ^{3}H α-toxin to membrane fragments from *Electrophorus* and *Torpedo* electric organs. II. *Molec. Pharmac.* **10**, 15–34.

3084. WEBER, M. & CHANGEUX, J. P. (1974). Binding of *Naja nigricollis* ^{3}H α-toxin to membrane fragments from *Electrophorus* and *Torpedo* electric organs. III. Effects of local anaesthetics on the binding of the tritiated α-neurotoxin. *Molec. Pharmac.*, **10**, 35–40.

3085. WEBER, W. & MARON, K. (1965). The effect of blastemal tissue on regeneration, II. Influence of blastema cell fractions on regenerative growth. *Folia biol.*, (*Praha*)., **13**, 383–396.

3086. WEBSTER, H. D. (1962). Transient focal accumulation of axonal mitochondria during the early stages of Wallerian degeneration. *J. Cell Biol.*, **12**, 361–383.

3087. WEBSTER, H. deF. (1965). The relationship between Schmidt-Lanterman incisurs and myelin segmentation during Wallerian degeneration. *Ann. N.Y. Acad. Sci.*, **122**, 29–38.

3088. WECHSLER, W. & HAGER, H. (1960). Elektronenmikroskopische Befunde am atrophischen quergestreiften Skelettmuskel der Ratte nach Nervdurchtrennung. *Naturwissenschaften*, **47**, 185–186.

3089. WECHSLER, W. & HAGER, H. (1962). Elektronen-mikroskopische Befunde zur Feinstruktur von Axonveränderungen in regeneriertenden Nervenfasern des Nervus ischiandicus der weissen Ratte. *Acta neuropath.*, **1**, 489–506.

3090. WEEDS, A. G. & LOWEY, S. (1971). Substructure of the myosin molecule. II. The light chains of myosin. *J. biol. Chem.*, **61**, 701–725.

3091. WEEDS, A. G. & POPE, B. (1971). Chemical studies on light chains from cardiac and skeletal muscle myosin. *Nature, Lond.*, **234**, 85–88.

3092. WEHRLI, E., MÜHLETHALER, K. & MOOR, H. (1970). Membrane structure as seen with a double replica method for freeze fracturing. *Expl Cell Res.*, **59**, 336–339.

3093. WEIGHT, F. F. & VOTAVA, J. (1970). Slow synaptic excitation in sympathetic ganglion cells: evidence for synaptic inactivation of potassium conductance. *Science*, **170**, 755–758.

3094. WEILL, C. L., McNAMEE, M. G. & KARLIN, A. (1974). Affinity labeling of purified acetylcholine receptor from *Torpedo Californica Biochem biophys. Res. Commun.*, **61**, 997.

3095. WEININGER, O. (1953). Mortality of albino rats under stress as a function of early handling. *Can. J. Psychol.*, **7**, 111–114.

3096. WEININGER, O. (1956). The effects of early experience on behaviour and growth characteristics. *J. comp. physiol. Psychol.*, **49**, 1–9.

3097. WEIS, J. S. (1969). The effects of nerve growth factor on the spinal ganglia of amblystoma maculatum. *J. exp. Zool.*, **170**, 481–488.
3098. WEIS, J. S. (1972). The effects of nerve growth factor on bullfrog tadpoles (Rana catesbiana) after limb amputation. *J. exp. Zool.*, **180**, 385–392.
3099. WEIS, J. S. & WEIS, P. (1970). The effect of nerve growth factor on limb regeneration in amblystoma *J. exp. Zool.*, **174**, 73–78.
3100. WEISENBERG, R. L. (1972). Changes in the organization of tubulin during meiosis in the eggs of the surf clam *Spisula solidissima. J. Cell Biol.*, **54**, 266–278.
3101. WEISENBERG, R. C. (1972). Microtubule formation *in vitro* in solutions containing low calcium concentrations. *Science, N.Y.*, **177**, 1104–1105.
3102. WEISENBERG, R. C., BORISY, G. G. & TAYLOR, E. W. (1968). The colchicine-binding protein of mammalian brain and its relation to microtubules. *Biochemistry, N.Y.*, **7**, 4466–4479.
3103. WEISENBERG, R. C. & TIMASHEFF, S. M. (1970). Aggregation of microtubule subunit protein. Effects of divalent ions, cochicine and vinblastine. *Biochemistry, N.Y.*, **9**, 4110–4116.
3104. WEISS, B., FERTEL, R. & UZUNOV, P. (1974). Selective alteration of the activity of the multiple forms of adenosine 3′5′-monophosphate phosphodiesterase of the rat cerebrum. *Molec. Pharmac.*, **10**, 615–625.
3105. WEISS, L. (1970). The cell periphery. *Int. Rev. Cytol.*, **26**, 63–105.
3106. WEISS, P. A. (1972). Neuronal dynamics and axonal flow. V. The semisolid state of the moving axonal column. *Proc. natn. Acad. Sci. U.S.A.*, **69**, 620–623.
3107. WEISS, P. & FERRIS, W. (1954). Electron microscopic study of the texture of the basement membrane of larval amphibian skin. *Proc. natn. Acad. Sci. U.S.A.*, **40**, 528–540.
3108. WEISS, P. & HISCOE, H. B. (1948). Experiments on the mechanisms of nerve growth. *J. exp. Zool.*, **107**, 315–395.
3109. WEISS, P. & HOAG, A. (1946). Competitive reinnervation of rat muscles by their own and foreign nerves. *J. Neurophysiol.*, **9**, 413–418.
3110. WEISSMAN, A. (1967). Drugs and retrograde amnesia. *Int. Rev. Neurobiol.*, **10**, 167–198.
3111. WELLER, M. & RODNIGHT, R. (1970). Stimulation by cyclic AMP of intrinsic protein kinase in ox brain membrane preparations. *Nature*, **225**, 187–188.
3112. WENDELL-SMITH, C. P., BLUNT, M. J., BALDWIN, F. & PAISLEY, P. B. (1965). Axon-satellite cell relationship in optic nerve. *J. Anat.*, **99**, 946.
3113. WERBOFF, J. & CORCORAN, T. B. (1961). Effects of sex hormone manipulation on audiogenic seizures. *Am. J. Physiol.*, **210**, 830–832.
3114. WERSÄLL, J., FLOCK, Å. & LUNDQUIST, P. G. (1965). Structural basis for directional sensitivity in cochlear and vestibular sensory receptiors. *Cold Spring Harb. Symp. quant. Biol.*, **30**, 115–132.

3115. WESTMAN, J. (1969). The lateral cervical nucleus in the cat. III. An electron microscopical study after transection of spinal afferents. *Expl Brain Res.*, **7**, 32–50.
3116. WESTRUM, L. E. (1966). Electron microscopy of degeneration in the prepyriform cortex. *J. Anat.*, **100**, 683–685.
3117. WESTRUM, L. E. (1969). Electron microscopy of degeneration in the lateral olfactory tract and plexiform layer of the prepyriform cortex of the rat. *Z. Zellforsch. mikrosk. Anat.*, **98**, 157–187.
3118. WESSELLS, N. K., SPOONER, B. S., ASH, J. F., BRADLEY, M., LUDUENA, M. A., TAYLOR, E. L., WRENN, J. T. & YAMADA, K. M. (1971). Microfilaments in cellular and developmental processes. *Science, N.Y.*, **171**, 135–143.
3119. WESSELLS, N. K., SPOONER, B. S. & LUDUENA, M. A. (1973). Surface movements, microfilaments and cell locomotion. *In Locomotion of tissue cells.* CIBA Fdn Symp. Elsevier, North Holland; Amsterdam. pp. 53–82.
3120. WESTRUM, L. E. & BLACK, R. G. (1971). Fine structural aspects of the synaptic organization of the spinal trigeminal nucleus (pars interpolaris) of the cat. *Brain Res.*, **25**, 265–288.
3121. WHITE, D. C. S. & THORSON, J. (1973). The kinetics of muscle contraction. *Prog. Biophys. molec. Biol.*, **27**, 175–258.
3122. WHITNEY, G. (1973). Vocalization of mice influenced by a single gene in a heterogeneous population. *Behav. Genet.*, **3**, 57–64.
3123. WHITTEMBURY, G. (1965). Sodium extrusion and potassium uptake in guinea pig kidney cortex slices. *J. gen. Physiol.*, **48**, 699–717.
3124. de WIED, D. (1969). *In*, Ganong, W. F. & Martini, L., *Frontiers in neuroendocrinology.* Oxford Univ. Press; pp. 97–140.
3125. de WIED, D., van DELFT, A. M. L., GISPEN, W. H., WEIJNEN, J. A. W. M. & GREIDANUS, W. T. B. (1972). The role of pituitary-adrenal system hormones in active avoidance conditioning. *In* Levine, S., *Hormones and behaviour.* Acad. Press; N.Y.
3126. WIESEL, T. N. & HUBEL, D. H. (1963). Single cell responses in striate cortex of kittens deprived of vision in one eye. *J. Neurophysiol.*, **26**, 1003–1017.
3127. WIESEL, T. N. & HUBEL, D. H. (1963). Effects of visual deprivation on morphology and physiology of cells in the cat's lateral geniculate body. *J. Neurophysiol.*, **26**, 978–993.
3128. WIESEL, T. N. & HUBEL, D. H. (1965). Extent of recovery from the effects of visual deprivation in kittens. *J. Neurophysiol.*, **28**, 1060–1072.
3129. WIESEL, T. N. & HUBEL, D. H. (1965). Comparison of the effects of unilateral and bilateral eye closure on cortical unit responses in kittens. *J. Neurophysiol.*, **28**, 1029–1040.
3130. WIESEL, T. N., HUBEL, D. H. & LAM, D. (1974). Autoradiographic demonstration of ocular dominance columns in the monkey striate cortex by means of transsynaptic transport. *Brain Res.*, **79**, 273–279.
3131. WIKLUND, R. & ALLISON, A. C. (1972). The effects of anaesthetics on

the motility of *Dictyostelium disciodeum:* evidence for a possible mechanism of anaesthesia. *Nature, Lond.*, **239**, 221–222.

3132. WIKSWO, M. A. & NOVALES, R. R. (1969). The effect of colchicine on the migration of pigment granules in the melanophores of *Fundulus heteroclitus. Biol. Bull. mar. biol. Lab., Woods Hole,* **137**, 228.
3133. WILKINS, M. H. F. (1972). X-ray studies of membranes and model systems. *Ann. N.Y. Acad. Sci.*, **195**, 291–292.
3134. WILLIAMS, C. M. & SCHNEIDERMAN, H. A. (1952). The necessity of motor innervation for the development of insect muscles. *Anat. Rec.*, **113**, 560.
3135. WILLIAMS, J. A. (1970). Origin of transmembrane potentials in nonexcitable cells. *J. theor. Biol.*, **28**, 287–296.
3136. WILLIAMS, P. L. & HALL, S. M. (1971). Prolonged *in vivo* observations of normal peripheral nerve fibres and their acute reactions to crush and deliberate trauma. *J. Anat.*, **108**, 397–408.
3137. WILLIAMS, T. H. & JEW, J. (1970). Collateral nerve sprouts produced experimentally. *Nature, Lond.*, **228**, 862–864.
3138. WILLIAMS, T. H., JEW, J. & PALAY, S. L. (1973). Morphological plasticity in the sympathetic chain. *Expl Neurol.*, **39**, 181–203.
3139. WILLIAMSON, A. R. & SCHWEET, R. (1965). Role of the genetic message in polyribosome function. *J. Molec. Biol.*, **11**, 358–372.
3140. WILSON, D. F. (1974). The effects of dibutyryl cyclic adenosine 3′5′-monophosphate, theophylline and aminophylline on neuromuscular transmission in the rat. *J. Pharmac. exp. Ther.*, **188**, 447.
3141. WILSON, J. (1968). Genetically determined neurological diseases in children. *In* Cumings, J. N. & Kremer, M., *Biochemical aspects of neurological disorders.* Blackwells; Oxford. pp. 250–281.
3142. WILSON, L. (1970). Properties of colchicine binding protein from chick embryo brain. Interactions with Vinca alkaloids and podophyllotoxin. *Biochemistry, N.Y.*, **9**, 4999–5007.
3143. WILSON, L., BRYAN, J., RUBY, A. & MAZIA, D. (1970). Precipitation of protein by vinblastine and calcium ions. *Proc. natn. Acad. Sci. U.S.A.*, **66**, 807–814.
3144. WILSON, L. D., WEINBERG, J. A. & BERN, H. A. (1957). The hypothalamic neurosecretory system of the tree frog *Hyla regilla. J. comp. Neurol.*, **107**, 253–271.
3145. WINDLE, W. F. & BAXTER, R. E. (1935). Development of reflex mechanisms in the spinal cord of albino rat embryos. Correlations between structure and function, and comparisons with the cat and the chick. *J. comp. Neurol.*, **63**, 189–209.
3146. WINDLE, W. F. & CHAMBERS, W. W. (1950). Regeneration in the spinal cord of the cat and dog. *J. comp. Neurol.*, **93**, 241–257.
3147. WINDLE, W. F., MINEAR, W. L., AUSTIN, M. F. & ORR, D. W. (1935). The origin and early development of somatic behaviour in the albino rat. *Physiol. Zool.*, **8**, 156–185.
3148. WINE, J. J. (1973). Invertebrate central neurons: orthograde degener-

ation and retrograde changes after axotomy. *Expl Neurol.*, **38**, 157–169.

3149. WINE, J. J. (1973). Invertebrate synapse: long-term maintainance of postsynaptic morphology following denervation. *Expl Neurol.*, **41**, 649–660.

3150. WINNICK, M., FISH, I. & ROSSO, P. (1968). Cellular recovery in rat tissues after a brief period of malnutrition. *J. Nutr.*, **95**, 623–626.

3151. WINNICK, M. & NOBLE, A. (1966). Quantitative changes in ribonucleic acids and protein during normal growth of rat placenta. *Nature, Lond.*, **212**, 34–35.

3152. WINNICK, M. & NOBLE, A. (1966). Cellular response in rats during malnutrition at various ages. *J. Nutr.*, **89**, 300–306.

3153. WINNICK, M., ROSSO, P. & BRASEL, J. A. (1972). Malnutrition and cellular growth in the brain; existence of critical periods *In Lipids, malnutrition and the developing brain.* Elsevier, North Holland; Amsterdam.

3154. WINSTON, H. D. (1964). Heterosis and learning in the mouse. *J. comp. physiol. Psychol.*, **57**, 279–283.

3155. WINSTON, H. D. & LINDZEY, G. (1964). Albinism and water escape performance in the mouse. *Science, N.Y.*, **144**, 189–191.

3156. WINZLER, R. J. (1970). Carbohydrates in cell surfaces. *Int. Rev. Cytol.*, **29**, 77–114.

3157. WIRTZ, K. W. A., KAMP. H. H. & van DEENEN, L. L. M. (1972). Isolation of a protein from beef liver which specifically stimulates the exchange of phosphatidyl choline. *Biochim. biophys, Acta,* **274**, 606–617.

3158. WIRTZ, K. W. A. & ZILVERSMIT, D. B. (1968). Exchange of phospholipids between liver mitochondria and microsomes in vitro. *J. biol. Chem.*, **243**, 3596–3602.

3159. WIRTZ, K. W. A. & ZILVERSMIT, D. B. (1969). Participation of soluble liver proteins in the exchange of membrane phospholipids. *Biochim biophys. Acta,* **193**, 105–116.

3160. WIRTZ, K. W. A. & ZILVERSMIT, D. B. (1970). Partial purification of phospholipid exchange protein from beef brain. *FEBS Lett.*, **7**, 44–46.

3161. WISLOCKI, C. B. & LEDUC, E. H. (1952). The cytology and histochemistry of the sucommisural organ and Reissner's fibre in rodents. *J. comp. Neurol.*, **97**, 515–544.

3162. WISLOCKI, G. B. & PUTNAM, T. J. (1920). Note on the anatomy of the area postrema. *Anat, Rec.*, **19**, 281–285.

3163. WISNIEWSKI, H. & MORELL, P. (1971). Quaking mouse: ultrastructural evidence for arrest of myelinogenesis. *Brain Res.*, **29**, 63–73.

3164. WISNIEWSKI, H. & OLSZEWSKI, J. (1963). Vascular permeability in the area postrema and hypothalamus, a study using iodinated radioactive albumin. *Neurology, Minneap.*, **13**, 885.

3165. WITMAN, G. B., KARLSON, K. & ROSENBAUM, J. L. (1972). Chlamydomonas flagella. II. The distribution of tubulins 1 and 2 in the outer doublet microtubules. *J. Cell Biol.*, **54**, 540–555.

3166. WOELK, H., KANIG, K. & PEILER-ICHIKAWA, K. (1974). Incorporation of ^{32}P into the phospholipids of neuronal and glial cell enriched fractions isolated from rabbit cerebral cortex: effect of norepinephrine. *J. Neurochem.*, **23**, 1057–1063.
3167. WOLF, M. K. (1970). Anatomy of cultured mouse cerebellum. II. Organotypic migration of granule cells demonstrated by silver impregnation of normal and mutant cultures. *J. comp. Neurol.*, **140**, 281–298.
3168. WOLF, M. K. & HOLDEN, A. B. (1969). Tissue culture analysis of the inherited defect of central nervous system myelination in jimpy mice. *J. Neuropath. exp. Neurol.*, **28**, 195–213.
3169. WOLF, M. K. & MERRILL, K. (1964). Differentiation of neuronal types and synapses in myelinating cultures of mouse cerebellum. *J. Cell Biol.*, **22**, 259–271.
3170. WOLFF, D., CANESSA-FISCHER, M., VARGAS, F. & DIAZ, G. (1971). The molecular organization of nerve membrane. V. Properties of mono- and bimolecular films formed with lipids isolated from an axolemma-rich preparation from squid retinal axons. *J. mem. Biol.*, **4**, 304–314.
3171. WOLFF, D., FISCHER, M. C., VARGAS, F. & DIAZ, G. (1971). The molecular organization of nerve membranes. *J. mem. Biol.*, **6**, 304–314.
3172. WOLFF, J. (1965). Untersuchungen über Struktur und Gestalt von Astrzytenforsätzen *Z. Zellforsch. mikrosk. Anat.*, **66**, 811–828.
3173. WOLFGRAM, F. (1966). A new proteolipid fraction of the nervous system. I. Isolation and amino acid analysis. *J. Neurochem.*, **13**, 461–470.
3174. WOLFGRAM, F. & KOTORII, K. (1968). The composition of the myelin proteins of the central nervous system. *J. Neurochem.*, **15**, 1281–1290.
3175. WOLFGRAM, F. & KOTORII, K. (1968). The composition of the myelin proteins of the peripheral nervous system. *J. Neurochem.*, **15**, 1291–1295.
3176. WOLFGRAM, H. & ROSE, A. S. (1957). The morphology of neuroglia in tissue culture with comparison to histological preparations. *J. Neuropath. exp. Neurol.*, **16**, 514–531.
3177. WOMBACHER, H. & WOLF, H. U. (1971). Regulation of membrane bound acetylcholinesterase activity by bisquaternary nitrogen compounds. *Molec. Pharmac.*, **7**, 554–566.
3178. WOOD, J. G. & KING, N. (1971). Turnover of basic protein in rat brain. *Nature, Lond.*, **229**, 56–58.
3179. WOODBURY, D. M. & WOODBURY, J. W. (1963). Correlation of microelectrode potential recordings with histology of rat and guinea-pig thyroid glands. *J. Physiol.*, **169**, 553–567.
3180. WOODWARD, D. L., REED, D. J. & WOODBURY, D. M. (1967). Extracellular space of rat cerebral cortex. *Am. J. Physiol.*, **212**, 367–370.
3181. WOOLARD, H. H. (1924). Vital staining of the leptomeninges. *J. Anat.*, **58**, 89–100.
3182. WORTHINGTON, C. R. (1969). Structural parameters of nerve myelin. *Proc. natn. Acad. Sci. U.S.A.*, **63**, 604–611.

3183. WORTHINGTON, C. R. (1969). The interpretation of low angle X-ray data from planar and concentric multilayer structures. *Biophys. J.*, **9**, 222–234.

3184. WORTHINGTON, C. R. (1970). On the interpretation of X-ray diffraction intensities from chemically treated frog sciatic nerve. *Biophys. J.*, **10**, 675–677.

3185. WORTHINGTON, C. R. (1971). Structure of photoreceptor membranes. *Fedn Proc. Fedn Am. Socs exp. Biol.*, **30**, 57.

3186. WORTHINGTON, C. R. (1971). X-ray analysis of nerve myelin *In* Adelman, W. J., *Biophysics and physiology of excitable membranes*, Van Nostrand Reinhold Co.; N.Y. pp. 1–46.

3187. WORTHINGTON, C. R. (1972). X-ray diffraction studies on nerve and photoreceptors. *Ann. N.Y. Acad. Sci.*, **195**, 293–308.

3188. WORTHINGTON, C. R. & BLAUROCK, A. E. (1969). A low-angle X-ray diffraction study of the swelling behaviour of peripheral nerve myelin. *Biochim. biophys. Acta*, **173**, 427–435.

3189. WORTHINGTON, C. R. & McINTOSH, T. J. (1974). Direct determination of the lamellar structure of peripheral nerve myelin at moderate resolution (7Å). *Biophys. J.*, **14**, 703.

3190. WREN, J. T. & WESSELLS, N. K. (1970). Cytochalasin B: effects upon microfilaments involved in morphogenesis of estrogen-induced glands of oviduct. *Proc. natn. Acad. Sci. U.S.A.*, **66**, 904–908.

3191. WRIGHT, M. R. (1946). Experiments on the lateral line system of anurans. *Proc. Soc. exp. Biol.*, **62**, 242–243.

3192. WRIGHT, M. R. (1947). Regeneration and degeneration experiments on lateral-line nerves and sense organs in anurans. *J. exp. Zool.*, **105**, 221–257.

3193. WRIGHT, M. R. (1951). The lateral line system of sense organs. *Q. Rev. Biol.*, **26**, 264–280.

3194. WUERKER, R. B. (1970). Neurofilaments and glial filaments. *Tiss. Cell*, **2**, 1–9.

3195. WUERKER, R. B. (1973). Changes in muscle morphology and histochemistry produced by denervation, 3,3′iminodipropionitrile and epineurial vinblastine. *Am. J. Anat.*, **136**, 221–234.

3196. WUERKER, R. B. & KIRKPATRICK, J. B. (1972). Neuronal microtubules, neurofilaments and microfilaments. *Int. Rev. Cytol.*, **33**, 45–75.

3197. WUERKER, R. B. & PALAY, S. L. (1969). Microfilaments in nerve cells. *Tiss. Cell*, **1**, 387–402.

3198. WURTZ, R. H. & O'FLAHERTY, J. J. (1967). Physiological correlates of steady potential shifts during sleep and wakefulness. I. Sensitivity of the steady potential to alterations in carbon dioxide. *Electroenceph. clin. Neurol.*, **22**, 30–42.

3199. WYCKOFF, R. W. G. & YOUNG, J. Z. (1956). The motorneurone surface. *Proc. R. Soc. B*, **144**, 440–450.

3200. YAMADA, K. M., SPOONER, B. S. & WESSELLS, N. K. (1970). Axon growth: roles of microfilaments and microtubules. *Proc. natn. Acad. Sci.*

U.S.A., **66**, 1206–1212.

3201. YAMADA, K. M., SPOONER, B. S. & WESSELLS, N. K. (1971). Ultrastructure and function of growth cones and axons of cultured nerve cells. *J. Cell Ciol.*, **49**, 614–635.

3202. YAMADA, K. M. & WESSELLS, N. K. (1971). Axon elongation. Effect of nerve growth factor on microtubule protein. *Expl Cell Res.*, **66**, 346–352.

3203. YANNET, H. (1940). Changes in the brain resulting from depletion of extracellular electrolytes. *Am. J. Physiol.*, **128**, 683.

3204. YASUI, T. & WATANABE, S. (1965). A study of "superprecipitation" of nyosin B by the change in turbidity. *J. biol. Chem.*, **240**, 98–104.

3205. YELLIN, H. (1967). Neural regulation of enzymes in muscle fibres of red and white muscle. *Expl Neurol.*, **19**, 92–103.

3206. YELLIN, H. (1967). Muscle fibre plasticity and the creation of localized motor units. *Anat. Rec.*, **157**, 345.

3207. YELLIN, H. & GUTH, L. (1970). The histochemical classification of muscle fibres. *Expl Neurol.*, **26**, 424–432.

3208. YINON, U. & ERICKSON, R. P. (1970). Adaptation and the neural code for taste. *Brain Res.*, **23**, 428–432.

3209. YNTEMA, C. L. (1959). Regeneration in sparsely innervated and aneurogenic forelimb of *amblystoma* larvae. *J. exp. Zool.*, **140**, 101–124.

3210. YNTEMA, C. L. (1959). Blastema formation in sparsely innervated and aneurogenic forelimbs of *amblystoma* larvae. *J. exp. Zool.*, **142**, 423–440.

3211. YNTEMA, C. L. (1962). Duplication and innervation in anterior extremities of *amblystoma* larvae. *J. exp. Zool.*, **149**, 127–146.

3212. YOON, M. (1971). Reorganization of retinotectal projections following surgical operations on the optic tectum in goldfish. *Expl Zool.*, **33**, 395–411.

3213. YOON, M. (1972). Reversibility of the reorganization of retinotectal projection in goldfish. *Expl Zool.*, **35**, 565–577.

3214. YOUNG, A. B. & SNYDER, S. H. (1973). Strychnine binding associated with glycine receptors of the central nervous system. *Proc. natn. Acad. Sci. U.S.A.*, **70**, 2832–2836.

3215. YOUNG, A. B. & SNYDER, S. H. (1974). Strychnine binding in rat spinal cord membranes associated with the synaptic glycine receptor: cooperativity of glycine interaction. *Molec. Pharmac.*, **10**, 790–809.

3216. YOUNG, A. B. & SNYDER, S. H. (1974). The glycine receptor: evidence that strychnine binding is associated with the ionic conductance mechanism. *Proc. natn. Acad. Sci. U.S.A.*, **71**, 4002.

3217. YOUNG, D. (1972). Specific reinnervation of limbs transplanted between segments in the cockroach, *Periplaneta americana*. *J. exp. Biol.*, **57**, 305–316.

3218. YOUNG, J. Z. (1945). History of the shape of nerve fibres. *In* d'Arcy Thompson, *Essays on growth and form.* Clarendon; Oxford.

3219. YOUNG, R. W. (1967). The renewal of photoreceptor cell outer segments. *J. Cell Biol.*, **33**, 61–72.
3220. ZACKS, S. I. & BLUMBERG, J. M. (1961). The histochemical localization of acetylcholinesterase in the fine structure of neuromuscular junction of mouse and human intercostal muscle. *J. Histochem. Cytochem.*, **9**, 317–324.
3221. ZADUNAISKY, J. A. & CURRAN, P. F. (1963). Sodium fluxes in isolated frog brain. *Am. J. Physiol.*, **205**, 949–956.
3222. ZALEWSKI, A. A. (1969). Neurotrophic-hormonal interaction in the regulation of taste buds in the rat's vallate papilla. *J. Neurobiol.*, **1**, 123–132.
3223. ZALEWSKI, A. A. (1969). Combined effects of testosterone and motor, sensory or gustatory reinnervation on the regeneration of taste buds in the rat. *Expl. Neurol.*, **24**, 285–297.
3224. ZALEWSKI, A. A. (1969). Regeneration of taste buds after reinnervation by peripheral or central fibres of vagal ganglia. *Expl Neurol.*, **25**, 429–437.
3225. ZALEWSKI, A. A. (1969). Regeneration of taste buds in the lingual epithelium after excision of the vallate papilla. *Expl Neurol.*, **26**, 621–629.
3226. ZALEWSKI, A. A. (1970). Trophic influence of *in vivo* transplanted sensory neurons on taste buds. *Expl Neurol.*, **29**, 462–467.
3227. ZALEWSKI, A. A. (1971). Prolonged survival and trophic function of neurons in homologous transplants of sensory ganglion *in vivo*. *Expl Neurol.*, **30**, 510–524.
3228. ZAMBRANO, F., CELLINO, M. & CANESSA-FISCHER, M. (1971). The molecular organization of nerve membranes. IV. The lipid composition of plasma membrane from squid retinal axons. *J. mem. Biol.*, **6**, 289–303.
3229. ZAMENHOF, S., MARTHENS, E. van, & GRAVEL, L. (1971). Prenatal cerebral development: effect of restricted diet, reversed by growth hormone. *Science, N.Y.*, **174**, 954–955.
3230. ZAMENHOF, S., van MARTHENS, E. & MARGOLIS, F. L. (1968). DNA (cell number) and protein in neonatal brain: alteration by maternal dietary protein restriction. *Science, N.Y.*, **160**, 322–323.
3231. ZAMENHOF, S., MOSLEY, J. & SCHULLER, E. (1966). Stimulation of the proliferation of cortical neurons by prenatal treatment with growth hormone. *Science, N.Y.*, **152**, 1396–1397.
3232. ZANELLA, J. & RALL, T. W. (1973). Evaluation of electrical pulses and elevated levels of potassium ions as stimulants of adenosine 3′5′-monophosphate accumulation in guinea pig brain. *J. Pharmac. exp. Ther.*, **186**, 241.
3233. ZECHMEISTER, L. (1944). *Cis- trans* isomerization and stereochemistry of carotenoids and diphenylopolyenes. *Chem. Rev.*, **34**, 267–344.
3234. ZELENA, J. (1970). Ribosome-like particles in myelinated axons of the rat. *Brain Res.*, **24**, 359–363.
3235. ZELENA, J., LUBINSKA, L. & GUTMANN, E. (1968). Accumulation of

organelles at the ends of interrupted axons. *Z. Zellforsch. mikrosk. Anat.*, **91**, 200–219.

3236. ZEMP, J. W., WILSON, J. E. & GLASSMAN, E. (1967). Brain function and macromolecules. II. Site of increased labelling of RNA in brains of mice during a short-term training experience. *Proc. natn. Acad. Sci. U.S.A.*, **58**, 1120–1125.

3237. ZEMP, J. W., WILSON, J. E., SCHLESINGER, K., BOGGAN, W. O. & GLASSMAN, E. (1966). Brain function and macromolecules. I. Incorporation of uridine into RNA of mouse brain during short-term training experience. *Proc. natn. Acad. Sci. U.S.A.*, **55**, 1423–1431.

3238. ZIEGLANSBERGER, W. & REITER, C. (1974). Interneuronal movement of procion yellow in cat spinal neurones. *Expl Brain Res.*, **20**, 527.

3239. ZIMMER, J. (1971). Ipsilateral afferents to the commisural zone of the fascia dentata, demonstrated in decommisurated rats by silver impregnation. *J. comp. Neurol.*, **142**, 393–416.

3240. ZIMMER, J. (1973). Extended commisural and ipsilateral projections in postnatally deentorhinated hippocampus and fascia dentata demonstrated in rats by silver impregnation. *Brain Res.*, **64**, 293–311.

3241. ZIMMER, J. (1973). Changes in the Timms sulfide staining pattern of the rat hippocampus and fascia dentata following early postnatal deafferentation. *Brain Res.*, **64**, 313–326.

3242. ZIMMER, J. (1973). Aberrant axonal growth (sprouting) in hippocampus. *Acta physiol scand.*, Suppl. 396, 32.

3243. ZORN, M. & FUTTERMAN, S. (1971). Properties of rhodopsin dependent on associated phospholipid. *J. biol. Chem.*, **246**, 881–886.

3244. ZUCKERMAN, J. E., HERSCHMANN, H. R. & LEVINE, L. (1970). Appearance of a brain-specific antigen (the S-100 protein) during foetal development. *J. Neurochem.*, **17**, 247–252.

3245. van ZUTPHEN, M. van DEENEN, L. L. & KINSKY, S. C. (1966). The action of polyene antibiotics on bilayer lipid membranes. *Biochem. biophys. Res. Commun.* **22**, 393.

3246. ZWILLING, E. (1961). Limb morphogenesis. *Adv. Morphogen.*, **1**, 301–330.

3247. ØYE, I. & SUTHERLAND, E. W. (1966). The effect of epinephrine and other agents on adenyl cyclase in the cell membrane of avian erythrocytes. *Biochim. biophys. Acta*, **127**, 347–354.

INDEX